Vegetation of Central Asia and Environs

Dilfuza Egamberdieva • Münir Öztürk
Editors

Vegetation of Central Asia and Environs

Springer

Editors
Dilfuza Egamberdieva
Faculty of Biology
National University of Uzbekistan
Tashkent, Uzbekistan

Münir Öztürk
Vice President "The Islamic World Acadmey of Sciences" Amann-Jordan and Centre for Environmental Studies-Botany Department
Ege University
Izmir, Turkey

ISBN 978-3-319-99727-8 ISBN 978-3-319-99728-5 (eBook)
https://doi.org/10.1007/978-3-319-99728-5

Library of Congress Control Number: 2018960829

This Springer imprint is published by the registered company Springer Nature Switzerland AG
The registered company address is: Gewerbestrasse 11, 6330 Cham, Switzerland

Foreword

Vegetation activity is regarded as one of the most important indicators for evaluating interactions between climate and different ecosystems. The vegetation is highly sensitive to climate change, especially in dry areas. Last two decades have been the warmest in nearly half a century in Central Asia, located in the hinterland of the Eurasian continent. This unique landscape has expansive but fragile mountain-oasis-desert ecosystems, being one of the driest areas in the world and characterized by low vegetation cover.

The geomorphological patterns of Kazakhstan, Kyrgyzstan, Tajikistan, Turkmenistan, and Uzbekistan, commonly known as Central Asia, start from the east of Caspian continuing to the south of Aral with towering glaciers and basins. All rivers in the region are inland rivers. The average temperatures are in a state of high variability, with decreases in precipitation during the last few decades, later potentially and strongly affecting the natural ecosystems of the region. Several publications reveal NDVI has increased at a rate of 0.004 per decade for some time but has started decreasing at a rate of 0.003 per decade lately. The shrub cover has increased significantly including their encroachment on grasslands. Nearly 10% of the grasslands have got converted to shrubland.

The increase in the melting of glaciers due to rising temperatures is resulting in the increase in runoff from mountain catchments. Aridity is growing in the region. The negative effects of changing climate as the major force of natural vegetation dynamics on the ecology of the region are also growing. Shrub encroachment has resulted in a mosaic landscape, with shrubs distributed among the grass patches.

The land use/cover classification data covering the last three decades reveals that land use in the region has experienced considerable changes, most notable being the grassland degradation, forest reduction, cultivated land expansion, and shrub encroachment. The grassland has decreased by nearly 26%, but the area of shrublands has gone up by 14%. This has changed the species interactions which can lead to a loss of biodiversity and at the same time effect ecosystem functioning. Vegetation in the region is mainly composed of mesophytes and xerophytes; the former are turning into xerophytes due to shrub encroachment, thus creating highly fragile ecosystems in the region.

The knowledge of the current changes and dynamics of different types of vegetation in relation to climatic changes and anthropogenic activities is critical for developing adaptation strategies to address the challenges posed by climate change and human activities for ecosystems. Vegetation pixels have significantly decreased for shrubs and sparse vegetation as compared to other vegetation types. The degradation of vegetation is serious in the Karakum and Kyzylkum deserts, the Ustyurt Plateau, the Aral Sea, and around the Caspian, triggered by exploitation of water resources in the upstream areas of Amu Darya basin and oil and natural gas extraction in the southern part of the Karakum Desert and southern Ustyurt Plateau. After independence, the abandoned pastures have got covered by dense vegetation in some parts of the region as the abandoned croplands reverted to grasslands. A wide range of mammal species are found here including the endangered snow leopard and ibex. Overgrazing and natural resource extraction are major threats to this ecoregion.

Deciduous forests are distributed between 1200 and 1700 m altitudes. *Malus sieversii* is one of the notable trees which grows along the river valleys of Tian Shan. Other temperate tree species are *Prunus armeniaca* and *Acer semenovii*, representing the remnants of broad-leaved temperate forests from the Tertiary period. Other taxa are the species of *Rosa* and *Berberis*, which are abundant in the region. Grass diversity is higher including several endemic species like *Atraphaxis muschketovii, Tulipa* spp., *Eremurus* spp., and *Ligularia macrophylla. Niedzwedzkia semiretschenskia* is an oldest relict endemic. *Artemisia cina* is a steppe endemic. Other endemics are *Tulipa greigii* and *Juno orchioides*.

In the deciduous forests at higher altitudes, one finds *Populus tremula* and *Celtis caucasica* together with economically important plants like *Dictamnus turkestanicus*, *Betula tianshanica*, and *Hippophae rhamnoides*.

Common mammals include red fox, corsac fox, wolf, steppe cat, jungle cat, weasel, Altai ferret, ferret, marbled polecat, badger, saiga antelope, arkhar, Tolai hare, Indian porcupine, various jerboas, birch mouse, sousliks, gerbils, water vole, vole, long-eared hedgehogs, larks, doves, wheateaters, Egyptian vulture, saker falcon, hawks, long-legged buzzard, kite, falcons, buntings, warblers, shrikes, lizards, runners, skinks, geckos, and rat snakes. Turanian tiger lived in this region but was hunted to extinction about 100 years ago. Some rare and endemic species are goitered gazelle, corsac fox, arkhar, saiga antelope, black vulture, imperial eagle, short-toed eagle, lesser kestrel, Central Asian tortoise, kulan, or onager which is an endangered subspecies of Asian wild ass.

There is a big gap of systematic and consistent information. This book will fill these gaps in Central Asia and its environs. It includes chapters on the Vegetation Classification and Habitat Mapping of Dachigam National Park, Kashmir; Ecosystem-Based Adaptation of Central Asian Rangelands and Their Role in the Resilience of Drought Stress: An Assessment Using Remote Sensing Data with Vegetation and Drought Indices (1982–2011); Ecology and Environmental Aspects of "Makmalzoloto" Gold Mining Area, Kyrgyzstan; Spatiotemporal Assessment of Trends in the Post-Soviet Central Asia; Medicinal Plants of Tajikistan; Causes and Impacts of Land Degradation and Desertification: Case Study from Kazakhstan; Assessment of the Current Plant Diversity Status in Kazakhstan; Landscape-Ecological Zoning of Agricultural Areas in South Kazakhstan; Plant Diversity of Ala Archa National Park in Kyrgyzstan with Emphasis on Its Economical Potential; Potential Impacts of Climate Change on Plant Diversity of Sary-Chelek Biosphere Reserve in Kyrgyzstan; Landscape-Ecological Zoning of Agricultural Areas in South Kazakhstan Region; Assessment of the Current Plant Biodiversity Status in the Territory of the Republic of Kazakhstan; and Current State and Prospects for Studies on the Diversity of Medicinal Flora in Kazakhstan.

It will be useful for the people involved at the government level, foresters, agricultural engineers, environmental activists, and researchers at all levels.

Nazim Mamedov
Stockbridge School of Agriculture
University of Massachusetts at Amherst
Amherst, MA, USA

Preface

Central Asia with a continental climate is one of the most arid areas in the world and includes Kazakhstan, Kyrgyzstan, Tajikistan, Turkmenistan, and Uzbekistan, as well as arid region of Northwest China. It lies west of the Helan Mountains, east of the Caspian Sea and Volga River, and south of the Aral Sea and Irtysh River. The geomorphological patterns of the five countries in Central Asia are towering glaciers, widespread deserts like Karakum, Kyzylkum, Gurbantünggüt, Badain Jaran, Tengger, and the famous Taklamakan, together with vast grasslands, sporadic oases, numerous mountains, and river basins. The region as the hinterland of the Eurasian continent has a unique landscape with highly sensitive ecosystems to global climate change, as well as increases in temperatures and evaporation. The areas in general are rich in solar energy receiving more sunlight, resulting in a greater annual evapotranspiration; the latter are the key processes of soil moisture loss than annual precipitation. Temporal as well as spatial changes in the soil moisture result from the integrated influence of warming, precipitation, increased solar radiation, and decreased surface-specific humidity. Both enhance the soil water consumption followed by a significant decrease in water storage and shallow groundwater levels which in turn effect the plants leading to shallow root formation in desert plants followed by a reduction in species diversity and vegetation cover, all rendering the ecosystems increasingly fragile.

Average temperatures show high variability; precipitation has remained relatively stable, but from the beginning of twenty-first century, increases in precipitation show a downward trend; average soil moisture is low. The climate changes can potentially have a strong impact on the region's natural ecosystems. One of the most important indicators for evaluating interactions between climate and terrestrial ecosystems is the vegetation activity, because it is highly sensitive to climate change, especially in arid to semiarid regions with a persistent moisture deficit and low relative humidity. The Caspian and Aral do not attenuate these extreme conditions beyond the directly adjacent areas. Ecological environment is fragile and unstable, and the species are extremely poor. The rising temperatures end up with an acceleration of glacier melting leading to an increase in runoff from mountain catchments which form the major water source replenishing the soil water with several ecological benefits. In contrast, the decrease in soil moisture caused by increased evapotranspiration, driven by rising temperature and radiation, produces negative ecological impacts. The aridity is growing due to climate change producing negative effects on the ecology of the arid desert region. It is the driving factor of natural vegetation dynamics.

Normalized difference vegetation index of natural vegetation in the region is exhibiting ups and downs; the values are higher on latitude basis in the northern part and lower in the southern part. The shrub cover has increased significantly followed by an encroachment on grasslands. Nearly 10% of the grasslands have changed into shrubland during the last decade. In view of this, terrestrial vegetation dynamics monitoring underpins efforts to understand the relations between vegetation and the atmosphere. In an arid-semiarid region, water-based ecosystem is extremely fragile, making it highly sensitive to climate change and human activities. Vegetation growth is directly affected by the hydrothermal conditions, but degeneration of the ecosystem is exacerbated by the rising temperatures, leading to an intensification of the desertification process. Sparse vegetation cover, plants mainly depending on the shallow soil and groundwater, continues growing. Vegetation in the northwestern areas of this region is less sparse than southeastern areas. The vegetation reflects a response to the moisture gradient. A high vegetation cover is observed mainly in northern Kazakhstan, the Ili River Valley, the Altay region, and the Qilian Mountains in Northwest China, while the areas of low vegetation are mainly toward the Chinese border because of less precipitation and glacier/snow meltwater in these areas. Tajikistan and Kyrgyzstan are mountainous and possess a relatively small desert area. In northern Kazakhstan, east of the Aral Sea, around Balkhash Lake, and in the northern Altai region, degradation of vegetation is prominent with decreasing trends in the valley bottom of Kyrgyzstan, southern Turkmenistan, the central Tian Shan Mountains, and the southern slope of the Altai Mountains. Stable vegetation shows an increasing trend in the eastern Balkhash Lake, southwest of Kazakhstan, and south of the Junggar Basin.

The land use in Central Asia based on the land use/cover classification data from the past three decades is experiencing considerable changes, especially during the last decade. Most notable being grassland degradation, forest reduction, cultivated land expansion, and shrub encroachment. Shrub encroachment has resulted in a mosaic landscape, with shrub patches interspaced with grass patches. It is the result of many interrelated factors influencing each other in multiple spatial and temporal scales. Grassland area in northern Kazakhstan has dramatically gone down. The open shrublands in Uzbekistan and Turkmenistan have increased to a large extent, highlighting shrub encroachment into grassland. These regions are characterized by high variability warming and shallow soil moisture. The warming accelerates the evapotranspiration of the shallow soil layer. The grasslands have decreased by more than 25%, and open shrublands have increased by nearly 14%.

Shrub encroachment into grasslands is changing the species interaction, which may enhance biodiversity loss and effect various aspects of ecosystem functioning. This encroachment is not equated with desertification; vulnerable terrestrial ecosystems are generally thought to be highly susceptible to degradation and desertification caused by climatic fluctuations.

The vegetation in general has two types of plants: the mesophytes and the xerophilous plants. Main water sources are precipitation, groundwater, and runoff from mountain areas. Shrub encroachment emphasizes that some mesophytes are changing into xerophytes, leading to water-based ecosystem becoming much more fragile in Central Asia. The vegetation dynamics are highly sensitive to alterations in water availability indicating that water conveyance is necessary for protecting important vulnerable ecological regions. The soil, carbon dioxide concentration, and human activities are important aspects affecting natural vegetation changes because of shrub encroachment.

The grazing on natural grasslands of Nalati, Tianchi, and Bayinbuluke and some others have been prohibited on a large scale, but some herdsmen settlements and livestock populations are still growing in some areas, producing negative impacts on vegetation diversity in arid ecosystems with serious degradation in grasslands. The vegetation pixels have decreased much for shrubs and sparse vegetation compared with other vegetation types. The sparse vegetation degradation is more serious in the Karakum and Kyzylkum deserts, the Ustyurt Plateau, and the wetland delta of the Aral Sea than in other regions.

The excessive exploitation of water resources in the Amu Darya basin and oil and natural gas extraction in the southern part of the Karakum Desert and the southern Ustyurt Plateau has led to an abandonment of pastures. In Kyrgyzstan and Kazakhstan, the population is denser and industrialization more advanced. Oil, coal, iron, and copper deposits are exploited, and contamination from such extraction affects air and water quality. Habitat in the foothill steppe areas has been negatively impacted by grazing and hunting. Overgrazing, oil and mineral extraction, and poaching are the major threats to this ecoregion. Livestock raising is also a dominant activity. A number of rare plant species are of a high value as food or medicinal plants.

Tashkent, Uzbekistan — Dilfuza Egamberdieva
Izmir, Turkey — Münir Öztürk

Acknowledgments

The editors would like to express their gratitude to the contributing authors for their noble academic efforts by spending their valuable time in writing the chapters included here. We highly appreciate their patience. Our special thanks are due to the staff at the Springer for full collaboration.

Contents

Contributors

Khursheed Ahmad Division of Wildlife Sciences, Faculty of Forestry, Sher-e-Kashmir University of Agricultural Sciences and Technology of Kashmir, Srinagar, Jammu and Kashmir, India

Volkan Altay Biology Department, Faculty of Science and Arts, Hatay Mustafa Kemal University, Antakya, Hatay, Turkey

Dildora Aralova Dresden Technology University (TU-Dresden), Institute of Photogrammetry and Remote Sensing, Dresden, Germany

Centre of Hydrometeorological Service at Ministry of Emergency Situations of the Republic of Uzbekistan (Uzhydromet), Tashkent, Uzbekistan

Mehmet Arslan Department of Geography, Land Management and Cadastre, Al-Farabi Kazakh National University, Almaty, Kazakhstan

Maulken Askarova Department of Geography, Land Management and Cadastre, Al-Farabi Kazakh National University, Almaty, Kazakhstan

Ermek Baibagyshov Department of Ecology, Naryn State University, Naryn, Kyrgyzstan

Florian Betz Faculty of Mathematics and Geography, Catholic University Eichstaett-Ingolstadt, Ingolstadt, Germany

Syeda Fatma Hassan Bukhari Atta-ur-Rahman School of Applied Biosciences, National University of Sciences and Technology, Islamabad, Pakistan

Nazima Chukunkyzy Kyrgyz State University, Bishkek, Kyrgyzstan

Akylbek Chymyrov Department of Geodesy and Geoinformatics, Kyrgyz State University of Construction, Transport and Architecture, Bishkek, Kyrgyzstan

Bernd Cyffka Faculty of Mathematics and Geography, Catholic University Eichstaett-Ingolstadt, Ingolstadt, Germany

Javaid M. Dad Central Institute of Temperate Horticulture, Srinagar, Jammu and Kashmir, India

Olena Dubovyk Center for Remote Sensing of Land Surfaces (ZFL), University of Bonn, Bonn, Germany

Dilfuza Egamberdieva Faculty of Biology, National University of Uzbekistan, Tashkent, Uzbekistan

Majid Farooq Department of Ecology, Environment and Remote Sensing, Jammu and Kashmir Forest Department, Srinagar, Jammu and Kashmir, India

Bukhtawar Fatima Atta-ur-Rahman School of Applied Biosciences, National University of Sciences and Technology, Islamabad, Pakistan

Dilshod Gafurov Cotton Breeding, Seeding and Cultivation Agro-technologies Scientific Research Institute, Kibray, Uzbekistan

Nadezhda G. Gemejiyeva Ministry of Education and Science, Institute of Botany and Phytointroduction, Almaty, Kazakhstan

Lyudmila M. Grudzinskaya Ministry of Education and Science, Institute of Botany and Phytointroduction, Almaty, Kazakhstan

Alvina Gul Atta-ur-Rahman School of Applied Biosciences, National University of Sciences and Technology, Islamabad, Pakistan

Plant Breeding-Genetics Section, School of Integrative Plant Science, Cornell University, Ithaca, NY, USA

Umut Halik Key Laboratory of Oasis Ecology, College of Resources & Environmental Sciences, Xinjiang University, Urumqi, Xinjiang, China

Sameen Ruqia Imadi Commission on Science and Technology for Sustainable Development in the South (COMSATS), Islamabad, Pakistan

Nazgül Imanberdieva Biology Department, Kyrgyzstan-Turkiye Manas University, Bishkek, Kyrgyzstan

Bazaraliev Imankul Sary-Chelek Government Biosphere Reserve, Jalal-Abad Oblast, Kyrgyzstan

Dilfuza Jabborova Institute of Genetics and Plant Experimental Biology, Academy of Sciences of Uzbekistan, Tashkent, Uzbekistan

Jahan Kariyeva Alberta Biodiversity Monitoring Institute, Edmonton, AB, Canada

Faria Khan Atta-ur-Rahman School of Applied Biosciences, National University of Sciences and Technology, Islamabad, Pakistan

Tahira Khan Atta-ur-Rahman School of Applied Biosciences, National University of Sciences and Technology, Islamabad, Pakistan

Timur Khujanazarov Water Resources Research Center, Disaster Prevention Research Institute, Kyoto University, Uji, Japan

Shnar Kairova Lyazat Kiyassova Department of Geography, Land Management and Cadastre, Al-Farabi Kazakh National University, Almaty, Kazakhstan

Zheenbek Kulenbekov American University of Central Asia, Bishkek, Kyrgyzstan

Alishir Kurban Xinjiang Institute of Ecology and Geography, Chinese Academy of Sciences, Beijing, China

Gulbubu Kurmanbekova Biology Department, Kyrgyzstan-Turkiye Manas University, Bishkek, Kyrgyzstan

Fatima Malik Atta-ur-Rahman School of Applied Biosciences, National University of Sciences and Technology, Islamabad, Pakistan

Akhmetkal Medeu Institute of Geography, Almaty, Kazakhstan

Alikhan Medeu Institute of Geography, Almaty, Kazakhstan

Gulnara Nyussupova Department of Geography, Land Management and Cadastre, Al-Farabi Kazakh National University, Almaty, Kazakhstan

Münir Öztürk Vice President "The Islamic World Acadmey of Sciences" Amann-Jordan and Centre for Environmental Studies-Botany Department, Ege University, Izmir, Turkey

Kanat N. Sarsenbayev Biological Sciences, Eurasian National University named after N. Gumilyov, Astana, Kazakhstan

William N. Setzer Department of Chemistry, University of Alabama in Huntsville, Huntsville, AL, USA

Zeki Severoğlu Biology Department, Kyrgyzstan-Turkiye Manas University, Bishkek, Kyrgyzstan

Biology Department, Faculty of Science and Arts, Marmara University, Istanbul, Turkey

Farukh Sharopov Department of Pharmaceutical Technology, Avicenna Tajik State Medical University, Dushanbe, Tajikistan

Zaineb Sohail Atta-ur-Rahman School of Applied Biosciences, National University of Sciences and Technology, Islamabad, Pakistan

Kristina Toderich ICBA-CAC (International Center for Bio-saline Agriculture in Central Asia and Caucasus), ICARDA Regional Office (The CGIAR Facilitation Unit) for CAC, Tashkent, Uzbekistan

Aigul Tokbergenova Department of Geography, Land Management and Cadastre, Al-Farabi Kazakh National University, Almaty, Kazakhstan

About the Editors

Dilfuza Egamberdieva graduated with a degree in biology from the National University of Uzbekistan and received her PhD in Agricultural Sciences from the Humboldt University of Berlin, Germany, in 2000. She conducted her postdoctoral studies at the Helsinki University of Finland, University of Florence, Manchester Metropolitan University, and Leiden University of Netherlands. She has a long-term experience in plant and soil biotechnology research and has collaborated widely with scientists across the world. She served as international project coordinator in her home country on a variety of topics, including soil improvement, crop productivity, food security, and sustainable management of natural resources. She is member of the Asian PGPR Society for Sustainable Agriculture. Dr. Egamberdieva is the recipient of numerous fellowships and awards, including the UNESCO-L'Oreal Fellowship for Women in Science and Alexander von Humboldt Fellowship. In 2012 she received the TWAS Prize in Agricultural Sciences for her contributions to the study of plant-microbe interactions in stressed environments. She is a member of several journal editorial boards, the author of 4 books, and coauthor of over 100 publications related to plant nutrition, agricultural biotechnology, and soil fertility.

Münir Öztürk has served at Ege University Izmir, Turkey, for 50 years. His fields of scientific interest are plant ecophysiology, medicinal and aromatic plants, conservation of plant diversity, biosaline agriculture and crops, and pollution and biomonitoring. He has over 450 publications to his credit, including 42 books, 68 book chapters, and nearly 200 papers in impact factor journals. He currently serves as Vice President of the Islamic World Academy of Sciences. He has received fellowships from the Alexander von Humboldt Foundation, Japanese Society for the Promotion of Science, and National Science Foundation (USA). Dr. Ozturk has served as Chairman of the Botany Department and Founding Director of the Centre for Environmental Studies at Ege University in Izmir, Turkey. He has also served as Consultant Fellow for the Faculty of Forestry at Universiti Putra Malaysia, Malaysia, and as Distinguished Visiting Scientist for the ICCBS, Karachi University, Pakistan.

Chapter 1
Spatiotemporal Assessment of Vegetation Trends in the Post-Soviet Central Asia

Olena Dubovyk

Abstract Currently there is a gap of spatially and temporally explicit information on vegetation cover dynamics and trends in the post-Soviet Central Asia at spatial scales sufficient to support decision-making in the region. Insufficient information also exists concerning vegetation variability across climatic gradients as well as vegetation response across different land uses, from natural rangelands to intensively irrigated croplands. We analyzed vegetation cover changes in five Central Asian countries in this study. This analysis included trends in key vegetation phenological parameters derived from 250 m Moderate Resolution Imaging Spectroradiometer (MODIS) normalized difference vegetation index (NDVI) time-series data for 2000–2011. In order to follow the vegetation changes over time, we calculated trends in phenometrics using a robust trend analysis method. The results showed that inter-annual vegetation dynamics followed precipitation patterns only outside irrigated areas, while clearly differentiated winter and summer seasons were observed throughout the study area. Specifically spatial patterns of long-term vegetation trends allowed defining areas characterized by decrease in overall vegetation greenness and peak greenness as well as revealing the shifts in timing of occurrence of peak greenness over the monitoring period. The information obtained will prove as a useful guide in the selection of field sites for detailed vegetation surveys and land rehabilitation interventions as well as improvement of overall understanding of vegetation dynamics and variability in the remote regions of Central Asia.

Keywords Vegetation · Phenology · Remote sensing · Land degradation · Central Asia

O. Dubovyk (✉)
Center for Remote Sensing of Land Surfaces (ZFL), University of Bonn, Bonn, Germany
e-mail: odubovyk@uni-bonn.de

D. Egamberdieva, M. Öztürk (eds.), *Vegetation of Central Asia and Environs*,
https://doi.org/10.1007/978-3-319-99728-5_1

1.1 Introduction

Countries in post-Soviet Central Asia (CA) have experienced drastic socioeconomic and environmental changes after the collapse of Soviet Union in 1991. This has led to extensive changes in land use and land cover (LULC) due to ongoing various transformations within the region (Mueller et al. 2014). For example, extensive irrigated areas established within CA, particularly in the downstream areas of the Amu Darya and the Syr Darya rivers, are reported to have almost disappeared specifically the marginal lands due to land degradation, poorly maintained irrigation and drainage infrastructure, and frequent droughts (Karthe et al. 2014). Ongoing socioeconomic processes, such as labor migration, combined with cropland degradation have caused widespread abandonment of not only irrigated but also rainfed cropland (Löw et al. 2015). In the light of predicted increasing scarcity of water resources resulting from climate change and reinforced by constant increase of population in the region (2014), understanding of vegetation dynamics has become a very important issue for the region.

Vegetation dynamics is a function related to the integrated indicator of vegetation responses to environmental factors (temperature, precipitation, topography) as well as factors related to human activities. Optical remote sensing provides the most cost- and resource-effective technique for assessing and monitoring vegetation dynamics and degradation over extensive areas. Alternative methods which rely on extensive field surveys are very resource demanding.

Vegetation phenology and vegetation cover trends in dryland ecosystems have not been intensively studied when compared to the more temperate regions (Siegfried et al. 2012). Often a patchy distribution of semiarid and arid vegetation is caused by erratic and scattered rainfall events, while micro-site variations in biophysical characteristics of the landscape further cause differences in vegetation patterns (Fensholt and Proud 2012). In the case of CA countries, the complexity of the landscape is not only a result of harsh climatic conditions but also human activities that have drastically transformed natural environment, thereby contributing to the variability and complexity of the regional landscapes. There is an urgent need for vegetation studies based on medium to high spatial resolution satellite images.

Frequent synoptic data observations from earth observation satellites are increasingly used to monitor spatially explicit vegetation cover patterns (Landmann and Dubovyk 2014). The Terra-Moderate Resolution Imaging Spectroradiometer (MODIS) was specifically designed to monitor land surface characteristics (Klein et al. 2014) that allows capturing vegetation dynamics at a relatively high level of spatial detail (250 m pixel size) (Bajocco et al. 2012). Considering the current call for improved spatial and temporal resolution of satellite time series to assess vegetation cover and trends (Walker et al. 2015) and given the availability of 250 m time-series data from MODIS spanning since 2000, advances in the mapping of spatial and temporal vegetation patterns are becoming easier to follow. Moreover, the use of remote sensing image time-series data is currently the only means to provide a synoptic view on environmental phenomena at different spatial and temporal scales.

Recently, a number of studies have focused on analysis of LULC at local to regional scales in CA. The studies undertaken by Fensholt and Proud (2012), Dubovyk et al. (2015), Zhou et al. (2015), and Kariyeva and van Leeuwen (2012) have analyzed the vegetation based on long-term normalized difference vegetation index (NDVI) data from the Advanced Very High Resolution Radiometer (AVHRR). De Beurs et al. (2015) assessed LULC changes in Kazakhstan using AVHRR-NDVI satellite time series. Land degradation and relations between degradation and possible triggers (soil quality, land use, groundwater level and salinity, irrigated infrastructure) have been observed at local scale in Uzbekistan using Moderate Resolution Imaging Spectroradiometer (MODIS) time-series data (Gessner et al. 2013). There are, however, only a few studies in CA that have explicitly focused on analysis of simultaneous changes in vegetation phenology and cover degradation. In addition, so far only a limited number of studies have applied medium spatial resolution satellite data time series in the region. Such datasets are likely to be more applicable and accurate for monitoring fragmented drylands landscapes of this region (Propastin et al. 2008).

Our aim in this chapter is to present an analysis of the vegetation dynamics and vegetation cover trends across the post-Soviet CA region using time series of 250 m MODIS-NDVI data. We hypothesize that (1) vegetation dynamics in CA varies across environmental and management gradients, (2) spatiotemporal dynamics of these parameters within and between years can be detected by the 250 m MODIS-NDVI data, and (3) it is possible to interpret spatial patterns of vegetation dynamics and trends with respect to climate- and/or human-induced land transformations.

1.2 Study Area

The study area consisted of five post-Soviet CA countries: Kazakhstan, Kyrgyzstan, Tajikistan, Turkmenistan, and Uzbekistan (Fig. 1.1). The climate is sharply continental with a distinct humidity gradient increasing south to north of the region. The mean summer temperatures lie between +2 °C and +31 °C, and the mean winter temperatures usually vary between −25 °C and +7 °C. At the same time, in the mountain areas the minimum temperatures can reach −45 °C, and in the desert areas the maximum temperatures can be as high as +50 °C (de Beurs and Henebry 2004). The precipitation has an uneven distribution and varies considerably (Dubovyk 2013). The mean annual precipitation is reported to lie between 60 mm in the driest areas in northern Turkmenistan and southern Uzbekistan and CA. 1000 mm in the mountainous areas of Tajikistan and Kyrgyzstan. Maximum precipitation occurs during winter and early spring, whereas summers are usually hot and dry (Dubovyk et al. 2013).

The study area includes three distinct subregional landscapes based on the Köppen climate classification (de Beurs et al. 2015): (1) northern semi-steppe and steppe subregion, (2) mountainous subregion, and (3) central semidesert/desert subregion. Rainfed cropland predominates in the northern and mountainous subregions, while irrigated agriculture is located in the central region.

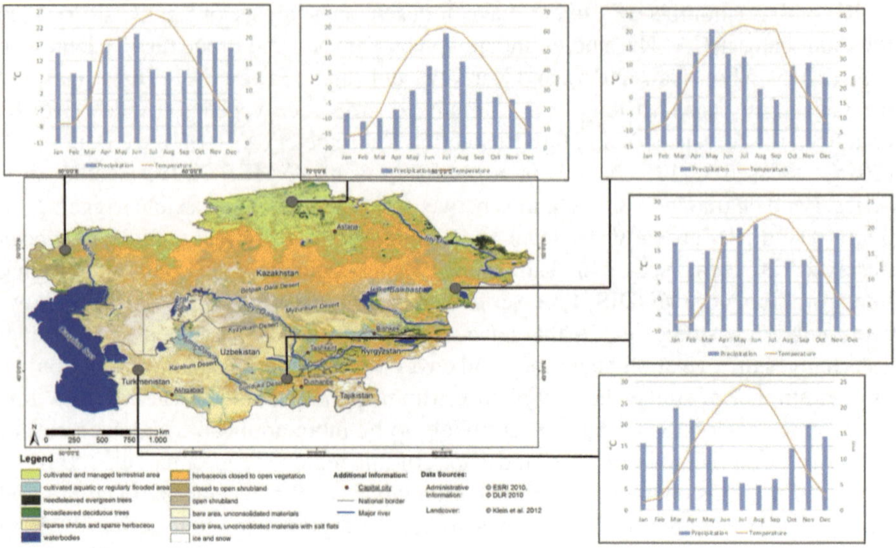

Fig. 1.1 The study area: post-Soviet Central Asia and the Walter-Lieth climate diagrams for different locations in the region

1.3 Methods

1.3.1 Satellite Data Processing

The NDVI time-series data for 2000–2011 from the MODIS 250 m MOD13Q1 (collection 5) product served as a main input for this analysis. The MOD13Q1 images are atmospherically corrected (Allison 2009) and composed of optimal observations acquired during a 16-day period. Optimal observations were calculated by evaluating overall pixel quality in terms of aerosol content, low view angle, and absence of clouds and/or cloud shadows. The further preprocessing step included time series smoothing with a Whittaker smoother algorithm (Walker et al. 2014) to reduce data noise levels resulting from cloud and cloud shadow contamination.

Vegetation activity is strongly linked to temperature and precipitation (Löw et al. 2015; Tüshaus et al. (2014). We used the updated high-resolution grids of monthly climatic observations (CRU TS3.10 Dataset), as these datasets are the best available time series of temperature and precipitation available for this region and for the observation period (Ganguly et al. 2010). Monthly aggregated CRU observations are gridded at a resolution of 0.5°. The CRU TS3.10 Dataset is derived from archives of climate station records that have been subject to extensive manual and semiautomated quality control measures.

1.3.2 Derivation of Seasonal Vegetation Metrics fro[illegible] and Climatic Data

We used harmonic regression to extract key seasonal vegetation metrics (or phenometrics) from the NDVI, precipitation, and temperature data. The formula for the generalized seasonal curve as derived from harmonic regression is as follows (Shuai et al. 2013):

$$y = a_0 + a_n \sin\left(\frac{2\pi nt}{T} + \varphi_n\right) \tag{1.1}$$

where α_n are amplitudes and the ϕ_n are phase angles ranging from 0° to 360°, n is a harmonic.

Amplitude 0 (a_0) is an overall metric, representing the mean annual value for NDVI/temperature/precipitation. Amplitude 1 (a_1) represents the magnitude or peak annual temperature and precipitation. Phase 1 (φ_1) refers to the timing of the annual peak defined by the position of the starting point of the corresponding sine wave of annual greenness. Phase image values potentially range from 0° to 360° such that each 30° represents a shift of approximately one calendar month.

1.3.3 Seasonal Trend Analysis

The computed seasonal metrics from NDVI, T, and P were used as an input, and from this we calculated trends and their significances using the Theil-Sen (TS) estimator (Mirzabaev et al. 2016; Xu et al. 2015) and also the Mann-Kendall (MK) test (Lioubimtseva and Henebry 2009; Kariyeva et al. 2012). This technique is robust to short-term inter-annual variability up to 29% of the length of the series and specifically rejects high-frequency sub-annual noise (Klein et al. 2012). The TS estimator is appropriate for noisy and short time series because it is robust to outliers and has a breakdown value of 29% (le et al. 2012). This means that outliers do not affect the values of the TS slopes when they are not in excess of 29% of the observations used in the time series. The significance of TS slopes was evaluated using the nonparametric MK test that is also robust to outliers (Harris et al. 2013).

1.4 Results and discussion

1.4.1 Observations

A 12-year distribution of vegetation cover of CA is presented in Fig. 1.2. It reflects a distinct climatic gradient in the region. The higher overall greenness (annual mean NDVI) values are located in the north and southeast parts of the region during

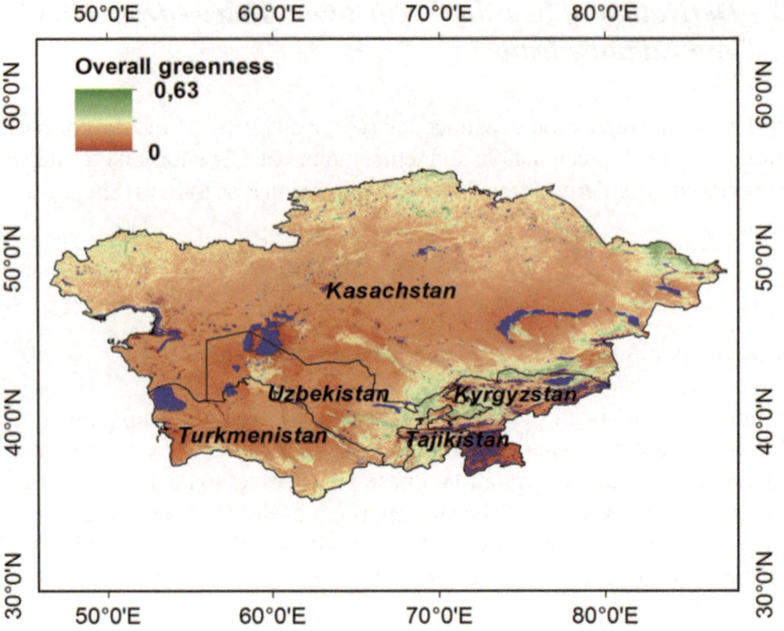

Fig. 1.2 Spatial patterns of vegetation distribution as described by overall greenness calculated from the 16-day time series of 250 m MODIS-NDVI for the time period of 2000–2011

2000–2011, whereas the lowest overall greenness values are located in the south (Fig. 1.2b, d). In general, higher overall greenness was observed in the northern semi-steppe and steppe subregion of CA that covers most of Kazakhstan. Vegetation dynamics in this subregion is driven primarily by climatic regimes that follow a latitudinal gradient (Zhou et al. 2015). Due to favorable precipitation conditions, almost all of the rainfed croplands of CA are located within this zone of the region (Mirzabaev et al. 2016). In 1954–1961, the Virgin Lands Campaign converted the rangelands located here that were most suitable for agriculture; into the cultivated cropland. Thereafter, land cultivation gradually expanded further southward to the lands less suitable for agriculture. Agricultural expansion to these marginal areas has resulted in severe land degradation and subsequent large-scale cropland abandonment, particularly visible in this area of CA.

The mountainous subregion portions of Tajikistan and Kyrgyzstan also experienced higher than average values of overall greenness. Vegetation growth patterns in this part follow a distinct latitudinal gradient, while they are constrained mainly by temperature thresholds rather than precipitation availability. This part of CA is reportedly most affected by the gradually increasing temperatures associated with global climate change (Glantz 2015). The lowest vegetation activity was detected in central and southern parts of Turkmenistan and Uzbekistan. These lands are located within the central desert subregion, which is the most arid part of the CA. Total annual precipitation within this subregion is about 100 mm. This area also includes most of the large-scale irrigated agricultural lands in the region with the main crops of cotton and winter wheat cultivated under the state procurement system (Dietz et al. 2013). In the decades of often-unsustainable irrigated water and land management, cropping on marginal lands unsuitable for agriculture has led to a widespread land degradation and abandonment to the cropland, like in the northern semi-steppe and steppe subregion (Siegfried et al. 2012).

1.4.2 Spatial Patterns of Vegetation, Precipitation, and Temperature Trends

The spatial patterns of the significant NDVI trends including trends in (a) overall greenness, (b) peak greenness, and (c) timing of peak greenness are shown in Fig. 1.3. The positive and negative values indicate the presence of increasing and decreasing trends over the monitoring period, respectively. In the case of timing of peak greenness, an increasing or decreasing phase angle indicates a shift to earlier

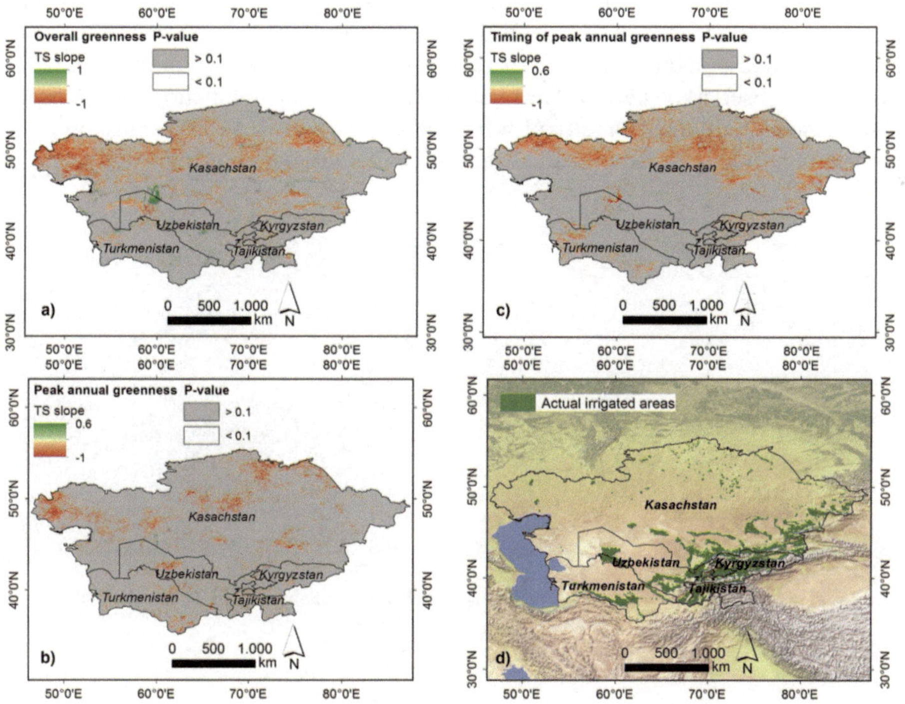

Fig. 1.3 Spatial patterns of vegetation trends calculated from the 16-day 250 m MODIS-NDVI time-series for 2000–2011: (**a**) overall greenness, (**b**) peak greenness, (**c**) timing of peak greenness, and (**d**) distribution of irrigated croplands

or later time of the year. Figure 1.3d shows the distribution of the irrigated cropland in the CA (Dietz et al. 2015) as additional reference information for interpretation.

Results of trend analysis indicate that the negative trends prevailed throughout the region with the positive trends manifesting only on a minor part of the total land area of the region. The spatial patterns of the negative trends show spatial coherence for overall greenness and peak greenness (Fig. 1.3a, b). Trends in peak greenness were similarly distributed as were overall greenness trends, though the area characterized by significant trends ($p < 0.1$) of peak greenness were slightly larger compared to those in overall greenness. In terms of the trends in timing of peak greenness (Fig. 1.3c), there is a distinctive pattern revealing a shift of the peak greenness to a later time (negative trends in Fig 1.3c) in the northern part of the study area (mainly in the northern and eastern part of Kazakhstan). In contrast, a tendency toward the shift of peak greenness to an earlier time during the monitoring period was not detected in the region within the monitoring period.

Large clusters of the negative trends for both overall peak and timing of peak greenness were clearly observed in the northern semi-steppe and steppe subregion located within the administrative borders of Kazakhstan. During the observation period, the clusters of the negative trends in the northern part of Kazakhstan were located within the areas that experienced increasing temperature trend and decreasing rainfall trends (compare Fig. 1.3a, b with Figs. 1.4 and 1.5). Although the trends in temperature and rainfall were mostly insignificant across the region, their negative or positive sign indicates the general tendency in the analyzed climatic variables during the observation period. The overall existing climate prediction for CA report the foreseen increasing scarcity of water resources, as well as an increase in mean temperature and extreme climatic events, such as droughts, resulting from global

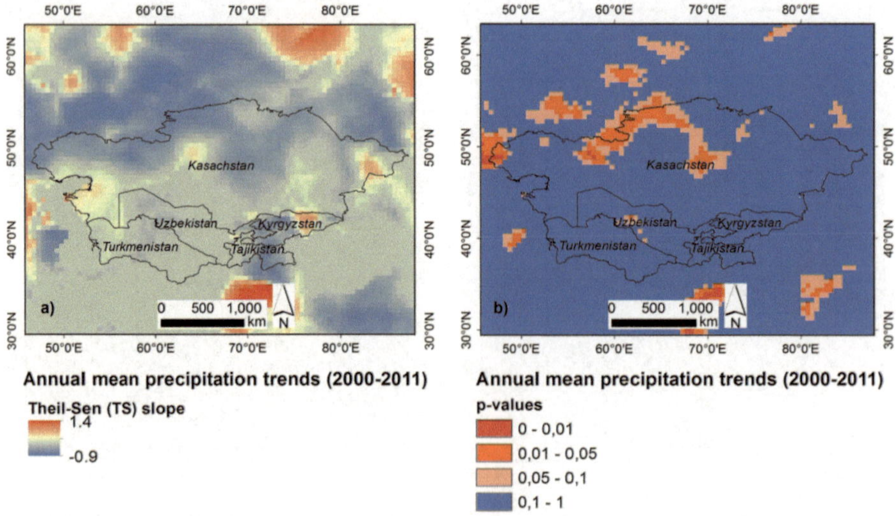

Fig. 1.4 Spatial patterns of precipitation trends calculated from CRU TS3.10 Dataset for 2000–2011: (**a**) Theil-Sen slope and (**b**) *p*-value

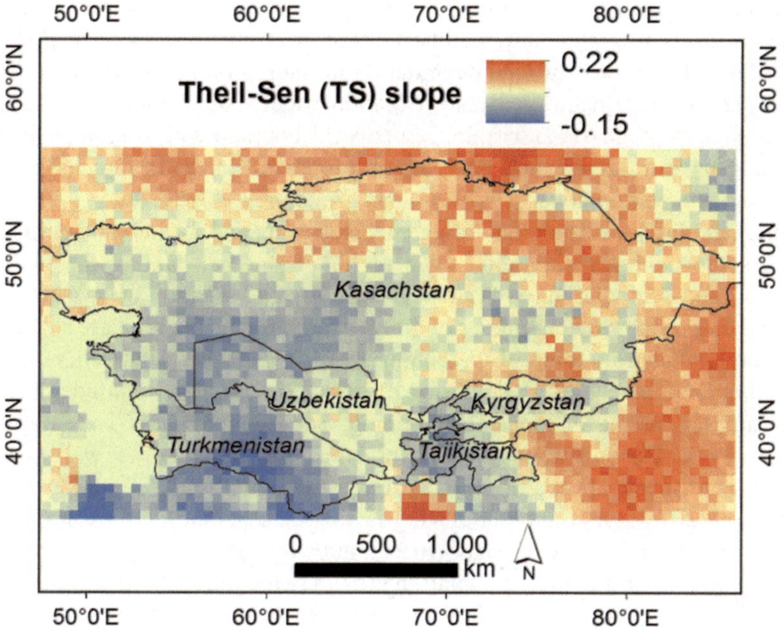

Fig. 1.5 Spatial patterns of temperature trends calculated from CRU TS3.10 Dataset for 2000–2011: Theil-Sen slope

climate change (Dubovyk et al. 2013). The latter was also to certain extent reflected in the results of the trend analysis conducted here (Figs. 1.4 and 1.5). The decreasing precipitation trends were found across the CA, while the increasing temperature trends were observed mainly in the northern part of the region. The corresponding *p*-values for temperature trends are not shown in Fig. 1.5 as *p*-values exceeded the threshold level of 0.1 across the region.

In addition to the climatic drivers of vegetation trends, the socioeconomic factors play an important role in defining the spatiotemporal trends in vegetation cover. For example, the processes of cropland abandonment, though widespread throughout the region, have been specifically acute in the northern Kazakhstan, probably leading to the herein detected clusters of the negative vegetation trends (Dubovyk et al. 2015), After the collapse of the Soviet Union in 1991 and due to the process of transition to a market economy, the sharp decline in cropland area took place across CA (Dietz et al. 2015). Among the factors that have led to such results are the loss of previously guaranteed markets, disintegration of value chain supplies, and failing price relationships between inputs and outputs during the transition period (Eastman et al. 2009).

Another important trigger of negative vegetation trends is an ongoing land degradation in the region (Freedman 2006), while within rainfed cropland areas and rangelands, land abandonment and overgrazing are the main triggers of the vegetation decline (Fig. 1.3a, b). The land degradation and reduced irrigation water supply

AG

are among the main factors of the reduced vegetation activity within the irrigated croplands of CA, which are mainly located in Uzbekistan and Turkmenistan (Fig. 1.3d). The process of land degradation in the region is governed by different factors. The unsustainable land and irrigation water management and inadequate maintenance of irrigation and drainage networks coupled with fragile environmental setting have triggered land degradation processes which in turn have led to a decrease in vegetation cover (Fig. 1.3a) (Siegfried et al. 2012).

Simultaneously, institutional arrangements, current and former land policies, and overall socioeconomic situation in the region are named among the underlying causes of land degradation (de Beurs and Henebry 2004; Bradley et al. 2007; Neeti et al. 2012). Former Soviet policies of wheat and cotton production are still continuously implemented in some of the countries, such as Uzbekistan. These have had resulted in an unprecedented expansion of irrigated cotton and rainfed wheat cultivation areas, often not suitable for cropping. Such policies have ended up with a lack of incentives for sustainable land practices in such areas as well as lack of maintenance of irrigation and drainage under the conditions of market economy (Dietz et al. 2015). Among other driving factors of vegetation trends in the region, the land tenure insecurity, breakdown of land/water management-related institutions, poverty, migration, and population growth (Mueller et al. 2014; de Jong et al. 2011; Kerlinger 1964; Kariyeva and van Leeuwen 2012) are adding to the preexisting problems. These suggest that in parallel to environmental factors that have a direct effect on the vegetation trends in the region, a special attention should be given to the analysis of socioeconomic and political factors. These factors have an indirect impact on the environmental processes but should not be neglected if a focus is on a comprehensive understanding of land dynamics and trends in the region. The incorporation of the full range of such casual factors in a quantitative manner is, however, currently constrained by availability of corresponding secondary datasets as well as by current limited access to existing socioeconomic information in the region (Siegfried et al. 2012).

1.5 Conclusions

The vegetation trend analysis over the post-Soviet Central Asia using 12-year (2000–2011) time series of medium spatial resolution (250 m) MODIS-NDVI data has been presented here. Robust seasonal trends have been derived using key phenometrics as overall greenness, peak greenness, and timing of peak greenness. The applied approach has enabled us to track spatiotemporal dynamics of these phenometrics across CA. Negative trends in overall greenness are more common than positive trends and have been identified throughout the study area with the biggest clusters located within rainfed cropland and rangelands in the northern Kazakhstan and within irrigated croplands of Uzbekistan and Turkmenistan. Trends in peak greenness are similarly distributed, as were the trends in overall greenness, though the number of pixels with significant trends is higher compared to trends in overall greenness. The timing of peak greenness tends to occur later within the year mainly in the northern parts of the study area and in the eastern part of Kazakhstan. Trend analysis of precipitation and temperature datasets have yielded mostly insignificant trend results, but these have allowed for a general understanding of trend direction in these key climatic variables. We conclude that both climatic and socioeconomic factors have derived vegetation trends in the region during 2000–2011. Non-irrigated croplands and naturally vegetated areas within northern and central subregions have been largely affected by (mainly insignificant) trends in both temperature and precipitation. Variables related to temperature have the greatest effects in irrigated areas and within mountainous landscapes. In addition, vegetation trends in CA are caused by numerous factors related to socioeconomic and political situation. The unsustainable land and water management practices, inadequate maintenance of irrigation and drainage networks, overgrazing, increasing poverty, and migration cause directly and indirectly such processes as land degradation and land abandonment which, in turn, have an impact on vegetation cover and productivity of the region. Finally, we can say that 250 m MODIS data, when appropriately processed and aggregated in a time series, is suitable for monitoring key vegetation phenometrics (i.e., overall greenness, peak greenness, and timing of peak greenness) and vegetation trends across large spatial extent. The long-term time series of the MODIS product allow us to derive information that is very useful for:

1. Assessment of vegetation dynamics and trends at regional level
2. Detection of the impact of precipitation and temperature trends and management driven changes (e.g., land abandonment and land degradation) on vegetation
3. Mapping of spatial patterns of vegetation cover trends as required for setting up the link to socioeconomic drivers

Acknowledgments We would like to thank Gohar Ghazaryan for his advice on the preparation of figures and to Francesco Vuolo for the MODIS data processing.

References

Allison PD (2009) Fixed effects regression models. SAGE, London, UK

Bradley BA, Jacob RW, Hermance JF, Mustard JF (2007) A curve fitting procedure to derive inter-annual phenologies from time series of noisy satellite NDVI data. Remote Sens Environ 106:137–145

Bajocco S, Angelis A, Perini L, Ferrara A, Salvati L (2012) The impact of land use/land cover changes on land degradation dynamics: a Mediterranean case study. Environ Manag 49:980–989

de Jong R, de Bruin S, de Wit A, Schaepman ME, Dent DL (2011) Analysis of monotonic greening and browning trends from global NDVI time-series. Remote Sens Environ 115:692–702

de Beurs KM, Henebry GM (2004) Land surface phenology, climatic variation, and institutional change: analyzing agricultural land cover change in Kazakhstan. Remote Sens Environ 89:497–509

de Beurs KM, Henebry GM, Owsley BC, Sokolik I (2015) Using multiple remote sensing perspectives to identify and attribute land surface dynamics in Central Asia 2001–2013. Remote Sens Environ 170:48–61

Dietz AJ, Kuenzer C, Conrad C (2013) Snow-cover variability in Central Asia between 2000 and 2011 derived from improved MODIS daily snow-cover products. Int J Remote Sens 34:3879–3902

Dietz AJ, Kuenzer C, Dech S (2015) Global snowpack: a new set of snow cover parameters for studying status and dynamics of the planetary snow cover extent. Remote Sens Lett 6:844–853

Dubovyk O (2013) Multi-scale targeting of land degradation in Northern Uzbekistan using satellite remote sensing. Ph.D. thesis, University of Bonn, Bonn, Germany

Dubovyk O, Menz G, Conrad C, Kan E, Machwitz M, Khamzina A (2013) Spatio-temporal analyses of cropland degradation in the irrigated lowlands of Uzbekistan using remote-sensing and logistic regression modeling. Environ Monit Assess 185:4775–4790

Dubovyk O, Landmann T, Erasmus BFN, Tewes A, Schellberg J (2015) Monitoring vegetation dynamics with medium resolution MODIS-EVI time series at sub-regional scale in Southern Africa. Int J Appl Earth Obs Geoinf 38:175–183

Eastman R, Sangermano F, Ghimire B, Zhu H, Chen H, Neeti N, Cai Y, Machado EA, Crema SC (2009) Seasonal trend analysis of image time series. Int J Remote Sens 30:2721–2726

Freedman D (2006) Statistical models: theory and practice. Cambridge University Press, Cambridge, UK

Fensholt R, Proud SR (2012) Evaluation of earth observation based global long term vegetation trends—comparing GIMMS and MODIS global NDVI time series. Remote Sens Environ 119:131–147

Ganguly S, Friedl MA, Tan B, Zhang X, Verma M (2010) Land surface phenology from MODIS: Characterization of the Collection 5 global land cover dynamics product. Remote Sen Environ 114:1805–1816

Gessner U, Naeimi V, Klein I, Kuenzer C, Klein D, Dech S (2013) The relationship between precipitation anomalies and satellite-derived vegetation activity in Central Asia. Glob Planet Change 110:74–87

Glantz MH (2015) Water, climate, and development issues in the Amu Darya Basin. Mitig Adapt Strat Glob Change 10:23–50

Harris I, Jones PD, Osborn TJ, Lister DH (2013) Updated high-resolution grids of monthly climatic observations—the CRU TS3.10 dataset. Int J Climatol 34:623–642

Karthe D, Chalov S, Borchardt D (2014) Water resources and their management in Central Asia in the early twenty first century: status, challenges and future prospects. Environ Earth Sci 73:487–499

Kerlinger FN (1964) Foundations of behavioral research. Holt, Rinehart & Winston, New York

Kariyeva J, van Leeuwen WJD, Woodhouse CA (2012) Impacts of climate gradients on the vegetation phenology of major land use types in Central Asia (1981–2008). Front Earth Sci 6:206–225

Kariyeva J, van Leeuwen WJD (2012) Phenological dynamics of irrigated and natural drylands in Central Asia before and after the USSR collapse. Agric Ecosyst Environ 162:77–89

Klein I, Gessner U, Künzer C (2014) Generation of up to date land cover maps for Central Asia. In: Mueller L, Saparov A, Lischeid G (eds) Novel measurement and assessment tools for monitoring and management of land and water resources in agricultural landscapes of Central Asia. Springer, Cham, Switzerland, pp 329–346

Klein I, Gessner U, Kuenzer C (2012) Regional land cover mapping and change detection in Central Asia using MODIS time-series. Appl Geogr 35:219–234

Lioubimtseva E, Henebry GM (2009) Climate and environmental change in arid Central Asia: impacts, vulnerability, and adaptations. J Arid Environ 73:963–977

Landmann T, Dubovyk O (2014) Spatial analysis of human-induced vegetation productivity decline over Eastern Africa using a decade (2001–2011) of medium resolution MODIS time-series data. Int J Appl Earth Obs Geoinf 33:76–82

Le QB, Tamene L, Vlek PLG (2012) Multi-pronged assessment of land degradation in West Africa to assess the importance of atmospheric fertilization in masking the processes involved. Glob Planet Change 92:71–81

Löw F, Fliemann E, Abdullaev I, Conrad C, Lamers JPA (2015) Mapping abandoned agricultural land in Kyzylorda, Kazakhstan using satellite remote sensing. Appl Geogr 62:377–390
Mirzabaev A, Goedecke J, Dubovyk O, Djanibekov U, Le Q, Aw-Hassan A (2016) Economics of land degradation in Central Asia. In: Nkonya E, Mirzabaev A, von Braun J (eds) Economics of land degradation and improvement—a global assessment for sustainable development. Springer, Cham, Switzerland, pp 261–290
Mueller L, Suleimenov M, Karimov A, Qadir M, Saparov A, Balgabayev N, Helming K, Lischeid G (2014) Land and water resources of Central Asia, their utilisation and ecological status. In: Mueller L, Saparov A, Lischeid G (eds) Novel measurement and assessment tools for monitoring and management of land and water resources in agricultural landscapes of Central Asia. Springer, Cham, Switzerland, pp 3–59
Neeti N, Rogan J, Christman Z, Eastman JR, Millones M, Schneider L, Nickl E, Schmook B, Turner BL, Ghimire B (2012) Mapping seasonal trends in vegetation using AVHRR-NDVI time series in the Yucatán Peninsula, Mexico. Remote Sens Lett 3:433–442
Propastin PA, Kappas M, Muratova NR (2008) Inter-annual changes in vegetation activities and their relationship to temperature and precipitation in Central Asia from 1982 to 2003. J Environ Inf 12:75–87
Siegfried T, Bernauer T, Guiennet R, Sellars S, Robertson A, Mankin J, Bauer-Gottwein P, Yakovlev A (2012) Will climate change exacerbate water stress in Central Asia? Clim Chang 112:881–899
Shuai Y, Schaaf C, Zhang X, Strahler A, Roy D, Morisette J, Wang Z, Nightingale J, Nickeson J, Richardson A, Xie D, Wang J, Li X, Strabala K, Davies JE (2013) Daily MODIS 500 m reflectance anisotropy direct broadcast (DB) products for monitoring vegetation phenology dynamics. Int J Remote Sens 34:5997–6016
Tüshaus J, Dubovyk O, Khamzina A, Menz G (2014) Comparison of medium spatial resolution ENVISAT-MERIS and Terra-MODIS time series for vegetation decline analysis: a case study in Central Asia. Remote Sens 6:5238–5256
Walker JJ, de Beurs KM, Henebry GM (2015) Land surface phenology along urban to rural gradients in the US Great Plains. Remote Sens Environ 165:42–52
Walker JJ, de Beurs KM, Wynne RH (2014) Dryland vegetation phenology across an elevation gradient in Arizona, USA, investigated with fused MODIS and Landsat data. Remote Sens Environ 144:85–97
Xu L, Zhou H, Du L, Yao H, Wang H (2015) Precipitation trends and variability from 1950 to 2000 in arid lands of Central Asia. J Arid Land 7:514–526
Zhou Y, Zhang L, Fensholt R, Wang K, Vitkovskaya I, Tian F (2015) Climate contributions to vegetation variations in Central Asian drylands: pre- and post-USSR collapse. Remote Sens 7:2449–2470

Chapter 2
Drought Variability and Land Degradation in Central Asia: Assessment Using Remote Sensing Data and Drought Indices

Dildora Aralova, Jahan Kariyeva, Timur Khujanazarov, and Kristina Toderich

Abstract The regional resilience of a landscape to climate change in water-scarce areas is one of the core environmental problems nowadays for Central Asian countries. Responses to increasing temperature and high evapotranspiration (ET_0) regimes have contributed to biodiversity loss and altered vegetation dynamics and changed the land use and management in these landscapes. Extremely dry conditions and droughts are recognized as an important factor that triggers land degradation in Central Asia. The aim of this study is to conduct attribution analysis to assess drought trends that are quantified using the Standardized Precipitation-Evapotranspiration Index (SPEI) and effects of other biophysical factors on the region and at a country level. The kriging (geostatistics) method was utilized to predict the status of vegetation change trends and generalize additive smoothed parameters to provide response factors for changes of land cover status. Specific

D. Aralova (✉)
Dresden Technology University (TU-Dresden), Institute of Photogrammetry and Remote Sensing, Dresden, Germany

Centre of Hydrometeorological Service at Ministry of Emergency Situations of the Republic of Uzbekistan (Uzhydromet), Tashkent, Uzbekistan
e-mail: aralovad@daad-alumni.de

J. Kariyeva
Alberta Biodiversity Monitoring Institute, Edmonton, AB, Canada
e-mail: kariyeva@ualberta.ca

T. Khujanazarov
Water Resources Research Center, Disaster Prevention Research Institute, Kyoto University, Uji, Japan

K. Toderich
ICBA-CAC (International Center for Bio-saline Agriculture in Central Asia and Caucasus), ICARDA Regional Office (The CGIAR Facilitation Unit) for CAC, Tashkent, Uzbekistan
e-mail: ktoderich@cgiar.org

D. Egamberdieva, M. Öztürk (eds.), *Vegetation of Central Asia and Environs*,
https://doi.org/10.1007/978-3-319-99728-5_2

objectives of the study were (a) *to assess drought trends and their effects on climate–vegetation trends at the regional and local level*; (b) *identify the main affected regions among five countries* (Kazakhstan, Kyrgyzstan, Tajikistan, Turkmenistan, and Uzbekistan) and *characterize their patterns for monitoring land tenures*; and (c) *define appropriate ecological risk zones*, especially trends of spatial changes over time with drought trends. The simulated and predicted maps with kriging dependence terms indicated that the climate–vegetation-driven dataset will suffer substantial losses of vegetation health [normalized difference vegetation index (NDVI)] in precipitation-driven regions of Turkmenistan, Uzbekistan, and Tajikistan, and that these areas, especially, Ahal and Lebap Provinces in Turkmenistan, Kyzylorda in Kazakhstan, Karakalpakstan Autonomous Republic in Uzbekistan, and Gorno-Badakhsan Autonomous Region (GBAR) in Tajikistan, are very sensitive to droughts, which might alert us to the fragility of this ecosystem.

Keywords Central Asia · Droughts · SPEI · NDVI-GIMMS3g · Precipitation · CRU-TS, kriging method

2.1 Introduction

Approximately 75% of the land in Central Asia (Kazakhstan, Kyrgyzstan, Tajikistan, Turkmenistan, and Uzbekistan) is occupied by rangelands (Bedunah et al. 2006); the total area of pastures in Central Asia is 262 million hectares (ha) (Squires 2012). Sustainable management of rangelands is a key goal for Central Asian countries, and the main challenge today is to conserve and restore ecosystems that are vital to human well-being. The effects of shrinking the Aral Sea Basin (Toderich et al. 2013) coupled with the USSR (Soviet Union) collapse have caused changes in land use practices (Kariyeva 2011), such as uncontrolled grazing, which lead to salinization and further deterioration of rangeland ecosystems (Aralova et al. 2016) in the region. During the USSR period, most of the agricultural policies were directed by the centralized government's decisions, which contributed strongly to degradation of dryland areas. A longer drought events and subsequent soil salinization are threats that have major impacts on land cover and land use change (LCLUC) patterns in the agricultural zones of Central Asia. The anthropogenic impact of converting vast dryland areas to irrigated lands, coupled with the accelerated increase of average temperature (as drivers) and precipitation anomalies observed in past decades, has been changing LCLUC dynamics in those areas. According to Nicholson et al. (1998) and Pickup (1998), drought has been considered to be a major factor in triggering land degradation and increasing desertification categories. However, some studies have argued that drought cannot be a factor explaining degradation trends (Vicente-Serrano et al. 2015a, b).

Central Asian countries still attribute 10–25% of their gross domestic product (GDP) to agriculture (FAOstat 2015). Large areas of steppe rangelands are used as pastures for breeding cattle, where it is the main income source for the rural population; also, 20–50% of the total employment rate is acquired in the agricultural sector (FAOstat 2015). Agriculture remains is an important sector in the economy of Central Asia, contributing 5.2% of the GDP in Kazakhstan, 7.5% in Turkmenistan, 18.5% in Uzbekistan, 20.8% in Kyrgyzstan, and 23.3% in Tajikistan (Abdullaev 2014; Bobojonov and Aw-Hassan 2014). The financial capacity of most countries is not robust enough to allocate adequate funds for drought management and mitigation, or for large-scale relief operations in the agriculture sector, especially in the developing or least developed countries. However, a number of remote sensing–based vegetation and drought indices have been developed for use in estimating vegetation status (such as the normalized difference vegetation index, NDVI) and degradation level, and Vegetation Drought Response Index (VegDRI) studies have begun to investigate the use for a drought-monitoring system in the United States (Wardlow et al. 2012a, b). The VegDRI targets the effects of drought on vegetation with applying general vegetation conditions (soil, elevation, etc.) and ratings of dryness extended by applying climate-based drought indices. Different algorithms have been proposed in drought research (Paulik et al. 2014; Ciabatta et al. 2016; Musyimi 2011; Kerr et al. 2012), but these are generally based on the exploitation of the available datasets (satellite derived), and suitable implementation for Central Asian landscapes has not yet been done. Severe to extreme drought conditions are causing a serious issue in dryland ecosystems, and the effects of drought on vegetation patterns in Central Asia were evident several times in the past two decades, and observations in various regions (Kariyeva and van Leuwen 2010) also affirmed these assumptions. In drylands, during a period with a high-temperature trend (June–August), drought days will increase more than three times in most areas of Central Asia at the end of the twenty-first century (Touge et al. 2015). Therefore, contradictory extremes (high rainfall ratings and hot temperature) are also negative influences on this landscape. As mentioned, there are still large gaps in managing or forecasting with contributing/creating an early warning system in the region level. In Central Asia, approximately 80% of rural people are involved in the agricultural sector, and the landscape of this area is vulnerable to droughts. As well, as can be drawn from the episode years of 2000–2001, such droughts hit in all five countries, resulting in rainfall levels decreasing below average (approximately 60% to 40%); and river flows dropping by 35–40% from those in a normal period. According to the accounts of Mirzabaev et al. (2015), the cost of land degradation categorized between 2001 and 2009 in Central Asia was estimated to be about $6 billion annually, as the result of rangeland degradation ($4.6 billion), followed by desertification ($0.8 billion), deforestation ($0.3 billion), and abandonment of croplands ($0.1 billion). As a result, most regions of the area were unable to manage disaster risks and to effectively recover from disasters.

A few analyses and studies provide for accurately quantifying drought trends and anomalies and finding relationships with vegetation pattern dynamics for these landscapes. Many researchers have analyzed vegetation sensitivity to climate trends (Propastin et al. 2008; Kariyeva 2011) between precipitation episodes (Gessner et al. 2012).

Drought indices such as the Standardized Precipitation Evapotranspiration Index (SPEI) allows drought severity to be compared through time and space, as it has estimated over a wide range of climates. A different scientific discipline enables detection, monitoring, and analysis of droughts and drought indices (Vicente-Serrano et al. 2015a, b). In this chapter, various satellite-based datasets were used and evaluated under kriging methodology, with the aim to forecast upcoming warming regions or sensitive areas in upcoming drought years or vegetation health threats in the case of Central Asia. As examples, we selected Lebap (Turkmenistan), Navoi (Uzbekistan), Kyzylorda (Kazakhstan), the Gorno-Badakhshan Autonomous Region (GBAR; Tajikistan), Ulytau (Kazakhstan), and Naryn (Kyrgyzstan).

The SPEI was calculated with the Penman–Monteith method following the work of Vicente-Serrano et al. (2017) and linking-based practices for drought risk reduction. The kriging method might be able to evaluate and understand the strength of NDVI drought–climate relationships and further reducing or warning of disaster risks in Central Asia. Rangeland diversity in Central Asia remains one of the important tasks of the Central Asian countries, understanding past and current conditions (Gintzburger et al. 2003) and forecasting future status. Also, it is important to provide measures for delay periods as reported by Udelhoven et al. (2009) and Gasparrini (2011), and to establish an adaptation mechanism for arid ecosystems. Especially between boundary countries and remote areas, adaptation strategies with action plans must be evaluated in various biodiversity loss areas of Central Asia.

2.2 Materials and Methods

2.2.1 Location

Central Asia includes a range of landscapes from mountains to steppes and deserts (Turkmenistan, Kazakhstan, and Uzbekistan) and is isolated by mountain ranges (Kyrgyzstan, Tajikistan) and the Caspian Sea (the borders with Turkmenistan, partly with Kazakhstan). The spatial extent of dryland rangelands in Central Asia is vast. The vegetation trends in this area (Fig. 2.1a) are mostly driven by precipitation and temperature dynamics; for example, warming temperature trends after spring lead to increasing NDVI values. As evaluated, non-cropland area is dominated by rangelands and mostly used in the agricultural sector for short-term crop rotation (Fig. 2.1b). As described on the mapped part of Central Asia (Fig. 2.2), the annual ascending ratings of precipitation are highly distributed mostly throughout the mountainous zones of Central Asia (Tajikistan and part of Kyrgyzstan), and more precipitation occurs on the eastern side of the regions than on the other sides of Central Asia.

In general, the objectives of this research are to assess and identify possible decreasing vegetation trends with quantifying SPIE trends and may enable under-

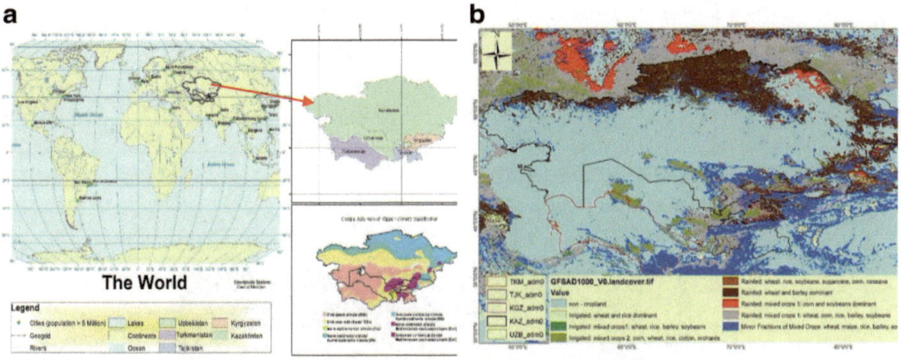

Fig. 2.1 Target area description: location of the middle zone of Asian continent (left) and climate classifications based on assumptions of Koeppen methodology (right). (**a**) Location of Central Asia and five neighbor countries and functional relationships with potential climate attributes in regions. Reference system: WGS84. Central Asian map with Köppen climate classification (right, down) with extraction permanent water attributes. (**b**) Central Asian Cropland Extent and Crop Dominance, modified after GFSAD1000 V0.0 Classes. Image resolution is aggregated to 500 m and projected to the Central Eurasian part from global cropland extent project. Accessed Google Earth Engine, May 2017 (Thenkabail et al. 2012)

standing the mechanism of land degradation in Central Asia or the regional scale of degradation within implementing various drivers, for example, Köppen classification, land use dynamics, and land cover dynamics.

2.2.1.1 Climatic Parameters and Description.

Precipitation in Central Asia Between 1982 and 2015

The analysis of spatial resolution observed by Gong et al. (2017) mentioned that high-value centers of precipitation appear in Kyrgyzstan, and then in Tajikistan. Our findings show similarity, the only exceptions being Turkmenistan, in that Turkmenistan has high precipitation accumulation ratings only for the eastern side. Accumulated high–moderate rating areas are the western part of Uzbekistan, the northern and northwestern part of Kazakhstan, and the northern part of Turkmenistan. Generally, the Central Asia precipitation trend has increased in some regions, especially in the month of May, 2015, in Kazakhstan; the opposite contradictions happened also in the southern areas such as Uzbekistan (eastern side) and Turkmenistan (northeastern side), where the precipitation trends were decreased from November months until February (trends are seen in Fig. 2.2). Accounting yearly values of precipitation also described that in that part of Kazakhstan, Kyrgyzstan and Uzbekistan have raised precipitation ratings for the year 2015. Even a peak of precipitation trends was seen in Kazakhstan in 2015, which was troubled with waterlogging in the north, and on the northwestern side of

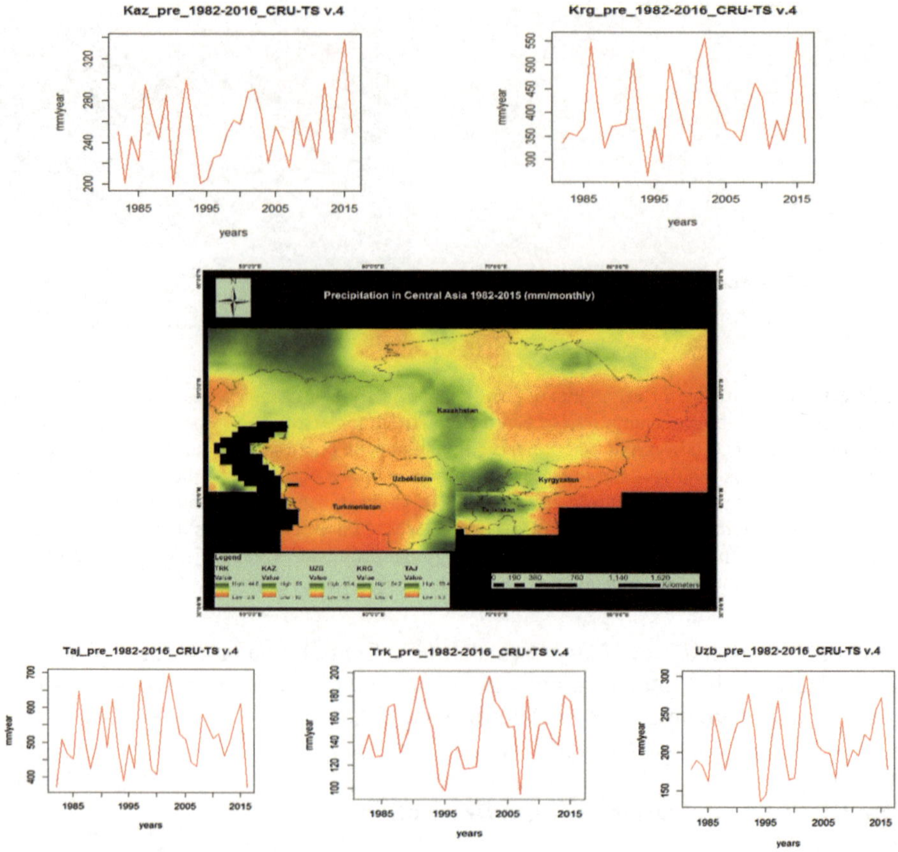

Fig. 2.2 Precipitation trends in the five countries estimated from the period 1982–2015 and illustrated yearly contributions (up side and down side) and monthly mapping of precipitation (in the middle) in Central Asia for the same period. All datasets based on CRU-TS 4 version

Kazakhstan. At the same time, the rate of ET_0 is increased (not shown) and continued at least 90 days per year in these regions. In those periods, land surface precipitation reached 0–10 mm/month. Results show that under climatic conditions in which low annual variability of precipitation predominates (Turkmenistan, Uzbekistan, part of Kazakhstan), the drought indices respond mainly as higher negative trends. Also, annually trends of precipitation is responding on high productivity of the rainfed zones, and these areas are generally for Kazakhstan, Tajikistan, and Kyrgyzstan, where drought indices contributing positively on mostly estimated periods and as well as, some years also contributed as negative. These selected and targeted areas are dependent on rainfall rates. But, controversy also occurs. At the same time, in these areas the hottest period of summer was

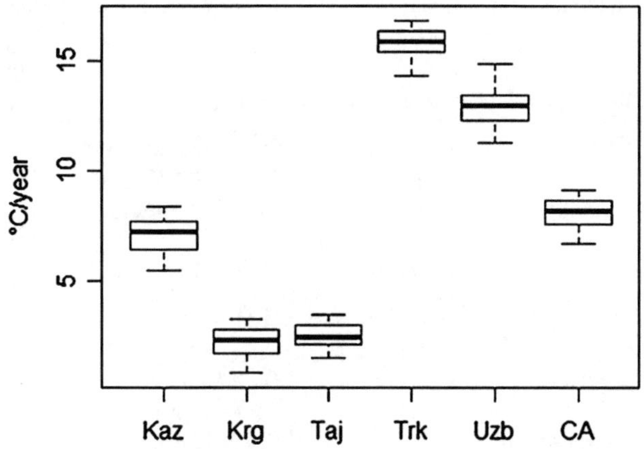

Fig. 2.3 Boxplot of annual average temperature in Central Asian countries based on monthly dataset CRU-TS ver.4 and generally for Central Asia

observed, after heavy rainfall (in 2015). However, the temperature ratings estimations are based on annual average calculations (Fig. 2.3), and, in fact, the sum of annual rainfall was greater than <500 mm/yearly; the environment or temperature is cooler in this country, similar to Tajikistan and Kyrgyzstan (Fig. 2.2).

2.2.2 *Classification of Targeted Area*

As shown on Fig. 2.1a, three main categories apply in the south and southwestern parts of Central Asia (Table 2.1): the categories with letter "*B*" are classified as *desert zones* and the main criteria for this category are MAP $<5 \times P^{1}_{\text{threshold}}$; obviously the Köppen categories are described as a high ET_0 and zones of less precipitation (Fig. 2.1a) and occupy the south and southwestern parts (Turkmenistan, Uzbekistan, and in part the southwestern side of Kazakhstan) of Central Asia (see Fig. 2.1b). The "BWk" category is mostly contributed in Central Asia. Multidisciplinary analysis of climatic drivers that are modified by the Köppen classification provides a visual interpretation of vegetation–climate anomalies dynamics, especially in the targeted area, and offers a clear idea of the biotic and abiotic stresses that are the response to temporal vegetation syndrome (Fig. 2.1a).

Theoretically, we have modified the target area based on the climate gradient with Köppen classification to estimate the SPIE dataset (Table 2.1). For instance, in the category *BWk* and *Csa* are located the following regions: Lebap (Turkmenistan)

[1] Mean annual precipitation (70%) accumulated during winter period.

Table 2.1 Köppen climate symbols and related criteria for Central Asia (modified after Peel et al. 2007)

1st	2nd	3rd	Description	Regions	Criteria[a]	SPIE selected category
B			Arid		$MAP < 10 \times P_{threshold}$	
	W		–Desert	Uzb,Trk,	$MAP < 5 \times P_{threshold}$	
	S		–Steppe	Kzk, Uzb, Trk	$MAP \geq 5 \times P_{threshold}$	BSk-Ulytau, Naryn
		h	Hot	Trk, Uzb	$MAT \geq 18$	
		k	Cold	Trk, Uzb, Kzk	$MAT < 18$	BWk-Lebap, Navoi, GBAR
C			Temperate		$T_{hot} > 10$ & $0 < T_{cold} < 18$	
	s		–Dry summer		$P_{sdry} < 40$ & $P_{sdry} < P_{wwet}/3$	
	w		–Dry winter		$P_{wdry} < P_{swet}/10$	
	f		–Without dry season		Not (Cs) or (Cw)	
		a	Hot summer	Taj	$T_{hot} \geq 22$	Csa-Lebap, Navoi
		b	–Warm summer		Not (a) & $T_{mon10} \geq 4$	
		c	–Cold summer		Not (a or b) & $1 \leq T_{mon10} < 4$	
D			Cold		$T_{hot} > 10$ & $0 < T_{cold} \leq 0$	
	s		–Dry summer	Uzb, Taj, Krg, Kaz	$P_{sdry} < 40$ & $P_{sdry} < P_{wwet}/3$	
	w		–Dry winter		$P_{wdry} < P_{swet}/10$	
	f		–Without dry season		Not (Ds) or (Dw)	
		a	–Hot summer	Kaz, Krg	$T_{hot} \geq 22$	Dsa-GBAR
		b	–Warm summer	Krg	Not (a) & $T_{mon10} \geq 4$	
		c	–Cold summer		Not (a, b or d)	
		d	–Very cold summer		Not (a or b) & $T_{cold} < -38$	

[a]*MAP* mean annual precipitation, *MAT* mean annual temperature, T_{hot} temperature of the hottest month, T_{cold} temperature of the coldest month, T_{mon10} number of months where the temperature is above 10°, P_{dry} precipitation of the driest month, P_{sdry} precipitation of the driest month in summer, P_{wdry} precipitation of the driest month in winter, P_{swet} precipitation of the wettest month in summer, P_{wwet} precipitation of the wettest month in winter, $P_{threshold}$ varies according to the following rules (if 70% of MAP occurs in winter then $P_{threshold} = 2 \times MAT$; if 70% of MAP occurs in summer then $P_{threshold} = 2 \times MAT + 28$, otherwise $P_{threshold} = 2 \times MAT + 14$). Summer (winter) is defined as the warmer (cooler) 6-month period of ONDJFM and AMJJAS

Table 2.2 Summary of input datasets for parameterizing methodology of kriging

Data	Indices	Temporal scale	Time span (extracted)[a]	Spatial scale	Data source
NDVI	Vegetation	Bimonthly	1982–2015	8 km	AVHRR-GIMMS (NDVI 3g)
SPIE	Drought	Monthly	1982–2015/1950–2017	0.5 × 0.5⁰	SPIEbase
Average temperature	Climate	Monthly	1901–2015/1982–2015	0.5 × 0.5⁰	CRU-TS (ver. 4)
Average precipitation	Climate	Monthly	1901–2015/1982–2015	0.5 × 0.5⁰	CRU-TS (ver. 4)

[a]For simulation data, and for kriging methodology it has been extracted years between 1982 and 2015, originality data available or occupied past (before 1982) and present (after 2011) periods

and Navoi (Uzbekistan); in *BWk* are Kyzylorda (Kazakhstan) and GBAR (Tajikistan); and in the *BSk* category are Ulytau (Kazakhstan), Naryn (Kyrgyzstan); all estimated by their past and ongoing occurrences.

2.2.2.1 Datasets and Methods

To determine whether and how vegetation dynamics in Central Asian drylands are associated with climate patterns and NDVI, we used bimonthly 8-km GIMMS AVHRR data (1982–2015), and compared with time series data [CRU TS (v4.23)] for precipitation and temperature effects (1982–2015) during the selected period of time (Jones and Harris 2008) and calibrated with SPIE time series data (1982–2017). In Table 2.2, a detailed explanation is provided for each dataset and resolution scales. However, correction of the forecasting (prediction map) was estimated for the period 1982–2015 because of the availability of the NDVI dataset for this period. Originally, some drivers have longer periods, such as SPIE (1950–2017) and CRU-TS (1901–2015). Environmental variables (temperature, precipitation, and NDVI; SPIE) have been analyzed with the geostatistics method: a detailed description is illustrated on the flowchart (Fig. 2.4).

Environmental variables are indicated as a certainty dataset because of the utilization of various factors for estimation parameters and their dependency on each other. To account for surface biophysical properties of various habitats (Fig. 2.1a, b) and to assess the temporal movement dynamics of vegetation patterns in these cold desert and semi-desert ecosystems, we utilized a simple kriging methodology that was developed to classify further the vegetation index, which is identifying as index certainty associated factors (Prec/Temp/NDVI/SPIE) influencing the solid earth.

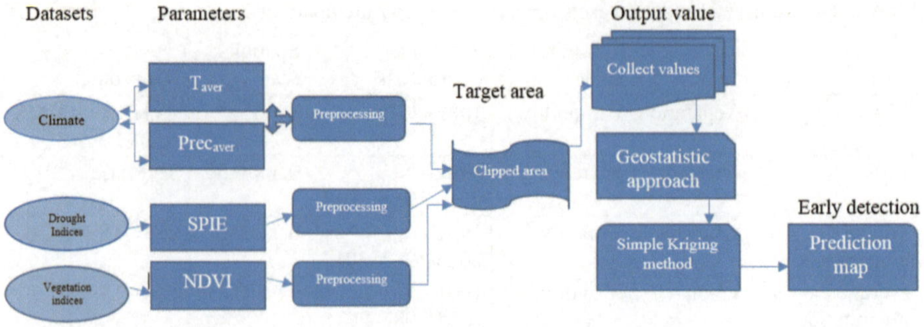

Fig. 2.4 Flowchart represents the steps of composing datasets and data processing for target area explored among Kazakhstan, Kyrgyzstan, Tajikistan, Turkmenistan, and Uzbekistan

2.2.2.2 Climatic Variables

The CRU time series (0.5 × 0.5 degree) grid datasets were extracted for 1982–2015 to assess month-by-month variation for climate/precipitation on a larger scale (Harris et al. 2014). Mapping climatic and precipitation data need to consider that some gaps are available during/after 1982–1991; it is clear that after collapse of the USSR some stations have ceased their missions, affecting the quantity of data. Climate data visualized over this time show quite high temperatures (since 1982–1992), which later becomes stabilized (after 2000).

2.2.2.3 A Drought Index: The Standardized Precipitation-Evapotranspiration Index (SPEI)

The SPEI is a multiscalar drought index based on climatic data and estimated based on the CRU-TS dataset. The SPEI-drought monitor offers near real-time information about drought conditions at the global scale, with a 0.5° spatial resolution and a monthly time resolution. Available to be downloaded on the following https://climatedataguide.ucar.edu/data-type/climate-indices/drought/spei, the SPEI is obtained from the monthly climatic water balance [precipitation minus reference evapotranspiration, $\left(ET_{0^-}\right)$], which is adjusted using a three-parameter log-logistic distribution. The values are accumulated at various time scales and converted to standard deviations with respect to average values (Vicente-Serrano et al. 2015a, b). The calibration period for the SPEI is January 1950 to December 2017 (last access: March 2018). For long-term analysis and calculation of selected target zone (Central Asia) habitats, we have scaled with a 1-month period with the same available calibration time (1950–2017) and as requested a robust understanding of the interrelations of the drought rankings in Central Asia (Fig. 2.5b). Determining the onset and duration of drought conditions with respect to normal conditions in a variety of natural and managed systems such as crops is best suited for drought monitoring and early warning purposes. Within this purpose, we have explored five countries

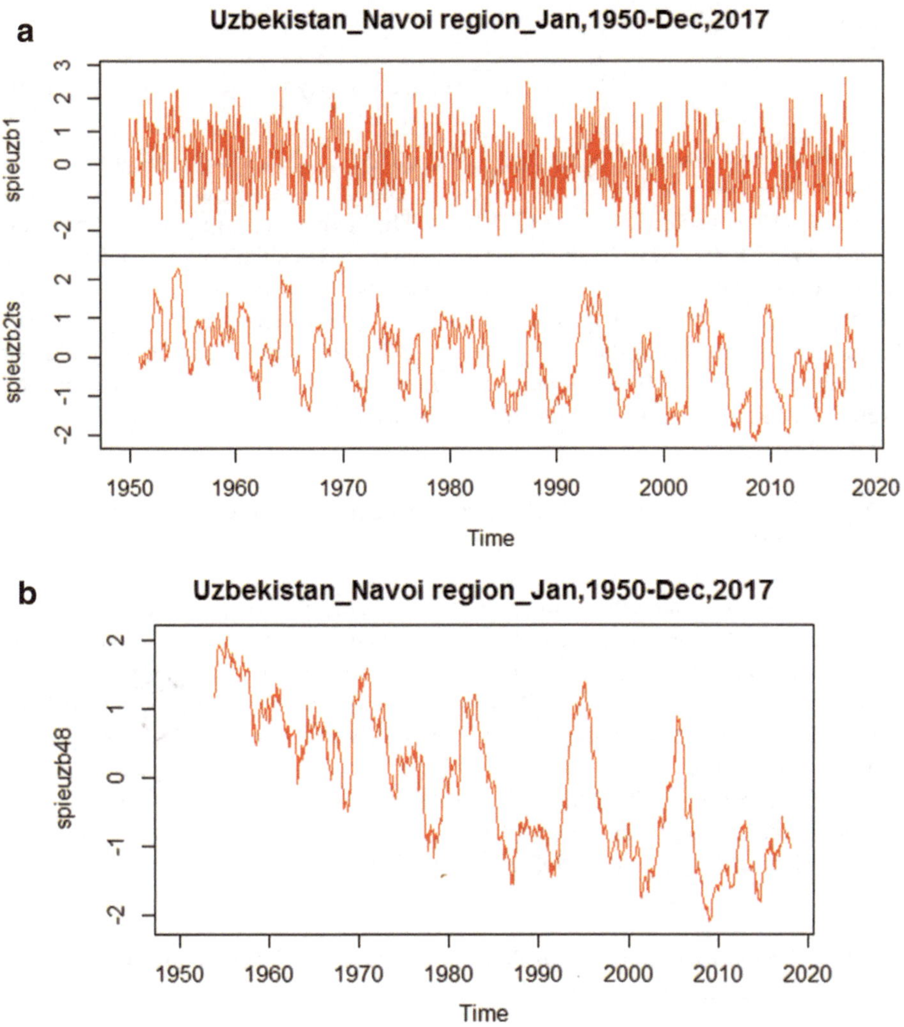

Fig. 2.5 Main drought episodes in Uzbekistan for Navoi region as occurred by SPIE dataset: monthly and annually distributions of drought (**a**) and 4 years cycling the drought trends (**b**). (**a**) SPIE_1: monthly and SPIE_12: annual dataset for Uzbekistan (Jan 1950–Dec 2017). (**b**) SPIE_48: counting and averaging for each 4 years

separately to visualize SPIE ratings for the past 70 years, within this purpose to be able compare with NDVI values (1982–2015). The process is determined by applying R programming language (within "spie" dataset) and this set of function (dataset) computing SPIE dataset (Beguería et al. 2014) and following the classical approximation as estimated and updated by Abramowitz and Stegun (1965).

$$\mathrm{SPEI} = W - \frac{C_0 + C_1 W + C_2 W^2}{1 + d_1 W + d_2 W^2 + d_3 W^3}, \tag{2.1}$$

where

$$W = \sqrt{-2\ln(P)} \quad \text{for} \quad P \leq 0.5 \tag{2.2}$$

and P is the probability of exceeding a determined d value, $P = 1 - F(x)$.

If $P > 0.5$, then P is replaced by $1 - P$ and the sign of the resultant SPEI is reversed (Vicente-Serrano et al. 2010). The constants are $C_0 = 2.515517$, $C_1 = 0.802853$, $C_2 = 0.010328$, $d_1 = 1.432788$, $d_2 = 0.189269$, and $d_3 = 0.001308$. The equations are cited at http://spei.csic.es/home.html and upgraded by Vicente (Vicente-Serrano et al. 2010). An R package is available for calculating the SPEI from user-selected input data using either the Thornthwaite, Penman-Monteith, or Hargreaves methods. For mapping, we have used the SPIEbase with based on NetCDF format, what was upgraded after the CRU-TS dataset.

2.2.2.4 Vegetation Indices: Normalized Difference Vegetation Index (NDVI)

In the framework, datasets from the Global Inventory Monitoring and Modeling System (GIMMS) project are carefully assembled from different AVHRR sensors and account for various deleterious effects, such as calibration loss, orbital drift, and volcanic eruptions (Tucker and Pinzon 2013). Bimonthly with 8-km resolution GIMMS AVHRR-NDVI3g data (1982–2015) were utilized to analyze NDVI (vegetation) status for Central Asia. The vegetation models were developed and generated subsequently with using climatic variables [temperature (average) and precipitation (average)] and drought index (SPIE) and NDVI, respectively, in the same period for kriging methodology approaches.

2.2.2.5 Probalistic Methods to Predict and Monitor Further Status of Landscapes (Kriging)

The assessment based on application of satellite images (NDVI 3g) with certainty datasets ($Prec_{monthly}$/$Temp_{monthly}$/$SPIE_{monthly}$) was developed on the base requirements of geostatistics to provide probability of further status within prediction of negative patterns or land degradation areas in Central Asia. Within this aim, we have used the kriging method with variograms to detect degradation categories, and calculated several environmental variable parameters for better understanding of forthcoming years conditions resulting from scarcity/limitations of water resources and fast land triggering issues. A major factor also is a warming temperature that influenced mismanagement basins (abiotic factors) by the overuse of its tributary rivers. More accuracy was found when we utilized geostatistical methodology for a large dataset (the territory of Central Asia is ~4,002,900 km^2). Another method using Empirical Bayesian Kriging (EBK) is more suitable, and also this methodology is reliable for

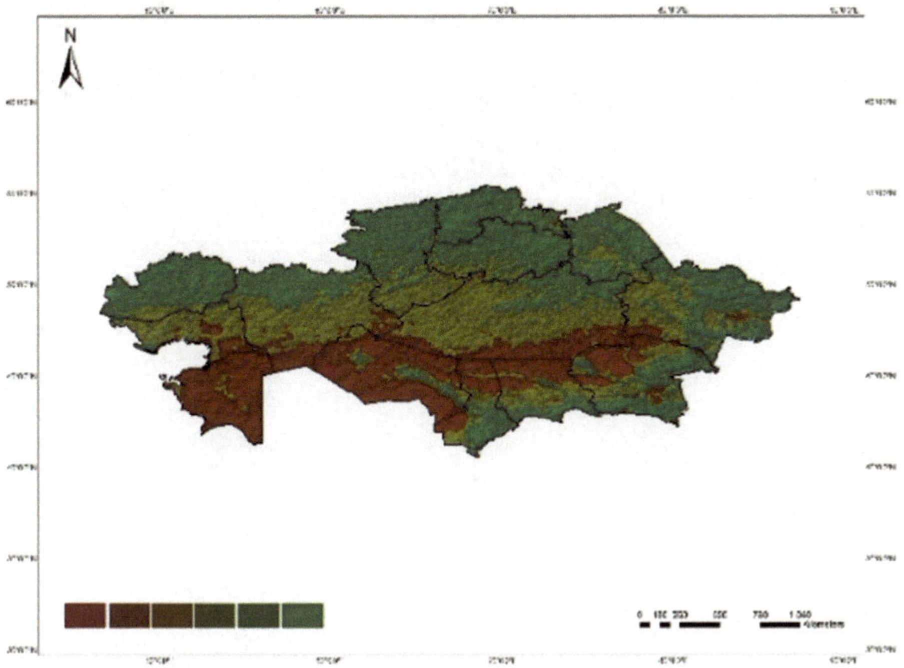

Fig. 2.6 SPIE-based mapping of drought episodes in Kazakhstan (1950–2017)

interpolation of the target area. This methodology provides greater accuracy for estimating forest or cropland status (areas with a high greenness index), but for drylands low to moderate vegetation values are better simulated with the kriging (simple and ordinary) method (see results on Figs. 2.6 and 2.7). The predicted map with the kriging dependence terms visualized more realistic means to classify vegetation patterns (low values of NDVI) and particularly with application of SPIE datasets. A probalistic method is required to provide information on prediction uncertainty limits and the choice of interpolators to statistical ones.

2.3 Results

We are intending to demonstrate the SPIE dataset as responsible for loss of energy in the balance (vegetation patterns) resulting from an outgoing high ET_0 in the drylands of CA. On the open rangelands, incoming high radiation is the source of high ET_0 ratings. Centrally, Turkmenistan and Uzbekistan, and partly Tajikistan (western), and farthest in Kazakhstan (southern side), this is a main source of losing a high energy balance from vegetation patterns.

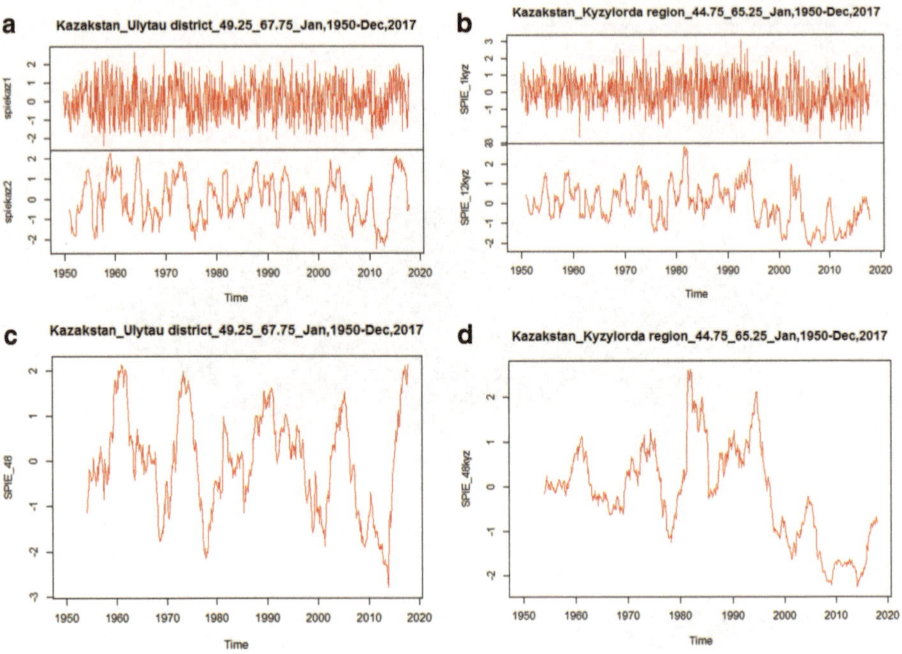

Fig. 2.7 Main drought episodes in Kazakhstan for Ulytau and Kyzylorda regions occurred by SPIE dataset (**a, b**) monthly and annual distributions of drought and (**c, d**) 4 years cycling of drought trends. (**a**) SPIE_1: Monthly and SPIE_12: annually dataset for Ulytau district, Kazakhstan (Jan 1950–Dec 2017). (**b**) SPIE_1: Monthly and SPIE_12: annually dataset for Kyzylorda, Kazakhstan (Jan 1950–Dec 2017). (**c**) SPIE_48: counting and averaging for each 4 years in case of Ulatau, Kazakhstan. (**d**) SPIE_48: counting and averaging for each 4 years in case of Kyzylorda, Kazakhstan

2.3.1 Long-Term Trends of SPIE Data (1950–2017) for Selected Areas

Mapping and determining monthly trends, annually drought variability (1950–2017) and multiplicative year (each 4 years) of trends (1950–2017) was found in targeted areas with positive and negative trends of SPIE (ratings between 2 and −2). Drought categories were derived based on Charusombat and Niyogi (2011) and modified after Ta et al. (2018) (listed in Table 2.3). Inserted shapes of the mapping part indicated alterations in the selected regions (green is normal, red is negative). Trends modified and updated on the database SPIEbase 2, and observation, show evidence that a drought is triggering the land degradation process, and that long-term drought is stressful for crops, explaining the major degradation trends (Figs. 2.5, 2.8, 2.9, 2.10, and 2.11).

Table 2.3 Standardized Precipitation-Evapotranspiration Index (SPEI) category for estimation of drought variety

Drought ranking	SPIE category
Extreme drought	SPEI ≤ −2.0
Severe drought	−1.99 < SPEI < −1.5
Moderate drought	−1.49 < SPEI ≤ −1.0
Mild drought	−0.99 < SPEI ≤ −0.5
Non-drought	SPEI ≥ −0

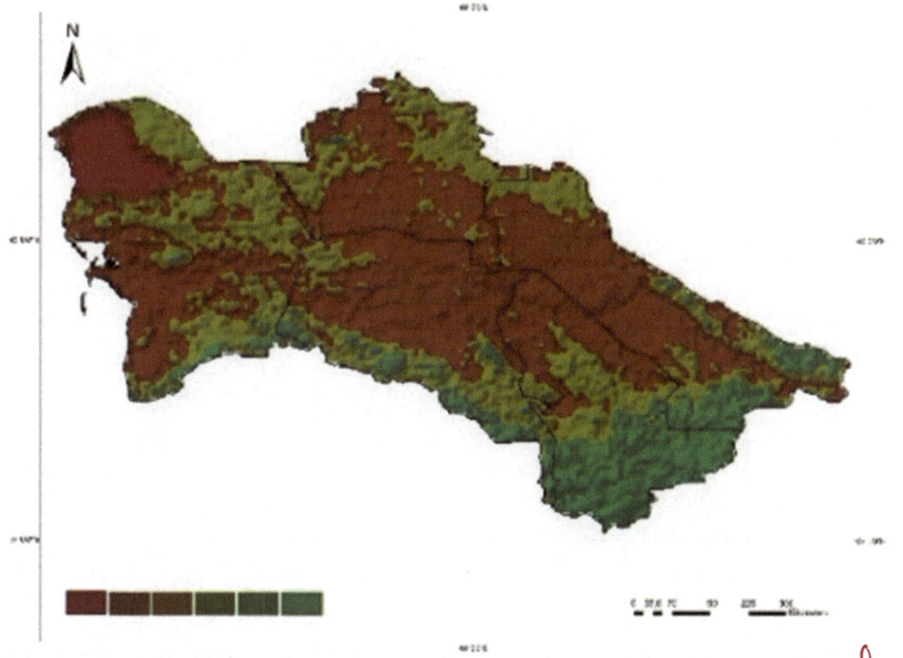

Fig. 2.8 SPIE-based mapping of drought episodes in Turkmenistan (1950–2017)

The main drought episodes occurred in the late 1950s, and unfavorable occasions were observed in late 2010 for all five countries within six regions. For the time series dataset we have added extra basic requirements for better understanding the anomaly in decades of time (*Y* coordinate values reached ≥−1 and ≥−2, which means a serious drought period of years or month in the region, and 0 trends mean no changes during the annual period), and the mapping part is inserted with shapes where time series are analyzed on pixel-based coordinate data (Figs. 2.4, 2.12, 2.13, 2.14, and 2.15). We are quickly able to see a high drought period for each 4 years (Figs. 2.4, 2.12, 2.13, 2.14, and 2.15, right side, with trends); 48 SPIE datasets were analyzed, demonstrating the averaging drought value for each 4 years based on average values.

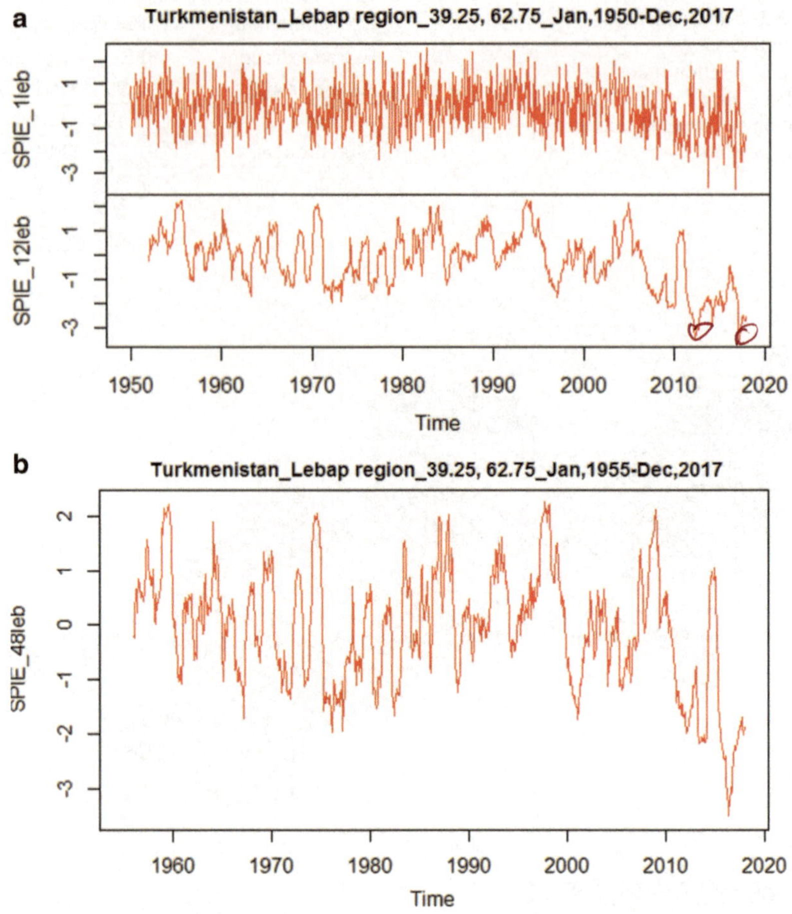

Fig. 2.9 Main drought episodes in Turkmenistan for Lebap region as occurred by SPIE dataset: monthly and annual distributions of drought (**a**) and 4 years cycling of drought trends (**b**). (**a**) SPIE_1: Monthly and SPIE_12: annually dataset for Lebap region, Kazakhstan (Jan 1950–Dec 2017). (**b**) SPIE_48: counting and averaging for each 4 years in case of Lebap, Turkmenistan

2.3.1.1 Uzbekistan

Drought is a common occasion in Central Asia, especially for the parts of Turkmenistan and Uzbekistan, and is more influenced by climate or landscape distributions. for the part of Uzbekistan, the Navoi region was selected (coordinates 40°25′–65°25′), located in the middle zone of the country. Originally, the area was classified as cold desert zone (BWk, Koeppen classification), and drought frequency is the usual issue on this area because of high temperature and low precipitation. However, how frequently drought is observed in this area is demonstrated in

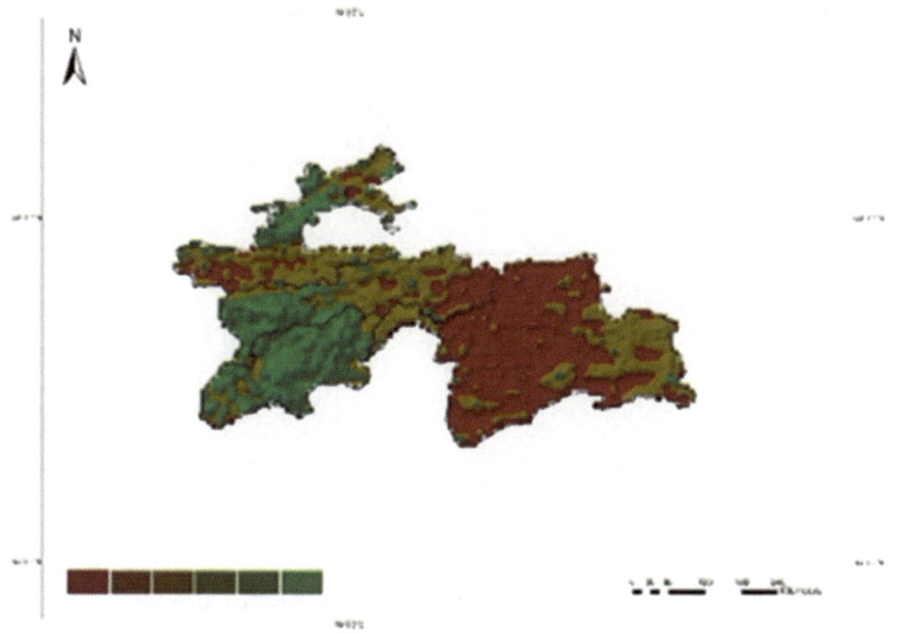

Fig. 2.10 SPIE-based mapping of drought episodes in Tajikistan (1982–2015)

Fig. 2.16, and for each 4 years it possible to observe strong droughts that correspond to above >−2 (Fig. 2.16).

Based on annual data, the strongest droughts were observed in the middle 1960s and the early 1970s. In past decades, during the 2000s, 2009 was recorded as the strongest one (Fig. 2.4a, left side). The importance of this selected point is related to estimating regions that are not a source to water basins and to get an idea about drought occurrences for areas far away from water resources. Based on vegetation classification, it is classified as *Artemisia* spp., a shrubland zone and therefore our first point (Navoi region) classified as being the desert type of vegetation (Fig. 2.1b).

2.3.1.2 Kazakhstan

Almost 70 years observation (1950–2017) estimated the strong droughts in Ulytau, and the last years negatively with further negative trends for Kyzylorda region (Fig. 2.12), whereas the mapping part shows that the northern part illustrated a positive scale (low drought) and the middle of the country a moderate level (values) of drought (Fig. 2.8). The difference between Ulytau and Kyzylorda is related to an accumulation on minimum values of drought indices, such as strongest or peak of drought (max >−2) observed in the Kyzylorda region (see mean values with minus),

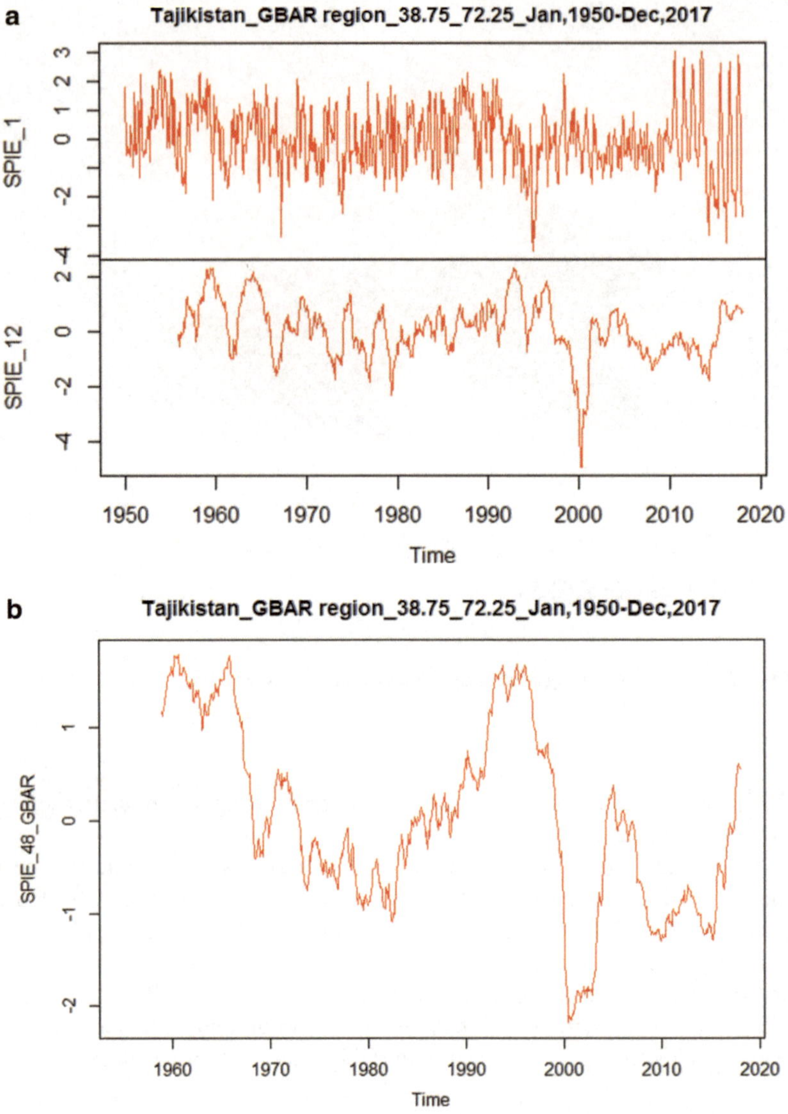

Fig. 2.11 SPIE based mapping of drought episodes in Tajikistan (1982–2015). The selected point located on the GBAR region, the area where covered partly with mountains. The main drought episodes in Tajikistan for Lebap region occurred by SPIE dataset, (**a**) monthly and annually distributions of drought and (**b**) 4 years cycling the drought trends. (**a**) SPIE_1: Monthly and SPIE_12: annually dataset for GBAR region, Tajikistan (Jan 1950–Dec 2017). (**b**) SPIE_48: counting and averaging for each 4 years in case of GBAR, Tajikistan

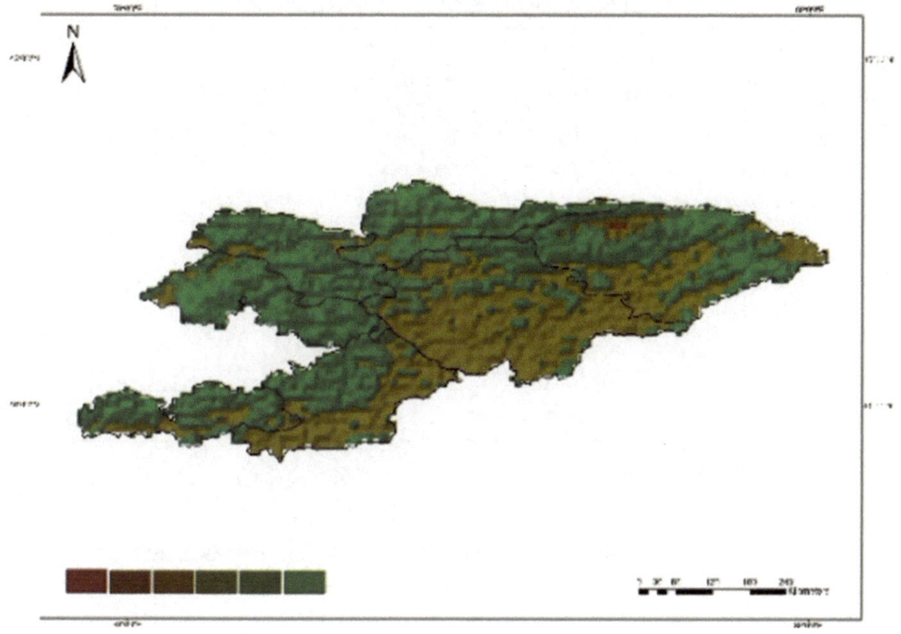

Fig. 2.12 SPIE-based mapping of drought episodes in Kyrgyzstan (1982–2015)

while frequently a positive median among null (no droughts) in Ulytau occurred with greater frequency.

2.3.1.3 Turkmenistan

The main drought episodes in Turkmenistan occurred in the past decades; also, SPIE_48 for the Lebap region (Fig. 2.13a); Fig. 2.13b illustrates that among 4 years calculations the strongest one is between 2013 and 2017.On the monthly dataset, the strongest or extreme strong drought values are illustrated between 2015 and 2017 (Fig. 2.13b).

2.3.1.4 Tajikistan

In this area, the last years are categorized mostly with strong drought trends that reached SPIE >−4 in some months of 2010–2017 (Fig. 2.14a, b), whereas in further upcoming years for the GBAR region might be also observed no drought trends as counted by linearity forecasting (Fig. 2.14c). The strongest SPIE >−3 is observed in

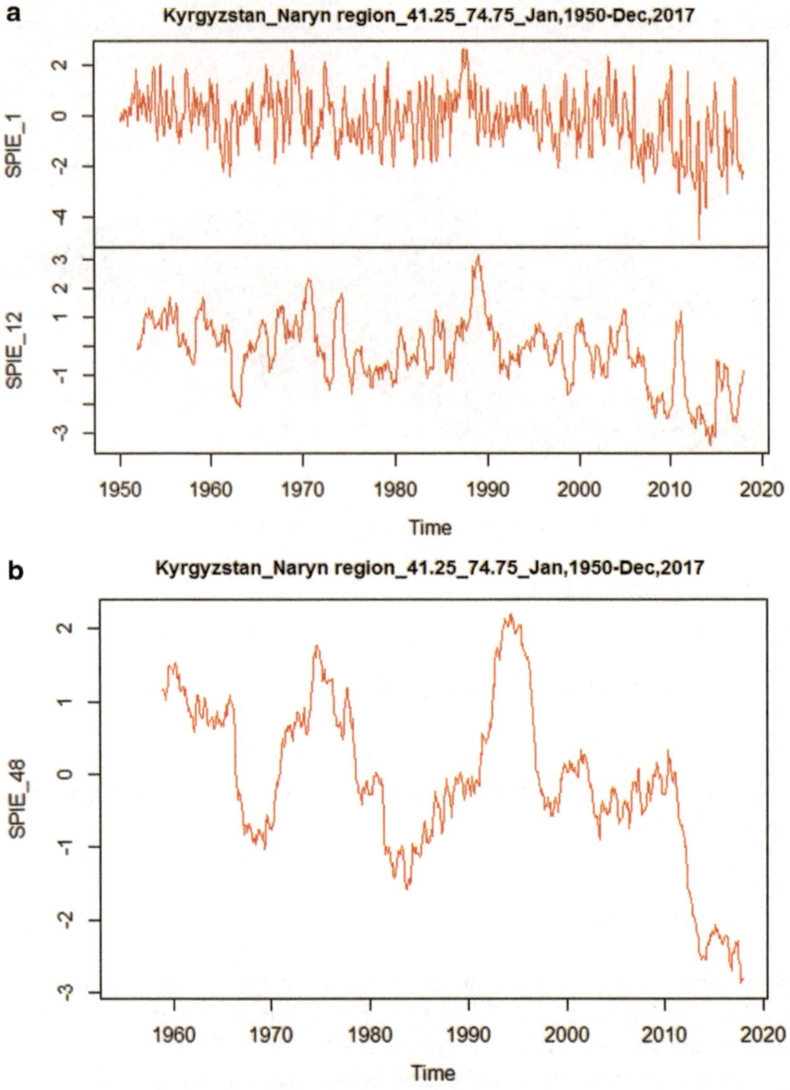

Fig. 2.13 The main drought episodes in Kyrgyzstan for Naryn region occurred by SPIE dataset, (**a**) monthly and annually distributions of drought and (**b**) 4 years cycling the drought trends. (**a**) SPIE_1: Monthly and SPIE_12: annually dataset for Naryn region, Kyrgyzstan (Jan 1950–Dec 2017). (**b**) SPIE_48: counting and averaging for each 4 years in case of Naryn, Kyrgyzstan

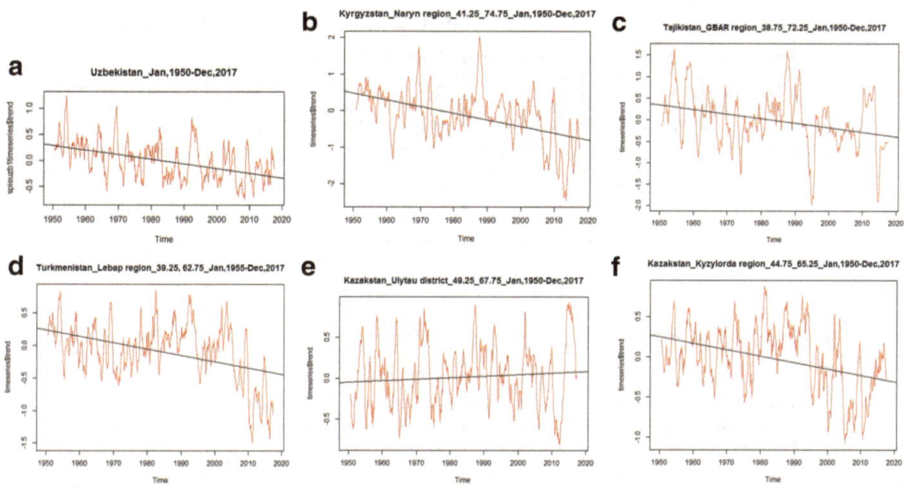

Fig. 2.14 Annual drought trends in Central Asia (1950–2017) and linear prediction to 2020. (**a**) Annual drought trends in Uzbekistan (1950–2017) and linear prediction to 2020. (**b**) Annual drought trends in Kyrgyzstan (1950–2017) and linear prediction to 2020. (**c**) Annual drought trends in Tajikistan (1950–2017) and linear prediction to 2020. (**d**) Annual drought trends in Turkmenistan (1950–2017) and linear prediction to 2020. (**e**) Annual drought trends in Ulytau, Kazakhstan (1950–2017) and linear prediction to 2020. (**f**) Annual drought trends in Kyzylorda, Kazakhstan (1950–2017) and linear prediction to 2020

the 2000s, while annually is mostly on positive level (Fig. 2.14d, e). According to 4-year estimations, the period of 2010 was also a moderate drought period (Fig. 2.14f).

2.3.1.5 Kyrgyzstan

In this area, past years show mostly strong drought trends, and it is estimated that it reached SPIE >−3 in some months after 2010 (Fig. 2.15); for further upcoming years for the Naryn region, drought trends might be also observed (as SPIE >−1) as counted by linearity forecasting (Fig. 2.17b). Results indicated that negative trends of drought severity index (red color) are related to low accumulation of NDVI values, and a positive drought index included high or moderate values of NDVI, as in Central Asia, where high values of NDVI ranged between 0.35 and 0.50. Kyrgyzstan is dominant with high values of NDVI, and no drought according to drought mapping (Fig. 2.11), and mostly positive trends of drought severity index indicated in Naryn region (except for 2015).

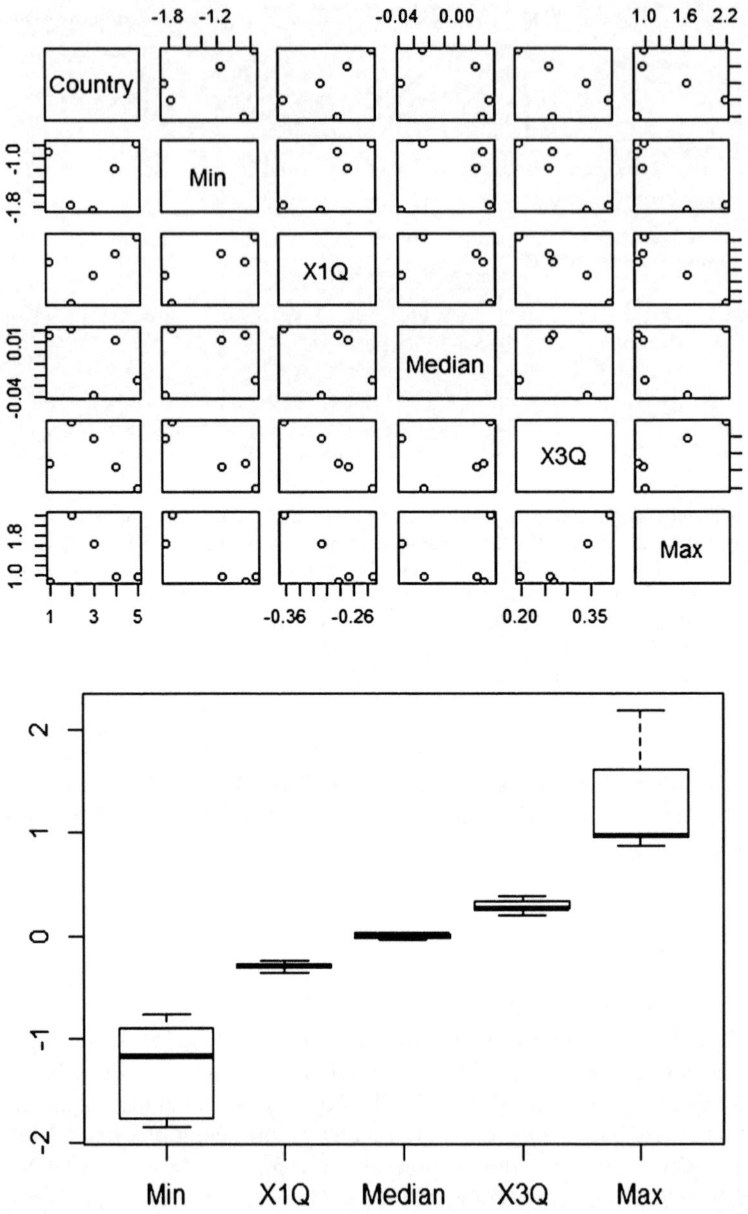

Fig. 2.15 Scatterplot and boxplot analysis for drought residuals in five areas of Central Asia. On the column level: *1* Uzbekistan, *2* Kazakhstan, *3* Kyrgyzstan, *4* Turkmenistan, *5* Tajikistan

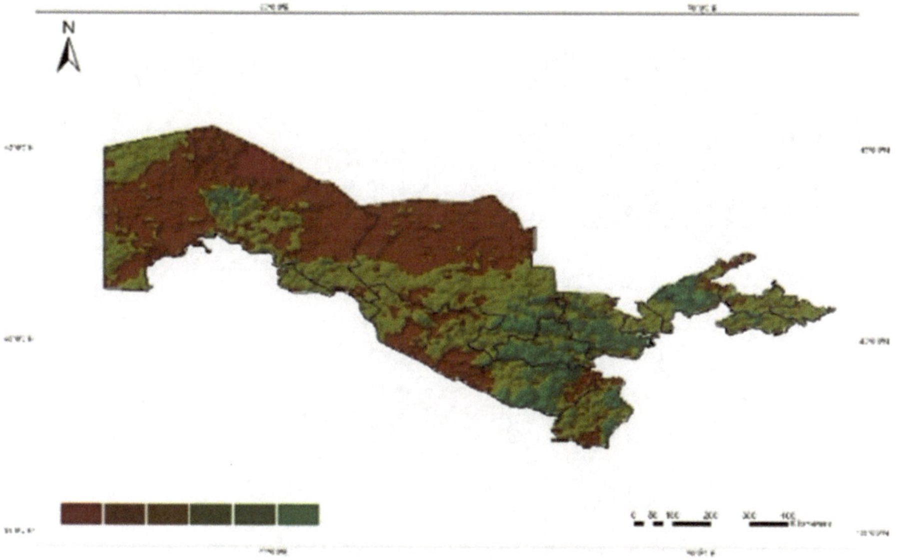

Fig. 2.16 Mapping annual SPIE dataset for Uzbekistan based on available period (1950–2017)

This is a preliminary conclusion of results, and all highlights are a complexity of drought activity processes to describe generally.

2.3.2 Summary About Droughts for Central Asia

On the basis of the results, each country has observed a very extreme drought period at least two times during the observation period (1950–2017); occasionally, it is not similar for each country in addition to region. However, mostly the strongest drought was observed for 2010 in all countries. These phenomena, which vary with time, related to develop a particular span of time. As demonstrated results and drought coefficients, the negligible condition is observed in Lebap (Turkmenistan) and Ulytau (Kazakhstan) regions. Based on observed results, the SPEI is a good indicator to predict further drought anomalies or alternatively to be able develop crop failure or less productive zones under a statistical approach (Table 2.4).

2.3.3 Statistical Description of Annual Trend Analysis of Droughts and Their Residuals

The selected method outputs describes the results of drought on annual trends based on monthly dataset. The seasonal fluctuations in a time series can be contrasted with cyclical patterns (see on top) (Tables 2.5, 2.6, and 2.7). To handle seasonality with amplitude and phase are a linear regression for drought trends and calculated with

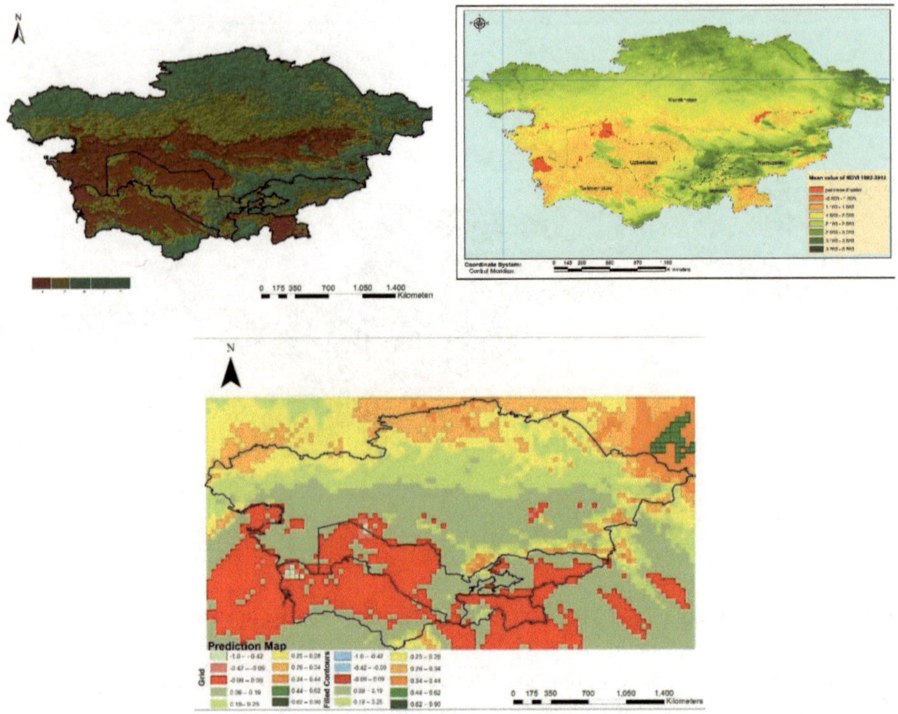

Fig. 2.17 Mapping prediction standard error on base kriging results (prediction made on the basis of regression equation). Results indicated similarity of SPIE index with Köppen classification (**a**, **b**) and average annual NDVI values (**c**) and prediction map with simple kriging method (**d**). (**a**) Drought indices of SPEI (1982–2015) and identified similarity Köppen classification. (**b**) NDVI values (1982–2015) determined on the high evaporation zones with low vegetation values. (**c**) Prediction map for changes of vegetation indices in Central Asia. A high loss of vegetation indices is indicated in the southern territories and occupied Turkmenistan, Tajikistan, and Uzbekistan. Red color is modified as early drought detection zones

Table 2.4 Basic statistical description monthly available SPIE dataset in the part of Uzbekistan (1957–2017)

Minimum	1st quarter	Median	Mean	3rd quarter	Maximum
−2.17	−0.83	−0.01	−0.05	0.65	2.42

Table 2.5 Basic statistical description of monthly available SPIE dataset in the part of Uzbekistan (1957–2017)

Ulytau					
Minimum	1st quarter	Median	Mean	3rd quarter	Maximum
−2.415740	−0.731258	0.000685	0.015693	0.748020	2.764420
Kyzylorda					
Minimum	1st quarter	Median	Mean	3rd quarter	Maximum
-2.78183	-0.73523	-0.04258	-0.02281	0.66500	3.15759

Table 2.6 Basic Statistical description monthly available SPIE dataset in the part of Uzbekistan (1957–2017)

Min.	1st Qu.	Median	Mean	3rd Qu.	Max.	NA's
−3.89508	−0.77346	−0.07035	−0.01224	0.76384	2.99554	1

Table 2.7 Basic statistical description monthly available SPIE dataset in the part of Uzbekistan (1957–2017)

Min.	1st Qu.	Median	Mean	3rd Qu.	Max.	NA's
−4.89750	−0.87012	−0.07526	−0.15617	0.61815	2.64364	2

decomposition method from time series analysis series. As we have seen, for the six cases (selected target areas), the annual trend values ranged between −2 and 2 (as minus "drought," with plus "low drought"). Based on residuals and forecasting in further drought status (Fig. 2.17) illustrates that only in the Ulytau region, Kazakhstan will have less drought, or positive trends will have affirmed it. Other resting areas are shown with negative trends (descending line); especially, Turkmenistan (Lebap) and Kazakhstan (Kyzylorda) will be stronger than other areas. Other statically available data for residuals is illustrated in Tables 2.8 and 2.9. The following residual analysis of annual drought trends is based on the monthly dataset accounted for each region (Table 2.9) and the following abbreviations meaning the value: for example, for Uzbekistan, minimum (Min): Min = −0.76; first quartile (1Q):1Q = −0.23; Median (Med): Med = −0.02; Third Quartile (3Q):3Q = 0.19; and Maximum (Max): Max = 0.98.

2.3.4 *Resilience of Ecosystem and an Assessment of the Consequences of Current Factors for Vegetation Trends*

Drought-persistent periods are various and dependent on annual rainfall data, and the following nonsmoothing parameters are developed for irregular patterns of precipitation and drought trends. Obviously, temperature raising trends are affected on Central Asian drylands and our preliminary studies has proved that land degradation is directly linked to Central Asian habitats. Using a semi-variogram as a model, new data are unconditionally simulated at each of the input locations in the subset; in the case of the large number of datasets are plotted together and measured for prediction. The indicator prediction values are calculated using the semi-variogram modeled from the binary (0–1) data, the creation dataset based on indicator transformations of original data (Prec/Temp/NDVI/SPIE).

Cross-validation sequentially omits a point and calculates indicator prediction values for each dataset. The simulating semi-variogram (Fig. 2.18a) with the cer-

Table 2.8 Drought coefficients in selection regions of Central Asia

Countries	Intercepts	Estimate	St. error	*t*-value	Pr(>|*t*|)	Multiple *R*-squared	Adjusted *R*-squared	*p*-value
Uzbekistan	sin(2* π*drought)	17.92	1.13	15.85	<2e−16 ***			
Uzbekistan	cos(2* π *drought)	−0.01	0.00	−15.86	<2e−16 ***	0.2387	0.2378	<2.2e−16
Kyrgyzstan	sin(2* π *drought)	35.99	2.26	15.86	<2e−16 ***			
Kyrgyzstan	cos(2* π *drought)	−0.01	0.00	−15.92	<2e−16 ***	0.2406	0.2397	<2.2e−16
Tajikistan	sin(2* π *drought)	20.76	2.14	9.668	<2e−16 ***			
Tajikistan	cos(2* π *drought)	−0.01	0.00	−9.673	<2e−16 ***	0.10	0.10	<2.2e−16
Turkmenistan	sin(2* π *drought)	19.31	1.43	13.46	<2e−16			
Turkmenistan	cos(2* π *drought)	−0.01	0.00	−13.52	<2e−16	0.1857	0.1847	<2.2e−16
Kazakhstan(Ulytau)	sin(2* π *drought)	−3.74	1.35	−2.76	0.00584 **			
Kazakhstan(Ulytau)	cos(2* π *drought)	0.00	0.00	2.77	0.00563 **	0.01	0.01	0.005631

Signif. codes: 0 '***'; 0.001 '**' ;0.01 '*'; 0.05 '.' ; 0.1 ' ' ; 1

Table 2.9 Summary for residual analysis of monthly drought values and outputs between trends and seasonality

Country	Min	1Q	Median	3Q	Max
Uzbekistan (Navoi)	−0.75964	−0.23210	−0.02444	0.19754	0.98208
Kazakhstan (Ulatau)	−0.88626	−0.28184	0.01544	0.27061	0.87385
Kyrgyzstan (Naryn)	−1.76631	−0.36260	0.02083	0.39100	2.18888
Turkmenistan (Lebap)	−1.1657	−0.2658	0.0114	0.2616	0.9576
Tajikistan (GBAR)	−1.85109	−0.30753	−0.03922	0.34162	1.61468

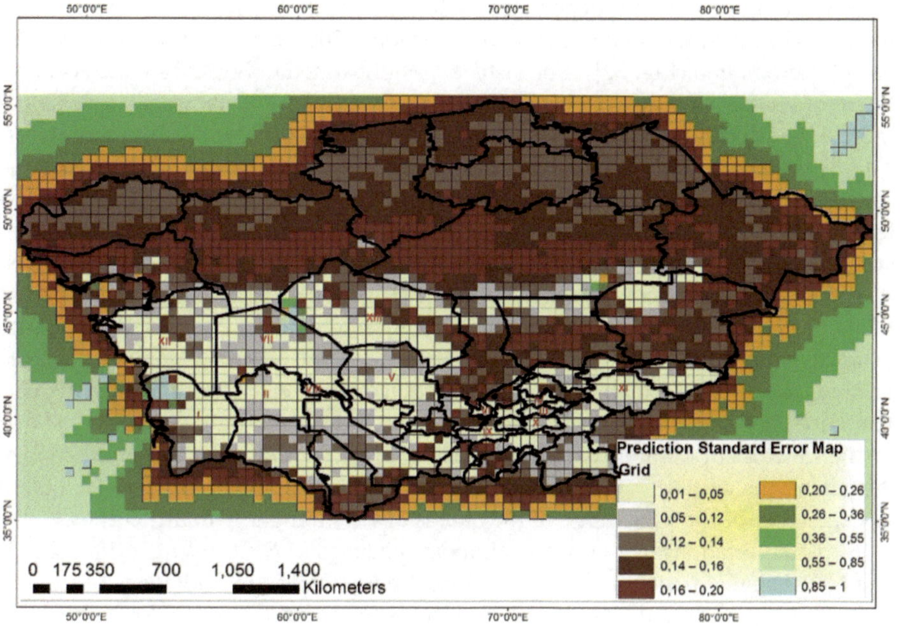

Fig. 2.18 Mapping prediction standard error on base kriging results (prediction made on the basis of regression equation)

tainty associated (*Prec/Temp/NDVI/SPIE*) dataset is constructed by calculating half the average squared difference of the values of all pairs (NDVI vs. Prec, SPIE vs. Temp) of measurements at locations separated by a given distance h (h*10^{-1}) and plotted as y (y*10^{-2}) against the separation distance h (Fig. 2.18a). In reality, it means the uncertainty prediction after measuring level 0.636 is various or diffused and a randomly occupied measured distance (Fig. 2.18b). Because of the accumulation of low vegetation as sparse vegetation, after 0.636 it was hard to predict after this level and variations of the dataset are occasionally diffused. An inserted shape indicates dominant values for these zones that are tightly accumulated and less extending (more stabilized) in both sides between planks.

The number of lags 12 monthly, lag size >12 for datasets, defines the weights that determine the contribution of each observed data point to the prediction of new values at unsampled locations. The selected number 12 is indicated as monthly contributions and their delay effects. This is an effective way to acquire vegetation covers because they are time consuming, and curvature (Fig. 2.18a) effects to lags are very small and averaged datasets (blue crosses) are nonsignificant to the semi-variogram. As illustrated in Fig. 2.18a, the semi-variogram is estimated from data to make optimal prediction among various distances (dataset as frequently diffusion from estimated blue line) and nonlinear correlation observed between SPIE and precipitation. Interpreting predictions with the semi-variogram provides a better understanding of possible dependency dataset to each others and the generalized outputs. However, the model makes an enormous difference if the function is to be used for prediction (Fig. 2.7). For instance, as illustrated in Fig. 2.18a, the accumulation of NDVI values on the beginning of the curvature line shows positively, and more or less correlation is available between pairs 0.05, 0.20, and 1 between values. This is a nice example of the distinction between statistical significance and the scientific importance of the dataset. The curvature is highly significant and given a high variance in *y*, the effect of curvature is less effective, just a few distance (measured *x*) periods is effectively correlated. The curvature has to pass to the averaged dataset. In the case of the curvature model, no information on non-linearity other than that contained within data, then parsimony suggests that errors will be smaller using the simpler, linear model prediction (Fig. 2.17). We have applied a simple kriging standard error map (Fig. 2.6c) to receive fewer errors within applying certain datasets for the prediction status of patterns (loss and gain productivity) and resilience areas of Central Asia. Both models (Figs. 2.6 and 2.7) are equally good at describing the data ($r^2 = 0.76$) (the power law model) and $y = 0.96 * x + 0.80$, but extrapolation beyond the range of the data is always fraught with difficulties.

Explored relationships between temporal changes of SPIE events (1982–2015) and NDVI spatial patterns (1982–2015) for the arid and semi-arid zones of Central Asia (Fig. 2.6a, b) and linear relationships between varies: $y = 0.96 * x + 0.80$ (Figs. 2.6 and 2.7); also linear/positive relationship observed between periods of precipitation anomalies and high evapotranspiration ($Prec_{low}$ /$SPEI_{max}$), as observed regionally drought periods in Central Asia. The positive relationship means that SPIE is getting a peak period (plus ratings) and, the lowest ratings of precipitation is accumulated on same time.

Generally, drought ratings continues more than 90 days, prolongation droughts and more than 90 days associated with raising temperature ratings. In contrast, the main drought episodes were identified by the SPEI (Fig. 2.7) and NDVI values that illustrated vegetation accumulation ratings (0.02–0.43) (Fig. 2.7), when SPIE had evaporated under following indices: 0.06–0.41; The further anomalies scenarios regarding on datasets (Fig. 2.7), quantity of samples are equal to 1863 (as dots) and indicated linear regression when applied four certainty datasets (Prec/Temp/NDVI/SPIE). Moreover, if temperature increased progressively by 2° or 4 °C on the global level, the reinforcement of drought severity is associated with higher water demand by potential evapotranspiration (Vicente-Serrano et al. 2015a, b). The similarity or progressive rise in temperature is already estimated in Turkmenistan, then in Uzbekistan and Kazakhstan.

2.3.5 Ongoing Process and Early Drought Detection with the Kriging Method

This research assessed the vegetation dynamics in Central Asian drylands to determine which associated anthropogenic pressures are versus climate anomalies. During high ET_0 is expected to continuous of decreasing levels of water resources, as shown in Fig. 2.17c, the rating of the negative vegetation values (Fig. 2.17) with high transpiration is forecast across the borders of Turkmenistan and Uzbekistan, Uzbekistan and Tajikistan, and Kyrgyzstan and small areas of Kazakhstan around the Kyzylkum Desert. The background of the prediction map (Fig. 2.17) illustrates improper low vegetation values in the southwestern part of the area (red color); it might be a landscape pattern these areas mostly under desert zones and vegetation patterns or coverage is frequently sparse.

Large-scale datasets are used to assess and measure relationships between global climate patterns and regional-scale vegetation responses to support land and water use and management in this drought-prone region with prediction map/trend. The prediction map (Fig. 2.18) illustrates a more complex situation for resiliency; among five countries only Kazakhstan and Kyrgyzstan are more resilient in this ongoing scenario, with projections suggesting losses in areas of Turkmenistan, Uzbekistan, Tajikistan, and partly Kyrgyzstan, and same time potential gains (increasing more greenness) in Kazakhstan lands. It is more important to see that a vast area of rangeland/grasslands are stabilizing or restoring in part of Kazakhstan. However, it is a vast country and on the regional or local level, we may find diverse problems related to anthropogenic effect or the climate change issue. A right example is that for Kyzylorda region; after rice paddying and high evapotranspiration in atmosphere, huge agricultural zones have been converted on saline and abandoned areas.

Some main affected areas among regions in Central Asia and mapping were identified with standard error for detection of change values or prediction standard errors (Fig. 2.17) quantify the uncertainty for each location in the surface that we created and developed criteria to illustrate vegetation loss areas. A simple rule of thumb is that 95% of the time, the true value of the surface will be within the interval formed by the predicted value. Appropriate phytoindicators for modifying and designing different ecological zones, especially trends of spatial changes of vegetation cover over time trends which are associated with climate patterns, assessed a better understanding vegetation movement dynamics and their mechanisms. Within prediction standard error surface that locations (five countries) near sample points generally have lower error and more accuracy for receiving further status of vegetation (Fig. 2.18).

On the basis of criteria to develop appropriate phytoindicators with fragile zones as lesser resilience of climate anomalies (regions of the country) are the following:

Turkmenistan: Balkan (I), Taschauz (II),

Uzbekistan: Ferghana (III), Namangan (IV), *Navoi* (V), Syrdaryo (VI), Karakalpakstan (VII), Khorezm(VIII)

Tajikistan: Khodjand (IX)

Kyrgyzstan: Batken (X), Naryn (XI),

Kazakhstan: Mangghystau (XII), Kyzylorda (XIII)

To resilience of this occasions or as a potential winner (Kazakhstan and partly Kyrgyzstan) and losers (Turkmenistan, Uzbekistan, and Tajikistan). This assumption will have to undertake proactive adaptation tools to reduce early drought damage for loser category countries. On the basis of results, we are able to forecast that two regions of Central Asia are detected as high potential risk zones: Turkmenistan and Tajikistan. Also, more than half the area of Uzbekistan is also indicated as same status.

2.4 Conclusion

The rising occurrence of drought events and following soil salinization are serious threats that have major impacts on land use and land cover (LULC) change patterns in agricultural zones of Central Asia. This research demonstrates that it is possible to estimate based on satellite image sources and effectively disseminate an early warning of disaster risk zones for upcoming months or years. For developing countries, there is increased awareness of the loss of biodiversity caused by abiotic stresses, especially drought events observed and characterized as a higher influence by agricultural sectors. Therefore, the drought is an occasional issue in Central Asia, and the minimum values have not reached the positive values, which means drought trends will be observed as longer or shorter term in all selected countries and regions. The strongest drought in past years was observed in Lebap (SPIE > −3) and in Naryn (SPIE > −3), also in GBAR region (SPIE > −2), while this region is famous with mountain ecosystems. Based on median values (Fig. 2.17, left side) in Central Asia drought or drought anomalies are observed usually in targeted areas (based on minus values). For selected areas, at least two times of peak drought periods were estimated, mostly in Turkmenistan, and after that in Uzbekistan. In general, arid and semi-arid regions particularly have high evaporation loss ratings, and therefore, high ET_0 ratings were observed, where the water supply is most limited and very valuable. Performance of utilization of satellite images within drought indexes gives affordable and visual information for current and past condition to analyze and develop information systems for early drought detection. Arid zones are prone to frequent seasonal droughts, and complex terrain that both hinders ground and satellite-based remote sensing is an applicability of conventional subgrid process to make parametrizations for detection of early drought anomalies with developing indicator approaches.

Acknowledgments The project work was financed by the DAAD Program, partly by GFF (TU-Dresden) through a research exchange between Central Asia and the European Union. We thank Prof. Elmar Csaplovics (TU-Dresden) for his potential interest for this area and supporting it. For the technical and statistical part we thank the colleagues of TU-Dresden, Institute Remote Sensing and Photogrammetry, for their expert assistance to simulate data with statistical approach and time series analysis.

References

Abdullaev I (2014) Water-energy-agriculture and environment nexus in Central Asia: current state and future. Presentation from the World Water Week, Stockholm, Sweden. http://programme.worldwaterweek.org/sites/default/files/water-energy food_and_environmental_nexus_in_central_asia.pdf

Abramowitz M, Stegun A (1965) Handbook of mathematical functions. Dover, New York

Aralova D, Toderich K, Jarihani B, Gafurov D Gismatulina L, Osunmadewa BA, Abualgasim R, Majdalin (2016) Environmental resilience of rangeland ecosystems: assessment drought indices and vegetation trends on arid and semi-arid zones of Central Asia. In: Proceedings of the SPIE 10005, earth resources and environmental remote sensing/GIS applications VII, 100050R (18 October 2016)

Bedunah DJ, McArthur ED, Fernandez-Gimenez M, eds (2006) Rangelands of Central Asia: proceedings of the conference on transformations, issues, and future challenges. 2004 January 27, Salt Lake City, UT. Proceedings RMRS-P-39. U.S. Department of Agriculture, Forest Service, Rocky Mountain Research Station, Fort Collins, 127 p

Beguería S, Vicente-Serrano SM, Reig F, Latorre B (2014) Standardized precipitation evapotranspiration index (SPEI) revisited: parameter fitting, evapotranspiration models, tools, datasets and drought monitoring. Int J Climatol 34(10):3001–3023. https://doi.org/10.1002/joc.3887

Bobojonov I, Aw-Hassan A (2014) Impacts of climate change on farm income security in central Asia: an integrated modeling approach. Agric Ecosyst Environ 188:245–255

Charusombat U, Niyogi D (2011) A hydroclimatological assessment of regional drought vulnerability: a case study of Indiana droughts. Earth Interact 15(26):1–65

Ciabatta L, et al (2016) Using Python® language for the validation of the Cci soil moisture products via Sm2Rain. 1–6. https://doi.org/10.7287/peerj.preprints.2131v1

FAOstat (2015) http://faostat3.fao.org/home/E. Accessed Mar 2015

Gasparrini A (2011) Distributed lag linear and non-linear models in R: The package dlnm. J Stat Softw 43:1–20

Gessner U, Naeimi V, Klein I, Kuenzer C, Klein D, Dech S (2012) The relationship between precipitation anomalies and satellite-derived vegetation activity in Central Asia. Glob Planetary Change 110:74–87. https://doi.org/10.1016/j.gloplacha.2012.09.007

Gintzburger G, Toderich K, Mardonov B, Mahmudov M (2003) Rangelands of the arid and semi-arid zones in Uzbekistan. CIRAD–ICARDA Montpellier, p 426

Gong Z et al (2017) Research on trend of warm-humid climate in Central Asia. IOP Conf Ser Earth Environ Sci 74:012017

Harris I, Jones PD, Osborn TJ, Lister DH (2014) Updated high-resolution grids of monthly climatic observations—the CRU TS3.10 Dataset. Int J Climatol 34(3):623–642

Jones PD, Harris IC (2008) Climatic Research Unit (CRU) time-series datasets of variations in climate with variations in other phenomena. NCAS British Atmospheric Data Centre, date of citation. http://catalogue.ceda.ac.uk/uuid/3f8944800cc48e1cbc29a5ee12d8542d

Kariyeva J (2011) Land surface phenological responces to land use and climate variation in a changing Central Asia. Dissertation. In: Johnson R (ed) Graduate College of University of Arizona, Tucson, p 179

Kariyeva J, van Leuwen WJD (2010) Environmental drivers of NDVI-based vegetation phenology in Central Asia. Remote Sens 3(2):203–246. https://doi.org/10.3390/rs3020203

Kerr YH et al (2012) The SMOS soil moisture retrieval algorithm. Geosci Remote Sens 50:1384–1403

Mirzabaev A, Guta D, Goedecke J, Gaur V, Börner J, Virchow D, Denich M, von Braun J (2015) Bioenergy, food security and poverty reduction: trade-offs and synergies along the water–energy–food security nexus. Water Int 40(5–6):772–790

Musyimi Z (2011) Temporal relationships between remotely sensed soil moisture and NDVI over Africa: potential for drought early warning? pp 1–49

Nicholson SE, Tucker CJ, Ba MB (1998) Desertification, drought, and surface 1055 vegetation: an example from the West African Sahel. Bull Am Meteorol Soc 79:815–829

Paulik C, et al (2014) Open source toolbox and web application for soil moisture validation. In: IEEE geoscience and remote sensing symposium, pp 3331–3333. https://doi.org/10.1109/IGARSS.2014.6947193

Peel MC, Finlayson BL, McMahon TA (2007) Updated world map of the Köppen-Geiger climate classification. Hydrol Earth Syst Sci 11(5):1633–1644

Pickup G (1998) Desertification and climate change—the Australian perspective. Clim Res 11:51–63

Propastin PA, Kappas M, Muratova NR (2008) A remote sensing based monitoring system for discrimination between climate and human-induced vegetation change in Central Asia. Manag Environ Qual Int J 19(5):579–596. https://doi.org/10.1108/14777830810894256

Squires V (2012) Rangeland stewardship in Central Asia: balancing improved livelihoods, biodiversity conservation and land protection. ISBN 978-94-007-5366-2. https://pasture.klink.asia/klink/9c1102/download?embed=true

Ta Z, Yu R, Chen X, Mu G, Guo Y (2018) Analysis of the spatio-temporal patterns of dry and wet conditions in Central Asia. Atmosphere 9(1). https://doi.org/10.3390/atmos9010007

Thenkabail PS, Knox JW, Ozdogan M, Gumma MK, Congalton RG, Wu Z, Milesi C, Finkral A, Marshall M, Mariotto I, You S, Giri C, Nagler P (2012) Assessing future risks to agricultural productivity, water resources and food security: how can remote sensing help? Photogramm Eng Rem S 78(8):773–782

Toderich K, Shuyskaya E, Taha FK, Matsuo N, Ismail S, Aralova D, Radjabov T (2013) Integrating agroforestry and pastures for soil salinity management in dryland ecosystems in Aral Sea basin. In: Developments in soil salinity assessment and reclamation, pp 579–602

Touge Y, Tanaka K, Khujanazarov T, Toderich K, Kozan O, Nakakita E (2015) Developing a water circulation model in the Aral Sea Basin based on in situ measurements on irrigated farms. J Arid Land Stud 25(3):133–136

Tucker CJ, Pinzon JE (2013) A 1981–2012 non-stationary AVHRR NDVI global 8-km data set AGU fall meeting abstracts

Udelhoven T, Stellmes M, del Barrio G, Hill J (2009) Assessment of rainfall and NDVI anomalies in Spain (1989–1999) using distributed lag models. Int J Remote Sens 30:1961–1976

Vicente-Serrano SM, Begueria S, Lopez-Moreno JI, Angulo M, El Kenawy A (2010) A new global 0.5 degrees gridded dataset (1901–2006) of a multiscalar drought index: Comparison with current drought index datasets based on the Palmer Drought Severity Index. J Hydrometeorol 11:1033–1043. https://doi.org/10.1175/2010JHM1224.1

Vicente-Serrano SM et al (2015a) Drought variability and land degradation in semiarid regions: assessment using remote sensing data and drought indices (1982–2011). Remote Sens 7:4391–4423

Vicente-Serrano SM, Cabello D, Tomas-Burguera M, Martin-Hernandez N, Beguera S, Azorin-Molina C, Kenawy AE (2015b) Drought variability and land degradation in semiarid regions: assessment using remote sensing data and drought indices (1982–2011). Remote Sens 7(4):4391–4423. https://doi.org/10.3390/rs70404391

Vicente-Serrano SM, Tomas-Burguera M, Beguería S, Reig F, Latorre B, Peña-Gallardo M, Luna MY, Morata A, González-Hidalgo JC (2017) A high resolution dataset of drought indices for Spain. Data 2:22

Wardlow B, Anderson M, Verdin J (eds) (2012a) Remote sensing of drought. CRC Press, Boca Raton

Wardlow BD, Anderson MC, Sheffield, J, Doorn, BD, Verdin, JP, Zhan X, Rodell M, Future Opportunities and Challenges in Remote Sensing of Drought (2012b) Drought Mitigation Center Faculty Publications. p 103. http://digitalcommons.unl.edu/droughtfacpub/103

Chapter 3
NDVI-Based Monitoring Long-Term Vegetation Change Dynamics in the Drylands of Central Asia

Dildora Aralova, Dilshod Gafurov, and Kristina Toderich

Abstract An understanding of the complex environmental systems at a regional scale is still a challenging problem in the Central Asian countries, which cover 399.4 million hectares (ha). This chapter focuses on one specific (specific regions are classified as *Oblast* and it is targeted as a province) region for each country of Central Asia. The time series data through remote sensing represents a promising resource for studying connectivity within dynamic ecosystems. Among diverse landscapes in Central Asia, the selected zones are classified on base elevation and major land use type as mountain zones—Jalalabad region (Kyrgyzstan), Gorno-Badakhshan Autonomous Region (Tajikistan), and flat zones—Lebap region (Turkmenistan), Navoi region (Uzbekistan), and Kyzylorda region (Kazakhstan). The seasonal variations of NDVI derived from AVHRR-GIMMS 3g data for the period 1982–2015 estimated differences of spectral profiles enacted to indicate the weakness and strength of vegetation patterns. In response to the medium spatial resolution of the freely available multispectral Sentinel-2 datasets (0.3 m) these were upgraded for specific areas for monitoring the surroundings. The outputs of breakpoint of NDVI in Navoi (Uzbekistan) were described as slightly changed after 2002, whereas in general decreasing NDVI trends of vegetation values in Uzbekistan and Turkmenistan were observed in the past 3–5 years. With application of Mann–Kendall (MK) monotonic trends analysis we are approaching estimation of

---

D. Aralova (✉)
Dresden Technology University (TU-Dresden), Institute of Photogrammetry and Remote Sensing, Dresden, Germany

Centre of Hydrometeorological Service at Ministry of Emergency Situations of the Republic of Uzbekistan (Uzhydromet), Tashkent, Uzbekistan
e-mail: aralovad@daad-alumni.de

D. Gafurov
Cotton Breeding, Seeding and Cultivation Agro-technologies Scientific Research Institute, Kibray, Uzbekistan

K. Toderich
ICBA-CAC (International Center for Bio-saline Agriculture in Central Asia and Caucasus), ICARDA Regional Office (The CGIAR Facilitation Unit) for CAC, Tashkent, Uzbekistan
e-mail: ktoderich@cgiar.org

D. Egamberdieva, M. Öztürk (eds.), *Vegetation of Central Asia and Environs*,
https://doi.org/10.1007/978-3-319-99728-5_3

statistically significant trends (*p* value, *MK-tau*) in each selected zone. The negative significance procedure of MK-tau outputs is described in the regions of Turkmenistan, Uzbekistan, and Kyrgyzstan. Others are described as a fast recovering or caused by the increase of levels of precipitation as observed after 2013–2015. This study introduces approaches to enable and combine high- and low-resolution datasets for monitoring large–scale rangeland habitats and estimates of the amount of the data needed to better interpret biodiversity loss levels.

3.1 Introduction

In Kyrgyzstan, Tajikistan, and Kazakhstan, pasture occupies 69.1–85.0% of agriculture land. The importance of this research is related to better estimating natural and crop parameters for analysis of future behavior of vegetational succession or surroundings. The last findings proved that unpalatable plant cover in rangelands is associated with dense greenness and with a higher seasonal greenness value and more correspondence to sequestering palatable plants. Vegetation successions are a response to climate anomalies, and by observing transect changes of vegetation in this region we decided to analyze irreversible landscape changes and measure the impacts of environmental and anthropogenic factors in arid and semiarid rangelands. A vast area of the land currently faces with salinization after the shrinking of the Aral Sea basin and the irreversible use of water resources. The northern part of Uzbekistan (Karakalpakstan and Khorezm regions), the southern part of Kazakhstan (Aral'sk and Kyzylorda regions), and the Turkmenistan side are faced with environmental threats. The Aral Sea is an example of the effects of water abstractions for irrigation from the rivers of Amu Darya and the Syr-Darya, the rivers that feed the Aral Sea, to grow cotton in the desert (Chapagain 2005). Land degradation has begun to be a large challenge in Central Asia; for example, most desert lands that had been converted to irrigated land in the past are currently abandoned (unfertile) because of high salinization and depletion of water resources. In the past, pressure from the water surface dampened the strength of the northerly and northeasterly winds; the loss of the Aral Sea basin means that this protective action of the wind was significantly reduced (Khatamov 2014). Satellite images have revealed that as many as ten major dust storms occur each year in the region, most occurring between the months of April and July (Glantz 1999). Another challenge related to the decrease of precipitation levels or precipitation anomalies is reduction of the rainfed zones in Central Asia. A decrease in groundwater levels and salt-moving effects (the shrinking of the Aral Sea rapidly increased the number of dust and salt storms in the area) explicitly cause declining vegetation productivity. Therefore, salinization and increasing length of drought periods may strongly influence the spatial distribution of ecosystems and their biodiversity and functioning.

Linked to this issue, we have examined pixel-based temporal trends in the Normalized Difference Vegetation Index (NDVI) in the Kyzylorda region (Kazakhstan) and its neighbors: Uzbekistan (Navoi) and Turkmenistan (Lebap).

Tajikistan and Kyrgyzstan are an important corridor because they are adjacent to Afganistan and China. Responses of surrounding vegetation in the drylands are related to a steady rise in temperature in the past decades, and these factors affect the slight changes of vegetation cover. Dynamics of vegetation trends associated with a time series analysis (e.g., linear regression, Mann–Kendall trends) and a positive correlation describes precipitation effects and the negative correlation between temperature analyses. During the hot periods of summer, because of low precipitation and high evapotranspiration ratings, the soil is exposed to dryness. Most desert vegetation, especially semi-shrubs and shrubs, drop their leaves to reduce evapotranspiration level, and this period accumulates a lower NDVI index that determines the low vegetation values. It is a challenge to extract desert vegetation information from satellite images because of the influence of the low vegetation values resulting from the low precipitation of the background and the high index of brightness. Therefore, this study extracts the rangelands that are close to the regions and associated as the "Turanian Depression Ecosystems": Navoi (*Uzbekistan*), Lebap (*Turkmenistan*), and Kyzylorda (*Kazakhstan*). A low-lying desert basin region called the Turan Lowlands or Turan Depressions stretches from Turkmenistan (the Dashoguz region is the major area) and southern part of Karakum desert through Uzbekistan (Urganch and Karakalpakstan regions) to Kazakhstan (western part). The Amudarya River runs through the lowlands, extending from the southeast and northwestern directions. Reduction in the volume of watersheds in Central Asia is creating a current shortage in agricultural lands, and at the same time precipitation anomalies are also causing declining yields of the rainfed and arable zones. In addition, a huge territory of these two countries (Uzbekistan and Turkmenistan) is occupied by deserts, and the majority of water for irrigation is soaked up in dry channels (aryk), blatantly wasting the water (between 25% and 75% of it, depending on the season).

The received rainfall per year is less than 380 mm in these areas. Nevertheless, some other desert zones receive even less rainfall, between 100 and 200 mm per year. The Aralo-Caspian Lowlands is a prime example of marginal and saline environments (Toderich et al. 2014); however, these two countries rely heavily on irrigated agriculture and are mostly contributing to the economics of rural livelihoods. The derived vegetation index images from different satellite image sensors provide the opportunity to estimate the spatial and temporal dynamics of vegetation trends/patterns and to better understand the concept of vegetation indices (Jackson and Huete, 1991), even in hyperarid ecosystems.

3.2 Study Sites

Central Asia is located on the middle of the Asian continent. The following coordinates define the selected zone: 45.4507°N, 68.8319°E. The selected zones stretch from the Caspian Sea in the west (near Turkmenistan) to China in the east

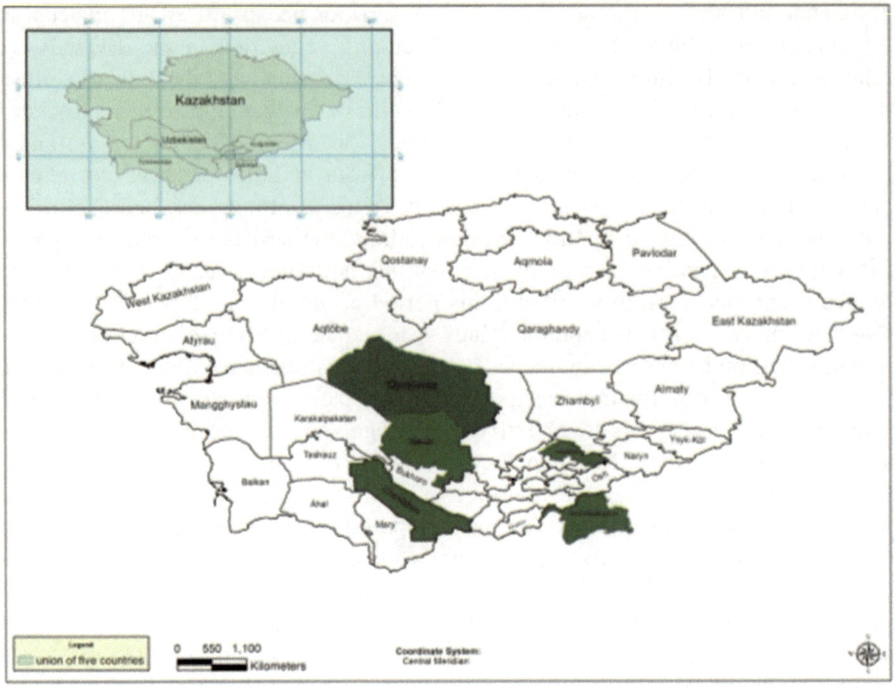

Fig. 3.1 Administrative boundaries of provinces in Central Asia and selected target areas (*green*). At the *top* of selected areas is located Kyzylorda (Kazakhstan), in the *middle*, Navoi (Uzbekistan), *below*, Lebap (Turkmenistan) and eastern part at the top located in Jalalabad (Kyrgyzstan), and *bottom*, Gorno-Badakhschon Autonomous Region (Tajikistan)

(Kyrgyzstan borders) and from Afghanistan in the south (Uzbekistan borders) to Russia in the north (Kazakhstan). The selected regions are shown in Fig. 3.1 and their districts with names are shown on Fig. 3.2. The corresponding geographic locations are at the top of the northern part, occupied by the Kyzylorda region (Kazakhstan), and in the middle part occupied by the Navoi region (Uzbekistan); in the eastern part the smallest one is the Jalalabad region (Kyrgyzstan) and below the Gorno-Badakhschan Autonomous Region (Tajikistan); the southern part is located in Lebap province (Turkmenistan), which borders with Uzbekistan.

3.2.1 Climatic Conditions in Central Asia

In these areas, climate increasing dynamics are already observed, and more change is expected again with rising trends (Fig. 3.2). Since 1957, there has been an observed trend of warming within Uzbekistan, Turkmenistan, and Kazakhstan as well. Past climate trends (precipitation and temperature) as below are described for

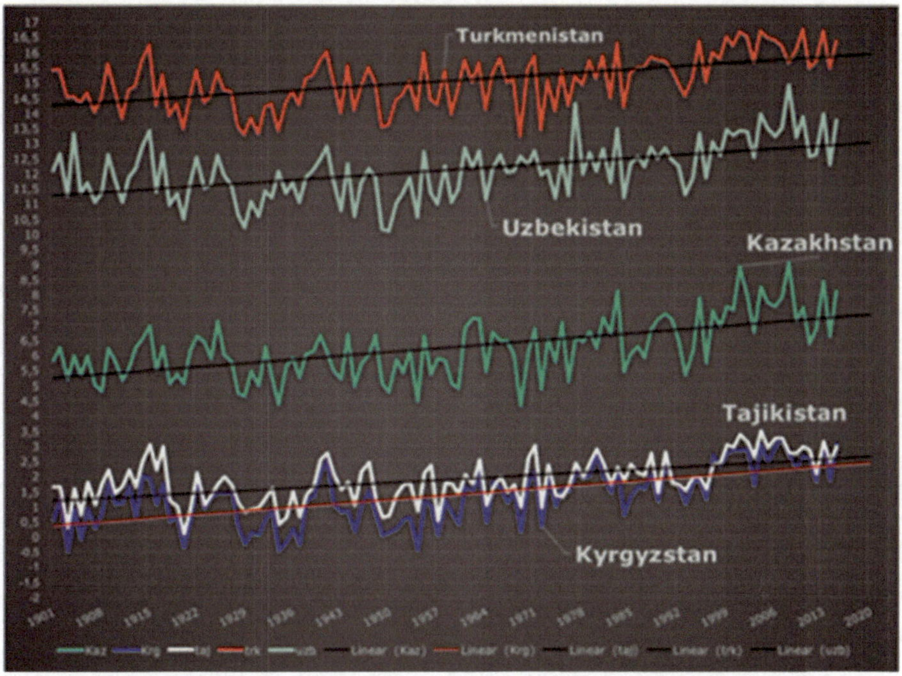

Fig. 3.2 Climate dataset in Central Asia, retrieved from CRU-TS 4 dataset and further raising trends described by linear trend (2016–2020)

each targeted area and with past datasets retrieved and modified after www.climatecharts.net from TU-Dresden[1]. The datasets (past) were developed based on USSR meteorological stations, originated and available from the Global Historical Climatology Network (GHCN). GHCN[2]-Monthly provides climatological observations for four elements: monthly mean maximum temperature, minimum temperature, mean temperature, and monthly total precipitation.

The annual average temperature has increased by 0.29°C for each 10 years since 1957, for example, with minimum temperatures increasing more than maximum temperatures. There are some significant exceptions to this trend, including (1) the Aral Sea, where the maximum temperature has increased more than the national average while the minimum temperature has remained constant; and (2) mountainous areas, where warming has been less than the national average.

[1] https://climatecharts.net/

[2] https://www.ncdc.noaa.gov/ghcnm/

Table 3.1 Selected regions: names of their districts and covered area (km^2)

Jalalabad (Kyrgyzstan)	Navoi (Uzbekistan)	Kyzylorda (Kazakhstan)	Lebap (former Chordjou) Turkmenistan	Gorno-Badakhschon Autonomous Region (Tajikistan)[a]
33,700 km^2	110,800 km^2	226,019 km^2	93,730 km^2	64,200 km^2
Chatkal	Uchkuduk	Aral'skiy	Magdanly City (formerly Gowurdak)	Vanj
Ala-Buka	Tamdy	Kazalinskiy	Seýdi City	Ishkashim
Aksyi	Konimekh	Karmakchinskiy	Turkmenabat City (formerly Charjew)	Murgab
Nooken	Nurota	Zhalagashskiy	Atamyrat	Roshtkala
Bazar-Korgon	Khatirchi	Terenozekskiy		Darvaz
Suzak	Karmana	Shieliyskiy		Rushan
	Qiziltepa	Zhankorganskiy		Shugnan
	Navbahor	Kyzylorda		

[a]The overall observed land is equal to 528,449 km^2 or 52,844.900 ha

3.2.2 *Detailed Information for Selected Eco-regions and Their Land Use and Land Cover Change Data*

Among different land cover transitions, the selected zones (Table 3.1) are classified as mountain zones—Jalalabad region (Kyrgyzstan), Gorno-Badakhshan Autonomous Region (Tajikistan), and flat zones—Lebap region (Turkmenistan), Navoi region (Uzbekistan), and Kyzylorda region (Kazakhstan). According to the classification of Köppen-Geiger (Köppen, 1936), the Central Asian selected zones are classified as Bwk, which gives a clear idea of temporal vegetation anomalies syndrome during precipitation anomalies periods. As shown on Fig. 3.1b, the south and southwestern parts of Central Asia are mainly occupied with three categories of Köppen classification Bwk: the categories with letter "B" are classified as arid zones, "W" are classified as a desert, and the last, "k," refers to cold.

Table 3.2 describes the selected zones coordinates and general information about total territory (km^2). Among Central Asian locations, the northern part is occupied by Kazakhstan and mostly covered with Kazakh steppe (Fig. 3.3), and the southern part includes Uzbekistan and Turkmenistan and the "southern desert" named as Caspian lowland deserts. The largest territory among the five countries is counted for Kazakhstan and the smallest is Tajikistan (Table 3.2).

3.2.3 *Kyzylorda Region–Kazakhstan*

This region is located to the east of the Aral Sea Basin in the lower reaches of the Syrdarya River, mainly within the Turan Lowlands, with altitude ranging between 50 and 200 m (Fig. 3.2). On the left bank of the Syrdarya River is a vast areas with

Table 3.2 Coordinates of observed zones and information about the total territory of Central Asia

Country	Area, km²	Coordinates of selected zones
Kazakhstan	2.724,900	**Aral'sk** 45°16′39″N,61°24′4.59″E **Shieliyskiy** 44°29′1.50″N, 65°54′56.59″E **Karmakchinskiy** 45°29′36.91″N, 64°25′23.28″E
Kyrgyzstan	199,900	**Aksyi** 41°54′51.49″N, 72°4′55.04″E **Ala-Buka** 41°11′33.59″N, 71°1′9.43″E **Bazar-Korgon** 41°18′32.28″N, 73°8′51.68″E
Tajikistan	143,100	**Darvoz** 38°35′39.02″N, 71°2′45.59″E **Murghob** 38°20′56.97″N, 72°44′4.46″E **Roshtqal´a** 37°27′33.85″N, 72°29′16.99″E
Turkmenistan	488,100	**Gazadzhak** 41°0′48.08″N, 60°56′36.86″E **No name** 37°35′13.33″N, 63°38′12.87″E **Seydi** 38°22′12.82″N, 62°31′13.59″E
Uzbekistan	447,400	**Konimekh** 40°16′46.40″N, 65°9′21.18″E **Nurata** 40°15′14.13″N, 65°56′52.27″E **Tamdy** 42°4′1.48″N, 64°22′52.25″E

hilly-ridge sands of Kyzylkum desert, and a considerable part of the territory is occupied by sands, almost devoid of vegetation. Dry channels of Zhanadarya and Inkardarya are located in the middle area of the region; shallow basins are occupied by solonchaks. In the floodplain of the Syrdarya, streams are often occupied by saline soils covered with meadow vegetation with rare tugai forests; in the delta and along the coasts are extensive thickets of reeds (*Phragmites* sp.). The climatic parameters of the area are shown on Fig. 3.4.

3.2.4 Gorno-Badakhschon Autonomous Region–Tajikistan

The Gorno Badakhshan Autonomous Region (GBAR) is one of the most mountainous in Tajikistan, sparsely populated, underdeveloped, and large areas located in the eastern part of the country. The selection criteria of this region is related to

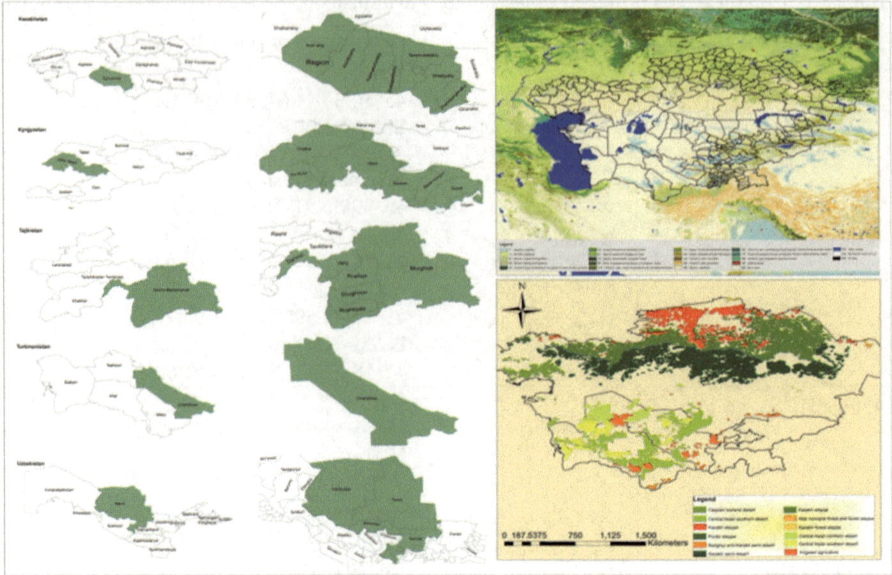

Fig. 3.3 Administrative boundaries of targeted areas and Turanian Lowland distributions in Central Asia (*right, below*)

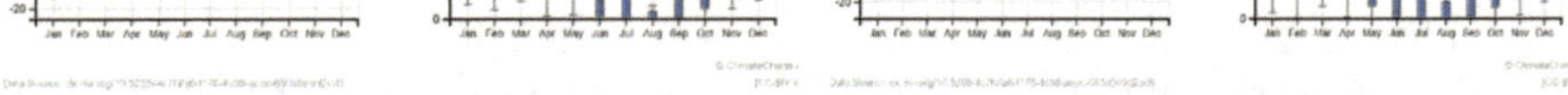

a) Shieli city, Kyzylorda region, Kazakstan *b) Kyzylorda city, Kyzylorda region, Kazakhstan*

Fig. 3.4 Past climatic trends for Kyzylorda region based on meteorological stations, Kazakhstan: (**a**) Shieli city, Kyzylorda region, Kazakstan; (**b**) Kyzylorda city, Kyzylorda region, Kazakhstan

demonstrating the biodiversity of the unpopulated region; at the same time it borders on the north with Kyrgyzstan, in the east with China, and in the south and west with Afghanistan. The GBAR area is 64,200 km^2 (Table 3.1), 44.9% of the country's territory. Irrigated lands of GBAR amount to 17.7 thousand ha, which equals 2.4% of Tajikistan's irrigated lands; rainfed lands are 6.4 million ha, and meadows 757 thousand ha. The climatic parameters of the area are shown in Fig. 3.5.

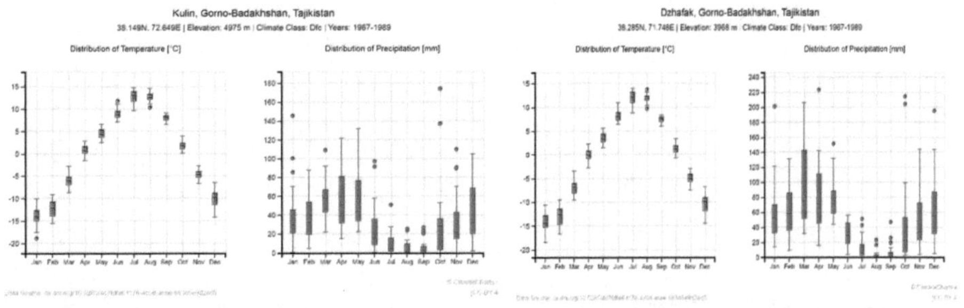

a) Kulin city, GBAR, Tajikistan b) Djafak city, GBAR, Tajikistan

Fig. 3.5 Past climatic trends for Gorno-Badakhschan Autonomous Region (GBAR) region based on meteorological stations, Tajikistan: (**a**) Kulin city, GBAR, Tajikistan; (**b**) Djafak city, GBAR, Tajikistan

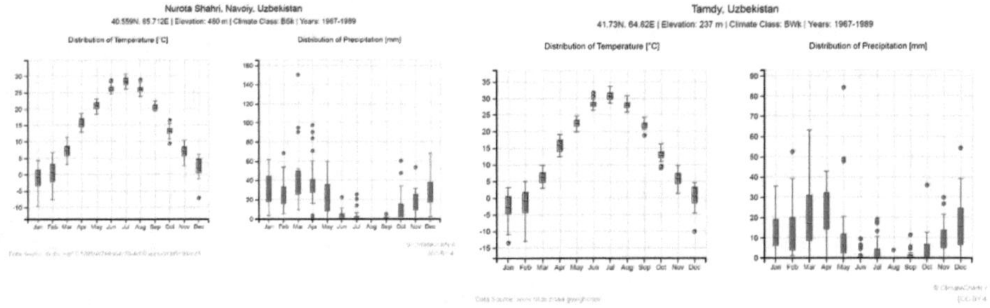

a) Nurota city, Navoi region, Uzbekistan b) Tamdy city, Navoi region, Uzbekistan

Fig. 3.6 Past climatic trends for Navoi based on meteorological stations in Uzbekistan: (**a**) Nurota city, Navoi region, Uzbekistan; (**b**) Tamdy city, Navoi region, Uzbekistan

3.2.5 Navoi Region–Uzbekistan

The Navoi province covers 24.8% of the total area of the country (Table 3.1). The northwestern part of the province is occupied by the Kyzylkum plateau; Nurata mountainous ridges stretch in the east and the southern part of the province is fringed with the Zarafshan River. Vast areas of the oasis are occupied by sands of Kyzylkum desert. The climatic parameters of the area are shown in Fig. 3.6.

3.2.6 Jalalabad Region–Kyrgyzstan

Jalalabad province covers 16.9% of the total area of the country. Jalalabad is located in the centralwestern part of Kyrgyzstan. The southern edge of the region is part of the Fergana Valley (Uzbekistan); the rest of the region is mountainous. The Jalalabad region is rich in ecological resources. Among strictly protected areas (IUCN

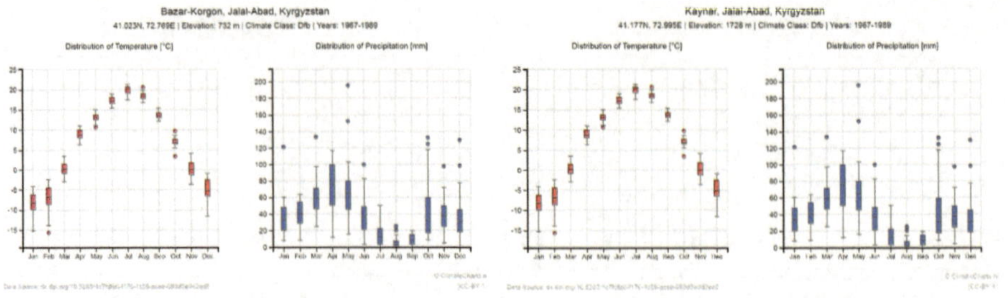

a) Bazar –Korgon city, Jalalabad region, Kyrgyzstan *b) Kaynar city, Jalalabad region, Kyrgyzstan*

Fig. 3.7 Past climatic trends for Jalalabad region based on meteorological stations in Kyrgyzstan (1967–1989): (**a**) Bazar—Korgon city, Jalalabad region, Kyrgyzstan; (**b**) Kaynar city, Jalalabad region, Kyrgyzstan

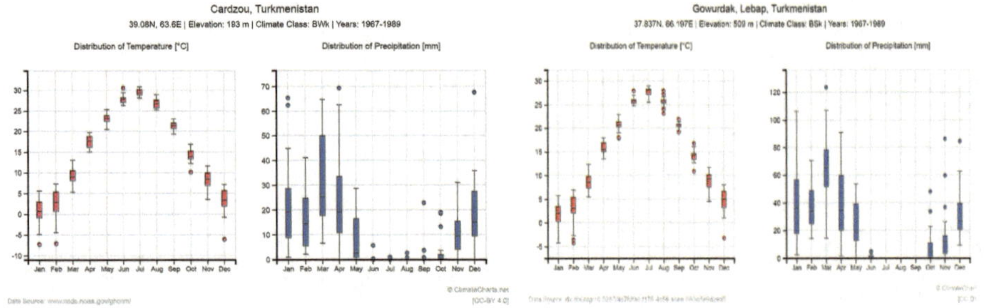

a) Chorjuy(former) Lebap region, Turkmenistan *b) Gowurdak city, Lebap region, Turkmenistan*

Fig. 3.8 Past climatic trends for both regions of Turkmenistan (1967–1989) based on available meteorological stations: (**a**) Chorjuy (former) Lebap region, Turkmenistan; (**b**) Gowurdak city, Lebap region, Turkmenistan

category) located in the region are Sary-Chelek State Biosphere Nature Reserve, Besh-Aral State Nature Reserve, and Padyshata State Nature Reserve. The climatic parameters of the area are shown in Fig. 3.7.

3.2.7 Lebap (Former Chordjou) Region–Turkmenistan

The downstream of Amudarya crosses this region. The region of Lebap is covered by a desert, and more than 90% of the territory located in a desert plain zone. On the left bank stretches the desert of Karakum, and northwest of the right bank is the desert of Kyzylkum. As well, this kind of landscape diversity is well influenced by climatic conditions: the summers are very hot and precipitation annually is estimated as about 50–60 mm. Also in this area is located the highest point of the country, Mount Ayrybaba (3139 m, a.s.l.). In the territory of the region there are two channels: Amu-Bukhara and Kashkadarya. In this region, there are three reserves: Amudarya, Kugitang, and Repetek. The Repetek Reserve is known for being the hottest place in Central Asia (Fig. 3.8).

3.3 Materials and Methods

3.3.1 Vegetation Datasets

3.3.1.1 GIMMS 3g

Bimonthly data with 8-km resolution GIMMS AVHRR (1982–2015) were utilized by the following source: https://ecocast.arc.nasa.gov/data/pub/gimms/3g.v1/00FILE-LIST.txt, to analyze NDVI status in Central Asia extracted shapefiles from DIVA source and masking it. The calculation is dynamically interpreted by Python language (ArcPy) and providing some specialized functions to prepare numerical datasets after retrieving GeoTIFF format.

3.3.1.2 Sentinel-NDVI

Sentinel-2 was launched in 2014, and is able to monitor land, vegetation, and water bodies. The sentinel NDVI ratio can be determined from the contribution of *visible* wavelength and *near-infrared* wavelengths (Eq. 3.1). The efforts of Sentinel-2 are related to vegetation in poor condition or thin areas such as Central Asian ecosystems:

$$\mathrm{NDVI} = \frac{B_8 - B_4}{B_8 + B_4} \tag{3.1}$$

The scene of Sentinel 2 downloaded by GitHub: https://scihub.copernicus.eu/dhus/#/home and utilized with Sen2Cor tool. Sen2Cor tool is a processor for Sentinel-2 products and converted jpeg2000 to GeoTIFF format; it performs the atmospheric, terrain, and cirrus corrections. Then Sentinel datasets used a resampling method for displays of properties. Because NDVI is defined as a ratio of difference over sum of bands 8 and 4 (NIR and R), Eq. 3.1 was used for available to precedent NDVI. The available datasets were 2016 of the period 11 September.

3.3.2 Trend Analysis

For the trend analysis itself, different methods are tested. Test procedures (Price 1987) are carried out for the assessment of quality of trend modeling, especially for time series approaches. It is useful to be able to turn a time series analysis for NDVI values into components, and a suitable tool for that is Python language. Another tool, developed under programming language R and a time series, divides into three components: the trend, seasonality, and remainder, available within R via the **stl** function. The time series methodology was presented by Robert Cleveland et al. (1990).

3.3.3 Time Series Decomposition

The frequency of seasonality is the initial point for the decomposition into trend, seasonal, and random components. All analyses are exemplarily performed for five regions as mentioned in Table 3.1 and carefully selected locations (8 × 8 km) in Central Asia. To analyze the interannual trend, it is necessary to remove this seasonal component from the input data (Neeti et al. 2012; Osunmadewa et al. 2015).

In the first step, the time series (*Y*) must be separated into a trend component (*T*), a season component (*S*), and an additional component that contains the random noise (*e*) of the input signal. An additive or multiplicative model can describe the relationship among these three components. The decomposition for the time series of the study area (Central Asian regions) was achieved using the "decompose" function implemented in the open-source software called R programming language, 23, that was also used for the entire analyses presented in this chapter.

$$Y_t = T_t + S_t + e_t \tag{3.2}$$

For the moving average, a window size of frequency of seasonality was used. After removing the trend from the original time series, the seasonal figure (S_t) was computed by averaging, for each time unit, over all periods and centered in a second step. For one period (year), the seasonal component S_i is

$$\underset{t_{\max}-\frac{n}{2}}{\overset{t_{\min}+\frac{n}{2}}{T_t}} = \frac{\sum_{i=t-\frac{n}{2}}^{i=t+\frac{n}{2}} Y_i}{n+1} \tag{3.3}$$

$$\underset{i=1}{\overset{n}{S_i}} = \frac{\sum_{j=0}^{m-1} Y_{j\cdot n+i}}{m} \tag{3.4}$$

where *m* is the number of years. If both trend and seasonal figures are removed from the input data, the error component (random) remains.

$$e_t = Y_t - \left(T_t + S_t\right) \tag{3.5}$$

3.3.4 Seasonal Mann–Kendall Trend Test

The Seasonal Mann–Kendall trend test is a test for monotonic trend in a time series with seasonal variation (Kendal and Stuart, 1983). Hirsch et al. (1982) developed such a test by computing the Kendall score separately for each month. The separate monthly scores are then summed to obtain the test statistic. The variance of the test statistic is obtained by summing the variances of the Kendall score statistic for each month. The τ coefficient is defined by

$$\tau = \frac{\sum_{i=1}^{S} S_i}{\sum_{i=1}^{S} D_i} \tag{3.6}$$

where S_i, D_i, $i = 1, \cdots, s$ denote the Kendall scores and denominators for the ith season and s is the seasonal period.

3.4 Results

3.4.1 *Monitoring NDVI Characteristics and Dynamics for Selected Arid and Humid Regions During 1982–2015 in Central Asia*

For better explanation and interpretation of grid-based values of the NDVI, each selected zone has been divided into three parts (cities, districts) for each region of Central Asia. The selected three parts of the regions describe the spectral profile of the vegetation trends (Figs. 3.9, 3.10, 3.11, 3.12, and 3.13).

3.4.2 *Gorno-Badakhschon Autonomous Region-Tajikistan*

The selected subregions (Fig. 3.9) are selected based on geographic location and described in Fig. 3.2, such as Darvoz district located in the western part of selected zone (3); Murghob district (2) occupied the eastern part and southern part named as Roshtqala district (1). The largest one is Murghob district (38,442 km^2) and the smallest is Darvoz district (2824 km^2). Pixel-based NDVI values are observed for each zone (see Figs. 3.9, 3.10, 3.11, 3.12, and 3.13) as dense vegetation, in the western part of the region in Darvoz distict (3), according to the spectral profile of values indicated that ranges between 0.60 and 0.80; and sparse vegetation described in Murghob (2) districts that ranges between 0.05 and 0.07. In the southern part of the region (Roshtqala region), also described as sparse vegetation, it ranges between 0.16 and 0.19. Other regions are not interpreted but suitable images are demonstrated on Figs. 3.10, 3.11, 3.12, and 3.13.

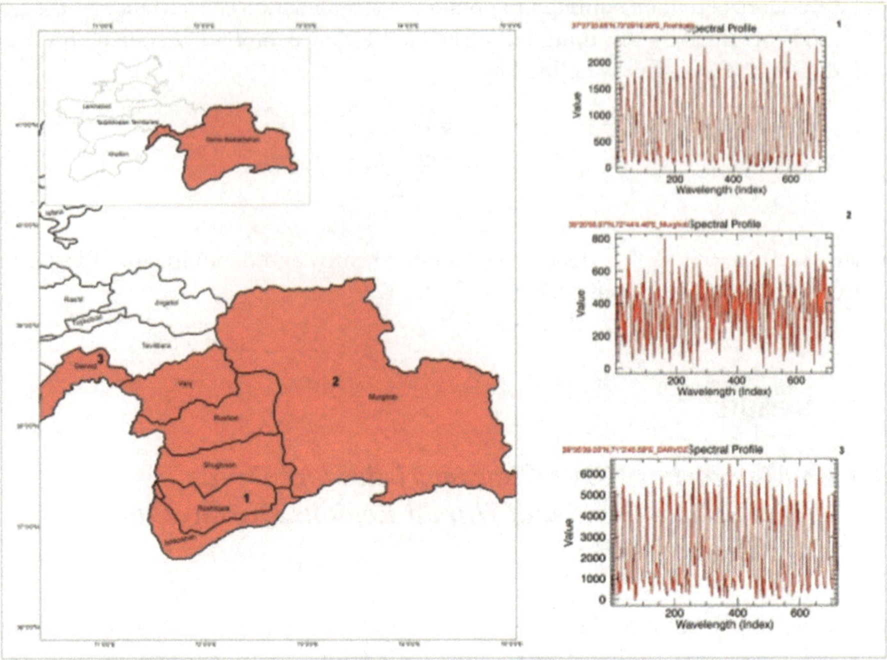

Fig. 3.9 Selected districts from Gorno-Badakhschon Autonomous Regions (GBAR): *1*, Roshtqala, *2*, Murghob, *3*, Darvoz

3.4.3 Sentinel-2 NDVI Scenes for Efficiency Description of Crop and Shrublands as Examples for Kazakhstan and Uzbekistan

In this part, we have described the new satellite images to better interpret and understand the relevance of landscape diversity with utilization of the NDVI indexes. The selected scenes are clearly explained: the crop zones as irrigated lands (top scenes in Fig. 3.14) and shrubland zones between the boundaries in Kazakhstan and Uzbekistan (lower scenes in Fig. 3.14).

3.4.4 Decomposition and Seasonal NDVI Accumulation in Selected Provinces of Central Asia

The seasonal changes on breakpoint (red line) are observed in Navoi region (Fig. 3.15), while other countries are kept in stable position. The abrupt changes of NDVI in Navoi province might be related to increasing greenness in the eastern part of the districts (Nurata, Sarmysh hills). However, the results of *MK-tau* demonstrated a positive significance on the eastern and southeastern parts of the area,

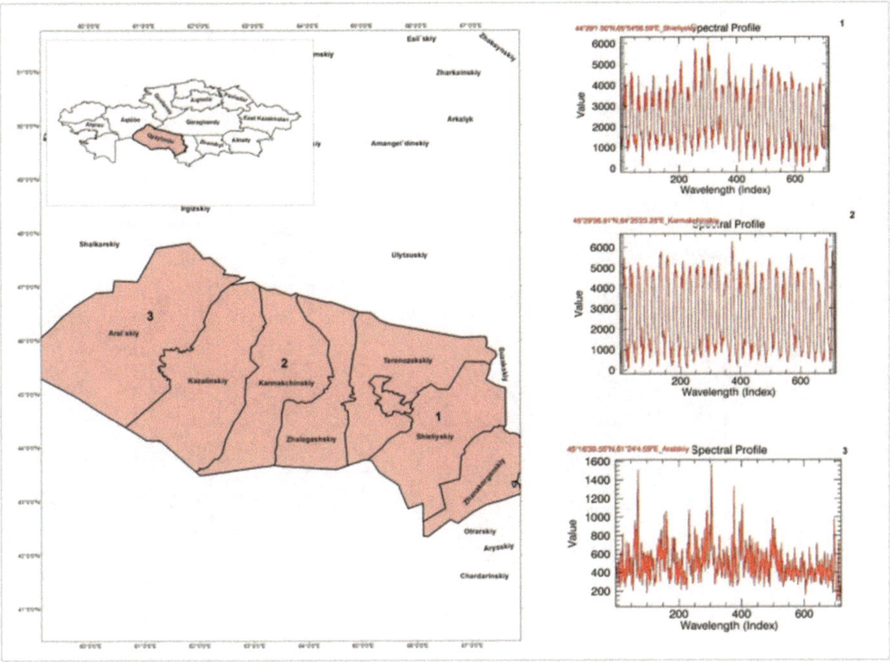

Fig. 3.10 Selected districts from Kyzylorda regions: *1*, Shielieyskiy, *2*, Karmakchinskiy, *3*, Aral'skiy

whereas a vast area maintained a negative significance of NDVI values. The maximum ratings of NDVI or NDVI increasing trends were observed in GBAR (Tajikistan) and Kyzylorda (Kazakhstan) in the last observed years (2014–2015). In Kyzylorda, the past years observed the recreation of dams near the Aral Sea Basin, and the establishment of dams has more influence on the level of greenness in the western part of Kyzylorda. The results are also shown on the spectral profile in Aral'sk (Kyzylorda) (Fig. 3.16): the ratings of spectral profile are increasing in the last observed 4 years (2011–2015), related to stabilizing riparian ecosystems near Aral Sea Basin zones.

3.4.5 *Kyzylorda Region–Kazakhstan*

A large variety of detected changes were found during aggregation over the Kyzylorda region (Kazakhstan), and this might be related to the effects of salt movement after shrinking of the Aral Sea Basin and the agricultural intensification of rice paddies among the streams of the Syrdarya River basins. In addition, for the difference of the trends compared to the other countries, it was estimated under a vast area occupied by temporal drying lakes and lakes around the Aral'sk districts (Figs. 3.10 and 3.16). At the same time, the anthropogenic factors have much influence in this

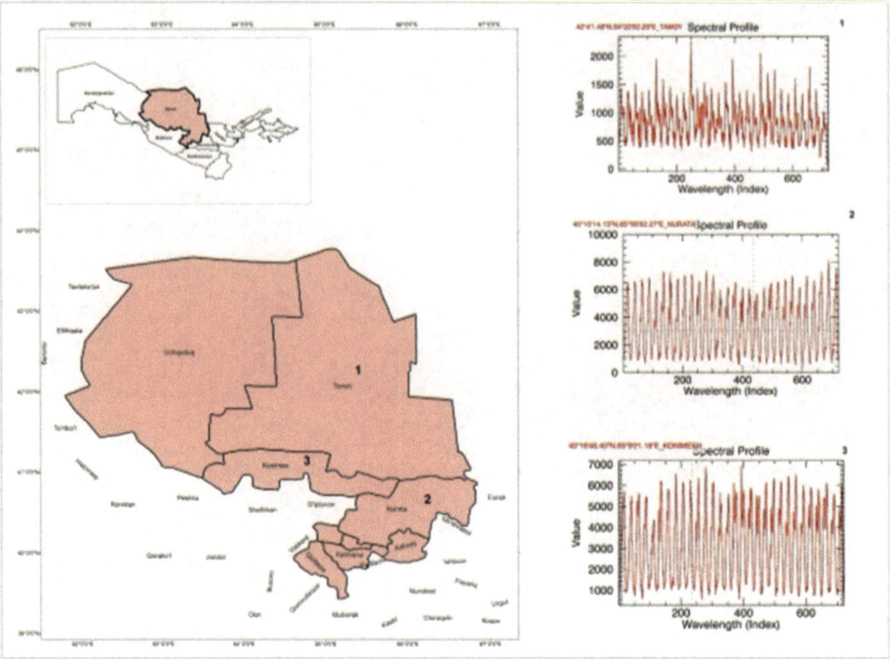

Fig. 3.11 Selected districts from Navoi regions: *1*, Tamdy, *2*, Nurata, *3*, Konimekh

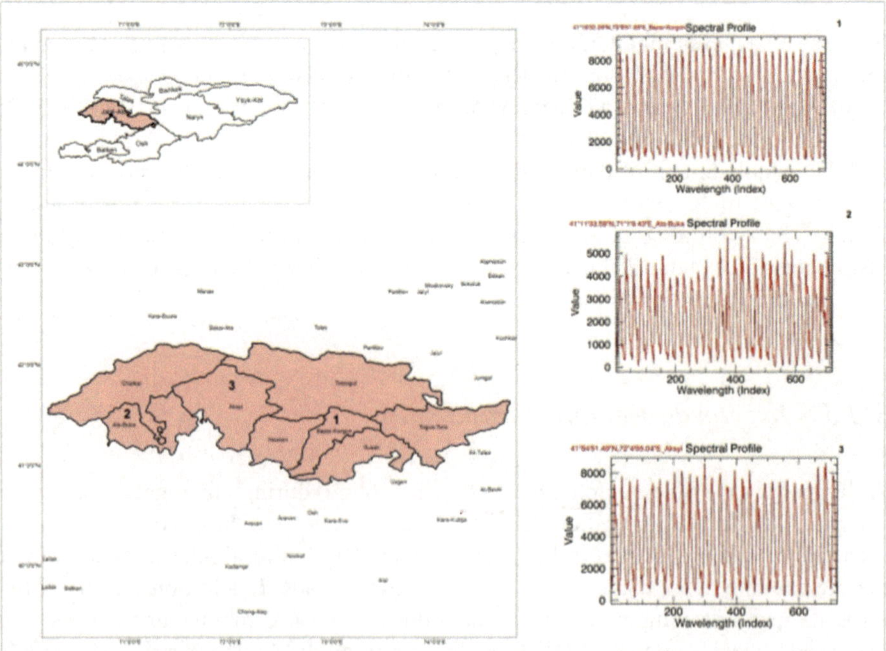

Fig. 3.12 Selected districts from Jalalabad regions: *1*, *Bazar-Korgon*, *2*, *Ala-Buka*, *3*, *Aksyi*

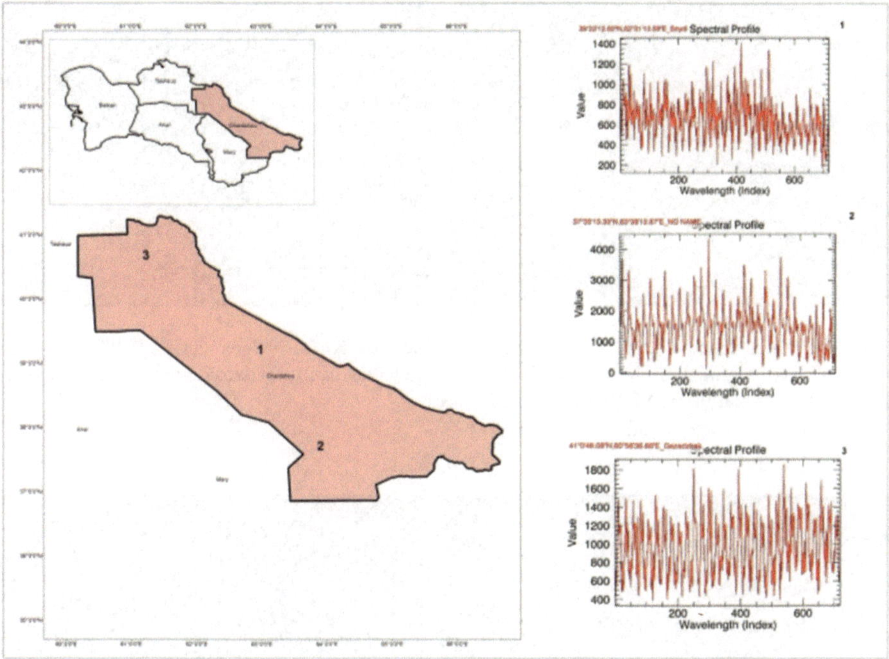

Fig. 3.13 Selected districts from Lebap regions: *1*, *Seydi*, *2*, *No name*, *3*, *Gazadzhak*

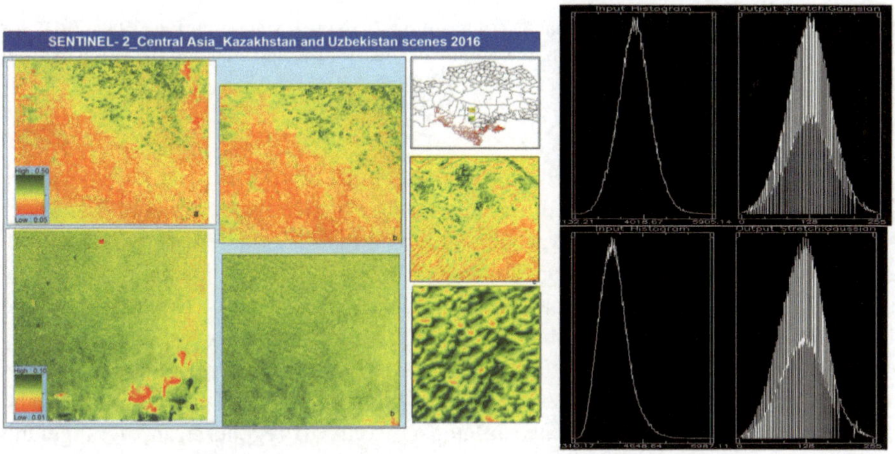

Fig. 3.14 Description of crop and shrubland zones in the southern part of Central Asia and estimation of Normalized Difference Vegetation Index (NDVI) indices in the selected zones: *top*, part of Kazakstan part; *below*, between Kazakhstan and Uzbekistan. Spectral profiles of Sentinel-2 datasets were examined in two scenarios: Kazakhstan and Uzbekistan

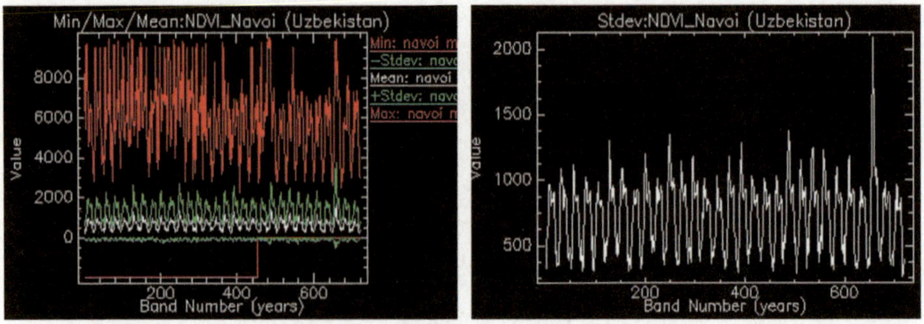

Fig. 3.15 Accumulated and seasonal decomposition of NDVI values during 1982–2014 for selected Navoi regions: minimum, maximum average values, and standard deviation (±) (*left side*) and standard deviation of NDVI values (*right side*)

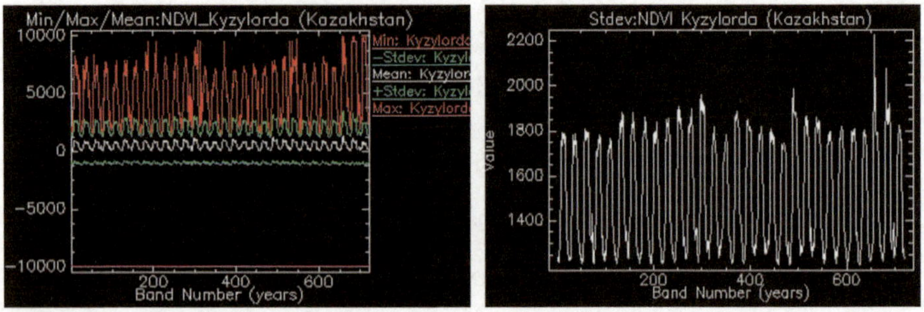

Fig. 3.16 Accumulated and seasonal decomposition of NDVI values during 1982–2014 for selected Kyzylorda regions: minimum, maximum average values, and standard deviation (±) (*left side*) and standard deviation of NDVI values (*right side*)

area, related to the intensification of use of water resources for agriculture purposes in the past; currently it is related to the extension of rice paddy plots on the lower streams of Syrdarya, where the vast abandonment lands expanded after salinization. Abandoned land has resulted from shifting cultivation after high salinization of fertile land, and at the same time a southern part of the territory is occupied by the Kyzylkum desert. The overestimating of salinization lands correspond to further occupying the lowlands of Kyzylkum, resulting from the high impact of dust storms and topographic location, and expanding to the borders of all five countries. The increases of NDVI values are also related not only to regeneration of rangeland greenness values, but mostly related to the expansion of agricultural lands along river channels in the past 7–8 years in the regions of Kyzylorda and totally in Kazakhstan (Fig. 3.16).

3.4.6 Navoi Region-Uzbekistan

The geographic coordinates of Kyzylorda and Navoi regions and the selected zone are 42–45° (lat.) and 62–67° (long.); estimation (measures) of the area are approximately 113,764 km^2. The overall area for both countries is 336,819 km^2. The main interest related to observations between the two neighbor countries is how it has changed vegetation patterns; however, the distance among them is more than a hundred thousand kilometers, but at the same time landscape and topographical diversity are more or less similar. Results revealed that, since 1982, in the Navoi region NDVI values ranged between 0.06 and 0.10, and after 1995 the indicated drop in rating of NDVI values might be related to the procession of cotton fields in this area.

Also, as already mentioned, in these areas are observed an altered seasonal breakpoint of NDVI after 2002–2003, and as a fact, in these years more precipitation and snow cover were observed in rangelands. However, during 1999, 2000, and 2001 was observed a drought anomaly (peak) in the past century; northwest Uzbekistan experienced drought and water shortage, raising of salinity levels in fertile lands, and depletion of groundwater levels. All these factors are detected to change the seasonal breakpoint of NDVI. In addition, one practice observed is that farmers and shepherds are estimated to use arable land and rainfed zones for temporal crops or short-term crops in seasonality.

3.4.7 Gorno-Badakhschon Autonomous Region–Tajikistan

The relative changes of NDVI in GBAO (Fig. 3.17) are mostly related to expansion of agricultural zones after the breakup of the USSR in these regions, and also, totally, to the observed increased food productivity in past decades. However, the Tajik-Pamirs, with only 240 km^2 of arable land, will never be able to achieve local self-sufficiency.

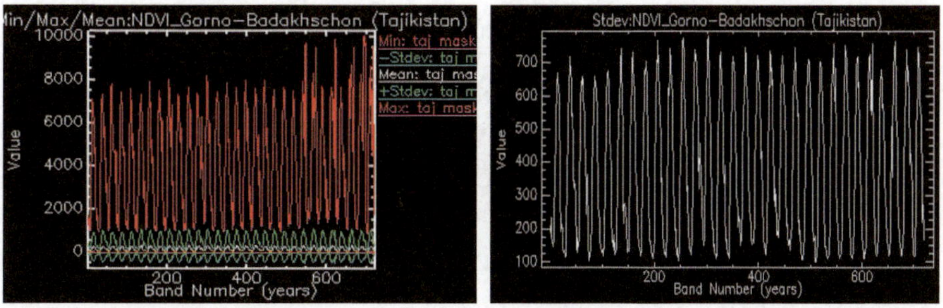

Fig. 3.17 Accumulated and seasonal decomposition of NDVI values during 1982–2014 for selected GBAR regions: minimum, maximum average values, and standard deviation (±) (*left side*) and standard deviation of NDVI values (*right side*)

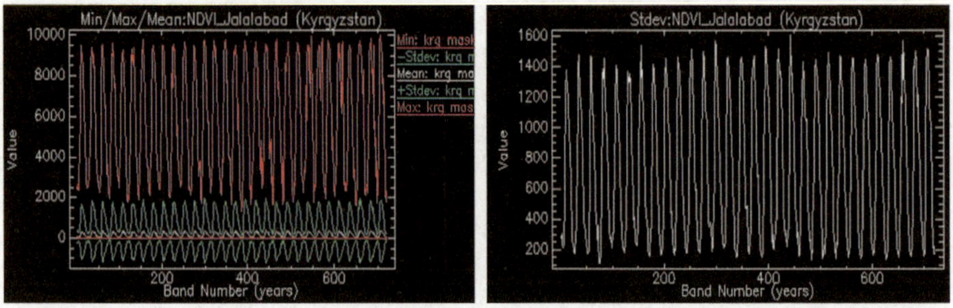

Fig. 3.18 Accumulated and seasonal decomposition of NDVI values during 1982–2014 for selected Jalalabad regions: minimum, maximum average values, and standard deviation (±) (*left side*) and standard deviation of NDVI values (*right side*)

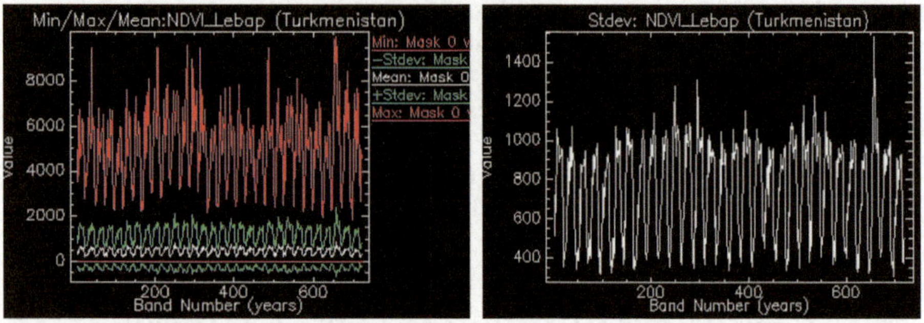

Fig. 3.19 Accumulated and seasonal decomposition of NDVI values during 1982–2014 for selected Jalalabad regions: minimum, maximum average values, and standard deviation (±) (*left side*) and standard deviation of NDVI values (*right side*)

3.4.8 Jalalabad Region–Kyrgyzstan

The smooth results observed only in Jalalabad province accord with the results estimated for the whole region (Fig. 3.18). The area has stability parameters (standard deviation, greenness indices, etc.). On the basis of agricultural expansions, it was estimated that lines were dropped; however, these areas are mostly overpopulated zones and mostly expecting the increasing level of the trend in agriculture and livestock sector in these zones.

3.4.9 Lebap Region–Turkmenistan

Low indices of NDVI were observed in this area (Fig. 3.19). The area has stability parameters (standard deviation, greenness indices, etc.). On the base of agricultural expansions, it was estimated as dropped lines also.

3.5 Discussion

Within this aim, we concentrated on finding a natural vegetation succession process in areas where anthropogenic factors have less influence; a bright example is the GBAR region, and reversing version, as well as the factors where humans introduced changes and destroyed the natural succession process in the ecosystem: Kyzylorda (Kazakhstan), Lebap (Turkmenistan, and Navoi (Uzbekistan). But the GBAR area was partly destroyed during the civil war of Tajikistan, and nowadays this area is also faced with serious environmental threat issues. Results indicated (GIMMS 3g) on Kyzylorda Province (Kazakhstan) estimated several unstable vegetation change trends, and because of the shrinking of the Aral Sea or expansion of high-salinity areas, at the same time the neighbor country Uzbekistan faced precipitation anomalies and the greening trend was reduced.

Climatic stress contributes to an overall loss of valuable biodiversity on a large scale. The potential amount of interaction that the vegetation–soil–atmosphere approaches gives a clear idea to develop ecosystem services with valuation, but still that is always less actual in these regions and this is a limiting factor to characterizing conditions (increasing levels of soil salinization) annually. Characterizing natural plant characteristics or their resistance to climate anomalies and the uptake of carbon from the atmosphere are important issues nowadays in developing further pasture strategies, and it will take more attention to investigate perennial plants that are less sensitive to extreme drought conditions. Assessing the grazing potentials of degraded rangelands by mapping zonal halophytic vegetation allowed us to identify salt pioneer plant species for each studied zone of Central Asia to initiate the reclamation process of saline-prone soils and to stabilize sand dunes.

3.5.1 Expansion of Agricultural Lands in Central Asia After 1992

The datasets modified under source and the results are categorized into two subjects: expansion (Fig. 3.20) and decrease (Fig. 3.21) of agricultural areas in Central Asia. Agricultural lands occupied a vast territory in Central Asia, and this category

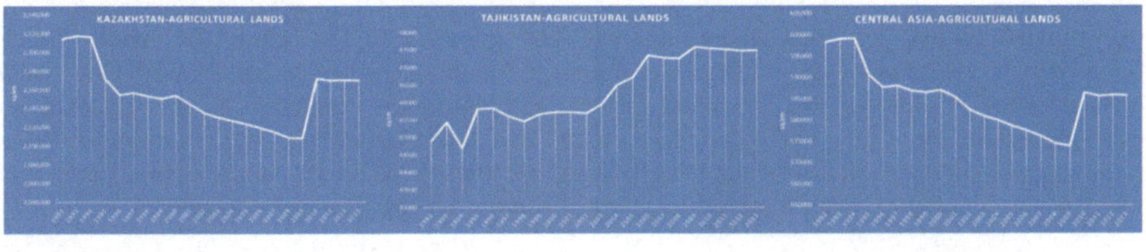

Fig. 3.20 Agricultural land occupation in Central Asia: Kazakhstan (2,180,000 km^2), Tajikistan (47,500 km^2), and Kyrgyztan (106,000 km^2). Generally, the average in Central Asia (590,000 km^2) has expanded in the past 3 years (2011–2013). [Modified after https://knoema.com/atlas (last data updated 2016)]

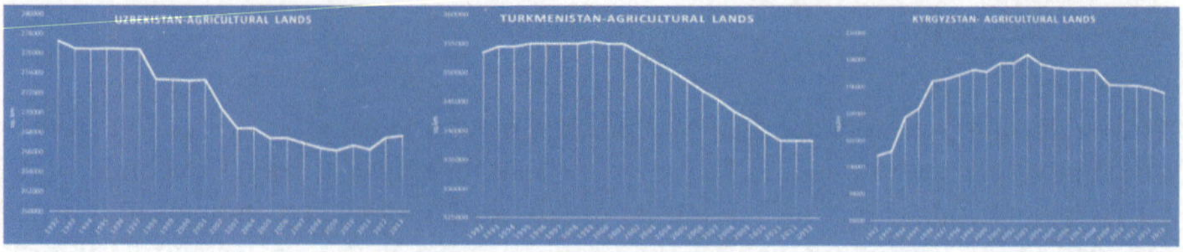

Fig. 3.21 Recession of agricultural lands in Central Asia after 1992. [Modified after https://knoema.com/atlas (last data updated 2016)]

refers to the share of land area that is arable, under permanent crops, and under permanent pastures. Permanent pasture is land used for 5 or more years for forage, including natural and cultivated crops also categorized as agricultural land. In the category of expansion, lands are demonstrated in Kazakhstan, Tajikistan, and generally, in Central Asia as well. The average total amount in Central Asia is similar to Kazakhstan, because a very large amount of land in Kazakhstan is indicated as similar.

3.5.2 Recession/Decline of Agricultural Lands in Central Asia

The decline of agricultural lands in these areas is related to salinization and declining cotton productivity each year. Irrigated agriculture in these regions is the continuous challenge of land salinization (Rozema and Flowers 2008). Reduction in watershed volume in Central Asia is currently causing water shortage in agricultural lands and at the same time precipitation anomalies are also decreasing the rainfed zones. Also, a huge territory of these two countries is occupied by deserts and the majority of water for irrigation is soaked up in dry channels (aryk) and the water is blatantly wasted[3] (between 25% and 75% of it, depending on the period). In Central Asia, the highest rate of salinization is caused by irrigation, with higher ratings in Turkmenistan, whereas the lowest percentage of the area affected by salinization is in Tajikistan[4]. In Turkmenistan, the Karakum Desert occupies 80% of the territory, whereas in Uzbekistan 85% of the land consists of arid and semi-arid desert. As mentioned earlier, the result is crop desiccation.

Acknowledgments The lead authors express their gratitude to Dr. Roman Bohovich (Czech University) for sharing of eco-region descriptions after validation data; to Mr. Siavush Ghiasvand (TU-Dresden) for assisting to download datasets with automatic programming, and Dr. Jarihani (Australia) with modifying Python language; to Dr. Babatunde Osunmadewa (TU-Dresden) for his assistance to simulate Mann–Kendall analysis among the five countries; to regional experts who helped with field experiences and interpreting results in Uzbekistan and Kazakhstan countries.

[3] http://www.columbia.edu/~tmt2120/introduction.htm

[4] http://ecfs.msu.ru/en/news/soil-salinization-central-asia-737

We are grateful to the NASA group for providing GIMMS 3g dataset at the latest version (1981–2014) https://ecocast.arc.nasa.gov/data/pub/gimms/3g.v1/ [accessed on 15 Sep 2016], and also the SENTINEL group as well as https://scihub.copernicus.eu/dhus/#/home [accessed in April 2017]. The research work was supported by DAAD Program, under the number CA 91527284.

References

Chapagain A K, Gautam R, Hoekstra A Y, (2005) "The Water footprint of cotton consumption," UNESCO- IHE, The Institute of Water Education

Glantz M H (1999) Creeping environmental problems in the Aral Sea basin. NATO Advanced Research Workshop on Critical Scientific Issues of the Aral Sea Basin: Cambridge University Press, Cambridge 320 p.

Jackson R, Huete A (1991) Interpreting vegetation indices. Prev Vet Med 11:185–200

Kendall M, Stuart A (1983) The advanced theory of statistics, vol 3. Charles Griffin, London, pp 410–414

Khatamov JB (2014) The Aral Sea problem in Central Asia. Young Scientist, pp 190–191

Köppen, W. Das geographische system der klimate. In Handbuch der Klimatologie; Köppen, C.W., Geiger, R., Eds.; Gebrüder Bornträger: Berlin, Germany, 1936; vol. 3.

Neeti N, Rogan J, Christman Z, Eastman JR, Millones M, Schneider L, Ghimire B (2012) Mapping seasonal trends in vegetation using AVHRR-NDVI time series in the Yucatán Peninsula, Mexico. Remote Sens Lett 3(5):433–442

Osunmadewa BA, Wessollek C, Karrasch P (2015) Linear and segmented linear trend detection for vegetation cover using GIMMS normalized difference vegetation index data in semiarid regions of Nigeria. J Appl Remote Sens 9(1):096029

Price J (1987) Calibration of satellite radiometers and the comparison of vegetation indices. Remote Sens Environ 21:15–27

Ray T (1994) FAQ on vegetation in remote sensing. http://www.yale.edu/ceo/Documentation/rsvegfaq.html. Updated 13 Oct 1994

Rozema J, Flowers T (2008) ECOLOGY: Crops for a Salinized World. Science 322 (5907):1478–1480

Robert M. Hirsch, James R. Slack, Richard A. Smith, (1982) Techniques of trend analysis for monthly water quality data. Water Resources Research 18 (1):107–121

Sustainable Livelihoods in a Fragile Environment—Case Studies from the Tajik Pamirs. From web site: http://www.pamirs.org/Sustainable-Livelihoods.pdf

Toderich K, Shuyskaya E, Taha FK, Matsuo N, Ismail S, Aralova D, Radjabov T (2014) Integrating agroforestry and pastures for soil salinity management in dryland ecosystems in Aral Sea basin. In: Developments in soil salinity assessment and reclamation, pp 579–602

Chapter 4
Deforestation: A Continuous Battle—A Case Study from Central Asia and Other Countries

Faria Khan, Zaineb Sohail, Tahira Khan, Bukhtawar Fatima, Fatima Malik, Syeda Fatma Hassan Bukhari, Sameen Ruqia Imadi, and Alvina Gul

Abstract Deforestation is a plague that is not new to the earth, but it has certainly accelerated in the past few decades. A reduction in the number of forest canopies has increased at an alarming rate. Forestry is important because of its eminent use as biofuels, as source of food and earnings, and in mitigating climatic changes. This global problem needs to be addressed, controlled, and coordinated in an efficient manner. The chapter focuses on the policies practiced in the Central Asian states and other countries in the world being practiced to eradicate and control deforestation rates. Forest management practices focus on balancing rate of deforestation and growth, increasing product yield, and enhancing services obtained from forests. Measures such as legality of forest lands, establishment of public frontier institutions, forest management certification, and provision of incentives can result in implementation of forest management practices on ground level.

Keywords Deforestation · Climate change · Degradation · Implementation · Conservation

---

F. Khan · Z. Sohail · T. Khan · B. Fatima · F. Malik · S. F. H. Bukhari
Atta-ur-Rahman School of Applied Biosciences, National University of Sciences and Technology, Islamabad, Pakistan

S. R. Imadi
Commission on Science and Technology for Sustainable Development in the South (COMSATS), Islamabad, Pakistan

A. Gul (✉)
Atta-ur-Rahman School of Applied Biosciences, National University of Sciences and Technology, Islamabad, Pakistan

Plant Breeding-Genetics Section, School of Integrative Plant Science, Cornell University, Ithaca, NY, USA

D. Egamberdieva, M. Öztürk (eds.), *Vegetation of Central Asia and Environs*,
https://doi.org/10.1007/978-3-319-99728-5_4

4.1 Introduction

Ever since the invention of agriculture, forests have been cut down for thousands of years. In many regions over the world, crops cannot be grown, and neither can livestock graze until forests have been cleared away. Not only does agriculture require the removal of shrubs and weeds, but the trees also need to be cleared away. Regardless of various benefits of forests, fundamentally, there has always been a conflict between forests and agriculture (Ozturk et al. 1991, 2010a, b, c, 2017a; Olagunju 2015). In this era, deforestation continues, even though the societies in which it occurs have shown great changes. Satellite imaging has changed our views on deforestation, as they can clearly show the clearing of forests and allow detailed analysis and comparison. When this information is combined with detailed on-the-ground study, a very vivid picture of deforestation is brought forward. This has also led to the coining of a new term "forest degradation."

Deforestation, strictly speaking, is the complete clearance of a piece of land from trees. This is easy to detect via satellite imaging. But there are also some other important changes that are hard to detect through satellite imaging as most of the tress and canopy remain, even though the forest below has been depleted, hence the term forest degradation. Elimination of trees and loss of carbon occur, but the forest canopy isn't removed. Various factors can lead to forest degradation. It can be caused by a fire that sweeps through the understory, livestock grazing, or trampling over saplings and seedlings. Selective logging can also be a reason, where most tree species lack commercial value, so only a few large trees are removed, leaving the forest canopy unharmed (Asner et al. 2005; Olagunju 2015).

Degradation proves to be more challenging to study as compared to deforestation as it is not only hard to keep the track of rate of degradation, but the possible outcomes are also hard to predict. On one hand, if the forests are left undisturbed, the trees that were cut can be replaced hence neutralizing the loss of carbon and canopy depletion over time (Rice et al. 1997). On the other hand, forest degradation can continue, eventually widening the canopy and gradually turning degradation into eventual deforestation. Moreover, degraded forests have been shown to have a higher likelihood of being completely deforested as compared to intact forests in subsequent years (Foley et al. 2007). Thus, the impact of forest degradation, which is already difficult to predict, can range from mild to devastating.

Deforestation and forest degradation are the primary causes of climate change, responsible for approximately 15% of global warming pollution (www.ipcc.ch). The simple reason to this inference is that trees contain extremely large quantities of carbon, up to almost 50% of the wood's weight. On cutting the tree, this carbon is released into the atmosphere as carbon dioxide. This can be a result of burning, rotting, or even conversion to paper; the only difference is the time it takes for the carbon to be converted to carbon dioxide (DiNicola 1997; Dolgikh et al. 1997; FAO 2010; Ozturk et al. 2010c; Mirzabaev 2013; Li et al. 2015).

4.2 Deforestation: Past and Present

A common misconception regarding deforestation is that it's a recent occurrence, gaining momentum in the tropical regions of the world since around 1950. However, in reality, its history stretches far back into the corridors of time when humans first occupied the earth and learned how to control fire—probably half a million years ago. All that has changed is the rate of acceleration of this ancient practice resulting in environments being irreversibly damaged as compared to the old ages. Possibly as much as nine tenths of all deforestation occurred before 1950 (Williams 2001). The Carboniferous rainforest collapse (Sahney et al. 2010) was an event that occurred 300 million years ago. Tropical rainforests were devastated by climate changes, leading to the extinction of many plant and animal species. The change was abrupt and skewed the climate toward cooler and drier conditions that were not favorable for the growth of rainforests and most of the biodiversity they harbored. Rainforests now became fragmented and greatly shrunk in size (Ebeling and Yasué 2009; Syuharni et al. 2014; www.wrm.org.uy).

Rainforests once covered around 14% of the earth's surface; now they cover a mere 6%, and experts have estimated that if forests continue to be cleared at the same alarming rate, the last remaining rainforest could be consumed in less than 40 years (Taylor 2004). Small-scale deforestation is an ancient practice of various societies, even before proper civilization began (Flannery 1994). The first evidence of deforestation comes from the Mesolithic period (Brown 1997). It was probably carried out to transform closed forests into more open ecosystems, which would prove to be more favorable for game animals (Flannery 1994). With the dawn of agriculture, larger areas underwent deforestation, as fire became the primary tool to clear land for cultivation of crops. Mesolithic foragers utilized fire to create clearings for red deer and wild boars. In Great Britain, according to the pollen records, shade-tolerant species of trees such as oak and ash were replaced by brambles, hazels, grasses, and nettles. The removal of forests led to decreased transpiration; hence upland peat bogs started to form. There was a widespread decrease in elm pollen in Europe between 8400–8300 BC and 7200–7000 BC, which may represent land clearing by fire at the dawn of the Neolithic agriculture. The Neolithic period also saw extensive clearing of forests for farming land. Advancement of tools used to cut trees also led to an increase in the rate of deforestation in this era. Evidence of deforestation has been also collected from the Minoan Crete, e.g., the Bronze Age saw the severe deforestation of the Palace of Knossos (Neolithic Age from 4.000 BC to 2.200 BC or New Stone Age) (www.en.m.wikipedia.org).

The pre-industrial era caused a further increase in the rates of deforestation, along with intermittent pulses of soil erosion that had not been experienced before. Easter Island is a classic example of a civilization that declined due to extensive deforestation and overexploitation of almost all the resources. The disappearance of the island's trees seems to coincide with the decline of its civilization, which was around the seventeenth and eighteenth century (Hogan 2007). This became a trend in most regions, where cities were often built in forested areas, which provided wood required by the inhabitants. However, when deforestation occurred without proper replanting, local wood supplies would become insufficient. This would then lead to the city's abandonment, as evident in case of Easter Island. From 1100 to 1500 AD, an increase in deforestation was observed in Western Europe. The main reasons were the expanding human population and the introduction of a new use of wood—as charcoal. The use of charcoal at an industrial scale has been a main accelerant of the rate of deforestation, as the demand of wood increased manifold (Cantor 1994).

The industrial era introduced steamboats around the nineteenth century, which became the major cause of deforestation of banks of major rivers. One of the environmental results of this was increased and more severe flooding. Wood was cut from the riverbanks on a regular basis to fuel the steam engines (Norris 1997). This process led to many rivers changing their lateral course. Moreover, crews attempted to improve navigation by the use of snag pullers, which often led to clearing large trees 100–200 ft. back from the banks. This caused the flooding and abandonment of several cities. The widespread clearance of woodland to create agricultural land can be seen in many parts of the world, such as the central forest-grasslands transition and other areas. Specific parallels are seen in the twentieth century when deforestation occurred in many developing and developed countries (www.en.m.wikipedia.org).

Deforestation has never been an irrational act, considering the contexts in which it occurs. The clearance of forests by people and organizations is often carried out for good or considerably beneficial reasons, usually economic in nature. However, this may not always lead to long-term benefit as the economy is embedded in a vortex of political and cultural aspects that may lead to clearance of forests even when it is not the most profitable choice. This matter because the agents of most deforestation today are various businesses. Deforestation has changed from being a "state-initiated" process to an "enterprise-driven" one (Rudel et al. 2009). Deforestation is now being considered as an economic alternative by corporations, which consider it as being advantageous in terms of cost. Therefore, reductions in deforestation in one area, by limiting supply and increasing prices, can increase the pressure for forest clearance elsewhere. Sometimes, the same corporations, like multinational timber companies, can move from one forested area to another, thus nullifying the net effect (Lambin and Meyfroidt 2011). This does not equate that deforestation never decreases, it simply moves from one region to the other. Nevertheless, success is not unattainable. The world can achieve development without deforestation. Despite the global spread of the drivers of deforestation, they can be defeated. Strong action by governments and civil societies can pressurize these corporations to choose alternatives for deforestation. With continuous efforts, the loss of forests can be eliminated altogether. After thousands of years of clearing, mankind is surely able to make deforestation history.

4.3 Global Trends in Deforestation

The rates as well as the causes of deforestation have been found to vary from one region to the other. In 2009, two thirds of the world's forests were in Russia, Brazil, Canada, the USA, China, Australia, Congo, Indonesia, Peru, and India (Al Gore 2009). Tropical dry forests experienced the highest rate of deforestation on planet earth (Peña-Claros et al. 2009; Janssen et al. 2018). The world's annual rate of deforestation is estimated to be 13.7 million ha, equal to the area of Greece (www1.american.edu). During 1990–2015, total forest area of the world decreased from 4.28 to 3.99 billion ha, with the percent global forest cover dropped from 31.85 to 30.85%. The area of planted forests increased from 167.5 to 277.9 million ha or 4.06–6.95% of total forest area (Payn et al. 2015). Barely half of this deforested area is compensated for by growing new forests. Major deforested regions of the world will now be discussed.

According to the United Nations Environment Programme (UNEP), Africa's rate of deforestation is twice the global deforestation rate. It has been noted that almost 90% of West Africa's original forests have been wiped out by deforestation (www.independent.co.uk). Deforestation has been accelerated in Central Africa (www.scidev.net/global). Among all the continents, the highest percentage of tropical forests was lost in Africa from the 1980s to the early 2000s according to the report of FAO (2008). Also, according to FAO's figures from 1996, over 87.2% of West Africa's moist forests have been degraded (www.mongabay.com). Within just 15 years (1990–2005), 81% of Nigeria's old growth forests have been lost. Massive deforestation has now started to pose a threat to the food security of a number of African countries (Mweninguwe 2010). A major contributor to the high rates of deforestation in African countries is the use of wood as fuel for heating as well as cooking (Yvonne 1998). Rates of illegal logging in Africa vary from 50% in Equatorial Guinea and Cameroon to 70% in Gabon and up to 80% in Liberia (www.rainforests.mongabay.com). There are three major reasons that have contributed to deforestation in the Democratic Republic of Congo: unregulated logging, mining, and demands of subsistence activities of the poor population. In the eastern side of the country, over three million people live within a short range of Virunga National Park. Most of these people use wood from the park as lumber for construction, firewood, and to produce charcoal. This is a major threat to the park in general but also poses a serious threat to the habitat of almost endangered mountain gorilla (Anon 2007). The main driving force of deforestation in Ethiopia is a growing population and its ever-increasing demand for agriculture, fuelwood, and livestock production (Sucoff 2003). Low education and inaction from the government (Mccann 1999) has also contributed to deforestation in this country, although the new government has taken steps to tackle this problem (Maddox 2006). Organizations like Farm Africa have joined hands with the federal and local governments to initiate a system to look after forest management. Shortage in rain and depletion of most of its natural resources are the cause of Africa's third largest country by population to be hit by famine many times. The already low rainfall has been further declined due to increased deforestation and thus has led to extensive erosion. The last 50 years have witnessed the loss of 98% of Ethiopia's forested region. Approximately 420.000 km^2 (35%) of Ethiopia's land was covered with forests when the twentieth century began. However, it has been indicated by recent reports that forests now cover a meager 11.9% of Ethiopia's land as of 2005 (www.rainforests.mongabay.com). A loss of 14% of the forests, or 21,000 km^2, has been observed between the years 1990 and 2005. Almost 94% of Madagascar's previously productive land has been affected with deforestation resulting from desertification, water resource degradation, as well as soil loss (www.mongabay.com). Humans first arrived in Madagascar 2000 years ago, and since then it has lost over 90% of its original forested regions. Most of this deforestation is a consequence of Madagascar's freedom from the French, as that was when people first started to use the slash-and-burn agriculture practice for subsistence (www.mongabay.com). Deforestation is a major factor contributing to the country's inability to provide its fast-growing population with adequate amount of food, freshwater, and sanitation facilities (www1.american.edu). As stated by FAO, the world's highest rates for deforestation of primary forests have been observed in Nigeria, where almost 90% of the original forests have been destroyed. Main causes of this high rate are the collection of wood for fuel, logging, and clearance of land for agriculture.

Approximately 12 million km^2 of boreal forests are distributed in Russia. These are the largest forested area compared to other places on our earth. These forests contain more than 50% of the total global conifers and contribute more than 10% to the biomass of earth. The deforested land in the country has been summed up to lie around be 20.000 km^2. The areas bordering China have been greatly affected by deforestation, as the main market for timber is in that area. The damage caused here by deforestation is tremendous when compared to other regions. The forests here have a short growing season due to harsh winters. Similarly Southeast Asia is the world's second greatest biodiversity hotspots; however, forest loss is tremendous. Vietnam is only second to Nigeria as far as the highest primary deforestation rates are concerned. More than 90% of the Philippine archipelago's old-growth forests are lost to deforestation. The mangrove forests have depleted at a rate of more than 0.15% a year during 2000–2012 (Donata et al. 2011; Syuharni et al. 2014; Richards and Friess 2016).

4.4 Economic and Environmental Perspective of the Forests

Forest provides solution to disasters (Sumithrabai et al. 2011) and can be considered as "safety nets." They play a great role in climate change adaptation and mitigation. They provide food when crops are damaged by floods and other natural calamities. They are also a source of food for livestock (East et al. 2007). In rural areas, their products are commercialized to earn money under unexpected circumstances. They are a source of self-insurance for rural people living in remote areas (Karahan et al. 2015; Ozturk et al. 2017b). Forests planted in crop areas maintain crop production during dried and wet conditions due to their characteristics like their extensive root system absorbing water and nutrients from soil, increased soil porosity leading to increased water retention during conditions of water stress, and trees having highest transpiration rate keeping soil fully aerated (Verchot et al. 2007). Tree-based systems usually generate crops of better value. Recently, species named *Melia volkensii* is being considered best species so far for agroforestry (www.worldagroforestry.org). Government should support small-scale farmers of developing countries to plant trees and promote agroforestation (Guariguata 2009).

Forests are best reservoirs of carbon and are reported to store about one third of carbon dioxide released by fossil fuels each year. Forests are best reservoirs of carbon only when they are allowed to grow properly to their age of maturity (Pan et al. 2011). Cutting down forests for different purposes can reduce the amount of carbon stored in them. Slow regeneration of forests can overcome loss of carbon but loss cannot be completely overcome as the rate of carbon reduction is reported to be more (Werf et al. 2009). Tropical forests, savannah systems are good reservoirs of carbon as carbon storage persists in them even when they are not fully mature. Forest sinks mitigate climate change caused primarily by release of CO_2 from fossil fuels affecting the global carbon cycle. Peat lands and mangroves have a higher carbon storage capacity than tropical forests. Their conservation should be given prime importance (Donata et al. 2011).

Wild animals are a source of bush meat for people living in rural areas (Nasi et al. 2011). Wild animals are a rich source of protein and are essential to overcome food deficiency in rural areas mostly across tropics. Deforestation has both direct and indirect impact on food security. It directly leads to loss of biodiversity, which is a source of food and indirectly causes soil degradation reducing the production of food (Olagunju 2015).

Animals such as tapir, duikers, deer, pigs, peccaries, primates, larger rodents, birds, and reptiles are mostly extracted from forest ecosystems and provide benefit to indigenous people in terms of food and livelihood (Nasi et al. 2011). Some species are near extinction from forest ecosystem due to excessive hunting, whereas other species are able to maintain harvesting pressure by adopting different strategies like altering their biological factors or occupying habitats of extinct species (Hurtado-Gonzales and Bodmer 2004). Bush meat is not always attributed to elimination of species from ecosystem. Alternative sources of nutrition like beef will require a massive deforestation to convert tropical forest in to grazing lands (Nasi et al. 2008). Data has shown that Brazilian beef production is responsible for 50 million ha of deforestation. Better substitution of bush meat is poultry or pig stocks in rural areas as these species can easily survive on remains of Kitchen ingredients and crops (East et al. 2007). Urban users have many sources of obtaining their protein meat, but their choice of protein food depend on their personal preference and the cost (Wilkie et al. 2005). Forests like mangroves grow near aquatic ecosystem and maintain fish species (Donata et al. 2011).

Forests play a role in agricultural zone as it maintains livestock survival by providing them fodder and shade. It is a place where many agricultural pollinators like bees and bats make their homes. A study has shown that coffee plants give low yields when planted in areas far away from forests due to absence of agricultural pollinators (Ricketts et al. 2004). Forests also regulate the amount of water and quality of water required to maintain agricultural production and control the process of sedimentation in agricultural areas (Guariguata 2009).

Forests provide a role in conservation of biodiversity (Gücel et al. 2008; Sunderland 2011; Altay et al. 2012; Ozyigit et al. 2015; Sezer et al. 2015; Ozturk et al. 2008, 2010a, b, c, 2011, 2017c). Destruction of biodiversity has led to loss of available genetically different forms of crop plant and has led to oversimplification of diet. A data showed that India and China has lost many rice landraces and Mexico has lost more than 80% of genetic diversity of maize crop. Loss of disease-resistant livestock species has also been increased at a very high rate globally. Availability of a few genetic forms of dietary crops makes society prone to disease, which ultimately lead to famine. The decrease in biodiversity of cereal crops in developing countries has led to deficiency of micronutrients. Worldwide, one billion people are reported to suffer from micronutrient deficiency due to genetic uniformity (Sunderland 2011).

Forests make up one third of household income of people living in rural areas near forests (Angelsen 2010; Karahan et al. 2015). Increased global trends toward deforestation to meet global agricultural needs have contributed a little in meeting agricultural needs globally. Recent data shows that increased deforestation only make up 0.3% of agricultural needs particularly in developing countries (Angelsen 2010). Deforestation in hectares of land in Brazilian Amazon by cattle farmers has resulted in raising only a single cow (CFIOR study). But a survey conducted in Indonesia showed that oil palm has taken the place of rubber and rice crops previously used for rural livelihood. Farmers make collaboration with companies and banks and earn their livelihood. Major limitations in this strategy are unawareness of policies, their rights, and rules and regulation. Farmers accustomed to grow rubber crops previously are also ignorant of growth conditions and growth requirements of oil palm crops (Feintrenie et al. 2010). Although improved livelihood of farmers has been observed in many instances, sometimes selling their land to big companies does not result in huge benefits for them. They may lose access to forest-based household income (Baños et al. 2011).

4.5 Forests and Biofuels: Source of Charcoal and Fuel Energy

Deforestation is the obliteration of natural forests and woodlands. Global warming, population explosion, and depleting energy reserves are diverting trends toward biofuels. Biofuels can reduce greenhouse gas (GHG) emission effect. First-generation biofuels are obtained from various crops like sugarcane, oil palm, and soya bean (Sumithrabai et al. 2011; Ozturk et al. 2017a). Forests suffer due to expansion of first-generation liquid biofuels in terms of land usage (Searchinger et al. 2008). As GHG emission effect increases due to frequent land usage, the emission of carbon or carbon debt resulting from conversion of forest land to agricultural lands takes approximately 100–1000 years depending upon particular ecosystem involved in this land conversion incident. However, recent evidences suggest that apart from agricultural expansion, there are also other factors supporting land usage effects. These factors are harvesting of timber, building of infrastructure, and converting land to promote agricultural growth. Both direct land use change (clearing specific land as a part of chain of biofuels production) and indirect land use change (clearing random land for biofuel purpose) are increasing GHG emissions. Application of suitable crop management practices can reduce GHG emission effect due to direct and indirect land use change. Among various practices, analysis showed that no-till practice with winter cover crops is so far the best strategy for reducing effect of land use change (Sumithrabai et al. 2011). Forests are sites of carbon storage, but upon disturbances like wildfire, forest harvesting can lead to the possible danger of carbon emission stored in them (Ozturk et al. 2010c; www.thegef.org).

However, carbon emission effects from forests are a matter of debate. Scientists suggest that existence of trees preserving carbon for more than 1400 years and their continual growth is depicting their suitability for carbon storage. Moreover, increased plant yield and decreased decomposition rates of plants in high CO_2 presence in various experiments are supporting their carbon storage capacity. Second-generation biofuels from forest and agricultural residues have not yet been commercialized. Forest residues can promote development of rural countries in developing countries. Utilization of other biomass sources require high production costs and need of large-scale facility which appears to be a major hindrance of their implications in developing countries (Searchinger et al. 2008). However, scientists consider second-generation biofuels as a sustainable source of energy (Sumithrabai et al. 2011).

Charcoal is a feasible source of energy as it is easily accessible. Forests have been used to produce charcoal to meet energy needs and reduce environmental effects caused by emission of greenhouse gases. The quality of charcoal depends mainly on factors like species of plant and type of kiln used and process of carbonization carried out. Cracked firewood obtained from Brazilian Kiln has shown to increase charcoal production. The international standard of charcoal production is to break charcoal before packing to obtain good quality of charcoal. Wood pellets and wood chips are the most convenient sources of biomass and fuel energy. These forms can easily be transported to long distances promoting trade among countries (Veronica et al. 2012). Forests are used at domestic level as a source of cooking and heating particularly in developing countries. Data suggest that more than 90% of wood energy is being utilized at domestic level mainly for heating and cooking purposes. The use of fuelwood energy in African countries, particularly as a source of carbon neutral energy, requires a massive financial investment (Kaimowitz 2003).

4.6 General Causes and Types of Deforestation

According to the United Nations Framework Convention on Climate Change, the major cause of deforestation globally is the removal of trees for the agricultural expansion purposes. In poor countries, the subsistence agriculture is responsible for 48% of deforestation, commercial agriculture for 32%, commercial logging only for 14%, and charcoal and other fuelwood cleaning for less than 6% (www.phys.org). Another cause of deprivation of forest ecosystems is for economic incentives, which make forest conversion more profitable than forest conservation. Substantial deforestation arises due to lack of property rights security and absence of a system ensuring the effective implementation of conservation policies. These factors are primarily seen in developing countries, and in some cases, corruption and terrorism are parallel factors in deforestation.

There are two main causes of deforestation. The primary and most common reasons for deforestation are known as the direct causes. Logging, overpopulation, urbanization, and dam construction are the direct causes. The other main cause of deforestation is natural since it is brought by the Mother Nature. Rapid population growth has resulted in the conversion of forest areas to non-forest lands for settlement and farming. Together with this are urbanization and residential area expansion. This takes a significant loss of forests for harvesting forest products as more people need more lumber to build their houses and for requiring greater area for building their houses, malls, and business centers. Forests are cleared to make way for the construction of the transport networks. Clearing of large areas of forest is also done for raising cash crops. Roads and railway tracks also make it easier for people and companies to enter the forest to extract resources. Dams and harvesting of hydroelectric power in forests such as the Amazon destroy trees through flooding, resettling of people to the area around the dam, and cutting down trees on the shoreline of the dam (www.eoearth.org). An increase in population also means an increase in produce consumption. Thus, rainforests are destroyed and converted to cattle pastures to supply the burgeoning demand for meat. In the Central America, almost half of rainforests have been slashed and burned for cattle farming in order to comply with the foreign demands. Twenty-five percent of the Amazon's forests have also been destroyed for cattle ranches (www.stopearthdestruction.com).

The lack of government legislation for land reforms has also cleared the forests especially in developing countries of South East Asia. The people in that region are among the poorest and are desperate for a piece of land. Unequal distribution of resources has led these people to find their way by exploiting the forests (Butler 2012). The majority of rural population has wood as the only fuel to cook food and provide heat in chilly winters. Firewood collection contributes much to the depletion of tree cover. The most important causes of deforestation are due to human activities (www.phys.org). Deforestation is a great threat to life worldwide. We can only hope that the natural forces causing deforestation would not do great damage. However, right decisions and appropriate actions must be taken to address the problems caused by the hands of people.

Cleaning of land for grazing purposes; Expanding the production of crop; Timber harvesting and commercial categorization (logging); Slash-and-burn forest cutting for farming subsistence and Natural disasters such as: Volcanic eruptions, Stand wind-throw from hurricanes, Catastrophic fires in forests, Changes in local climate, and Rainfall system. Volcanic eruption is one of the several natural forces capable of causing damage to forests (Butler 2012).

Wind damage to forests is determined by numerous factors. In July 1999, catastrophic windstorm hit the Boundary Waters Canoe Area. The levels of damage varied from 29.5 to 86.8% of basal area fallen and 23.3 to 63.4% of stems fallen. In all sites, the disturbance reduced mean trunk diameter of standing trees (Butler 2012). Drought initially weakens canopy plants by the reduction of local humidity and rainfall; storms can cause extensive damage in the rainforest through tree falls (www.climateandweather.net). Natural forest fires occur in rainforests, in smaller fires, ground vegetation, shrubbery, saplings, and smaller trees are eradicated (Butler 2012).

Estimates on the contribution of deforestation to carbon emissions vary but are commonly held to be around 19% of global emissions—greater than the emissions produced by the whole global transport sector. The bulk of emissions from deforestation arise when land is converted to agricultural production, particularly if forests are first cleared with burning (Butler 2012). Research indicates that deforestation results in warmer and drier conditions. Deforestation may pose the same effects to global warming as burning of fossil fuels (Lawrence and Vandecar 2015).

The potential for forests to become even greater sources of carbon emissions due to deforestation and degradation is massive (www.eoearth.org). Examples can be found from forest areas across the globe, including Russia's boreal forest, the forests of the Congo Basin, and Sumatra's peat swamp forests.

Deforestation is continuing at an alarming rate. Once distributed over half the planet, forests now cover only a quarter of its land surface—and forest loss, particularly in the tropics, is continuing at an alarming rate. Figures released by the UN FAO in 2005 indicate that the rate of natural tropical forest loss is about 13 million ha each year—equivalent to 36 football fields a minute (www.climateandweather.net). Deforestation is an important factor in global climate change. Climate change is because of a buildup of carbon dioxide in out atmosphere, and if we carry on cutting down the main tool, we have to diminish this CO_2 buildup, and we can expect the climate of our planet to change dramatically over the next decades (Butler 2012; Ozturk et al. 2015). It is estimated that more than 1.5 billion tons of carbon dioxide are released to the atmosphere due to deforestation, mainly the cutting and burning of forests, every year. Over 30 million acres of forests and woodlands are lost every year due to deforestation; causing a massive loss of income to poor people living in remote areas who depend on the forest to survive (Butler 2012).

4.7 Deforestation and Its Effects

Deforestation in the long run is causing climatic changes globally, where increase in global warming is seen with prevailing deforestation trends worldwide. Moreover, it is noted that the loss of habitat and forestry is inadvertently causing a loss of important animal, plant, and wildlife in important forests and basins such as Amazon, Congo Basin, and Russia's boreal forests. However, not only forests are affected, but this trend also results in soil erosion resulting in loss of arable lands for agricultural purposes. There is also some evidence about coral reef loss and decline due to siltation and soil erosion around mangroves.

Forests have a critical role to play in combating global warming as researches have clearly pointed out the significance of global warming incidences linked with worldwide deforestation trends. It must be noted that forests are the largest storehouse of carbon after the oceans. However, when forests are destroyed by activities such as logging, forest clearing for roads and settlements, and land conversion for agriculture, they also become a source of release of large quantities of CO_2 and other greenhouse gases into the atmosphere.

It is now widely recognized through extensive research that rising concentrations of greenhouse gases (GHGs) are driving changes in the earth's climate patterns, resulting in catastrophic weather events, such as hurricanes, global warming, heat waves, droughts and floods, and glacier melting, thereby threatening plant and animal life. Forests play an important role in protecting the earth from extreme climatic changes by regulating climate patterns: the trees—trunks, branches, and roots—and shrubs, grass, and plants in general, even soil, absorb and store CO_2, acting as a natural reservoir for GHG. It is recorded that the earth's vegetation and soils currently contain the equivalent of approximately 7500 gigatons of CO_2. It is more carbon than is contained in all the remaining oil stocks on the earth and more than double the total amount of carbon currently in the atmosphere. However, when forests are destroyed or degraded by human and natural activities, they release large quantities of CO_2 and other GHGs into the atmosphere, to become a significant (and for developing countries, a major) source of GHG emissions hence a contributor to climate change (wwf.panda.org).

Forests play an important role in recycling rainwater and groundwater on a global scale at continental and intercontinental levels (Ellison et al. 2012). Deforestation in one part of the world can reduce rainfall in another part. The water evaporated from forests of Eurasian continents is responsible for 80% of water sources in China. Similarly, rainfall in Sahel is caused by moisture evaporating from Congo Basin. The penetration of water in tree roots makes them giant reserves of water, and about 75% of water is stored globally. The water stored in roots of trees is slowly released to provide supplies during dry periods. They have a role in removing pollutants from soil and convert these into less harmful substances. They also reduce soil erosion and sediment formation, so the areas, where forests are removed, are subjected to more soil erosion and sediment formation. The water-storing capacity of forests makes them highly valuable. A study conducted on forests of China has shown that the water-storing capacity of forests in China costs 7.5 trillion Yuan (Rudi et al. 2010). Similarly, a study of Kenyan forests has shown that water-storing capacity of forests saved national economy for more than USD 20 million. The world's most developed areas including New York, Jakarta, and Singapore rely on forested areas for their water supply (Ellison et al. 2012).

4.8 Habitat for Species and Forestry

The importance of biodiversity and forest and its adjoining wetlands as a habitat for a diverse population of species can be seen by the following quote: "Certain species may live in a forest but depend on a nearby grassland or wetland for a food source" (Elizabeth Brown). Deforestation, as previously discussed has the eminent negative effects on the biotic aorta overall. The most dramatic impact of deforestation is a loss of habitat for millions of species. According to statistics, 70% of earth's land animals and plants live in forests, and many cannot survive the effects of deforestation that are destroying their natural habitats. Removing trees deprives the forest of major portions of its canopy, which serves to block the sunlight during the day and hold heat at night. This disruption leads to more extreme temperatures changes that can be harmful to plants and animals residing in the vicinity of these canopies (Ellis 2003–2012).

A single tree means thousands of species. Not only the plants, trees and forested area serve as a habitat but it also acts as a source of food to an enormous amount of plant and animal life. Any basic tree, anywhere in a wood, can provide high branches for birds, vegetation for insects to reside, and animals to eat. It also acts as a shelter for shade plants and burrows for animals such as squirrels, toads, birds, and foxes. It helps to maintain balance between the exchange of beneficial nutrients from the soil and species connected with a single tree in the vicinity. When deforestation occurs, a high percentage of local plants and animals disappear because the environment cannot support their existence. According to the study carried out by the International Union for Conservation of Nature and Natural Resources (IUCN), many species face extinction, and the primary reason is attributed to deforestation (www.wisegeek.com).

Many of the major areas of the world suffering from deforestation are the reservoirs of species habitats, also possessing the most life-filled environments on the planet. The Amazon rainforest in South America is considered to be one of the biggest hotbeds of species diversity on earth, yet this fertile and immense forest is a major target for lumber, agricultural land conversion, and industrial clearing. Experts from the IUCN, World Wildlife Fund, and other environmentalist groups believe that the statistical data show the planet has already lost hundreds of thousands of species largely due to the horrendous effects of deforestation and may lose thousands more (www.wisegeek.com).

In the middle of South America, amid the Amazon bowl to the north and west and the temperate grasslands of Argentina in the south, locale two of the most astonishing ecosystems on the planet. The Atlantic Forest and the Gran Chaco both prop an incredible diversity of fauna and flora that matches that of the far extra Amazon rainforest. Yet beyond South America, most people have never ever heard of them, nor do they understand that they are amid the most intimidated habitats on earth (www.pulsamerica.co.uk).

The Atlantic Forest stretches along the Atlantic coasts of southeastern Brazil and Uruguay inland to eastern Paraguay and the northernmost isthmus of Argentina. It is the world's most beautiful tropical and subtropical forest and known as one of the most biodiverse habitats on the planet and a designated world biosphere reserve. The forest includes a number of species found nowhere else such as the golden lion tamarin, the maned three-toed sloth, the woolly spider monkey, and the critically endangered Itatiaia highland frog, as well as hordes of bird and insect species. However, it is alarming to note that only 7% of the Atlantic forest remains today. This shocking statistic doesn't even begin to cover the full extent of the damage to its native fauna and flora. Instead of being one single, untouched stretch of wooded refuge, the forest consists of hundreds of small vulnerable fragments, which further functions as a competitor for survival of these extremely rare and precious species (www.d.umn.edu; www.pulsamerica.co.uk).

Extinction Debt, the Forests, Habitats, and Climate—Are All Linked Up?

The species extinction, like global warming and even nuclear blast effects, cannot be seen instantaneously. Loss of forest species in the past may not be apparent yet today; hence estimation of specie loss to deforestation might not be predicted now. Ward (1997) used the term "extinction debt" to explain such long-term extinction of species and their declining populations long after habitat changes. "Extinction debts are hence bad liabilities, and after they become a part of our system, the globe will become a terrible place" (www.rainforests.mongabay.com).

For example, the vanishing of critical pollinators will not cause the instant demise of tree species alongside existence cycles measured over course of centuries. Similarly, a study on West African primates resulted in findings that there is a liability of extinction, above 30% of the finished primate fauna as a consequence of significant deforestation (Ward 1997). This suggests that nowadays merely the protection of still conserved forests in these spans could not be sufficient to stop extinctions provoked by past habitat loss. At present, we could be able to forecast the results of the extinction of a little species; we understand little relative to the large bulk of species capitulated to deforestation to create reasonable projections. Hence, the unanticipated elimination of unfamiliar species will have a magnified result above the tiny protrusion of period in adjacent future. The intricacy of the rainforest creates additional hazard to anticipate how, which, and when species will disappear (www.rainforests.mongabay.com).

At present, it is noted and well acknowledged that the tropical species are not merely intimidated undeviatingly by deforestation but additionally by global meteorological conditions change. Even if species endure deforestation stress in protected reserves, they could perish as a consequence of rising marine levels and climactic changes. Countless tropical species are adapted to steady, year-round conditions of temperature, weather, aridity, and humidity. They are however not adapted to meteorological condition change even if it is as tiny as 1.8 °F (1 °C). Adjustments in seasonal length, rain, and intensity and frequency of great events that might transpire to the earth could powerfully affect biodiversity in seasonal tropical forests and cloud forests. Studies suggest that infrequent and fluctuating meteorological conditions can eventually result in populace variations of countless forest animals (www.rainforests.mongabay.com).We have often overlooked the consequence of increased temperatures in the spread of illness amid wild, feral animals. The spread of these illnesses to upland forest as a consequence of temperature fluctuation should plausibly mean the eventual extinction of countless endangered bird species (www.rainforests.mongabay.com).

Besides losing exceptionally rare species, we are losing an incredible pool of genetic diversity. We are perhaps now heading in the direction of a mass extinction of common wildlife and glorious beasts: ferocious tigers, armored rhinos, brilliant macaws, hunters and deers, pandas and penguins, and frogs and toads. As these species vanish from the globe, the globe will be a poorer place to live in (Butler 2012; www.rainforests.mongabay.com). The general pattern associated with deforestation and soil erosion and its impact on the habitat and species loss (with particular importance to research and science) is as follows:

Top soil accumulates slowly → erosion → unusable land → disastrous flooding → loss of scientific possibilities, i.e., identification of species and in them cures for deadly diseases. The loss of trees, which anchor the soil with their roots, causes widespread erosion throughout the tropics and forests. Only a minority of areas are left with good soils, which, after clearing of trees, are quickly washed away by heavy rains. Thus, crop yields decline, and people must spend income to import foreign fertilizers or clear additional forest. The rate of increase of soil loss after deforestation is proportional to the rate of forest clearing; a study in Ivory Coast established that forested gradient regions lost 0.03 tons of soil per year per hectare; cultivated gradients almost annually lost 90 tons per hectare, while bare gradients lost 138 tons per hectare. After heavy tropical rains fall on cleared pastures and forest lands, the runoff passes soil into local canals, creeks, and rivers. The rivers carry the eroded soils downstream, causing major problems. Siltation process also raises riverbeds, increasing the occurrence and severity of floods, and creates shoals, silt bars, and sandbars that make river movement far more problematic. The increased sediment rate of rivers smothers fish breeding and decreases fish eggs hatchery, causing lower hatch rates of prawns and other water life. Coastal fisheries are affected not just by the loss of coral reefs and their communities but also by the damage inflicted on mangrove forests by heavy siltation rates. Deforestation-induced erosion can have detrimental effect on roads and highways that cross through the forest (www.rainforests.mongabay.com).

4.9 Research on Deforestation

Deforestation has a huge impact on soil's physical and chemical properties (Hajabbasi et al. 1997). Quantification of soil quality adjustments pursuing deforestation by measurable soil qualities is vital to hold conservation process. A discovery was commenced in 1994 to assess the results of deforestation in Iran according to which deforestation and clear-cutting of forests in the central Zagros massifs resulted in a lower soil quality therefore cutting the usual productivity of the soil. Elevated population rate and consequent need for more food and fiber need extra earth to farm; consequently every year, hundreds of hectares of forests in northern and central Zagros of Iran are deforested and modified to croplands and agrarian areas. Deforestation results in a lower soil quality in nearby locations as determined by examining its physical and chemical properties. Cutting soil organic matter and aggregate size, rising soil bulk density, and changing the center rank of the soil were found to be inevitable and detrimental aftermath of the deforestation.

Cultivating and cropping in the stand forest is one more exercise of crop creation in the span that additionally cut the soil quality to a little extent, but not as much as completely deforested method. Several underlying indicators of deforestation include forest cover, per capita income, poverty, agricultural production, food, governance, and population growth (Khuc et al. 2018). Therefore, it is noted in the research that cropping amid the forest trees could be the most feasible and suggesting method of crop creation; by that the moderately elevated populated span will be nourished because the usual resources like forest trees and soil will moderately stay conserved.

GHG emission effect has been increased due to deforestation. Deforestation causes climate change leading to loss of stored carbon in trees. A scheme to reduce GHG emission effect due to deforestation is "compensated reduction (CR)" (Gitz and Ciais 2004). This scheme involves giving stored carbon in trees a financial value, creating an incentive to reduce deforestation and protection of forests from cutting (Santilli et al. 2005). Using CR, underdeveloped countries provide government agencies and other private investors their annual deforestation rate and carbon certificates. On gaining carbon credits, these underdeveloped countries release stored carbon and are bound not to go above and below the level of their annual deforestation rate (Gitz and Ciais 2004).

A set of practices known as SFM (sustainable forest management practices) has been suggested to reduce negative impacts caused by timber harvesting practices. The common layout of all these practices is to balance harvesting and growth, induction of proper plans of harvesting process to increase yield of products obtained from forests, and to enhance services obtained from forests. Most of the forest management practices are focused on tropical forests (Boscolo et al. 2009). Tropical forests are places where about 98% of species survive (large biodiversity) and which have large sources of stored carbon and also provide hydrological services. Worldwide data suggest that 96 million ha of tropical forests have management plans and 2.5 million are suggested to be well managed, while ten million are certified by third parties like Forest Stewardship Council (Siry et al. 2005).

These forest management practices are collectively called elements of reduced impact logging (RID). Common goal of policies is to extend concession rights, change criteria of stumpage fee collection, collect it on basis of area, and implement public bidding systems for using public forest lands. Some of the practices are implemented before harvesting like agreement of cutting trees, vine cutting, and planning of road, while some are implemented after harvesting including directional falling. These RID practices are estimated to reduce damage to forests and increasing profit from logging procedures. These practices improve the methods with which forest operations are carried out. Better law enforcement can lead to better adaptation of practices. Many organizations showed consensus to commercial tress after the enforcement of law in 1996 and up to 1998, about 80% of organizations adopted this practice. Bolivia forestry system is improved by implementation of SFM practices through law and by increasing awareness among timber producers regarding the benefits of SFM (Holmes et al. 2002). Apart from Bolivia forestry systems, forest management practices in many other tropical forests systems are lacking. There are many factors affecting implementation of these practices. One of the reasons is people indigenous to forest systems are more concerned with commercial benefits and are least interested in environmental concerns. A study revealed that, in the Brazilian Amazon, total GDP in 2006 was 28 billion of which 23 billion is obtained from timber industry. High cost associated with forest management practices is another major hindrance in adaptation of these practices by timber harvesters (Nasi et al. 2011). Forest management certification is a costly process, and these certificates are not long-lasting and get expired mainly due to hurricanes, low volumes of harvesting, and acquisition of very few commercial benefits by selling certified wood in markets of country (Barry and Taylor 2008). Lack of policies for giving land tenures to local communities and small-scale stakeholders is a problem. Giving land tenures to deserving communities might solve problem to some extent, but some small communities do not have a proper control over informal logging within their region. In Mexico, boundary disputes are also seen which are affecting forests operations in these regions. These policies in the long-term cause commercial logging (Larson et al. 2008). Complexity of these policies makes them hard to understand by local communities. These forest policies are unaware of ground problems faced by local communities and indigenous forest users and are generally being disregarded in Brazil, Peru and Mexico. Implementation of policies at public level is hard to monitor (Enneker 2008). Recently, a case was reported in Brazilian Amazon in which about 100 government personnel are arrested, who were involved in illegal logging practices (Keller et al. 2007). Another factor in hindrance of proper forest management practices is lack of proper trained staff ensuring compliance of forest management practices (Durst et al. 2006). According to an estimation, if national REDD+ targets have to be met in Brazilian forests, then approximately 27,000–33,000 trained forest officers are required (improving sustainable tropical forest management). Similarly, in Mexico, the trend of changing technical staff after every 3 years leads to organizational memory loss and causes repeated high cost of training individuals for forest operations (Bray et al. 2006). In addition, inefficient timber harvesting practices lead to time loss, and sufficient quantity of timber cannot reach markets. The data suggests that timber loss practices have increased to 30%, and only 50% of high-quality swan timber species are able to reach the market (Keller et al. 2007).

Many solutions have been proposed to overcome problems faced in forest management practices. Measures should be taken to improve legality of forest lands. Establishment of public institutions in frontier regions and improvement of forest management certification can also improve the situation. Forest products should be legally verified and funds should be given to region, where carbon mitigation effects have been observed (CIFOR's Poverty and Environment Network study). Another step that can improve situation is giving incentives to timber harvesters. These incentives are in the form of performance bonds. These bonds are refundable and are deposited to accounts of government. On completion of proper forest management plan by timber harvesters, the bond is returned to them. If some disobeying of practice is seen, then due fine is detected from bonds. These refundable bonds are a source of income for harvesters and can contribute to reduce harmful effects of deforestation. Forest certification gives many benefits to users including implementation of forest regulations by government, accessibility of land tenures, and availability of financial incentives, but this forest certification is not affordable by small-scale holders. REDD+ funds can be used to support forest certification for these small holders and local communities (Richards and Costa 1999). Degraded forests can be regenerated naturally after 30–40 years or artificially by active management to restore degraded areas. Some researchers use approaches promoting restoration of degraded forests. This approach involves factors like blocking illegal logging, promoting forest management practices, reducing damage from grazing animals, and protection from wildfire. An alternative approach to enhance process of natural regeneration of forests is by planting seedlings and to allow natural species participating in natural regeneration process (Peña-Claros et al. 2009). Giving forest areas to local communities which are dependent on forests for their livelihood can preserve forests (Ebeling and Yasué 2009). Public forest areas should be given with a right of long-term access to concessionaries. These concessionaires can exercise forest management practices even in areas where human population is scarce (Drigo et al. 2008). One approach to minimize timber wasting during harvesting procedures and to ensure proper quantity of timber reaching markets from harvest areas is to take taxes from timber harvesters. These taxes are calculated on the basis of total volume of standing trees or on the basis of volume felled. Either approach requires corrupt-free officers to monitor the implementation of these procedures (CIFOR's Poverty and Environment Network study). Reduced impact logging (RIL) is more beneficial than conventional logging practices as former leads to rapid degeneration of forests and reduce CO_2 emission by 0.58 GT per year, but RIL can lead to reduce timber yields. Research has shown that marking of trees before and after logging and elimination of marked trees from local competition lead to their rapid regeneration (Vanclay 2008). Rules and regulations regarding forest management should be simplified and should be in compliance with ground problems faced by timber harvesters. Proper training of staff monitoring forest management should be ensured (Putz et al. 2008).

Technology is playing an important role in agricultural lands near forests. Use of appropriate technology with encouraging political and economic environments helps farmers, policymakers, and small-scale holders to enforce forest management practices. New technologies involving elastic products enhance deforestation. These crops are exported, and increased supply of these crops decreases their cost in local markets, suppressing farmer's income. This aspect makes use of new technology an undesirable practice. New technologies increase productivity of agriculture products, which create employment avenues in many other sectors. These new opportunities of employment in other sectors protect forests near agricultural lands from massive deforestation. Technologies which are labor-free facilitate farmers to expand agricultural lands and cause migration of labor to other agricultural areas, whereas technologies which depend on labor limit deforestation as these labor-dependent technologies affect number of labor families and increase wages of labor (Murdiyarso et al. 2000). Farmers mostly prefer labor-free technologies, which is a harmful aspect of technology use. Farmers are financially strengthened by progress in technology and invest money in activities associated with agricultural expansion like deforestation. Some farmers promote forest conservation by saving money due to technological progress. They invest their money in forest conversion. Progress in technology in more labor intensive agricultural lands cause elimination of resources in small amount and make demands of agricultural lands to be simpler. These extra resources can then be applied to expansion of agricultural lands, which ultimately results in increasing demand of agricultural lands (Kaimowitz and Angelson 1998).

One of the important factors in reducing deforestation rate is the reduction in greenhouse gas emission rate. Deforestation is a global problem, and it needs to be addressed, controlled, and coordinated in an efficient manner. International agencies, national conservation resources, and government organizations need to take steps to control and eradicate deforestation.

1. To enhance dissemination, transparency, and effective use of deforestation data by government agencies and civil society
2. To develop and implement functional, credible market mechanisms that provide financial incentives for conservation and sustainable use of tropical forests
3. To contribute to the development of public policies that will scale up the incentives for conservation and sustainable use in deforestation (www.monitoramento.sema.mt.gov.br)

4.10 Diversity and Ecological Significance of Forests

4.10.1 Central Asia

There are four major zones: irrigated, rainfed, rangeland, and mountainous areas. Four sources have triggered land degradation processes in the region: abandonment of massive areas formerly under rainfed crop production in Kazakhstan; continued desiccation of the Aral Sea; conversion of a sizable share of barren lands into other land uses, mainly shrublands and grasslands; and increases in the forested area across the region but especially in Kazakhstan. Kazakhstan, Kyrgyzstan, Tajikistan, Turkmenistan, and Uzbekistan are strongly affected by land degradation (Fig. 4.1) (Ozturk et al. 1996a, b; Bekturova and Romanova 2007; Pender et al. 2009; Klein et al. 2012; Nkonya et al. 2016).

It is a major development challenge in the region. Annual cost of land degradation in the region is about six billion USD, mainly due to deforestation (0.3 billion dollars) and other factors. There is a lack of published studies identifying the extent

Fig. 4.1 A general map showing the Central Asian Republics (Image Google Maps)

of land degradation in the region using observed data at national or regional scales (Ji 2008). Existing studies are mainly on the extent of land degradation in Central Asia, but based on qualitative estimates (Gupta et al. 2009). But there are a growing number of localized case studies based on detailed soil surveys or remote sensing data (O'Hara 1997; Buhlmann 2006; Dubovyk et al. 2013; Akramkhanov et al. 2011; Akramhanov and Vlek 2012; de Beurs et al. 2015; Nkonya et al. 2016).

The drivers of land degradation in the region are numerous (Pender et al. 2009; Nkonya et al. 2016). The main areas affected are concentrated in the north of Kazakhstan, continuing toward Eastern Kazakhstan, also covering Kyrgyzstan, northwest of Tajikistan, and south of Uzbekistan and Turkmenistan. In Uzbekistan annual costs associated with land degradation are estimated to be about one billion USD (Sutton et al. 2007; Nkonya et al. 2016). Degraded areas are localized in the lowlands of Xorazm and Karakalpakstan and in Bukhara, Navoi, Kashkadarya, and Fergana provinces. Area of rainfed rangelands has considerably decreased due to overgrazing and deforestation (Pender et al. 2009; Nkonya et al. 2016).

Mountainous ecosystems in the region occupy 10% of the total territory and are ecologically diverse. Land degradation in such areas has their own characteristics (Nkonya et al. 2016). Bai et al. (2008) has analyzed land degradation as a negative linear trend and found that land degradation ranges from 0.3% of the territory in Turkmenistan to 17.9% of the territory in Kazakhstan. Le et al. (2014) reports relatively higher share of land in the region. The extent of degradation, according to Le et al. (2014), ranges between 8% in Turkmenistan and Uzbekistan and 60% of the total area of Kazakhstan. The land degradation hotspots are concentrated in the north of Kazakhstan and stretch over Eastern Kazakhstan to the southern part of Central Asia, covering Kyrgyzstan, the northwest of Tajikistan, and the southern parts of Uzbekistan and Turkmenistan. The major drivers of land degradation in mountainous areas are population pressure leading to cultivation of sloping, easily erodible lands without sustainable soil conservation protection technology, and poor extension services (Gupta et al. 2009; Pender et al. 2009; Nkonya et al. 2016). Other change includes doubling of forested areas, from 2.3 million to 4.5 million ha, mainly through shifts from woodlands and grasslands to forests in Kazakhstan mainly Almaty and Eastern Kazakhstan provinces. Deforestation is leading to about 0.32 billion USD in annual losses; other sources of land improvement include afforestation on additional 2.2 million ha (Nkonya et al. 2016).

4.10.2 Kazakhstan

The country is poorly forested, covering only 4.5% of the country. The forests are unevenly distributed, with nearly 80% of the wood stock in the northern and northeastern areas. Forests in Kazakhstan are a carbon sink, and three categories of forest management activities have been identified as potentially decreasing CO_2 emissions: management for conservation, management for storage, and management for substitutions. Possible mitigation measures include promoting improved logging practices to reduce the damage to residual trees and the soil; encouraging agroforestry activities to contribute to sustainable development; promoting forest expansion; improving legal and policy framework, using technical exchange programs on environmental impact assessment and mitigation, and improving environmental planning; encouraging use of long-lived forest products; providing financial incentives for new afforestation activities; and controlling air pollution effects on forests (Kushlin et al. 2003; Baitulin et al. 2010; FAO 2010b; www.fao.org).

Haloxylon spp. occupy over 60% in the general forest area in the country. Current standing wood stock is around 375.8 million m^3, including 140 million m^3 of mature and overmature wood; the cutting volume has been reduced from 2.5 to 1.2 million m^3 during the last few years (Mátyás 2010). The contribution to the economy has been only around 2.5% of the gross national product. The standing forests represent a positive carbon stock with conifers like pine, fir, cedar, juniper, and similar species; softwood deciduous trees like birch, aspen, alder, and poplar; and hardwood deciduous trees like oak, ash, maple, and elm. In addition we come across *Haloxylon* (saxaul) forests including both black and white saxaul. Other woody taxa are apricot, plum, apple, and few others. The bushes include dwarf birch, hawthorn, and others. *Haloxylon* spp. occupy more than 65% of the forest area represented by 14.9 million m^3. The conifers and softwood deciduous trees form over 90% of the total standing wood stock (Kushlin et al. 2003; FAO 2010b).

The coefficient for the average annual gain of woody biomass in the country equals to 1.6 m^3 per hectare. The quantity of carbon absorbed by the forests has been estimated to be 2335 Gg CO_2. GHG emissions from forest fires and woodcutting have been recorded to be 493.1 Gg CO_2. These values have been estimated to be 1961.3 Gg CO_2 absorption, 39.8 Gg CO_2 emission, and a net gain of 1961.3 Gg CO_2. Ban on cutting down of conifers and *Haloxylon* spp. plantings has proved positive for preservation and restoration of woods (Baitulin et al. 2010; Mátyás 2010).

Kazakhstan stands low in the global list in terms of percentage of forest land, although the area per person is 0.77 ha. With a possible move to the south in mountain regions, the resistance of forest ecosystems implies ecoclimatic zone boundary disturbances (Baitulin et al. 2010; Mátyás 2010). The temperature and humidity changes are causing unsuitable conditions for conifers. Lower limit of spruce in the mountain regions is moving up by 100–120 m; giving way to deciduous softwood species, the fir plantings may disappear from some areas and remain only in a small area of East Kazakhstan (Yesserkepova 2008; Mátyás 2010).

Forest ecosystems are highly vulnerable to climate change. It is explained as the conifers at the southernmost border of area are very sensitive to temperature and humidity regimes. Junipers grow on the northern border of area but are capable of reacting to changed climatic conditions (Mátyás 2010). Percentage of woodlands in Kazakhstan together with *Haloxylon* spp. and bushes is around 5%. Forests play an important role in soil protection, in climate and water regulation, water protection and in recreation. New rules plan to create forests on 145,180 ha in the country (Yesserkepova 2008; Mátyás 2010). Electronic databank for forest ecosystems can permit to monitor woods, define volumes for sustainable forest use and their reproduction as well as control of activities by establishing forest management projects. Carbon sequestration by the forests may be significant because huge territories with trees are outside the system of forest accounting. There is a need to improve the inventory of the forests in Kazakhstan (Mátyás 2010).

4.10.3 Kyrgyzstan

Not much work has been done on the climate change in forest functions like productivity, survival ability, or loss of plantations (Mátyás 2010). The forests in the country are under the status of protected natural areas. They form unified state forests including both areas actually covered by forest and areas uncovered by forest but designated for forestry needs. Recent official assessment reveals that the total area of the state forestry lands is around 3.533.100 ha, including 932.100 ha covered by real forests. Vertical zonation and varying climatic zones have resulted in considerable diversity in forest-forming species and lead to rather low forest coverage (Blaser et al. 1995; Orlov et al. 2003; Vengelovsky 2006).

The country has limited forest cover, which is unevenly distributed and dominated by walnut (41,000 ha), spruce (124,100 ha), juniper (archa tree) (303,500 ha), and other (48,300 ha) forests. The pistachio and almond plantations grow at altitudes between 1300 and 1800 m. However, some of the mountain ridges are covered by a solid mass of walnut forest. *Juglans regia* is the most valuable species among the great variety of tree species and is the dominant species in the walnut forests (Blaser et al. 1995; Orlov et al. 2003; Vengelovsky 2006; Mátyás 2010).

In the northern parts of the country over the slopes of the Kyrgyz ridge, the dominant forest species is Tian Shan spruce (*Picea schrenkiana*) (Imanderdieva et al. 2018a, b), with an area of 128,200 ha. This comes to around 14% of the entire forest cover in the country. The spruce forests together with archa forests cover the steep slopes of the mountain ranges regulating mountain river runoff and direct surface runoff into subsurface runoff, attenuate erosion processes, and stabilize the soil against mud-and-stone landslides which have been the cause of severe disasters and devastation (Blaser et al. 1995; Orlov et al. 2003; Vengelovsky 2006).

The juniper forests (locally known as archa) and the associated dwarf forms are very important. These forests are widespread on the dry and harsh conditions of the Altai crest, covering 303,500 ha, which is nearly 33% of the country's forests. The largest area is concentrated in the regions of Oshsky and Batkensky, on the slopes of Turkistan and Altai mountain ranges. It is also extensively found in some other regions (Orlov et al. 2003; Vengelovsky 2006).

Other forest species in the mountain regions are located along the bottomland and shores of large rivers as well as along many small rivers, occupying about 48,300 ha which is nearly 6% of the country's forests. They typically have water conservation functions. There are narrow, broken forest strips in some mountainous areas, frequently forming riparian woods (tugais) composed of *Populus nigra*, *P. diversifolia*, *Salix alba*, *S. cinerea*, *Elaeagnus angustifolia*, *Tamarix laxa*, and *Hippophae rhamnoides*. *Ulmus* spp. and poplar forests grow along the shores of Talas River (Orlov et al. 2003; Mátyás 2010).

Among a wide variety of forest ecosystems, archa and spruce are in the highlands, walnut species on the mid-mountains, and tugai species in the low terrains. The dominating tree and shrub vegetation has low biomass growth coefficient, and the country's carbon absorption potential is relatively low. If forest cover expands by 8%, the additional annual accumulation of carbon in the forest reserves will be around 784 Gg CO_2 (Mátyás 2010). The forests mainly include mature and declining stands. There is a gradual transition from one age group to an older group. The area of young plantations is stable; the mature and declining stands cover over 40%. During the next two to three decades of forest aging, mature and declining stands are expected to occupy over 50%, and tree losses in spruce forests will increase due to tree die-off (Podrezov and Titova 2002; Orlov et al. 2003; Mátyás 2010).

The spruce forests as moisture-loving plants are distributed mainly between 1600 and 2900 m of altitudes. The changes in temperature and precipitation patterns are dependent largely on altitudinal changes. For their vertical range, there is a significant vegetation distribution and lack of homogeneity. If temperature increases the lower boundary of these forests, they may ascend by 150–200 m (Orlov et al. 2003).

The Central Asian plains are rich in widespread archa forests, occupying a narrow, thin band, distributed between 1200 and 3200 m. The area of archa forests has decreased by approximately 20% during the last two decades, and the rate of degradation has reached nearly 1% a year (Mátyás 2010). Only the area of sparsely closed stands has increased by more than 30% nearly by 12,000 ha. Soil erosion is increasing, and avalanches and mudslides have resulted in high damage in the economy. Decrease in soil water is due to the reduction in mountain forests (Orlov et al. 2003; Mátyás 2010).

The forest degradation reasons are industrial wood harvesting and wildfires, heavy livestock grazing, population and livestock growth in the mountainous areas, and increased recreation. Increase in summer temperatures on the archa forest ecosystem will mean gradual move to higher places. It could be 150–200 m higher than present in 2100 (Orlov et al. 2003; Mátyás 2010).

Forest restoration plan covering walnut forests and their development has been developed in the light of global warming, which includes effects on the growing conditions of walnut forests and cultivating walnut trees as well as other useful species, such as apple, pear, quince, jujube, plum, and almond (Mátyás 2010). There is considerable increase in the evaporation in comparison with precipitation as per the analysis of bioclimatic potential results. This is worsening the natural humidity regime in the area. CO_2 increases in the atmosphere; the wetter zone borders are going to shift upwards, by 100–200 m and may be 400 m in certain areas. An increased bioclimatic productivity is observed in the well-watered areas of walnut forests at altitudes of 1400–2300 m in the southwestern region. In the dry steppes and in the semidesert areas covered by pistachio and almond plantations at 800–1400 m, productivity will change or may even go down due to anthropogenic impacts (Blaser et al. 1995; Podrezov and Titova 2002; Orlov et al. 2003; Vengelovsky 2006; Mátyás 2010).There is need for an improvement of forest rules, national adaptation measures to climate change, institutional reform in the forests, sustainable forest management criteria and indicators and improvement of public awareness (Mátyás 2010).

Kyrgyzstan's forest sector is expected to face several threats. Climate change will contribute to the altitudinal advance of the deserts toward the upper border by 400 m. Steppes may advance by 250 m, forests by 150 m, and the subalpine cover by 100 m, leading to changes in many plant types and the main forest-forming species (Mátyás 2010). The lower limit of the European walnut will rise by 100–150 m due to increased soil moisture requirements. This will be conditioned by the increase in the active temperature (Orlov et al. 2003; Mátyás 2010).

The area of drought-resistant bushes like rose hip, hawthorn, and honeysuckle will change in the ecological niche of walnut forest. The pistachio, almond, and jujube can adapt to the increased temperatures and move up by 100–200 m (Mátyás 2010). Degradation of lands will go up in the vicinity of populated areas, with increased possibility of landslides, caused by excessive livestock grazing on the pastures near the villages. Worsening of the survival ability of forest plantations is insufficient financing for use in reproduction, protection, prevention of illegal forest harvesting, and fighting pests and diseases (Orozumbekov et al. 2009; www.forclimadapt.eu). The "Tian Shan Ecosystems Development" project has started as a continuation of the Central Asian transborder study on the preservation of the Western Tian Shan's biodiversity. This will contribute to the improvement of ecosystems management and sustainable forestry in Kyrgyzstan. It will be helpful in the application of ecological objectives like preservation of biodiversity and mitigation of the climate change through greenhouse gases (accumulation and entrapment) in the forests (Orlov et al. 2003; Mátyás 2010). The Institute of Forest and Walnut Cultivation of Kyrgyzstan is studying bioecological conditions of the forests in order to increase productivity of forest resources by improved management, supporting forest preservation measures and forest protection measures, increasing public awareness, development of sustainable forest management criteria and indicators, and increasing forest resources productivity. Walnut forest upgrading can deliver greater crop yields and provide local populations with more income while reducing damage to forest plantations (Blaser et al. 1995; Vengelovsky 2006; Orozumbekov et al. 2009; Mátyás 2010).

4.10.4 Turkmenistan

Turkmenistan is located in the west of Central Asia, bordered by Kazakhstan, Uzbekistan, Iran, and Afghanistan with an area of 491,200 km^2. The climate is extremely continental, except for the Caspian Sea coast and in the mountains. Maximum temperatures reach 48–50° in the central and southeast Karakum Desert. Average rainfall is 398 mm, with lowest values in the Kara-Bogaz-Gol Bay, averaging 95 mm. One of the largest deserts in the world is Karakum. It occupies the whole central part of the country; four fifths of the country is flat, but mountains and hills are mostly in the south and southwest (Mátyás 2010). Nearly 282.420 km^2 are desert. Water resources are distributed unequally, with more than 90% coming from Amu Darya River. The remaining 5% come from other rivers, streams, and springs. The southern Murghab, Tedzhen, and Sumbar rivers, and the smaller rivers of the foothills of the Kopet Dag, are fully exploited for irrigation (Popov 1994; Gintzburger et al. 2005; Pomfret 2008; www.tm.undp.org).

More than four million hectares are covered by forests, all classified as primary. Forests in the arid climate are of great ecological importance, and their importance is increasing. The timber used is between 30 and 50.000 m^3 per year. There is a reduction in unauthorized felling due to natural gas supply. The paper factory produces around 50,000 tons of paper annually, using cotton plant material and corn chaff as raw materials. Cattle grazing is restricted in areas where it could damage forest growth and development. Unique conditions are conducive to the growth of a range of plants, particularly trees and shrubs, from the Caucasus and Mediterranean regions, as well as from the western Tien Shan. There are extreme continental climatic conditions leading to a range in diversity between the forests of the mountains and foothills, and those of the sandy desert areas. Turkmenistan forests although have suffered negatively and efforts to conserve and protect them began only within the last few decades. Forest conservation in natural areas has been made possible by limiting human use of forest resources and national traditions (Atamuradov and Karryeva 2005; Mátyás 2010).

Turkmenistan has gained experience in forest cultivation using introduced species. The hardy *Pinus eldarica* grows under adverse climatic conditions and continues to grow. Much efforts are made on the planting of coniferous seedlings, and recreational forest area is increasing every year. Forest parks around the capital and in the foothills of the Kopet Dag are noteworthy. All sectors have been mobilized to establish forest parklands in the foothills of the Kopet Dag. More than 30 million saplings and seedlings of nearly 100 species of evergreen and deciduous trees and shrubs have been used (Mátyás 2010). Forest fire prevention organization and tackling diseases and pests have been developed due to the alleviation of climatic conditions and the establishment of forests. There has already been an increase in the number of wild animals and birds in forest parkland areas. Regular irrigation, drinking, and foraging sources for wild animals have increased. These are helping as additional factor in preserving forest biodiversity. In Karakum Desert, Altyn Asyr Lake has been constructed, which collects all drainage water from across the country. Saxaul trees, saltwort, *Ephedra*, desert *Acacia*, and other sand-tolerant species are expected to create conditions for 12 months pasture source for cattle. It will improve the conditions for winter grazing. A national forestry program is being developed to organize forest inventories and organize international and regional exchanges of experience and new practices and technologies in forest management and other organizational issues (Popov 1994; Atamuradov and Karryeva 2005; Mátyás 2010).

4.10.5 Uzbekistan

The total forest area includes 8.661.200 ha, including forest-covered areas, open artificial plantings, sparse forests, fire sites, perished stands, cut sites, groves, and abandoned sites. Forests are divided into mountain, floodplain valley, and desert forests. Approximately 80% of the distribution is of aridity-tolerant species, like *Haloxylon persicum* and *H. aphyllum*, *Salsola richleri*, *S. paletzkiana*, and *Calligonum* spp., together with some desert-type forest vegetation. The mountainous coniferous species occupy 11% of all the forest area, but nuciferous and wild fruit trees occupy around 3% only. Juniper forests are composed of *Juniperus seravshanica*, *J. semiglobosa*, and *J. turkestanica* (Botman 1999, 2008, 2009; Mátyás 2010).

Tugais occupy periodically flooded floodplains and river deltas, dominated by *Elaeagnus angustifolia*, *Populus euphratica*, and *Tamarix* spp., forming 5% of the forest area. Almost 80% of forests are found in Karakalpakstan, Navoi, and Bukhara regions, less than 1% in Sirdarya, Samarkand, and Fergana valley. Forest productivity is very low. The stocking density per hectare of mature and overmature forests on average basis is at 6 m^3, coniferous forests at 29 m^3, hardwoods 6 m^3, and saxaul forests 3 m^3 (Botman 2009; Djanibekov et al. 2012a; Mátyás 2010).

The forests of the country have huge protective importance, as sustainable ecosystems, highly adapted to specific soil-climatic conditions, help in maintaining biological diversity of the fauna and flora. In the mountains, they prevent erosion and other adverse factors. The forest stands mitigate conditions of habitat in deserts, fixing moving sands, protecting economic objects from sand movement, serving as local sources of fuelwood, and increasing productivity of desert pastures. Tugai forests help and play a role in the bank and water protection. Forest stands serve as protection from adverse actions of water and wind erosion in the irrigated plain lands and hot dry winds. They are the source of non-wood forest products like walnut, pistachio, almond, apple, pear, cherry plum, apricot, hawthorn, barberry, as well as mushrooms and berries, herbal medicines, and tanning and dyeing agents. The impacts of climate change on basic forest-forming species like juniper, floodplain valley, and desert saxaul forests have been assessed. The desert forests are around 78% of all forest-covered area, and saxauls account for over 65% of all forests. If habitat conditions become arid around 2080, climate will be warmer and dryer with smaller fluctuation of climatic indices. Climate changes will not affect much the conditions of saxaul stands, but still a decrease in productivity and deterioration of saxaul stands can be expected (Botman 1999, 2008, 2009; Mátyás 2010).

Floodplain and valley forests include forest plantations on irrigated and conditionally irrigated lands. In Uzbekistan, these stands grow only on irrigated or conditionally irrigated lands. Main factor limiting forest growth is access to moisture, which is assured or not assured by human intervention. Tugais occupy periodically inundated floodplains; their remains can be observed along modern riverbeds on surface terraces, if they are flooded from time to time. The remains of degraded tugai vegetation are found in the depth of desert in old river zones where moistening of soil currently depends only on precipitation. However, this vegetation suffers mainly from uncontrolled felling, uprooting with the purpose of agricultural reclamation of such lands, lack of high waters in natural terms due to overregulation of river runoff (regulated flow), lowered groundwater levels, etc. (Mátyás 2010; Djanibekov et al. 2012b).

Mountain forests like *Juniperus* formations need to be studied well during climate change. They are expected to be shifted toward dryer and colder climates. More constricted will become the area of preferred habitat of juniper, especially by 2080. The lower boundary level of Zeravshan junipers (*Juniperus seravshanica*) is expected to rise more, 1000–1050 m, in 2080. Upper boundary of Zeravshan junipers will rise by smaller values, 800–900 m. Semiglobular junipers (*J. semiglobosa*) will go up by 650–750 m and the upper boundary of Turkestan juniper (*J. turkestanica*) by 500–650 m (Botman 1999, 2008; Mátyás 2010). Outcome of such shifts in juniper-growing zones will mean reduction in the breadth of the juniper belt. In general, the juniper belt extent will decrease by 350 m (2080). This area will contract because the higher the hypsometrical level of the soil surface, the smaller its area. Increasing altitude above sea level results in substantial worsening of soil conditions as there is more coarse material, stone screes, shallow soils, etc. This will substantially reduce productivity of woodlands in new habitats and complicate work for their establishment (Botman 1999, 2008; Mátyás 2010). The semiglobular juniper and Turkestan juniper in Gissar-Darvaz forest-planting district may be lost as forecast according to scenario for 2030 and 2050 (Botman and Agafonova 1999; Botman 1999, 2008; Mátyás 2010). At least younger age classes should be able to transform to tree-type form in future, if situation will not suffer a permanent change. Most vulnerable to climate change are mountain forest stands (Botman and Agafonova 1999; Botman 2008; Mátyás 2010).

Strategies for adaptation of forestry to climate change include national forestry characteristics, forest inventory, database, monitoring, climate change aspects, improvement of forest management, development of forest farms, enhancement of effectiveness of forestry planning, grazing regulation, mitigation of anthropogenic pressures, development of applied scientific research/programs using knowledge gained for interrelation of science with production, legislative initiatives, and institutional changes (Botman and Agafonova 1999; Botman 2008; Mátyás 2010).

Threats for forestry development expected in the near future are lack of objective and detailed information on the conditions of forests, planning of scientifically based forestry development, decrease in the survival rate in regeneration, conditions of existing forests because of lack of material and financial resources, application of truncated technologies, uncontrolled grazing, illegal tree felling, fires, pests, diseases, as well as new and increasing negative climate change impacts, change in forest planting conditions in mountainous areas under the influence of changing climate, spatial shift in current forest areas, a shift upwards of growing boundaries for trees and shrubs, biodiversity loss, reduction in the area of forest-covered lands, reductions in species composition-forest density-productivity, age structure changes in forests because of traditional reasons as well as global warming. Tugai forests are becoming endangered especially *Populus euphratica*. Global warming will lead to increased water consumption in all needs. The reduction of water yield and river sink regulation is also threatening as these unique forests depend on this. The field-protective forest stands may disappear except for coppice linear plantings of mulberry and willow (Botman 2009; Mátyás 2010).

During the last decade, the projects were started on the creation of forest plantations in two forest farms including reconstruction of young forest plantations in order to increase their density in the valley zone of Djambai forestry farm of Samarkand region as well as Zamin forestry farm of Djizzak region in collaboration with locals for better CO_2 sequestration and climate change mitigation, improving methods of reforestation, managing current forests, creation of foundation for future initiatives, and improving existing models of tree growth calculations and assessment models of CO_2 sequestration. It has an indirect connection to climate change but with good sequestration potential due to large afforestation volumes. This project is expected to add to the ecosystem stability on degraded lands in Karakalpakstan and the Kyzylkum Desert. Detailed information on project preparation documents, practical implementation of projects, distribution of benefits from implementation, and rights and obligations of participating parties (Mátyás 2010). Reorganization of forest regulation and statistics is needed. Currently due to unsatisfactory condition of the forest regulation services, available statistics are not trustworthy. There is need for reliable and real statistical data (Mátyás 2010). This, in particular, also refers to statistical data required for undertaking GHG inventory in forestry.

Training in modeling of forest stand productivity in changing climatic conditions can be followed. The forestry sector of Uzbekistan has little experience in modeling, but overgrazing can be observed, as a basic factor due to its impact on the degradation of forest ecosystems. Creation of sustainable models of forest utilization in the mountains and deserts as pastures should be given a priority. Impact of global warming knowledge in forestry among specialists and the general population is a must (Botman and Agafonova 1999; Mátyás 2010).

4.10.6 Tajikistan

Tajikistan covers a total area of 143.000 km^2. The territory stretches 700 km from east to west and 350 km from north to south. It is a mountainous, landlocked country that borders Afghanistan, Uzbekistan, Kyrgyzstan, and the People's Republic of China. Over 70% of the country is high mountains, more than half of which rise 3.000 m above sea level. The east is covered by the Pamir Mountain range; across the north stretches the Alay Range. Only in parts of Khatlon province and in the Ferghana Valley, near the border to Uzbekistan, are small portions of intensively farmed lowland areas (www.pamir.at; www.naturalresources-centralasia.org). Overgrazing and deforestation are disrupting high-altitude ecosystems. This will have disastrous results with erosion signs visible everywhere. Ecologically sound and sustainable pasture and forestry management is a must. Suitable instruments to combine the two forms of land utilization are available but not part of regional planning (Pomfret 2008; Kirchhoff and Fabian 2010; www.naturalresources-centralasia.org).

Pamir Mountain chain with Pik Ismoil Somoni, rising 7.495 m as the highest peak, lies in the east of Tajikistan. This chain is formed of Tian Shan, Karakoram, Kunlun, and Hindu Kush ranges. Most of the range in Tajikistan, in the Gorno-Badakhshan area. Average elevation is between 3.600 and 4.400 m. There are many glaciers such as "the Fedchenko glacier" 80 km long. It is one of the longest glaciers outside the polar region with a depth of 800 m and an enormous water reservoir. The glaciers contribute to the network of fast-flowing streams, most of which empty into Tajikistan's major rivers the Syr Darya and the Amu Darya. These are the main sources of the Aral Sea. Climate is continental, subtropical, semiarid, and arid. The country is highly vulnerable to global warming as confirmed in recent reports (Pomfret 2008; Kirchhoff and Fabian 2010; www.naturalresources-centralasia.org).

Agricultural production capacity of the country is not good, forestry can help increase the land value, as its potential has been neglected. Precipitation and forest vegetation grow to the timberline (about 3.700 m) almost everywhere in the country. The forest vegetation can be expanded, which officially is around 410.000 ha. The livestock numbers and management considerably influence forest composition, which is an essential factor in forest degradation in Tajikistan. Elaboration of forest management plans and silvicultural systems should mitigate the livestock factor through better pasture management and/or improvement of pastures. Sustainable access to grazing can be an incentive to protect forest areas; local communities can be drawn into forest management (Kirchhoff and Fabian 2010; www.naturalresources-centralasia.org).

Forestry needs to be considered as an important part of the economy due to its environmental and ecological impacts. The individuals rely on timber imports, which effect negatively on the country's economy. Most important energy source is firewood. The existing forest resources cannot meet the demand. A comprehensive afforestation program is needed which will meet the needs for production of wood for fuel and construction. Nontimber forest products are important for contributing to the livelihood of rurals. The country's forest resources are relatively small as more than two thirds of the population lives in rural areas. The forest resources are major economic, social, and environmental factors in the livelihood and well-being of rurals. More ecological-economic aspects of forests can alleviate and even stop different types of erosion. Removal of forest cover results in various types of erosion, for example, tugai forests are characterized by thick undergrowth that grow along rivers; if these are destroyed, dangerous mudflows occur more frequently, threatening human lives. Proper protection, preservation, and management of forest resources can reduce risk management. Commercial logging has been prohibited in Tajikistan long back as such, there is need for lowering the country's dependence on timber imports, increasing forest resources, these should be envisaged and will positively impact the trade balance. A nationwide afforestation program is necessary (Kirchhoff and Fabian 2010; www.naturalresources-centralasia.org).

There is no timber industry at present in Tajikistan, and none of the woodworking industries are operating. Tajikistan used to import 400.000 cu^3 of timber from Russia four decades back. Out of this 350.000 cu^3 was processed to create products with added value, while some 50.000 cu^3 was used for fuelwood. Timber for construction is a precious product with high demand, because building typical Pamiri style houses require a lot of wood and 4 × 4 cm wood strips of relatively poor processing quality (conifers) cost much, while 1 cu^3 of construction wood (*Populus* sp.) is very expensive (Kirchhoff and Fabian 2010; www.naturalresources-centralasia.org).

Cutting any commodity from forest areas do not only imply tree harvesting; they include game animals, furbearers, seeds and nuts, berries, mushrooms, oils, foliage, medicinal plants, peat, fuelwood, honey, and the production and sale of seedlings. Nontimber products play a major role in the rural economy. These are essential for daily lives of Tajiks. They contribute to their subsistence and are also traded for cash. Survival of state-owned forests depends on these nontimber forest products. 1. The lezkhoz in the country depends on the harvesting and sale of these products to survive financially. The production of fruit trees in the state-owned nurseries, which are sold to private costumers or planted on lezkhoz land, as nursery revenue is crucial. The demand for fruit trees is higher than that for forest trees, seedling production is of exotic tree species in state-owned nurseries which uses obsolete technology. Seedlings are not raised in containers and are usually sold fairly large and planted bare root. The fruit tree plantations are recorded as "forests" in Tajikistan's official statistics (Kirchhoff and Fabian 2010; www.naturalresources-centralasia.org).

Nearly 70% of the country's population depends on firewood, which is used for cooking and heating. It is in high demand in the country because fuelwood is the most important energy source for rural households and demand is higher than supply. The fuelwood is collected from the woody vegetation and used immediately without drying. However, properly dried fuelwood could significantly increase the efficiency of this energy source. Tajikistan needs 168,000 cu^3 of fuelwood a year; nearly five million Tajiks use fuelwood as their primary source of household energy leading to an average per-capita consumption of 0.03 cu^3. Because of the poorly insulated houses and inefficiency of many stoves, annual demand for fuelwood in the country is higher lying around 15–20 million cu^3 used during the long and cold winters (Kirchhoff and Fabian 2010; www.naturalresources-centralasia.org).

The existing forest cover is not enough to provide so much fuel; locals use dry wood and burn dung. The consumption of fuelwood per year per head is around 3–4 cu^3. A normal populated village will need a plantation area between 270 and 360 ha to meet the immediate fuelwood needs. The availability of good technological sources can reduce fuelwood consumption. The usable forest area in the country should be urgently increased to satisfy the energy needs of its rurals. The establishment of fuelwood plantations around the villages can be a solution to protect the remaining natural forests and reduce the negative impacts of wind and water erosion. A crucial point here is that primary economic value and practical use of forests is for fuel. This issue must be solved by the government, as economic and social importance can be integrated into national planning for forestry (Kirchhoff and Fabian 2010; www.naturalresources-centralasia.org).

Rosa canina juice plays a great role at some places interlinked with the animal diversity like markhor and chukar in these areas together with the forests. Similarly wild boar and the Bukhara deer prefer woodlands and forests. However, it is difficult to consider wildlife from the point of view of forestry. The urial is a species which uses habitats interlinked with forests (Kirchhoff and Fabian 2010; www.naturalresources-centralasia.org).

It is very difficult to get reliable scientific data on the forest areas, up-todate maps are lacking, stocking volumes, species composition, and annual forest-destruction rates in Tajikistan. The data is nearly two decades old. Authentic data has been found in two sources (Akhmadov 2005, 2008). The data is not confirmed through satellite images, aerial photographs, and/or ground inventories. Average annual deforestation rate should be taken into consideration. (www.naturalresources-centralasia.org).

Nearly 25% of Tajikistan was covered by forests nearly a century back. Tugai forests were cut for cotton production, which greatly reduced the forests of the country. Further cutting of forests is forbidden; these have been declared as protected nature and anti-erosion and anti-mudflow zones. The few remaining forests are managed and protected according to the annual plans of government. These plans were implemented for a decade and updated and served another decade as the most important instrument for forest management. These plans have been discontinued after independence leading to an increase in the human pressure on forest resources. Earlier rural households were not guaranteed coal, oil or gas for their daily energy needs, which led to the ecological disaster of the near-total destruction of forests in the country (Kirchhoff and Fabian 2010; www.naturalresources-centralasia.org).

The forests in Tajikistan are state property. FAO statistics state that the forest area in the country has increased from 408.000 to 410.000 ha during 15 years up to 2005. There is no central database for forests and forestry, because reliable information on forest resources is not available, monitoring of forest resources is inadequate, and all the cartographic material is outdated. Illegal cutting, conversion to agricultural land, fuelwood harvesting, and overgrazing are four major factors in deforestation. Looking at the annual deforestation rate of 2%, some 172.000 ha of forests have been destroyed. Tajikistan's current forest area is estimated at some 250.000 ha. The few remaining forests are heavily degraded, and the stocking volume does not exceed 30 cu^3 per hectare. They are limited to very small relicts in remote and sparsely populated areas (Kirchhoff and Fabian 2010; www.naturalresources-centralasia.org).

Major issues in the forestry are open access which triggers the exploitation of forest resources-for fuelwood and through overgrazing. Fuelwood is in high demand during the winter in rural areas. Inefficient heating and cooking devices in poorly insulated houses add to the pressure on forest resources. The land tenure security and forest ownership awareness are lacking, and legal framework is not clear. Lezkhoz has generally weak administrative and managerial capacities. These are inadequate to overcome these issues. The law enforcement capacities are weak, and forest policies need to be organized to actual development. Sustainable forest management schemes are long overdue, and such reforms trigger impacts on the global and local environment. The establishment of sustainable forest management schemes in Tajikistan cannot be separated from the issues of pasture and fuelwood (Kirchhoff and Fabian 2010; www.naturalresources-centralasia.org). All forests in the country are classified as "protective forests." These are essential part of the country's watersheds. Natural forests are divided into five types: broad-leaved mesophilous forests, hard-leaved xerophilous light forests (shibliak), small-leaved microthermous mountain forests, juniper forests, and tugai forests (Kirchhoff and Fabian 2010; www.naturalresources-centralasia.org).

Juglans regia, *Malus* sp., *Acer regeli*, *A. turkestanicus*, and *Platanus orientalis* comprise the broad-leaved forests. Their ecological optimum lies between 1.200 and 2.300 m. The species composition changes from place to place; in some stands, *Acer* sp. predominate, while in others *Juglans regia* prevails. The forests are interlarded with grassland at higher altitudes, growing alongside *Juniperus*. These forests are under threat, but overgrazing prevents natural regeneration and negatively influences the biodiversity of this important ecosystem. For a natural regeneration of these forests, a pasture management system and participation of the rural population are a must. Hard-leaved xerophilous light forests consist of deciduous trees and shrubs, mainly occurring in light stands with a pronounced grass cover. These "shibliak stands" are well adapted to long, dry summers with a brief hibernation period. *Amygdalus bucharica*, *Pistacia vera*, *Calophaca grandiflora*, *Cercis griffithii*, and *Rhus coriaria* are the dominant species in these forests. Pistachio formations occur extensively on slopes and foothills in the southwest of country. The fuelwood gathering and uncontrolled grazing destroy young seedlings and prevent natural regeneration. Small-leaved forest associations have a dominance of deciduous mesophytic and microthermophilous trees, widespread in the flood belts next to the rivers in all the mountain ranges, from 1.500 m to the timberline. *Betula*, *Populus*, *Salix*, *Hippophae*, and *Fraxinus* are found in the small-leaved forests, which are often invaded by some shrubs and other grassland vegetation forming fairly thick, impenetrable vegetation. These forests have great biodiversity, providing habitat for a wide range of mammals and birds. In the Western Pamirs, remnants of these forests still exist. Small-leaved forests are under heavy pressure, mostly from fuelwood gathering and overgrazing. Their potential to produce firewood is considered great because of high annual increment and the ability of tree species (*Salix* sp. and *Populus* sp.) to coppice (Kirchhoff and Fabian 2010; www.naturalresources-centralasia.org). These forests should be included in the priority list for a comprehensive forest rehabilitation program in Tajikistan (Kirchhoff and Fabian 2010; www.naturalresources-centralasia.org). These forest resources can stop water erosion and serve as the principal energy source for rural households.

Juniperus forests are dominated by conifer species. Some endemics are *Juniperus sibirica*, *J. turkestanica*, *J. seravschanica*, *J. semiglobosa*, and *J. schugnanica*. The stands do not grow taller than 15 m, forming dwarf forests. They prevent soil erosion and protect hillsides from landslides and soil from being washed away. These forests are drought resistant and light-loving, with either semi-sparse or dense stands. *J. turkestanica* stands are also seen in mixed forests with other locally adapted species (Kirchhoff and Fabian 2010; www.naturalresources-centralasia.org).

In the early nineteenth century, one million hectare of alluvial tugai forests in Tajikistan formed an unusual system of dense forests with thick undergrowth along rivers, but most have been converted to agricultural use with cotton cultivation, which are currently under heavy pressure due to overgrazing, firewood cutting, and illegal felling. These forests are the best natural instrument to reduce the risk of mudflows and flooding. The protection and sustainable management of this important ecosystem should be listed as a priority in regreening the country. Handing over tugai forests to communities residing in the immediate neighborhood is the best solution (Kirchhoff and Fabian 2010; www.naturalresources-centralasia.org).

The remaining forests are deteriorating due to poor management, and many of these are in a bad condition. They are unable to perform normal ecological functions as they are degrading from inside and loss of forested areas is increasing (www.naturalresources-centralasia.org). This poses the danger from natural disasters like floods, landslides, and droughts. The disappearance of vegetation is reducing the biodiversity. Both plants and animals are losing ground. The ecological experts see little future for forests due to poor economic situation of the public. They still rely on timber for heating and cooking in many areas of the region. In southern Jalalabad region, deep hillside logging could lead to landslides hitting at any moment and threatening human settlements. The best recommendation includes increasing the number and area of the nature reserves. Somebody should explain it to the people living in the vicinity of forests, why their preservation is in their own interest. The inhabitants in the region should understand that destroying forests is like burning their own home. The heavy rain and hail have caused mudslides and floods in southern Tajikistan. Farmlands, more than 3.000 ha of cotton plantations, and roads and bridges have got damaged (Kirchhoff and Fabian 2010).

Tajikistan's few remaining forests are under severe threat due to overexploitation and uncontrolled grazing. These continue to reduce the country's remaining forest cover. The purpose should be to identify principal sector constraints; evaluate the lessons learned; compile reliable data on the forestry; make recommendations regarding policy changes, institutional reforms, capacity building, and financial investments; and prepare the groundwork for more donor participation in the forestry (Kirchhoff and Fabian 2010; www.naturalresources-centralasia.org).

Sustainable, managed forests are a necessary part of the development strategy in Tajikistan. Reliable information on the state of forestry and its resources is indispensable for making decisions about policies, strategies, and programs, as well as developing forestry sector in a sustainable way. This will allow the government, international donors, and other stakeholders to make strategic decisions to stop forest degradation and destruction in the country, with the ultimate goal of sustainable management of countries forests (Kirchhoff and Fabian 2010; www.naturalresources-centralasia.org).

4.10.7 Karakoram

The forests of Karakoram Range bordering Central Asian states are under tremendous threats due to climate change and anthropogenic activities such as overgrazing, increase in population, demographic pressure, household dependency on fuelwood, unscientific pastures beyond capacity, forest fires and storms, lack of financial inputs, and decline in traditional knowledge of sustainable use of forest trees (Faiza et al. 2017). The government of Pakistan needs to develop policies for enhancing the awareness for sustainable harvest of plants, reducing the grazing pressure, sharing equitable resources, increasing the marketing of products, and effective management of trees (Shedayi et al. 2016).

4.10.8 Other Countries

Russia, with approximately 12 million km^2 of boreal forest, has the largest forested area as of any other state on earth. Russian forests contain 55% of the world's total conifers and contribute 11% to the earth's biomass. The estimated deforested land in Russia has been summed up to be 20.000 km^2 (Soja et al. 2007). The areas closer to China have been affected greatly by deforestation, as the main market for timber is in that area (Samarrai 2011). The damage caused by deforestation in Russia is greater as compared to other regions as forests have a short growing season owing to the harsh winters; therefore the forests take a longer time to recover.

As of 2008, if the present rate of deforestation sustains, Indonesia would be wiped clear of its tropical rainforests in 10 years (China is black hole of Asia's deforestation 2006). By 2005, Indonesia had lost over 72% of its intact forests and 40% of all forests completely. Thirty-seven out of 41 national parks are victims of illegal logging, which costs up to US$ 4 billion per annum (Alley 2011).

Ever since the Vikings settled in Iceland in the ninth century, Europe has undergone massive deforestation. This has led to the degradation of vast regions of vegetation and land, soil erosion, and even desertification. Overexploitation, overgrazing under harsh climatic conditions, and logging have destroyed as much as half of the forested regions. Twenty-five percent of Iceland was once used to be covered by thick forests, out of which over 95% has been lost. However, afforestation and revegetation have restored a small part of this land. Sicily is often quoted whenever man-made deforestation is mentioned. Deforestation has been occurring in Italy since Roman times, when it was made an agricultural region, and has never stopped till this day (Porter and Prince 2009). This gradually started to affect the climatic conditions of Italy, leading to a decrease in rainfall and the consequential drying of rivers. One of the direct consequences of this ever-increasing deforestation is that the entire central and southwest provinces of Italy are currently devoid of forests (Trabia 2002). This has also led to a major negative impact on the island's wild fauna, of what little is left inhabits the inland's pastures and crop fields (Porter and Prince 2009).

Southeast Asia is the second world's greatest biodiversity hotspots. However, forest loss is acute in this part of the world. According to the 2005 report of FAO, Vietnam is only second to Nigeria as far as the highest primary deforestation rates are concerned. Over 90% of the Philippine archipelago's old-growth forest has been lost to deforestation. From 2000 to 2012, mangrove forest depleted at a rate of 0.18% per year. Some of the major drivers of loss of mangrove forests are rice agriculture and palm oil expansion (Richards and Friess 2016).

Canada underwent an estimated 56.000 ha of deforestation in 2005, which affected around 0.02% of Canadian forests. The major causes for deforestation in Canada are clearing for agricultural land, pasture, urban development, infrastructure development, hydroelectric development, and recreation. About 75% of this deforestation occurred in the boreal forests of Canada, mainly in Saskatchewan, Alberta, and Manitoba (www.cfs.nrcan.gc.ca). Before 2000, less than 8% of the boreal forests were protected in Canada, and over 50% were allocated to logging companies (Global Forest Watch Canada 2000).

Before the European Americans arrived, over half of the land area of present-day USA was covered with forests, approximately a million km^2 (990 million acres) in 1600. Over the next 300 years, the land was cleared for agriculture, mostly, at a rate that paralleled the population growth. For one person's addition to the population, 1–2 ha of land was cleared of forests for agriculture. This trend faded in the 1920s when, despite of population growth, the amount of crop land had stabilized. In due course, abandoned farmland started to revert to forests, and the forested regions increased from 1952 and peaked in 1963 (approximately 3 million km^2). Since 1963, the forested area has decreased steadily, with the exception of 1991, when some gain was observed (Myers and Tucker 1987). Most Central American countries have witnessed various cycles of deforestation and reforestation. Intensive agriculture by the Mayan civilization had considerably thinned the forests of Central America in the fifteenth century. Before the arrival of the Europeans, approximately 90% of the region was covered (Myers and Tucker 1987) with thick forests. Eventually, they cleared the land to sustain their demands of wood for exportation of primary products. However, since the 1960s, the main reason for land clearing has become cattle ranching. Though the main driving force in Brazil is debatable, a wide consensus exists that the clearance of land for ever-expanding croplands and pastures played a major role (www.mongabay.com). Certain areas like the Atlantic Rainforest have been reduced to barely 7% of their initial size (Olson et al. 2001; Olson and Dinerstein 2002). Despite many efforts to conserve these forested regions, only a few national parks have truly been protected. Even after all these efforts, almost 80% of logging in the Amazon Forest is illegal (Toyne et al. 2002). In 2008, a record increase was seen in the deforestation rates in the Amazon, which spiked up by 69% of what it was only a year ago (www.foxnews.com). WWF's new report states that if the current deforestation rates are not controlled, almost 60% of the Amazon could be completely wiped out or severely damaged by the end of 2030 (www.guardian.co.uk).

Australia was colonized relatively recently, which is why it has undergone high rates of deforestation, especially for agricultural purposes. Most of the clearing in recent years has occurred in Queensland and Tasmania (McAlpine et al. 2009), but the rates are expected to decrease once the new legislation has been implemented. Deforestation is thought to be responsible for almost 12% of Australia's total carbon emissions in 1998 (www.fpa.tas.gov.au). The expansion of urban areas is an additional factor that has contributed to loss of forests in Australia. The Littoral rainforest has also been victimized due to urban development, especially due to ribbon development (www.pittwater.nsw.gov.au). For 800 years that humans have occupied New Zealand, it bid farewell to almost 75% of its forests. Deforestation was initially carried out through wholesale burning of the forests by the Maori and Europeans, but the remaining forests were cleared to suffice the lumber demands of the increasing population. However, by 2000 all logging of the native trees was stopped in the public land (McAlpine et al. 2009). One of the world's largest rainforests is found in Papua New Guinea. It had the highest rate of illegal logging in the world in 2007, out of which 70–90% were estimated to be carried out for timber export (Alley 2011).

4.11 Conclusions

Trees are a source of oxygen, and if they vanish, so does clean air. Not all deforestation is intentional. A small portion of it is due to usual catastrophes like forest fires and overgrazing that halt trees from growing. Deforestation can be resolved merely if every single individual cooperates to stop people from vandalizing animals homes. It is the individual effort with regard to halting deforestation trends that can ultimately save the planet. Deforestation can be controlled using drastic measures including demarcation of forest boundaries, land use control policy, and afforestation campaign (Nazir and Ahmad 2018). Food sufficiency depending on forests is to be reduced to protect the forests. Moreover, afforestation must be started with planting of man-made forests on a large scale (Erb et al. 2016). The rate of deforestation can be reduced by enhancing awareness in the general public about the benefits of forests and establishing regulations to protect forests (Getahun et al. 2017).

Deforestation is a major issue in all of the Central Asian countries, which also leads to the loss of habitat and biodiversity. Due to increasing anthropogenic pressures, forests are deteriorating. In the Central Asia, around 30% population of rural areas lives near forests and depends on forest products. Studies show a tremendous increase in deforestation in this region. As of 2006, Tajikistan has the least amount of forests, 3.9%, whereas Uzbekistan has the highest percentage of forest covered land, which is 10.1%. Kyrgyzstan, Kazakhstan, and Turkmenistan have 6.2, 7, and 8.8% of forests and woodland covers, respectively (Djanibekov et al. 2015). Welfare in Central Asia largely depends on the mountain ecosystems of Pamirs, Tien Shan, and Altai, and they are under a great threat. These areas experience increase in deforestation and erosion together with other environmental degradations. There is an uncontrolled livestock grazing, illegal logging, and the plowing of slopes. All these have increased number of landslides, mudflows, and erosion in general. Overexploitation of natural resources, unsustainable land practices, and deficiencies in forest management are the major causes in this connection. The degradation of ecosystems has and is leading to a significant loss in biodiversity. The number of species and plants, which have disappeared or are endangered to become extinct, is constantly growing with irreversible consequences in some cases (www.ec-ifas.org).

Ecologists warn that if logging continues unabated in the region, all of middle Asia will face dire consequences such as water scarcity, health problems, and more frequent natural disasters. With forests shrinking together with the rate at which glaciers melt, main source of water in Central Asia is facing serious threats. The reports mention that only 4.3% of the mountainous territory is now forested. It was between one half and two thirds of the area 50 years ago. Nearly five decades back, the area was 6–8% in Kyrgyzstan alone. The deforestation rate has been worst during 1940–1950. Those times people here used to cut down timber for heating and cooking. Over the last 15 years, deforestation has increased as a result of economic crisis. There has been no commercial logging industry; still the forests have been damaged badly. The official figures show that 50.000–55.000 m^3 of trees were cut per year till 2006. From then onward, a 3-year moratorium was imposed on felling trees in particular valuable tree species in virgin forests. As a result of this, the legal logging has got reduced to 15.000 m^3 per year. This has raised hopes for the survival of remaining woods in the region. However, in spite of all these, deforestation is still an acute problem.

old

References

Akhmadov K (2005) Global forest resources assessment. Country reports. Tajikistan. FAO, Rome

Akhmadov K (2008) Forest and forest products country profile: Tajikistan. United Nations Economic Commission for Europe/Food and Agriculture Organization of the United Nations, Switzerland

Akramhanov A, Vlek PLG (2012) The assessment of spatial distribution of soil salinity risk using neural network. Environ Monit Assess 184(4):2475–2485

Akramkhanov A, Martius C, Park SJ, Hendrickx JMH (2011) Environmental factors of spatial distribution of soil salinity on flat irrigated terrain. Geoderma 163:55–62

Alley P (2011) Corruption: a root cause of deforestation and forest degradation. Global corruption report: climate change extract. Forestry Governance, pp 299–311

Altay V, Serin M, Yarci C, Severoglu Z (2012) Phytoecological and phytosociological investigations of the vegetation of Golcuk (Kocaeli/Turkey). Ekoloji 21(84):74–89

Angelsen A (2010) Policies for reduced deforestation and their impact on agricultural production. Proc Natl Acad Sci U S A 107(46):19639–19644

Anon (2007) Virunga under attack. Gorilla. cd. Virunga National Park blog. March

Asner GP, Knapp DE, Broadbent EN, Oliveira PJC, Keller M, Silva JN (2005) Selective logging in the Australian Greenhouse Average share in a sample of 8000 households in 24 countries, included in CIFOR's Poverty and Environment Network study. www.cifor.org/pen. Accessed 17 Feb 2013

Atamuradov A, Karryeva S (2005) Global forest resources assessment: Turkmenistan country report, Rome

Bai ZG, Dent DL, Olsson L, Schaepman ME (2008) Global assessment of land degradation and improvement 1: Identification by remote sensing. Report 2008/01. FAO/ISRIC, Rome

Baitulin IO, Proskuryakov MA, Yesserkepova IB (2010) Ecological doctrine of vegetation mantle monitoring of Kazakhstan in a view of global climate change. Hydrometeorol Ecol (in Russian)

Baños R, Manzano AF, Montoya FG, Gil C, Alcayde A, Góme J (2011) Optimization methods applied to renewable and sustainable energy: a review. Renew Sustain Energy Rev 15(4):1753–1766

Barry D, Taylor PL (2008) An ear to the ground: tenure changes and challenges for forest communities in Latin America. Rights and Resources Initiative, Washington, DC

Bekturova G, Romanova N (2007) Traditional land management knowledge in Central Asia: resource pack. S-Print, Almaty, p 86

de Beurs KM, Henebry GM, Owsley BC, Sokolik I (2015) Using multiple remote sensing perspectives to identify and attribute land surface dynamics in Central Asia 2001-2013. Remote Sens Environ 170:48–61

Blaser J, Carter EJ, Gilmour DA (1995) Biodiversity and sustainable use of Kyrgyzstan's walnut-fruit forests. In: Proceedings of the seminar, Arslanbob, Jalalabad Oblast, Kyrgyzstan, 4–8 Sept 1995

Boscolo M, Snook LK, Quevedo L (2009) Adoption of sustainable forest management practices in Bolivian timber concessions: a quantitative assessment. Int Forestry Rev 11(4):514–523

Botman EK (1999) Assessment of climatic conditions and vulnerability of Juniper formations under climate change in the territory of the Republic of Uzbekistan. In: Regional climatic scenarios, possible climate change impacts on agriculture and forest vegetation. Bulletin No. 1. SANIGMI, Tashkent, pp 54–64

Botman EK (2008) Vulnerability of juniper formations to climate change on the territory of Uzbekistan. In: Bulletin climate change consequences in Uzbekistan, adaptation issues Information on fulfilment of commitments to the UNFCCC by Uzbekistan. Edition No. 7. SANIGMI, Tashkent, pp 68–75

Botman EK (2009) Forest rehabilitation in the Republic of Uzbekistan. IUFRO, Vienna, pp 253–299

Botman EK, Agafonova (1999) Inventory of greenhouse gases in forestry of Uzbekistan. In: Mitigation measures on climate change in Uzbekistan. Bulletin No. 3. SANIGMI, Tashkent, pp 27–34

Bray DB, Antinori C, Torres-Rojo JM (2006) The Mexican model of community forest management: the role of agrarian policy, forest policy and entrepreneurial organization. Forest Policy Econ 8(4):470–484
Brown T (1997) Clearances and clearings: deforestation in Mesolithic/Neolithic Britain. Oxford J Archaeol 16(2):133
Buhlmann E (2006) Assessing soil erosion and conservation in the loess area of Faizabad, Western Tajikistan. Diploma Thesis, Faculty of Natural Sciences, University of Bern, 91p
Butler R (2012) Impact of deforestation: local and national consequences. www.rainforests.mongabay.com
Cantor NF (1994) The civilization of the Middle Ages: a completely revised and expanded edition of Medieval history, the life and death of a civilization. Harper Collins, New York, p 564
China is black hole of Asia's deforestation (2006) Asia News, 24 March 2006
DiNicola A (1997) Steps in preparing forestry and biodiversity sections of climate change action plans. Development and evaluation of forestry and biodiversity mitigation measures. Hydrometeorol Ecol 3:150–154
Djanibekov U, Khamzina A, Djanibekov N, Lamers JPA (2012a) How attractive are short-term CDM forestations in arid regions? The case of irrigated croplands in Uzbekistan. Forest Policy Econ 21:108–117
Djanibekov N, Van Assche K, Bobojonov I, Lamers JPA (2012b) Farm restructuring and land consolidation in Uzbekistan: new farms with old barriers. Eur Asia Stud 64(6):1101–1126
Djanibekov U, Dzhakypbekova K, Chamberlain J, Weyerhaeuser H, Zomer RJ, Villamor G, Xu J (2015) Agroforestry for landscape restoration and livelihood development in Central Asia. ICRAF Working Paper 186. World Agroforestry Centre East and Central Asia, Kunming, China, 41p
Dolgikh SA, Eserkepova IB, Shamen AM (1997) Assessment of global climate warming contribution in the desertification process development in Kazakhstan. Hydrometeorol Ecol 3:43–49 (in Russian)
Donata DC, Kauffman JB, Murdiyarso D, Kurnnianto S, Stidham M, Canine M (2011) Mangroves among the most carbon-rich forests in the tropics. Nat Geosci 4(5):293–297
Drigo KA, Piketty IG, Singer M-G (2008) Regulating industrial forest concession in Central Africa and South America. Forest Ecol Manag 256(7):1498–1508
Dubovyk O, Menz G, Conrad C, Kan E, Machwitz M, Khamzina A (2013) Spatio-temporal analyses of cropland degradation in the irrigated lowlands of Uzbekistan using remote-sensing and logistic regression modeling. Environ Monit Assess 185:4775–4790
Durst PB, McKenzie PJ, Brown CL, Appanah S (2006) Challenges facing certification and eco-labeling of forest products in developing countries. Int Forest Rev 8(2):193–200
East T, Kumpel NF, Milner-Gulland EJ, Rowcliffe M (2007) Determinants of bush meat consumption and tradein Río Muni, Equatorial Guinea. Biol Conserv 126:73–91
Ebeling J, Yasué M (2009) The effectiveness of market-based conservation in the tropics: forest certification in Ecuador and Bolivia. J Environ Manage 90(2):1145–1153
Ellis J (2003–2012) Effects of deforestation. Retrieved from wise geek. Conjecture Corporation. www.wisegeek.com/what-are-some-of-the-effects-of-deforestation.html
Ellison D, Futter NM, Bishop K (2012) On the forest cover-water yield debate: from demand-to supply-side thinking. Glob Chang Biol 18(3):806–820
Enneker C (2008) The impact of institutions on the constitution and performance of Community Forest Enterprises (CFE) in the Lowlands of Bolivia. PhD thesis, Wageningen University, Wageningen
Erb K-H, Lauk C, Kastner T, Mayer A, Theurl MC, Haberl H (2016) Exploring the biophysical option space for feeding the world without deforestation. Nature Article number 7:11382
Faiza N, Weiguo J, Aijun Y, Wenxing S (2017) Giant deforestation leads to drastic eco-environmental devastating effects since 2000; a case study of Pakistan. J Anim Plant Sci 27(4):1366–1376
FAO (2010) Forests and climate change in Eastern Europe and Central Asia. FAO, Rome, p 188
FAO (2010b) Global Forest Resources Assessment 2010 country report: Kazakhstan. Food and Agriculture Organization of the United Nations, Rome

FAO (Food and Agriculture Organization of the United Nations) (2008) Forests and energy: regional perspectives: opportunities and challenges for forests and forestry. African Forestry and Wildlife Commission. Sixteenth Session. Near East Forestry Commission, Khartoum, pp 1–88
Feintrenie L, Chong WK, Levang P (2010) Why do farmers prefer oil palm? Lessons learnt from Bungo district, Indonesia. Small Scale For 9(3):376–396
Flannery T (1994) The future eaters. Reed Books, Melbourne
Foley JA, Asner GP, Costa MH, Coe MT, DeFries R, Gibbs HK, Howard EA, Olson S, Patz J, Ramankutty N, Snyder P (2007) Amazonia revealed: forest degradation and loss of ecosystem goods and services in the Amazon Basin. Front Ecol Environ 5:25–32
Getahun K, Poesen J, van Rompaey A (2017) Impacts of resettlement programs on deforestation of moist evergreen Afromontane Forests in Southwest Ethiopia. Mountain Res Dev 37(4):474–486
Gintzburger G, Saidi S, Soti V (2005) Rangelands of the Ravnina region in the Karakum desert (Turkmenistan): current condition and utilization. INCO-COPERNICUS/RTD Project: ICA2-CT-2000-10015. Centre de coopération internationale en recherche agronomique pour le développement, Montpellier (Cirad)
Gitz V, Ciais P (2004) Future expansion of agriculture and pasture acts to amplify atmospheric CO_2 levels in response to fossil-fuel and land-use change emissions. Clim Change 67(2–3):161–184
Global Forest Watch Canada (2000) Canada's forests at a crossroads - an assessment in the year 2000
Gore A (2009) Our choice, a plan to solve the climate crises. Chapter 9, Forests. Bloomsbury Publishing, London, pp 170–195
Guariguata MR (2009) Tropical forest management and climate change adaptation. Rev Estud Soc 32:98–112
Gücel S, Ozkan K, Celik S, Yucel E, Ozturk M (2008) An overview of the geobotanical structure of Turkish *Pinus sylvestris* and *Carpinus betulus* Forests. Pakistan J Bot 40(4):1497–1520
Gupta R, Kienzler K, Mirzabaev A, Martius C, de Pauw E, Shideed K, Oweis T, Thomas R, Qadir M, Sayre K, Carli C, Saparov A, Bekenov M, Sanginov S, Nepesov M, Ikramov R (2009) Research prospectus: a vision for sustainable land management research in Central Asia. ICARDA Central Asia and Caucasus Program. Sustainable Agriculture in Central Asia and the Caucasus Series No. 1. CGIAR-PFU, Tashkent, Uzbekistan, 84p
Hajabbasi MA, Jalalian A, Karimzadeh HR (1997) Deforestation effects on soil physical and chemical properties, Lordegan, Iran. Plant and Soil 190(2):301–308
Hogan CM (2007) Knossos field notes. The Modern Antiquarian
Holmes TP, Blate GM, Zweede JC, Perlira RA, Barretu P, Boltz F (2002) Financial and ecological indicators of reduced impact logging performance in the eastern Amazon. For Ecol Manag 5583(2001):1–18
Hurtado-Gonzales J, Bodmer RE (2004) Assessing the sustainability of brocket deer hunting in the Tamshiyacu-Tahuayo Communal Reserve, northeastern Peru. Biol Conserv 116(1):1–7
Imanderdieva N, Imankul B, Severoğlu Z, Altay V, Ozturk M (2018a) Potential ımpacts of climate change on plant diversity of Sary-Chelek Biosphere Reserve in Kyrgyzstan. Springer (In press)
Imanderdieva N, Severoğlu Z, Kurmanbekova G, Altay V, Ozturk M (2018b) Plant diversity of Ala-Archa National Park in Kyrgyzstan and its economical potential. Springer (In press)
Janssen TAJ, Ametsitsi GKD, Collins M, Adu-Bredu S, Oliveras I, Mitchard ETA, Veenendaal EM (2018) Extending the baseline of tropical dry forest loss in Ghana (1984-2015) reveals drivers of major deforestation inside a protected area. Biol Conserv 218:163–172
Ji C (2008) Central Asian Countries initiative for land management multicountry partnership framework support project. ADB, Tashkent
Kaimowitz D (2003) Not by bread alone… forests and rural livelihoods in sub-Saharan Africa. In: Forests in poverty reduction strategies: capturing the potential. EFI Proceedings, No. 47, pp 45–63
Kaimowitz D, Angelson A (1998) Agricultural technologies and Tropical deforestation. Center for International Forestry Research (CIFOR), Bogor, Indonesia and Department of Economics and Social Sciences Agricultural University of Norway As, Norway

Karahan F, Altay V, Keskin M (2015) An ethnobotanical study on woody plants benefits from handicrafts in Antakya District (Hatay-Turkey). Int J Sci Technol Res 1(1):1–18

Keller M, Asner GP, Blate G, Glocklin M, Merry J, Peña-Claros F, Zweede M (2007) Timber production in selectively logged tropical forests in South America. Front Ecol Environ 5(4):213–216

Khuc QV, Tran BQ, Meyfroidt P, Paschke MW (2018) Drivers of deforestation and forest degradation in Vietnam: an exploratory analysis at the national level. Forest Policy Econ 90:128–141

Kirchhoff JF, Fabian A (2010) Forestry sector analysis of the republic of Tajikistan. Eschborn: Deutsche Gesellschaft für Technische Zusammenarbeit (GTZ)

Klein I, Gessner U, Kuenzer C (2012) Regional land cover mapping and change detection in Central Asia using MODIS time-series. Appl Geogr 35:219–234

Kushlin AV, Schillhorn van Veen TW, Sutton W (2003) Kazakhstan: forest sector in transition: the resource, the users and sustainable use. World Bank, Washington, DC

Lambin EF, Meyfroidt P (2011) Global land use change, economic globalization, and the looming land scarcity. Proc Natl Acad Sci 108:3465–3472

Larson AM, Cronkleton P, Barry D, Pacheco P (2008) Tenure rights and beyond: community access to forest resources in Latin America. CIFOR, Bogor

Lawrence D, Vandecar K (2015) Effects of tropical deforestation on climate and agriculture. Nat Clim Chang 5:27–36

Le QB, Nkonya E, Mirzabaev A (2014) Biomass productivity-based mapping of global land degradation hotspots. ZEF-Discussion Papers on Development Policy, p 193

Li C, Zhang C, Luo G, Chen X, Maisupova B, Madaminov AA et al (2015) Carbon stock and its responses to climate change in Central Asia. Glob Chang Biol 21:1951–1967

Maddox GH (2006) Sub-Saharan Africa: an environmental history. Santabarbara

Mátyás C (2010) Forests and climate change in Eastern Europe and Central Asia. Forests and Climate Change Working Paper 8. FAO, Rome

McAlpine CA, Etter A, Fearnside PM, Seabrook L, Laurance WF (2009) Increasing world consumption of beef as a driver of regional and global change: a call for policy action based on evidence from Queensland (Australia), Colombia and Brazil. Glob Environ Chang 19(1):21–33

Mccann JC (1999) Green land, Brown land, Black land: An environmental history of Africa 1800-1990. Heinemann, Portsmouth

Mirzabaev A (2013) Climate volatility and change in Central Asia: economic impacts and adaptation. PhD thesis at Agricultural Faculty, University of Bonn

Murdiyarso D, Stole F, Fagi AM (2000) Agricultural intensification, deforestation, and the environment: assessing tradeoffs in Sumatra, Indonesia. In: Lee D, Barrett C (eds) Tradeoffs or synergies?: Agricultural intensification, economic development, and the environment. CAB International, Wallingford, pp 221–224

Mweninguwe R (2010) Massive deforestation threatens food security. Newsfromafrica.org

Myers N, Tucker R (1987) Deforestation in Central America: Spanish legacy and North American consumers. Environ Rev 11(1):55–71

Nasi R, Brown D, Wilkie D, Bennett E, Tutin C, van Tol G, Christophersen T (2008) Conservation and use of wildlife-based resources: the bush meat crisis. Secretariat of the Convention on Biological Diversity and Center for International Forestry Research (CIFOR), Bogor, pp 5–42

Nasi R, Taber A, Van Vliet N (2011) Empty forests, empty stomachs? Bush meat and livelihoods in the Congo and Amazon Basins. Int For Rev 13(3):355–368

Nazir N, Ahmad S (2018) Forest land conversion dynamics: a case of Pakistan. Environ Dev Sustain 20(1):389–405

Nkonya E, Mirzabaev A, von Braun J (2016) Economics of land degradation and improvement - a global assessment for sustainable development. Springer, Cham, p 686

Norris FT (1997) Where did the villages go? Steamboats, deforestation, and archaeological loss in the Mississippi Valley. Common fields: an environmental history of St. Louis

O'Hara SL (1997) Irrigation and land degradation: implications for agriculture in Turkmenistan, Central Asia. J Arid Environ 37:165–179

Olagunju TE (2015) Impacts of Human induced deforestation, forest degradation and fragmentation on food security. N Y Sci J 8(1):4–16

Olson DM, Dinerstein E (2002) The Global 200: Priority ecoregions for global conservation. Ann Mo Bot Gard 89:199–224

Olson DM et al (2001) Terrestrial ecoregions of the world: a new map of life on Earth: a new global map of terrestrial ecoregions provides an innovative tool for conserving biodiversity. Bioscience 51:933–938

Orlov OV, Dzhumadylova CK, Khudaibergenov AD, Rodni Stuparik (2003) Climate change and its impact on the forest ecosystems of the Kyrgyz Republic. Bishkek

Orozumbekov A et al (2009) Forest rehabilitation in Kyrgyzstan. IUFRO, Vienna, pp 131–182

Ozturk M, Gemici Y, Gork G, Secmen O (1991) A general account of high mountain flora and vegetation of mediterranean part of Turkey. Ege Univ Sci Fac J 13:51–59

Ozturk M, Secmen O, Gork G (eds) (1996a) Plant life in Southwest and Central Asia, vol I. Ege University Press, Izmir, p 499

Ozturk M, Secmen O, Gork G (eds) (1996b) Plant life in Southwest and Central Asia, vol II. Ege University Press, Izmir, p 1093

Ozturk M, Celik A, Guvensen A, Hamzaoglu E (2008) Ecology of tertiary relict endemic *Liquidambar orientalis* Mill. forests. For Ecol Manage 256:510–518

Ozturk M, Gucel S, Ozkan K, Kucuk M, Saxena AK (2010a) Forests-forestry-environment in Turkey. J Environ Biol 31(1/2):245

Ozturk M, Aksoy A, Gucel S, Altundag E (2010b) Biodiversity & Conservation-A Case Study from Europe & Turkey. In: Bhatti R, Soomro SA (eds) Proceedings of second International conference "Biodiversity is our Life" (IC Biour-Life), Center for Biodiversity & Conservation, Shah Abdul Latif University, Khairpur, pp 001–012

Ozturk M, Gucel S, Kucuk M, Sakcali S (2010c) Forest, climate change & forest fires in the Mediterranean Region of Turkey. J Environ Biol 31:1–9

Ozturk M, Uysal I, Karabacak E, Celik S (2011) Plant species microendemism, rarity and conservation of Pseudo-Alpine Zone of Kazdagi (Mt. Ida) National Park in Turkey. Proc Soc Behav Sci 19:778–786

Ozturk M, Hakeem KR, Faridah-Hanum I, Efe R (eds) (2015) Climate change impacts on high-altitude ecosystems, vol XVII. Springer, New York, 695p

Ozturk M, Saba N, Altay V, Iqbal R, Hakeem KR, Jawaid M, Ibrahim FH (2017a) Biomass and bioenergy: an overview of the development potential in Turkey and Malaysia. Renew Sustain Energy Rev 79:1285–1302

Ozturk M, Altay V, Gucel S, Altundağ E (2017b) Plant diversity of the drylands in Southeast Anatolia-Turkey: role in human health and food security. In: Ansari AA, Gill SS (eds) Plant biodiversity: monitoring, assessment and conservation. Chapter 5. CABI, London, pp 83–124

Ozturk M, Uysal I, Yucel E, Altay V, Karabacak E (2017c) Soil-plant interactions in the monumental plane trees (*Platanus orientalis*) grove-Canakkale-Turkey. J Environ Biol 38(5):1129–1137

Ozyigit S, Altay V, Ozyigit II, Yarci C (2015) Vegetation ecology of the Princes' Islands, Istanbul-Turkey. J Environ Biol 36(1):113–120

Pan Y, Birdsey RA, Fang J, Houghton R, Kauppi PE, AKurz W, Phillips OL, Shvidenko A, Lewis SL, Canadell JG, Ciais P, Jackson RB, Pacala SW, McGuire AD, Piao S, Rautiainen A, Sitch S, Hayes D (2011) A large and persistent carbon sink in the world's forests. Science 333:998

Payn T, Carnus J-M, Freer-Smith P, Kimberley M, Kollert W, Liu S, Orazio C, Rodriguez L, Silva LN, Wingfield MJ (2015) Changes in planted forests and future global implications. For Ecol Manage 352:57–67

Peña-Claros VZ, Mustached M, Alarcón B, Licona A, Leaño JC, Pariona JC, Choque W (2009) Silvicultural treatments enhance growth rates of future crop trees in a tropical dry forest. For Ecol Manage 258(6):971–977

Pender J, Mirzabaev A, Kato E (2009) Economic analysis of sustainable land management options in Central Asia Vol 168, Final report for the ADB. IFPRI/ICARDA

Podrezov OA, Titova LI (2002) The modern climate of Kyrgyzstan and its change scenario in the 21st century. Bishkek

Pomfret R (2008) Tajikistan, Turkmenistan, and Uzbekistan. In: Anderson K, Swinnen JFM (eds) Distortions to agricultural incentives in Europe's transition economies. World Bank, Washington, DC, pp 297–338

Popov K (1994) Biogeography and Ecology of Turkmenistan. In: Fet V, Atamuradov K (eds) Monographiae Biologicae, vol 75, Chap 10. Springer, Netherlands, pp 173–186

Porter D, Prince D (2009) Frommer's Sicily. Frommer's, p 268

Putz FE, Zuidema PA, Pinard MA, Boot RG, Sayer JA, Sheil D et al (2008) Improved tropical forest management for carbon retention. PLoS Biol 6(7):e166

Rice RE, Gullison RE, Reid JW (1997) Can sustainable management save tropical forests? Sci Am 276:44–49

Richards M, Costa PM (1999) Can tropical forestry be made profitable by internalizing the Externalities? Nat Resour Perspect 46:1–6

Richards DR, Friess DA (2016) Rates and drivers of mangrove deforestation in Southeast Asia, 2000-2012. Proc Natl Acad Sci 113(2):344–349

Ricketts T, Daily G, Ehrlich P, Michener C (2004) Economic value of tropical forest to coffee production. Proc Natl Acad Sci U S A 101(34):12579–12582

Rudel TK, DeFries R, Asner GP, Laurance WF (2009) Changing drivers of deforestation and new opportunities for conservation. Conserv Biol 23:1396–1405

Rudi J, van der Ent, Hubert HG, Savenije B, Susan S, Steele-Dunne C (2010) Origin and fate of atmospheric moisture over continents. Water Resour Res 40(9):W09525, https://doi.org/10.1029/2010WR009127

Sahney S, Benton MJ, Falcon-Lang HJ (2010) Rainforest collapse triggered Pennsylvanian tetrapod diversification in Euramerica. Geology 38(12):1079–1082

Samarrai F (2011) UVA Today: archaeologynewsnetwork.blogspot.com/2011/03/russian-boreal-forests-undergoing.html#.ULI3U-S-o-1. Accessed 25 Nov 2012

Santilli M, Moutinho P, Schwartzman S, Nepstad D, Curran L, Nobre C (2005) Tropical deforestation and the Kyoto Protocol. Clim Change 71:267–276

Searchinger T, Heimlich R, Houghton RA, Dong F, Elobeid A, Fabiosa J, Tokgoz S, Hayes D, Yu T (2008) Use of U.S. croplands for biofuels increases greenhouse gases through emissions from land-use change. Science 319(5867):1238–1240

Sezer Y, Altay V, Ozyigit II, Yarci C (2015) Woody vegetation of Sile and its environs (Istanbul/Turkey) and destruction of the area. J Environ Biol 36(1):163–170

Shedayi AA, Xu M, Hussain F, Sadia S, Naseer I, Bano S (2016) Threatened plant resources: distribution and ecosystem services in the world's high elevation park of the Karakoram ranges. Pakistan J Bot 48(3):999–1012

Siry JP, Cubbage FW, Ahmed MR (2005) Sustainable forest management: global trends and opportunities. Forest Policy Econ 7(4):551–561

Soja AJ, Tchebakova NM, French NH, Flannigan MD, Shugart HH, Stocks BJ et al (2007) Climate-induced boreal forest change: predictions versus current observations. Global Planet Change 56(3-4):274–296

Sucoff E (2003) Deforestation. In: Environmental encyclopedia. Gale, Detroit, pp 358–359

Sumithrabai K, Thirumarimurugan M, Gopalakrishnan S (2011) Bio-fuel from algae. Int J Adv Eng Technol 2(3):320–323

Sunderland TCH (2011) Food security: why is biodiversity important? Int For Rev 13(3):355–368

Sutton W, Whitford P, Stephens EM, Galinato SP, Nevel B, Plonka B, Karamete E (2007) Integrating environment into agriculture and forestry. Progress and prospects in Eastern Europe and Central Asia. World Bank, Kosovo

Syuharni AW, Hakeem KR, Faridah-Hanum I, Alias MS, Ozturk M (2014) Ecology of the Coastal Heath Forest flora - a case study from Terengganu, Malaysia. Emir J Food Agric 26(12):1114–1123

Taylor L (2004) The healing power of rainforest herbs. Square one Publishers, Inc. Tropical Plant Database

Toyne P, O'Brien C, Nelson R (2002) The timber footprint of the G8 and China. Making the case for green procurement by government. WWF International 40, Gland, Switzerland

Trabia C (2002) A Sicilian Desert? Best of Sicily Magazine

Vanclay JK (2008) Improved tropical forest management for carbon retention. PLoS Biol 6:1368–1369

Vengelovsky BI (2006) Biological specifics of Kyrgyzstan's walnut forests restoration and development. Bishkek

Verchot L, Noordwijik M, Kandji S, Tomich T, Ong C, Albrecht A, Meckensen J, Bantilan C, Anupama K, Palm C (2007) Mitigation and adaptation strategies for global change. World Agroforestry 12(5):901–918

Veronica B, Carrillo-Parra A, Gonzalez-Rodriguez H, Ramirez-Lozano RG, Corral-Rivas JJ, Garza-Ocanas F (2012) Evaluation of charcoal production process from forest residues of *Quercus sideroxyla* Hum & Bonpl. in a Brazilian beehive Kiln. Ind Crop Prod 42:169–174

Ward PD (1997) The call of distant mammoths: why the ice age mammals disappeared. Copernicus, New York

Werf GR, Morton DC, DeFries RS, Olivier JGJ, Kasibhatla PS, Jackson RB, Collatz GJ, Randerson JT (2009) CO_2 emissions from forest loss. Nat Geosci 2:737–738

Wilkie DS, Starkey M, Abernethy K, Effa EN, Telfer P, Godoy R (2005) Role of prices and wealth in consumer demand for bushmeat in Gabon, Central Africa. Conserv Biol 19(1):268–274

Williams M (2001) The history of deforestation. Hist Today 51:7

Yesserkepova IB (2008) Constructing model-based climate change scenarios for Southeastern Kazakhstan and the possibilities for impact assessment. In: Qi J et al (eds) Environmental problems of Central Asia and their economic, social and security impacts. Springer, New York, pp 163–172

Yvonne A (1998) Deforestation in Sub-Saharan Africa. Af Tech Forum 8(1)

Websites

wwf.panda.org
www1.american.edu
www.cfs.nrcan.gc.ca
www.climateandweather.net
www.d.umn.edu
www.ec-ifas.org
www.en.m.wikipedia.org
www.eoearth.org
www.fao.org
www.fpa.tas.gov.au
www.forclimadapt.eu
www.foxnews.com
www.guardian.co.uk
www.independent.co.uk
www.ipcc.ch
www.mongabay.com
www.monitoramento.sema.mt.gov.br
www.naturalresources-centralasia.org
www.pamir.at
www.phys.org

www.pittwater.nsw.gov.au
www.pulsamerica.co.uk
www.rainforests.mongabay.com
www.scidev.net/global
www.stopearthdestruction.com
www.thegef.org
www.tm.undp.org
www.wisegeek.com
www.worldagroforestry.org
www.wrm.org.uy

Chapter 5
Vegetation Classification and Habitat Mapping of Dachigam National Park, Kashmir, India

Khursheed Ahmad and Majid Farooq

Abstract This chapter presents a detailed account of the vegetation classification and habitat mapping of Dachigam National Park in the Zanskar mountain range of Kashmir, India. A total of 203 vegetation plots in the entire study area based on the optimal size obtained from species-area curve were sampled during 2001–2010 for vegetation and habitat stratification following Mueller-Dumboise and Ellenberg (Aims and methods of vegetation ecology, Wiley, New York, p 574, 1974) and Rikhari et al. (Pattern of species distribution and community characters along a moisture gradient within an Oak zone of Kumaun Himalaya. Proc. Indian National Science Academy, pp 431–438, 1989). Forty-four (44) woody species were recorded, out of which 22 were trees and 22 were shrubs. Vegetation of Dachigam National Park was classified, based on Twinspan analysis, into six broader communities such as mixed riverine, mixed woodland, pine *Parrotiopsis*, grassland/scrub, mixed coniferous and alpine scrub. These communities based on the predominant vegetation types present were further classified into nine habitat types growing in an altitudinal sequence. The tree and shrub densities/ha differed significantly between different habitat types. The maximum tree density/ha (442.78 ± 18.89 S.E.) was recorded in mixed oak and pine *Parrotiopsis* (430.99 ± 58.92 S.E.) habitats. Ground cover showed significant differences between different seasons and habitat types. The riverine and mixed woodland habitats of Dachigam National Park showed the highest values of diversity indices, whereas rarefaction values were highest in case of mixed woodland and mixed coniferous habitats. Based on the tree diversity and density, cluster analysis (Bray-Curtis) showed that alpine scrub and mixed oak habitats followed by pine *Parrotiopsis* and mixed coniferous habitats of Dachigam National Park are very similar to each other, whereas cluster analysis (Bray-Curtis) of shrubs showed maximum similarities between riverine and mixed coniferous habitats and maximum dissimilarities between mixed oak and grassland/scrub habitats.

K. Ahmad (✉)
Division of Wildlife Sciences, Faculty of Forestry, Sher-e-Kashmir University of Agricultural Sciences and Technology of Kashmir, Srinagar, Jammu and Kashmir, India

M. Farooq
Department of Ecology, Environment and Remote Sensing, Jammu and Kashmir Forest Department, Srinagar, Jammu and Kashmir, India

D. Egamberdieva, M. Öztürk (eds.), *Vegetation of Central Asia and Environs*,
https://doi.org/10.1007/978-3-319-99728-5_5

The NDVI change detection analysis indicated that over the decades, there has been significant change in the vegetation in the National Park with area under open scrub having grown by about 5.81%, whereas area under alpine pastures has significantly reduced by 6.02%, followed by the blue pine forests by 4.75% and birch forests by about 3.57% which is imperative for long-term conservation and sustenance of the Park biodiversity.

Keywords Dachigam · Vegetation · Twinspan cluster analysis · Kashmir

5.1 Introduction

Vegetation which is universally recognized as a component of major importance in site evaluation and classification (Barnes et al. 1998) is strongly controlled by macro- and microclimate. The condition of vegetation in a stand, landscape, or region is a product of the interplay of forces of disturbance and biotic development on a stage set by patterns and dynamics of climate, soil, and landforms (Hunter 1999). The distribution, abundance, coverage, biomass, and vertical layering or stratification of vegetation, which are always prone to changes by natural or man-made processes (Barnes et al. 1998), are of major significance in animal ecology and wildlife management (Hunter 1999).

Dachigam National Park is located 21 km northeast of Srinagar City of the Jammu and Kashmir state along the Zanskar mountain range of Northwest Himalayan biogeographic zone (2A) (Rodgers and Panwar 1988; Rodgers et al. 2000). Dachigam has more or less natural vegetation and assumes a great ecological and aesthetic significance for showing a strong vegetation in contrast to the areas just outside the Park, as they have been subjected to varied types of biotic interferences (Singh and Kachroo 1978). With the exception of some steep slopes, natural vegetation has been replaced in the Kashmir valley by cultivated plants along roadsides and stream sides and in orchards (Kurt 1978; Singh and Kachroo 1978). Besides, Dachigam National Park is inhabited by the last genetically viable population of the Kashmir red deer or hangul (Ahmad et al. 2015), a subspecies of European red deer endemic to Kashmir recently been classified as a separate sub-species Cervus hanglu hanglu under Tarim Red deer (Cervus hanglu) (Brook et al. 2017). It is presumed that hangul had diverged from the Bactrian red deer and migrated to India from Tajikistan approximately 1.2 MYA (Kumar et al. 2016).

Compared to the number of studies carried out on the vegetation and habitat mapping in other parts of the Western Himalaya (Rawat et al. 1993; Singh and Rawat 1999, 2000; Singh 1999a, b), very few studies have been conducted in the Kashmir region of the Northwest Himalayas in general and Dachigam National Park in particular. The information available on the vegetation of Dachigam National Park is only restricted to four brief studies by Singh and Kachroo (1978), Bhat (1985), Anonymous (1985), and Bhat et al. (2002).

The only detailed vegetation study carried out in the Lower Dachigam shows that Dachigam National Park exhibits a variety of vegetation types manifested by habitat, form, and density of dominant species and controlled by a number of factors including habitat conditions, exposure, altitude, and, above all, the degree of biotic interference (Singh and Kachroo 1978). However, that study did not give any indication of the species diversity and density and other aspects of vegetation stratification and habitat mapping.

Since the forest structure, composition, and ecological processes change over a vast range of spatial and temporal scales, therefore, landscapes are not merely unique in structure, composition, and spatial pattern but are also dynamic (Andrew et al. 1997). A detailed comparative vegetation study in Dachigam National Park to evaluate any temporal changes in the vegetation structure was carried out from 2001 to 2010, after a gap of nearly 25 years. One of the aims of this study was to describe the vegetation structure of Dachigam National Park in order to evaluate habitat use by species particularly the hangul. The aspects investigated were (1) to analyze the plant community classification and hierarchical classification of vegetation communities in Dachigam NP; (2) to evaluate the density, abundance, and diversity of trees and shrubs in different habitat types; (3) to find out the significant differences in the vegetation and ground cover (grass/herb, litter, bare ground, rock) between different seasons and in different habitat types; and (4) to prepare vegetation (land use/land cover) and habitat mapping of Dachigam NP.

5.2 Study Area

Dachigam National Park (DNP henceforth) besides harboring unique faunal species of Palearctic affinity such as the endangered Kashmir musk deer *Moschus cupreus*, endangered Kashmir gray langur *Semnopithecus ajax*, near-threatened Himalayan serow *Capricornis thar*, Asiatic black bear *Ursus thibetanus*, Himalayan brown bear *Ursus arctos isabellinus*, and leopard *Panthera pardus* and other diverse flora and fauna (Anonymous 1985; Ahmad 2006) is the only home to the last genetically viable population of the Kashmir red deer or hangul (Ahmad et al 2009, 2015), a subspecies of European red deer endemic to Kashmir recently been classified as a separate sub-species Cervus hanglu hanglu under Tarim Red deer (Cervus hanglu) (Brook et al. 2017). It is presumed to have diverged from the Bactrian red deer and migrated to India from Tajikistan approximately 1.2 MYA (Kumar et al. 2016).

DNP encompasses an area of 141 km^2 ranging from 1700 to 4700 m and lies between 34° 05″ 00′ N and 34° 10″ 32′ N and 74° 53″ 50′ E and 75° 09″ 16′ E in the Zanskar mountain range of Northwest Himalayan biogeographic zone (2A) (Rodgers and Panwar 1988; Rodgers et al. 2000) (Fig. 5.1). DNP is roughly rectangular in shape approximately 22.5 km long and 8 km wide and covers roughly half of the catchment area of Dal Lake (Holloway and Wani 1970). The Dachigam National Park is bounded by Sindh Valley to the northeast; Tarsar, Lidderwath, Kolhai of Lidder Valley, and Overa Aru Wildlife Sanctuary in the far east; Tral range in the southeast; and Harwan, Brain, and Nishat in the west and southwest (Holloway and Wani 1970; Kurt 1978).

The climate of the study area may be described as sub-Mediterranean to typically temperate with bixeric regime with higher degrees of variation in precipitation and

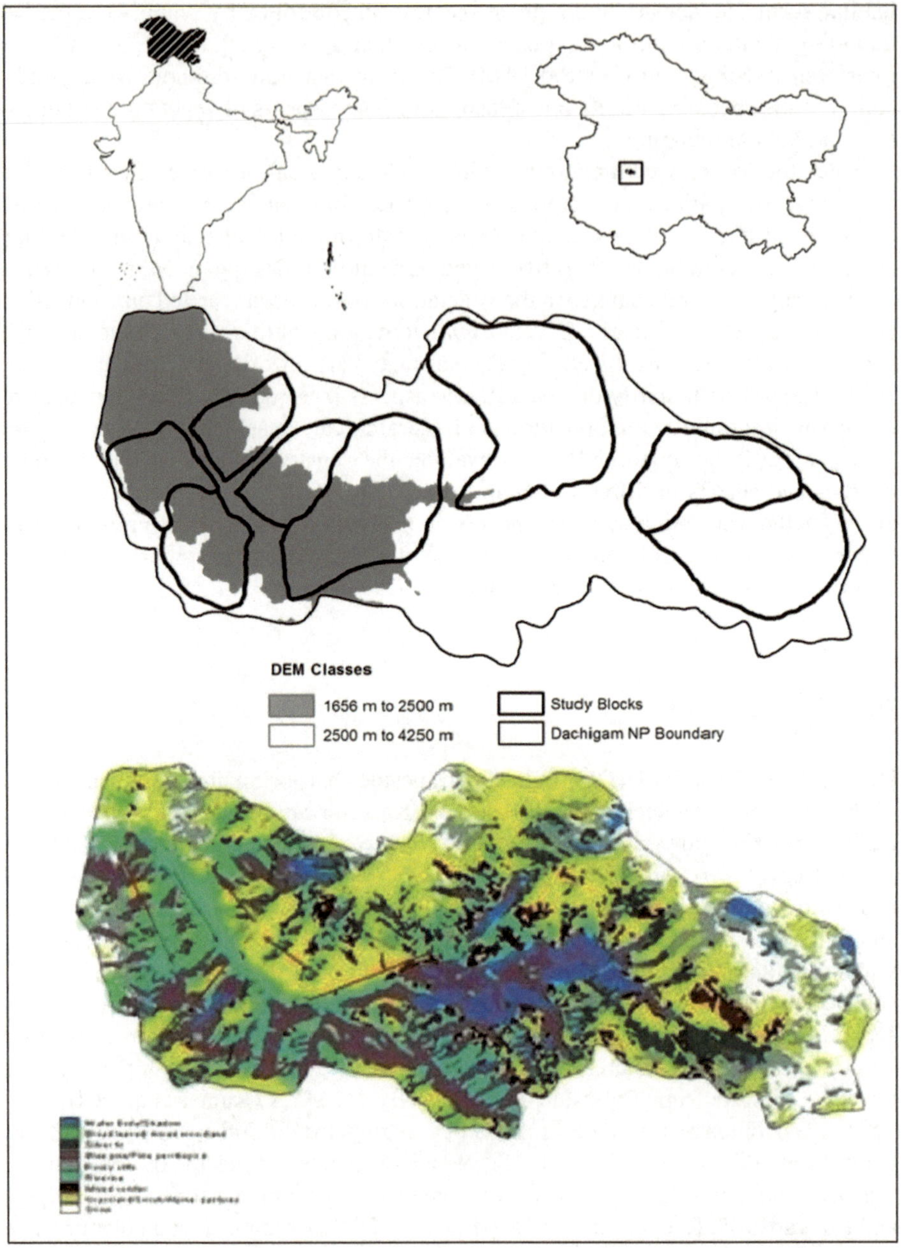

Fig. 5.1 Location of Dachigam National Park (top) that shows the land cover (lower right) and survey blocks (lower left) in the Jammu and Kashmir state of India

dryness (Singh and Kachroo 1978; Mani 1981). Generally, two spells of dryness are experienced, one in June and another in September to November: Snow is the main source of precipitation and in some parts melts till June. The annual minimum and maximum rainfall of Dachigam have been calculated ranging between 32 and 546 mm (Bhat 1985). Four distinct seasons occur in a year, viz., spring (March to May), summer (June to August), autumn (September to November), and winter (December to February). The monthly mean temperatures recorded during the study period ranged between a maximum of 32 °C in August (late summer) of 2002 and a minimum of −5.8 °C during January (midwinter) of 2003 (Ahmad 2006). The soil depth in the study area on the slope from lower to middle reaches is less than 25 cm and hence falls under the category of very shallow soils (Bhat 1985).

DNP is ecologically and administratively divided into two sectors: the Lower Dachigam (26 km^2) and Upper Dachigam (115 km^2). The Lower Dachigam altitude ranges from 1700 to 3500 m (Anonymous 1985) and thus has a complex mixture of vegetation types with broad-leaf mesophyll forest of *Acer caesium*, *Morus alba*, *Ulmus wallichiana*, *Ulmus villosa*, *Rhus succidiadiana*, *Juglans regia*, and *Parrotiopsis jacquemontiana* and a variety of conifers such as *Cedrus deodara*, *Pinus wallichiana*, *Picea smithiana*, and *Abies pindrow* growing in an altitudinal sequence (Holloway 1971; Ahmad 2006; Ahmad et al. 2009, 2015).

Upper Dachigam altitude ranges from 2000 m to c. 4700 m and comprises vegetation gradient of subalpine community of forest followed by scrub vegetation of birch *Betula utilis* and rhododendron *Rhododendron campanulatum* interspersed with meadows of herb-rich grasslands over 3300 m. This zone gradually merges in to the zone of permanent snow, which is above 3500 m (Holloway 1971; Ahmad 2006; Ahmad et al. 2009, 2015). The following methods as adopted by various workers in the Western Himalaya have been used (Rawat et al. 1993; Singh and Rawat 1999, 2000; Singh 1999a, b) for quantification of vegetation of Dachigam National Park.

5.3 Vegetation Sampling

Vegetation (trees, shrubs, and ground cover) and landscape parameters (altitude, aspect, and slope) were quantified both at the locations of animal sightings (animal focal plots), along with seven random but systematically laid line transects (1–2 km long), and in a random regular interval along seven stratified survey blocks in all the possible habitats and vegetation associations with an altitudinal sequence following Mueller-Domboise and Ellenberg (1974) and Rutledge (1982). Sampling of vegetation was done within 203 plots in the entire study area along 7 fixed study blocks, 7 line transects, and 52 (2 × 20 m) belt transects. The optimum size of vegetation sampling plots was chosen from the species-area curve.

At every sampling plot, data were collected on the vegetation and habitat parameters such as broad habitat/vegetation type (riverine, grassland/scrub, pine *Parrotiopsis*, mixed coniferous, mixed woodland, mixed *Morus*, mixed oak, and grassy/rocky slopes), altitude (in m) using altimeter and GPS, aspect (north, northeast,

east, south, southwest, west, northwest) using compass and GPS, and slope (flat 0–16^0, gentle 16–25^0, steep 25–34^0, very steep 34–50^0) based on visual estimation. In these plots, trees (the number of species and the number of individuals for each species) were quantified in circular plots (10.3 m radius), shrubs (the number of species and the number of individuals for each species) were quantified in 5 m circular plots, and ground cover (percentage of grass, herb, litter, bare ground rock, and snow) was quantified by line-intercept method (1 m tape) in four different directions following Mueller-Domboise and Ellenberg (1974) and Rikhari et al. (1989).

Habitat types were defined based on the predominant tree species or in other cases by the name that provided information about the main habitat features (such as riverine, grassland/scrub, etc.).

5.4 Analysis

Preliminary analysis was done to obtain the values of various parameters such as density, percentage of frequency, and abundance of vegetation (trees, shrubs, and ground cover) following Mueller-Domboise and Ellenberg (1974). The vegetation diversity indices, rarefaction (Krebs 1989), cluster analyses, and measurements of various diversity indices (Shannon-Weiner, Simpson, and Hill's number) were done using BioDiversity Pro (2000) windows-based software package developed by Neil McAleece for the Natural History Museum, London. Twinspan (Hill 1979) (a two-way indicator species analysis) was used for numerical and hierarchical classification of vegetation communities. Vegetation data were also analyzed for plant community classification (Mueller-Domboise and Ellenberg 1974) and hierarchical cluster analysis scheme of SPSS following Norris (1990). One-way analysis of variance (ANOVA) was used to find out the significant differences in the vegetation between different seasons and habitat types. All such statistical analyses were performed on computer program SPSS (Norris 1990).

5.5 Vegetation and Habitat Mapping

The Survey of India (SOI) topographic maps of Dachigam National Park of the series published on a scale of 1:50,000 was digitized using ArcInfo software. Delineation of the vegetation types based on hybrid classification (unsupervised, supervised, and rule-based classification) in the remotely sensed satellite data IRS-1D LISS-IV of 2012 into the micro-physiognomic units was done. Each category gave a unique combination of the tone, texture patterns, and DM values that helped in identification of various vegetation classes. The vegetation map of Dachigam National Park was generated using maximum likelihood classification approach of hybrid classification scheme. Digital data on contour and drainage were used to create digital elevation model (DEM) on the basis of interpolation. This elevation model was developed at 100 × 100 m spatial resolution and provided surrogate for altitude, aspect, and slope. Similarly, information on the availability and area proportion of altitude, slope, and aspect categories was extracted from a digital elevation model using ERDAS and ArcView 3.1 software program of GIS (Arc view Software 1996; Erdas 2000).

Land-use maps showing drainage, slope, aspect, altitude, study blocks, and transects were generated by using the toposheets and satellite imageries of IRS-1D LISS-IV of 2012 and identifying the ground features to position the sampling plots, study blocks, and transects. The location and directional layout of the study blocks and transects were denoted on the toposheet with the help of geographical positioning system (GPS) readings taken in the field. Further ground truthing was done for unsupervised classification and in vegetation types known apriority from Dachigam National Park.

Several steps were followed for the land-use classification and mapping. Firstly, the specific feature classes have been identified based on the visual interpretation of the satellite imagery, and then digital data analysis was carried out using ERDAS IMAGINE software. Land-use maps of 1972, 1981, 1992, 2001, and 2016 were prepared by employing supervised classification using maximum likelihood algorithm and parallelepiped nonparametric rule method. The land-use/land-cover classes include agriculture land, forest, settlements, rivers, pastures, scrub, rocky, snow, and built-up. A thorough field survey of the area was conducted to collect information about the land-use types prevailing in the area. Much information related to the agriculture, forest area, built-up, and other land-use practices have been obtained during the field visit. Global positioning system (GPS) was used to conduct ground verification. Later, change detection was done for the images to find out the changes that have taken place in the study. For monitoring the changes in land use/land cover and vegetation change, NDVI differencing method was used on Landsat-3 of 1981, Landsat-5 digital data of 1992, Landsat-7 ETM+ digital data of 2001, and IRS LISS-IV of 2016.

The effortlessness of calculation and interpretation of various types of satellite data has made NDVI a popular spectral vegetation index. Categorically, the NDVI is a function of two bands: the red band and the near-infrared spectral band. The NDVI differencing method is common and effective in change detection of vegetation changes. The main idea of this method is vegetation index is produced separately and then the second-date vegetation index is subtracted from the first-date vegetation index. This method has the advantage of emphasizing differences in the spectral response of different features and reduces impact of topographic effects and illumination, while the disadvantage is enhanced random noise or coherent noise. The NDVI, as mentioned earlier, is a function of two bands: the red band and near-infrared spectral band. It is calculated for both images using the following relationship:

$$(\text{NDVI}) = (\text{NIR} - \text{RED}) / (\text{NIR} + \text{RED})$$

where NIR is near-infrared band and RED is red band.

The annual rate of forest change was assessed using the compound interest formula that was used due to its explicit biological meaning [22]. This is as follows: $r = (1/(t2 - t1) \times \ln(a2/a1)) \times 100$ where r is the percentage of forest cover loss (annual rate of change), a2 is the area cover in current year, a1 is area in base year, t2 is the current year, t1 is the base year, and ln is natural logarithm.

5.6 Observations

During the study period, forty-four (44) woody species were encountered and sampled in the study area, out of which 22 species were trees and 22 species were shrubs. Prominent tree species recorded were *Pinus wallichiana*, *Morus alba*, *Celtis australis*, *Parrotiopsis jacquemontiana*, *Abies pindrow*, *Rhus succidiadiana*, and *Robinia pseudoacacia* with mean densities (no./ha) of 56.05 ± 10.45 S.E. (std. error), 31.20 ± 5.29 S.E., 28.84 ± 4.70 S.E., 23.66 ± 3.59 S.E., 19.08 ± 5.29 S.E., 16.71 ± 3.54 S.E., and 11.39 ± 3.98 S.E., respectively (Table 5.1) The dominant shrub species were *Indigofera heterantha*, *Isodon plectranthus*, *Prunus* spp., *Rhododendron campanulatum*, *Viburnum cotinifolium*, *Rosa webbiana*, and *Juniperus recurva* (Table 5.1).

Vegetation of Dachigam National Park was classified using Twinspan analysis into five broader communities (Fig. 5.2), viz., mixed riverine, mixed woodland, pine *Parrotiopsis*, grassland/scrub, and alpine scrub. These communities were further classified into nine habitat types growing in an altitudinal sequence based on the predominant vegetation types present, viz., (1) mixed broad-leaved riverine forest between 1700 and 1900 m; (2) pine *Parrotiopsis* forest comprising of blue pine (*Pinus wallichiana*) forest associated with *Parrotiopsis jacquemontiana* between 1900 and 2500 m; (3) grassland and open *Parrotiopsis* scrub between 1900 and 2500 m; (4) mixed coniferous forest comprised of coniferous association of blue pine, *Taxus wallichiana*, *Picea smithiana*, and silver fir *Abies pindrow* between 2500 and 3000 m; (5) subalpine forest of blue pine and silver fir 2900–3300 m; (6) birch (*Betula utilis*) and rhododendron (*Rhododendron campanulatum*) scrub 3000–3500 m; (7) alpine juniper (*Juniperus recurva*) scrub between 3300 and 3900 m; and (9) alpine meadows.

Of these, only the first three vegetation communities, viz., (1) mixed broad-leaved riverine forest between 1700 and 1900 m, (2) pine *Parrotiopsis* forest comprising of blue pine (*Pinus wallichiana*) forest associated with *Parrotiopsis jacquemontiana* between 1900 and 2500 m, and (3) grassland and open *Parrotiopsis* scrub between 1900 and 2500 m, represent the flora of Lower Dachigam, whereas the mixed coniferous forest community represents the edge community between Lower and Upper Dachigam, and the other communities represent the flora of Upper Dachigam.

Twinspan (a two-way indicator species analysis) has been used for numerical hierarchical classification of vegetation, and the technique is based on the concept that a group of samples, which constitutes a vegetation community type, would have a corresponding group of species characterizing the group. Since reciprocal averaging

Table 5.1 Mean density (per hectare) of tree and shrub species in Dachigam National Park (Feb. 01–Dec. 2010)

Tree species	Mean density/ha	Std. error	Shrub species	Mean density/ha	Std. error
Robinia pseudoacacia	11.39	3.98	*Indigofera heterantha*	1339.15	246.21
Prunus armeniaca	3.55	0.9	*Prunus cerasifera*	128.64	33.42
Salix babylonica	1.92	1.24	*Prunus persica*	10.04	6.68
Acer caesium	2.96	1.47	*Prunus armeniaca*	13.18	4.02
Morus alba	31.2	5.29	*Rosa webbiana*	47.69	15.35
Juglans regia	6.21	1.54	*Isodon plectranthus*	294.31	75.27
Celtis australis	28.84	4.7	*Prunus prostrata*	17.57	5.84
Ulmus wallichiana	7.25	1.77	*Vitis vinifera*	21.34	5.98
Ulmus laevigata	1.92	0.94	*Rosa macrophylla*	5.65	2.99
Aesculus indica	9.02	3.07	*Crataegus oxyacantha*	25.73	11.58
Parrotiopsis jacquemontiana	23.66	3.59	*Berberis lyceum*	81.58	17.76
Quercus robber	5.92	2.64	*Kerria japonica*	2.51	1.98
Platanus orientalis	1.63	1.36	*Indigofera* sp.	15.06	12.79
Pyrus malus	0.15	0.15	*Crataegus monogyna*	5.02	3.19
Rhus succidiadiana	16.71	3.54	*Cotoneaster nummularius.*	5.02	5.02
Pinus wallichiana	56.05	10.45	*Lonicera quinquelocularis*	6.28	3.05
Salix alba	7.1	3.28	*Viburnum cotinifolium*	48.32	20.3
Populus alba	0.59	0.47	*Rhododendron campanulatum*	111.07	36.99
Pyrus communis	3.84	1.36	*Juniperus recurva*	43.3	29.03
Taxus wallichiana	3.7	2.13			
Abies pindrow	19.08	5.29			
Betula utilis	7.99	4.01			

Average mean tree density/ha = 11.39 std. error 2.87, average mean shrub density/ha = 112.33 std. error 27.57

(RA) helped to arrange the species and samples in a way that could best express the relationship, RA has been used as the basis of Twinspan classification (Fig. 5.2).

The tree and shrub densities/ha differed significantly ($F = 20.02$, $P = 0.001$, $F = 10.36$, $P = 0.001$, respectively) between different habitat types. The maximum tree density/ha (442.78 ± 18.89 S.E.) was recorded in mixed oak and pine *Parrotiopsis* (430.99 ± 58.92 S.E.) habitats (Table 5.2; Fig. 5.3).

The mean tree densities in pine *Parrotiopsis* and mixed conifer were 430.99 ± 58.92 S.E. and 388.67 ± 42.73 S.E. compared to very low tree densities in grassland/scrub (59.55 ± 11.85) habitat. *Pinus wallichiana* was the most dominant tree species with density of 56.05 ± 10.45 S.E. This dominance of evergreens in the temperate zones of Himalayas has been shown to be due to their moist but cool and frost-prone climate that favors emergence and development of an evergreen vegetation (Saxena et al. 1982).

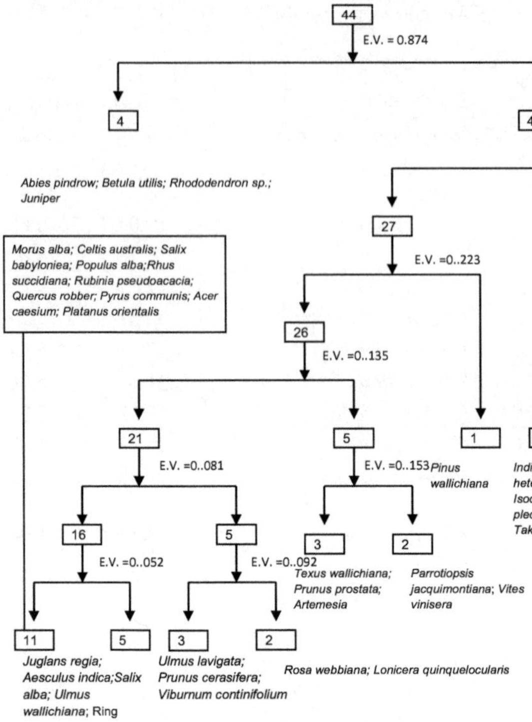

Fig. 5.2 Twinspan Classification of Vegetation of Dachigam National Park

The riverine and mixed woodland habitats of DNP showed the highest values of diversity indices (Table 5.3), whereas rarefaction values were highest in case of mixed woodland and mixed coniferous habitats (Fig. 5.4). The diversity indices of shrub vegetation in different habitat types of DNP are given in Table 5.4. Rarefaction values for shrubs are depicted in Fig. 5.5.

Maximum tree diversity indices (Table 5.5) were recorded in mixed woodland and riverine habitats, whereas maximum shrub diversity indices were recorded in pine *Parrotiopsis* and mixed coniferous habitats of DNP. The lowest tree and shrub diversity indices were recorded in grassy/rocky slope and alpine scrub habitats. Diversity indices actually explain both richness and evenness of species into a single value. The greater the chance that two randomly picked individuals in a community were of the same species, the less diverse is the community (Pielov 1975). The purpose of measuring habitat diversity was to judge its relationship with either to vegetation community properties or to the environmental conditions to which habitat was exposed. The changes in vegetation structure with the changing land-use patterns have been studied in the Himalayas. The heavily grazed habitats have been shown to have preponderance of therophytes (Yadava and Singh 1977).

The Shannon index (H) is the measure of average degree of uncertainty in predicting as to what species of individual chosen at random forms a collection of “S”

Table 5.2 Tree and shrub mean densities (no./ha) in different habitats of Dachigam National Park (February 1 to December 2010)

	Habitat type	Number of sample plots	Mean density/ha	Std. deviation	S.E.
Trees	Riverine	27	327.99	204.77	39.41
	Mixed woodland	35	391.96	152.36	25.75
	Grassland/scrub	62	59.55	93.28	11.85
	Pine *Parrotiopsis*	14	430.99	220.47	58.92
	Mixed coniferous	19	388.67	186.26	42.73
	Grassy/rocky slopes	9	326.87	262.32	87.44
	Alpine scrub	27	153.43	204.67	39.39
	Mixed oak	4	442.78	37.77	18.89
	Mixed *Morus*	6	410.26	199.73	81.54
	Total	203	251.24	223.16	15.66
Shrubs	Riverine	27	561.45	738.80	142.18
	Mixed woodland	35	676.98	930.37	157.26
	Grassland/scrub	62	5814.67	6401.47	812.99
	Pine *Parrotiopsis*	14	727.93	908.26	242.74
	Mixed coniferous	19	851.49	1462.38	335.49
	Grassy/rocky slopes	9	226.47	430.95	143.65
	Alpine scrub	27	901.16	1651.87	317.90
	Mixed oak	4	350.32	700.64	350.32
	Mixed *Morus*	6	530.79	472.66	192.96
	Total	203	2249.69	4341.10	304.69

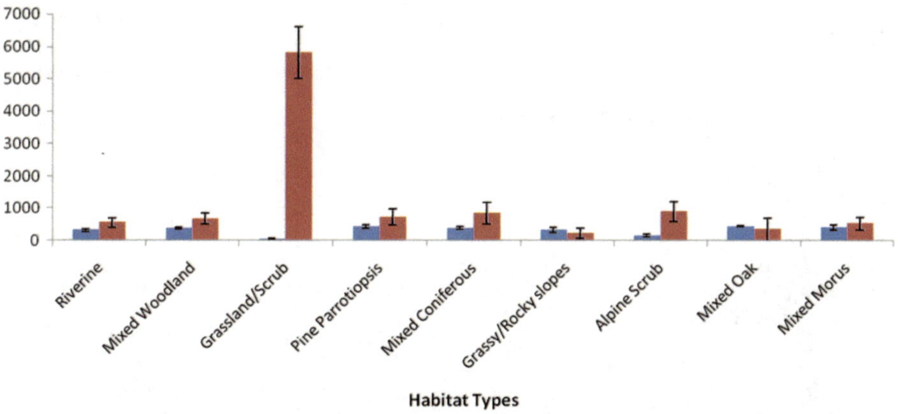

Fig. 5.3 Density of trees and shrubs in different habitat types in Dachigam National Park

species of "N" individuals. It was found that uncertainty and hence species richness were the highest in riverine and mixed woodland habitats for trees (Table 5.5) and riverine, grassland/scrub, and mixed woodland habitats for shrubs. Simpson index gives the probability that two individuals drawn at random from a population belong to the same species. If the probability is high that both individuals belong to the

Table 5.3 Diversity indices of trees in different habitat types of Dachigam National Park (February 1 to December 2010)

Index	Riverine	Mixed woodland	GL/ SC	Pine prt.	Coniferous	Grassy/ rocky slopes	Alpine scrub	Mixed oak	Mixed *Morus*
Shannon H′ log base 10	1.275	1.028	0.528	0.527	0.802	0.089	0.31	0.513	0.454
Shannon Hmax log base 10	1.362	1.301	1.114	1.146	1.114	0.301	0.477	0.845	0.903
Simpson diversity (1/D)	32.805	10.956	2.533	2.531	5.812	1.133	1.951	2.549	2.205
Hill's number H0	11	19	12	12	12	1	4	10	7
Hill's number H1	32.965	51.895	28.467	5.467	17.013	1.443	6.056	16.094	6.971

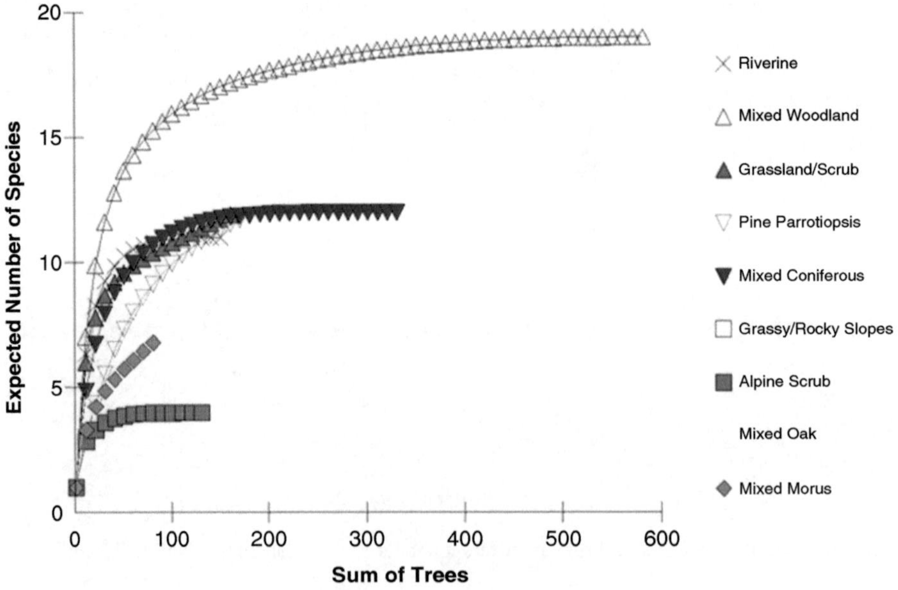

Fig. 5.4 Rarefaction plot for trees in Dachigam National Park

Table 5.4 Diversity indices of shrubs in different habitat types of Dachigam National Park (Feb. 01 to Dec. 2010)

Index	Riverine	Mixed woodland	GL/SC	Pine prt.	Coniferous	Grassy/rocky slopes	Alpine scrub	Mixed oak	Mixed *Morus*
Shannon H′ log base 10	1.28	0.72	0.49	0.77	0.64	0.28	0.45	0.22	0.38
Shannon Hmax log base 10	1.30	1.08	1.26	1.00	1.00	0.30	0.48	0.70	0.95
Simpson diversity (1/D)	32.36	7.36	2.03	6.63	3.64	2.06	3.04	1.29	1.59
Hill's number H0	6	11	17	9	9	1	2	4	8
Hill's number H1	13.763	12.117	5.701	21.402	19.923	1.443	3.585	8.64	21.423

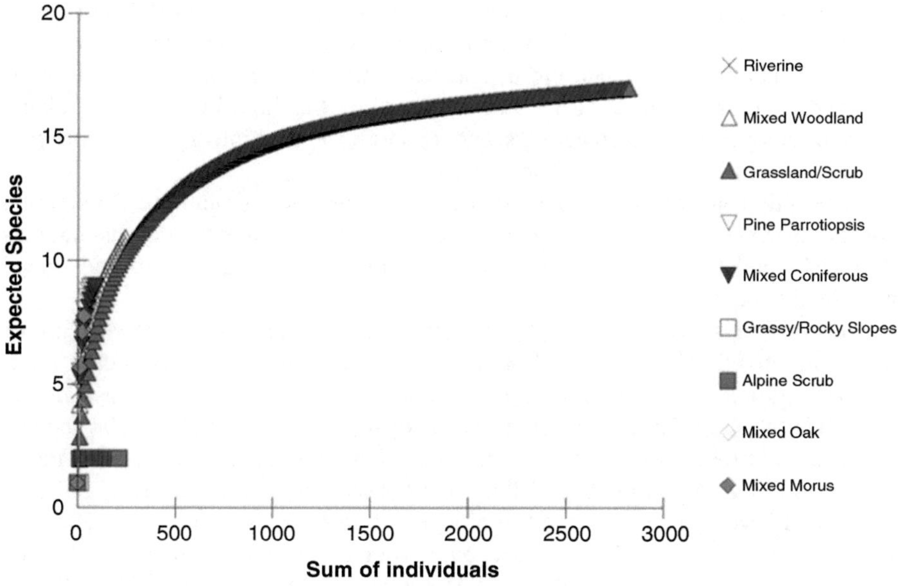

Fig. 5.5 Rarefaction plot for shrubs in Dachigam National Park

Table 5.5 Change detection of land use/land cover over the years of Dachigam National Park

	Area (km^2)					
Landuse/Landcover	2016	2001	1992	1981	1972	% Change
Agricultural area	1.82	1.75	1.79	1.62	1.77	2.75
Riverine forest	10.28	10.35	10.33	10.34	10.35	−0.68
Blue pine forest	32.56	32.56	33.1	33.2	34.1	−4.75
Blue pine/silver fir forests	14.25	14.25	14.29	14.32	14.38	−0.91
Pure silver fir forests	25.75	25.75	25.76	25.8	25.9	−0.58
Broad-leaved forests	11.18	11.25	11.29	11.35	11.37	−1.7
Birch forests	2.8	2.8	2.8	2.9	2.9	−3.57
Alpine pastures	22.92	23.32	23.4	23.9	24.3	−6.02
Open scrubs	28.25	28.25	28.12	27.9	26.61	5.81
Rocky cliffs	5.7	5.5	4.2	4.5	4.4	22.81
Permanent snow	1.68	1.5	2.3	2.5	2.44	−45.24
Waterways	3.25	3.25	3.25	3.25	3.25	0
Blank forest	2.33	2.25	2.14	1.19	1	57.08
Total	162.77	162.77	162.77	162.77	162.77	

same species (i.e., Simpson diversity value is high), then the diversity of community is low. However, Simpson diversity (1/D) is the reverse, and the higher the Simpson 1/D values, the higher is the community diversity. Simpson diversity (1/D) was high in riverine, mixed woodland for trees and riverine, mixed woodland and pine *Parrotiopsis* habitats for shrubs (Tables 5.5). Hill's effective number (H1 number) is a measure of the degree to which proportional abundance is distributed among the species. It is as such the measure of species in the sample where each species is weighed by its abundance. Hill's number for trees was high in mixed woodland, riverine, and grassland/scrub habitats, and for shrubs, it was high in pine *Parrotiopsis* and mixed coniferous habitats (Tables 5.5).

Rarefaction analysis represents the species richness in a community. Rarefaction curves for trees were high in mixed woodland and mixed coniferous habitats (Fig. 5.4), and for shrubs, the curves were high for grassland/scrub and mixed woodland habitats (Fig. 5.5).

Cluster analysis (Bray-Curtis) of trees showed close similarities between alpine scrub and mixed oak habitats followed by pine *Parrotiopsis* and mixed coniferous habitats and maximum dissimilarities between mixed woodland and grassy/rocky slope habitats (Fig. 5.6), whereas cluster analysis (Bray-Curtis) of shrubs showed maximum similarities between riverine and mixed coniferous habitats and maximum dissimilarities between mixed oak and grassland/scrub habitats (Fig. 5.7).

Ground cover (percent) showed significant differences {(grass/herb $F = 333.483$, $P = 0.001$; litter $F = 74.37$, $P = 0.001$; bare ground $F = 3.99$, $P = 0.001$; rock 4.89, $P = 0.002$; snow $F = 106.\ 06$, $P = 0.001$)} between different seasons (Fig. 5.8) and habitat types {(grass/herb $F = 13.18$, $P = 0.00$; litter $F = 12.93$, $P = 0.001$; bare ground $F = 4.74$, $P = 0.001$; rock 5.95, $P = 0.002$; snow $F = 29.01$, $P = 0.001$)} (Fig. 5.9).

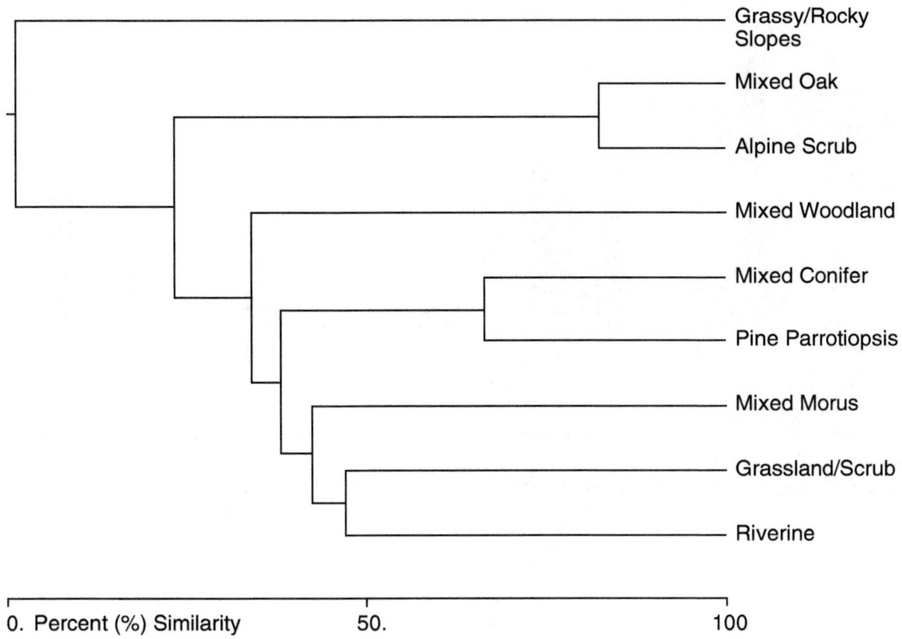

Fig. 5.6 Cluster analysis (Bray-Curtis) of Trees in Dachigam National Park

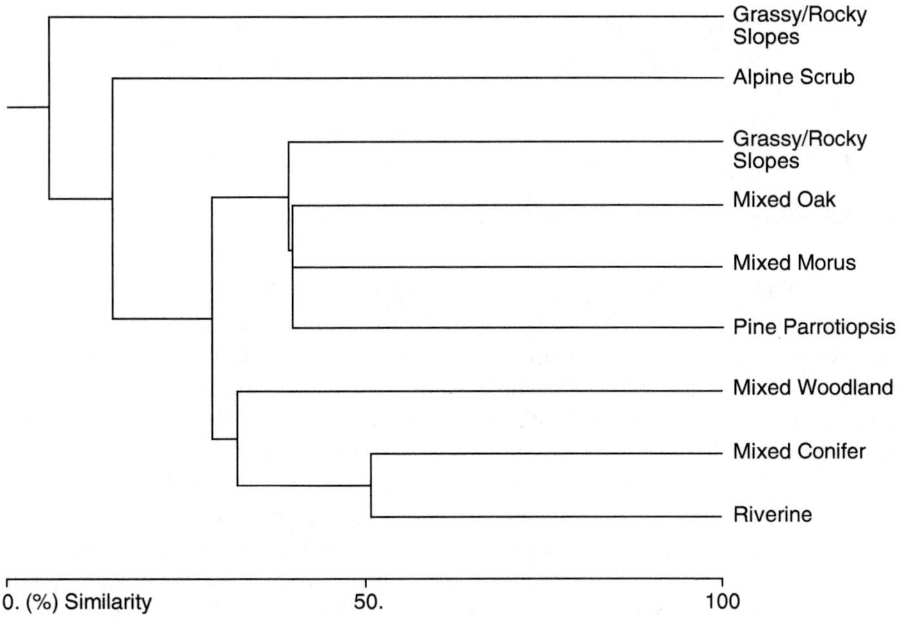

Fig. 5.7 Cluster analysis (Bray-Curtis) of shrubs in Dachigam National Park

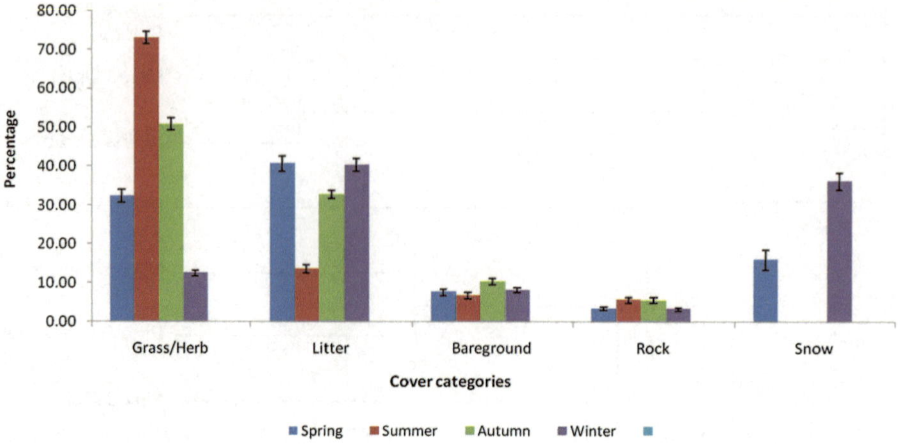

Fig. 5.8 Percentage ground cover in different seasons in Dachigam National Park

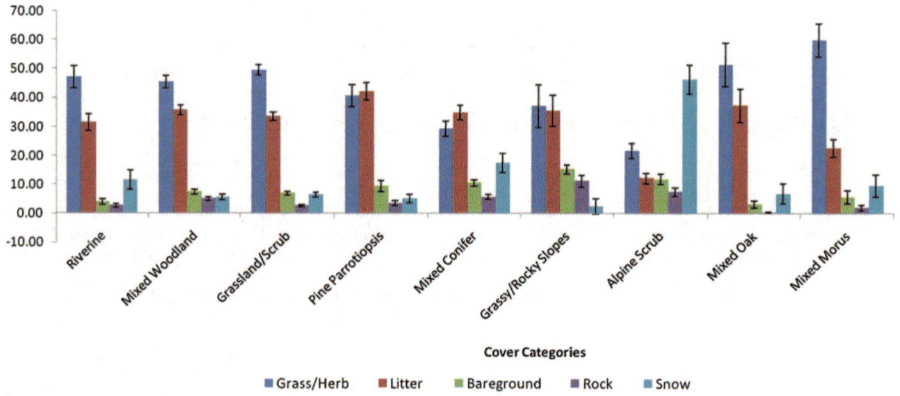

Fig. 5.9 Percentage ground cover in different habitat types of Dachigam National Park

The DEM (Fig. 5.10) slope (Fig. 5.11), aspect maps (Fig. 5.12), and vegetation and habitat map (Fig. 5.13) of Dachigam National Park were prepared in ERDAS and ArcView software using both satellite data and toposheets. Satellite data have been widely used for study of wildlife habitat and vegetation monitoring in many parts of the world, and the use of satellite data in India and abroad has been standardized with reliable accuracy in mapping (Botkin et al. 1984; Hilderbrandt 1986; Roy et al. 1991, 1992, 1995). Preparation of animal distribution and habitat maps is one of the prerequisites, which enable the Park management to implement conservation actions effectively. Spatially explicit maps are considered more advantageous than absolute numbers and help in easily advocating for required wildlife management and conservation implications (Buckland and Elston 1993; Cardillo et al. 1999; Boone and Krohn 2000).

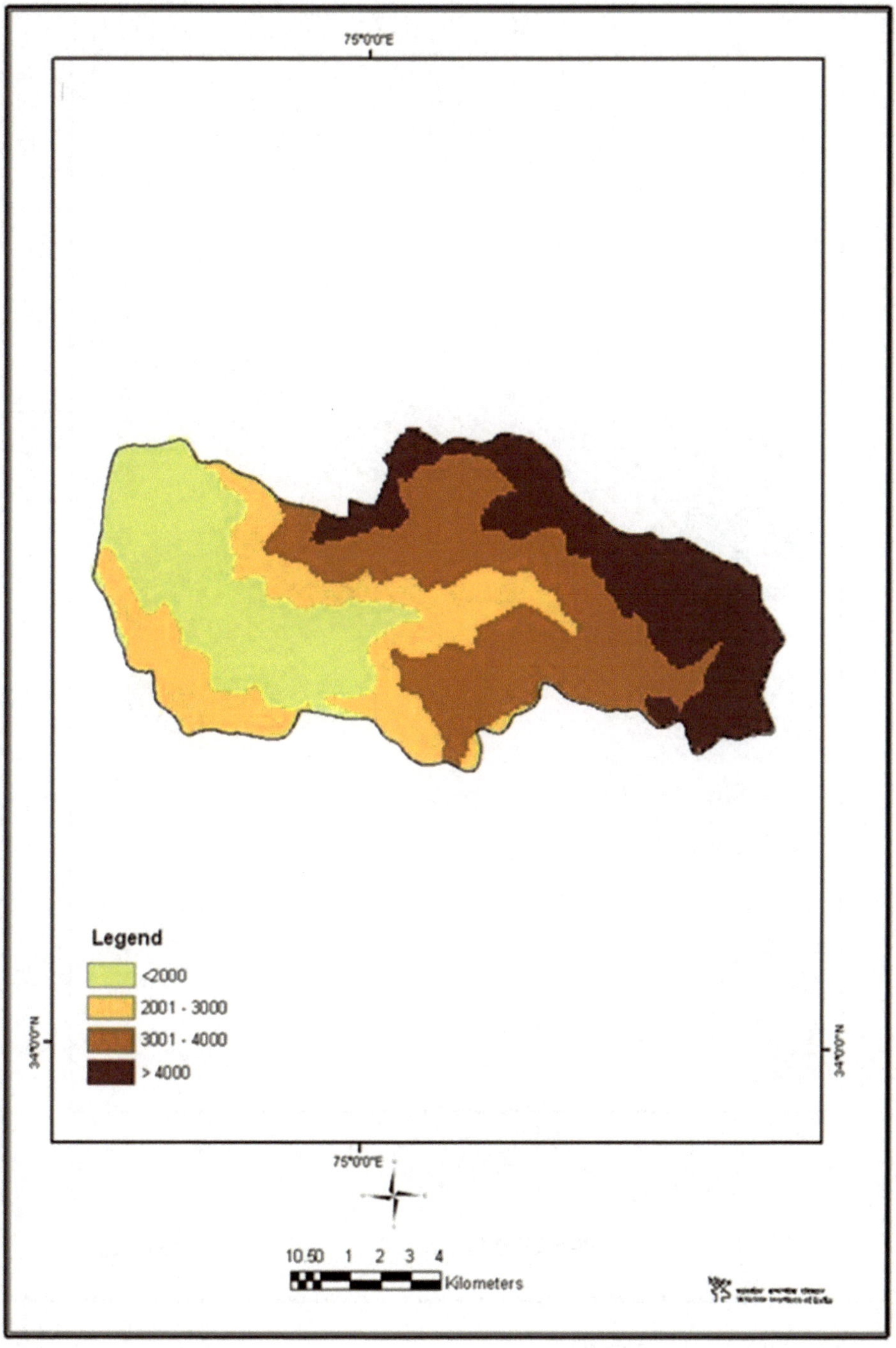

Fig. 5.10 DEM of Dachigam N.P

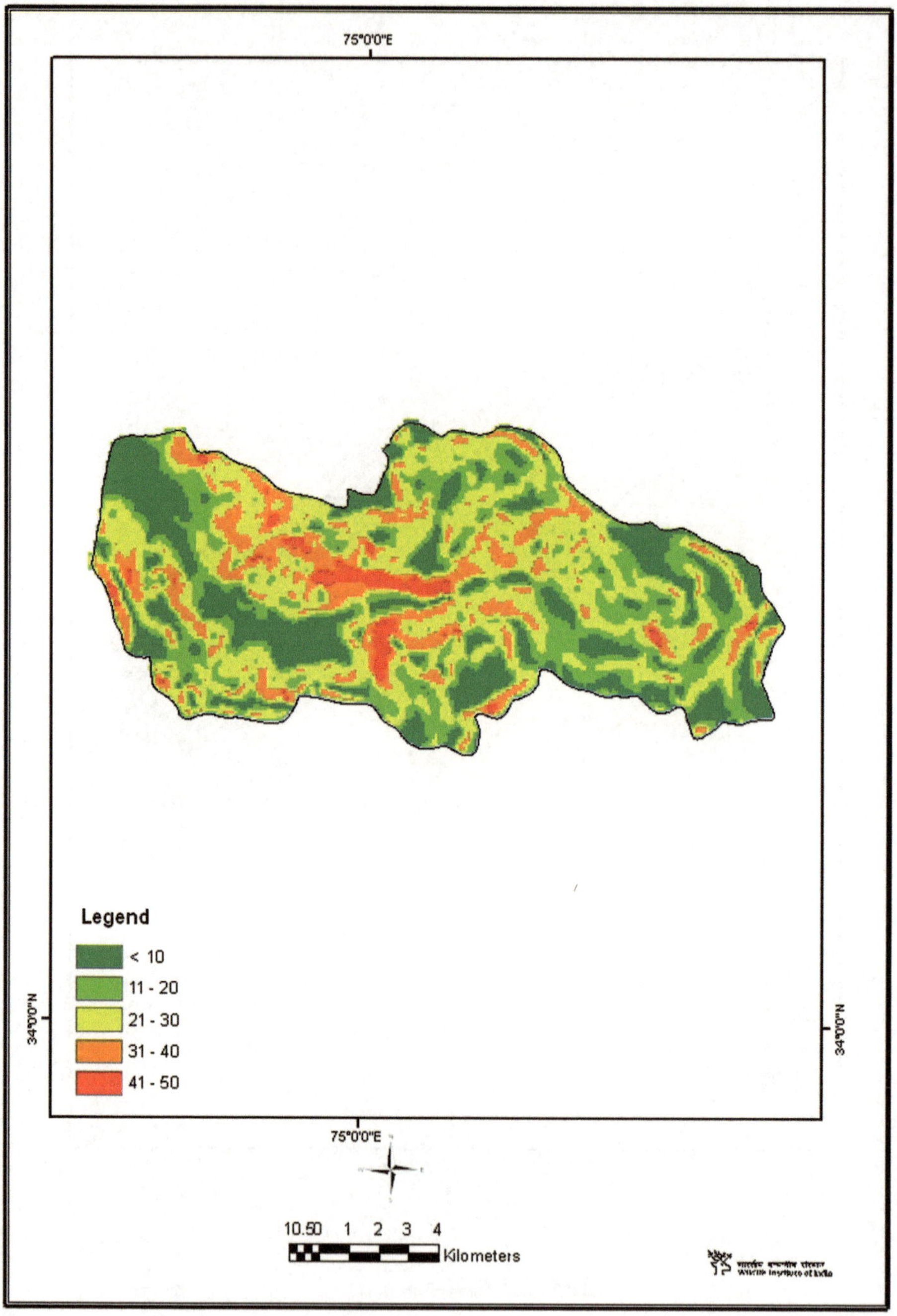

Fig. 5.11 Slope Map of Dachigam N.P

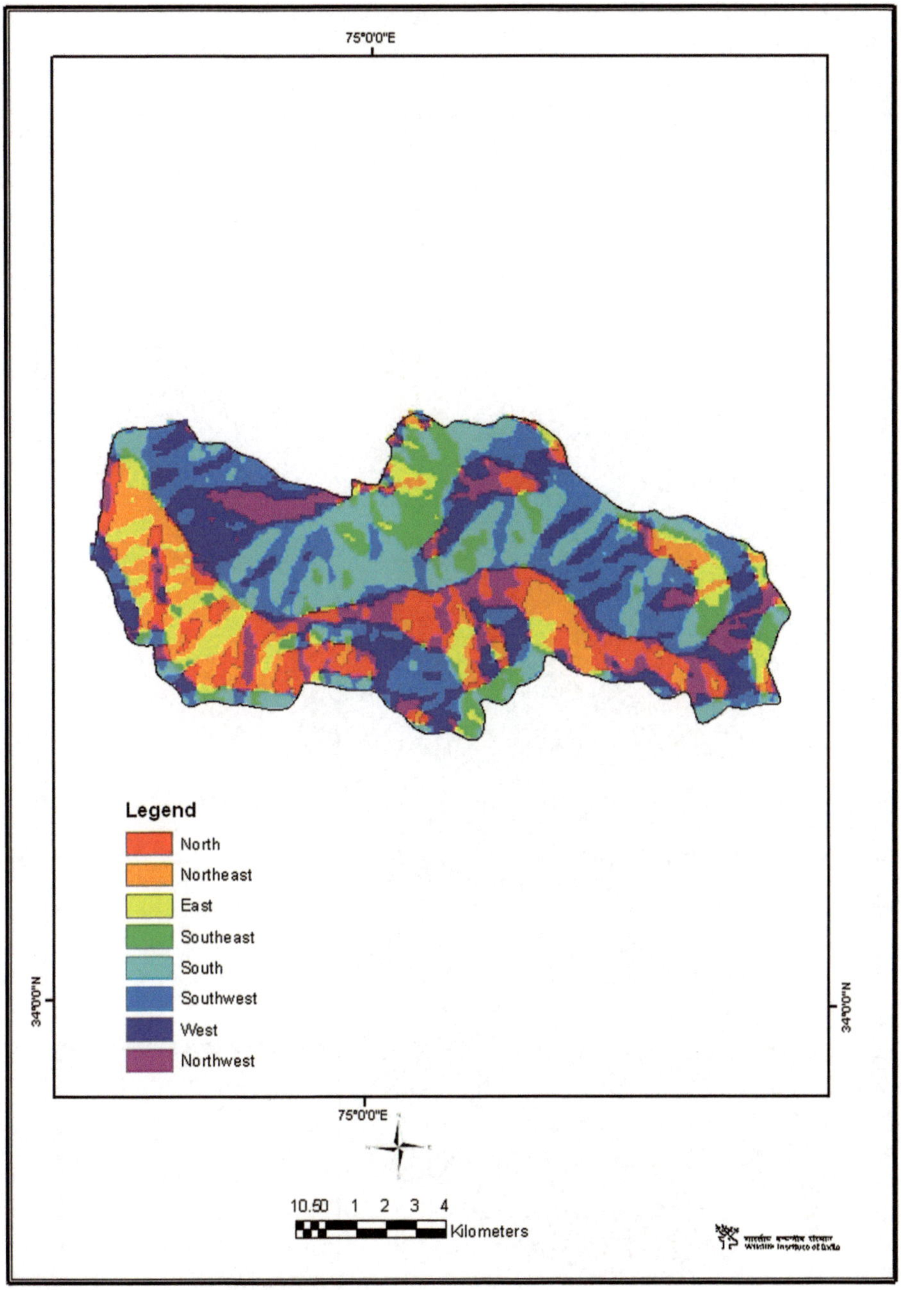

Fig. 5.12 Aspect Map of Dachigam N.P

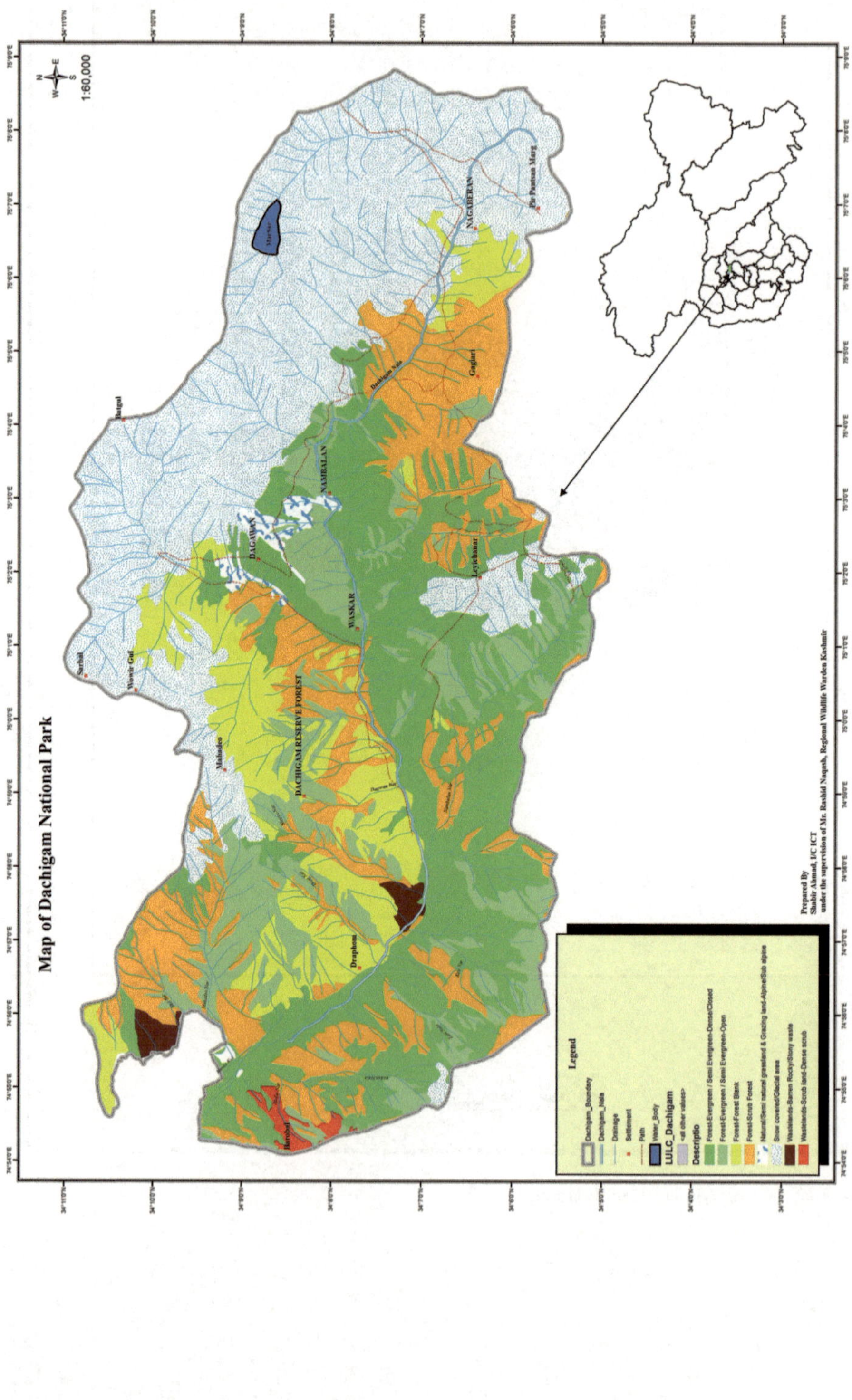

Fig. 5.13 Vegetation map of Dachigam National Park, Kashmir, India

The findings of this study are beyond the scope of comparisons with the earlier vegetation studies of Singh and Kachroo (1978) and Bhat (1985) as those studies do not present any information with regard to plant species density, diversity, hierarchal classification, vegetation, and habitat mapping in GIS domain.

The supervised classification and NDVI analysis of satellite images of 1972, 1981, 1992, 2001, and 2016 indicated 13 vegetation types in DNP (Table 5.5; Fig. 5.10). The result of this process was a set of 8-bit gray scale NDVI images (Fig. 5.14) representing the amount of vegetation present at each time. Examining a gray scale of the NDVI for each successive year was a visually simplistic way to analyze the progression of vegetation, the regrowth area, and the remover field area or planting field over the 4-year period. In the first two images in the third set of Fig. 5.14, areas with healthy vegetation are white, while areas that are gray have little or no vegetation, and areas where no vegetation exists are black. The white area which represents vegetated areas has stronger near-infrared reflectance. This means that most of the visible light was used for product biomass, thereby producing NDVI values ranging between 0.2 and 1. This represents regions of plants with good condition, high leaf biomass, canopy closure, and vegetation with high chlorophyll content. Conversely, negative NDVI values were recorded in a dark area. This is as a result of the fact that features reflect more in the visible band than they do in the near-infrared band, indicating regions of low vegetation, typical water, cloud, bare soil, and rock. The rectangular-shaped area immediately above the lake in the fourth set of Fig. 5.14 clearly depicts the boundary of agricultural field of the image acquired in 1986. The land was left without any activity on it for nearly 4 years, thereby giving way for natural vegetation to grow. It can be observed that the boundary is no longer visible from the image acquired in 1990. These areas were closed-off and abandoned for over 5 years which gives rise to a most likely change from agricultural land to noncultivation area rather than phonological changes in vegetation. To indicate more subtle details, the 8-bit gray scale NDVI results of both images were colored using a green color table. The black areas represent regions of plants with good condition, high leaf biomass, canopy closure, and vegetation with high chlorophyll content. On the contrary, the white areas indicate regions of low vegetation, typical water, cloud, bare soil, and rock.

Based on selective field checks, the overall classification accuracy of the LULC maps derived from the satellite images was determined, which ranged from 90 to 95%. The kappa coefficient ranged from 0.87 to 0.93. The area computations from temporal data indicated that the area under open scrub in 1972, 1982, 1992, 2001, and 2016 was 26.61 km^2, 27.90 km^2, 28.12 km^2, 28.25 km^2, and 28.25 km^2, respectively. Thus, over the years from 1972 to 2016, the area under open scrub has grown by about 5.81%. The results also revealed that area under alpine pastures has significantly reduced by 6.02%, followed by the blue pine forests by 4.75% and birch forests by about 3.57%. Gradual changes have also been found in riverine forests and broad-leaved forest. Agriculture land shows a definite though not significant increase in the area at 2.57%. The classes such as rocky cliffs, permanent snow, and blank forests were not considered as coverage of snow in images during different years can change its land cover within each other.

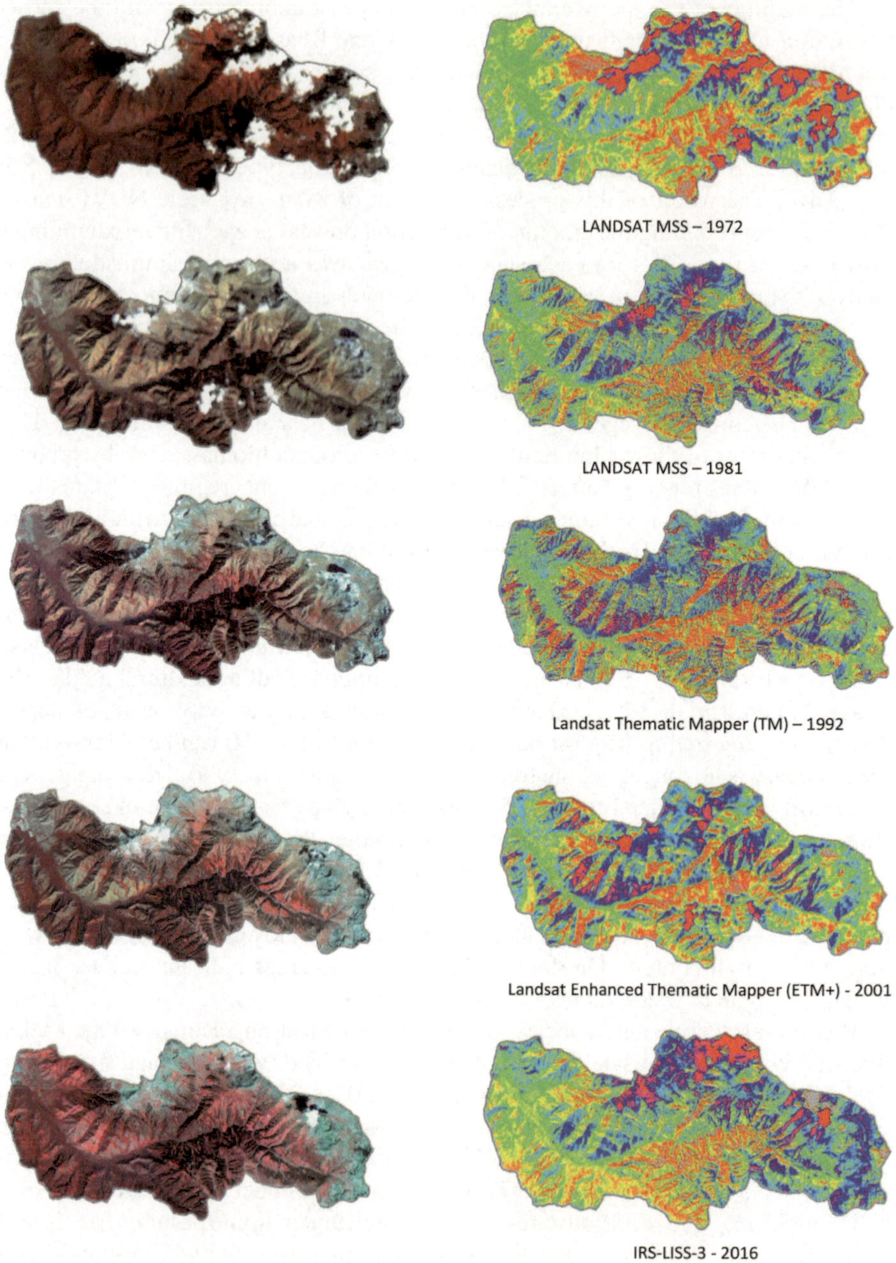

Fig. 5.14 NDVI of temporal multispectral data of Dachigam National Park

Acknowledgments We sincerely acknowledge the Department of Wildlife Protection, Jammu and Kashmir Government, and Ministry of Environment, Forests, and Climate Change, Government of India, New Delhi, for sponsoring the research studies in Dachigam National Park.

References

Ahmad K (2006) Aspects of Ecology of Hangul (Cervus elaphus hanglu) in Dachigam National Park, Kashmir, India. Ph.D. thesis, Forest Research Institute University, Dehradun. 220p

Ahmad K, Sathyakumar S, Qureshi Q (2009) Conservation status of the Last Surviving Wild population of Hangul or Kashmir Deer (*Cervus elaphus hanglu*) in Kashmir. India J Bomb Nat Hist Soc 106(1):63–71

Ahmad K, Qureshi Q, Agoramoorthy G, Nigam P (2015) Habitat use and food habits of Kashmir Red deer *Cervus elaphus hanglu* in Dachigam National Park, Kashmir, India. Ethol Ecol Evol 28(1):85–101

Andrew NG, Bleich VC, August PVT, Steven G (1997) Demography of mountain sheep in the east Chocolate Mountains, California. Calif Fish Game 83(2):68–77

Anonymous (1985) Ecological-cum-management plan for Dachigam National Park 1985-90. Jammu and Kashmir Wildlife Department, Directorate of Wildlife Protection Srinagar

Arc view Software (1996) Arc view 3.1. extension software. VSGS-BRD. Alaska Biological Science Centre

Barnes BV, Zak DR, Denton SR, Spurr SH (1998) Forest ecology, 4th edn. Wiley, New York

Botkin DB, Estes JE, MacDonald RM, Wilson MV (1984) Studying the earth, vegetation, forest space. Biosciences 34:508–511

Boone RB, Krohn WB (2000) Predicting broad—scale occurrence of vertebrates in patchy landscapes. Landsc Ecol 15:63–74

Bhat GA (1985) Biological studies of Grassland of Dachigam National Park, Kashmir. Ph.D. thesis, CORD, University of Kashmir, Srinagar. 521p

Bhat GA, Qadri MY, Zutshi DP (2002) An ecological survey of Dachigam National Park, Kashmir, with special emphasis on grasslands. In: Pandit AK (ed) Natural Resources of Western Himalaya. Valley Book House Publications, Srinagar, pp 341–376

Brook SM, Thakur M, Ranjitsinh MK, Donnithorne-Tait D. Ahmad K (2017). Cervus hanglu ssp. hanglu. The IUCN Red List of Threatened Species 2017: e.T113259123A113281791. http://dx.doi.org/10.2305/IUCN.UK.2017-2.RLTS.T113259123A113281791.en

Buckland ST, Elston DA (1993) Empirical models for the spatial distribution of wildlife. J Appl Ecol 30:478–495

Cardillo M, Macdonald DW, Rushton SP (1999) Predicting mammal species richness and distribution: testing the effectiveness of satellite derived land cover data. Landsc Ecol 14:423–435

Erdas (2000) Erdas software. Erdas Inc.

Hill MO (1979) TWINSPAN—a FORTRAN Programme for arranging multivariate data in ordered two-day by classification of the individuals and attributes. Cornell University, Itahaca, New York. 90p

Hilderbrandt G (1986) Potential and limitations of remote sensing for forest inventory and mapping. Reutsch Gesclschatt for Luff and faunfahrt, Bonn, pp 165–185

Holloway CW, Wani AR (1970) Management Plan for Dachigam Sanctuary. 1971-75. Cyclostyled (Mimeo) 26p

Holloway CW (1971) The Hangul in Dachigam: a census. Oryx 10(6):373–382

Hunter MLJ (1999) Maintaining biodiversity in forest ecosystems. Cambridge University Press, 698p

Krebs CJ (1989) Ecological methodology. Harper and Row Publication, New York, NY, p 654

Kumar VP, Thakur M, Rajpoot A, Joshi BD, Nigam P, Ahmad K, Kumar D, Goyal SP (2016) Resolving the phylogenetic status and taxonomic relationships of the Hangul (Cervus elaphus hanglu) in the family *Cervidae*. Mitochondrial DNA A DNA Mapp Seq Anal 28(6):835–842

Kurt F (1978) Threatened Deer. Proceedings of IUCN threatened deer programme. Kashmir deer (*Cervus elaphus hanglu*) in Dachigam. IUCN Specialist Group Publications, pp 87–109

Mani A (1981) The climate of Himalaya. In: Lall JS (ed) The Himalaya: aspects of change. Oxford University Press, Indian International Centre, Delhi, pp 3–15

Mueller-Domboise D, Ellenberg H (1974) Aims and methods of vegetation ecology. Wiley, New York, p 547

Norris MJ (1990) SPSS/PC + Statistics 4.0 for IBM PC/XT/AT and PS/2 SPSS International Br. The Netherlands

Pielov EC (1975) Ecological diversity. Wiley, New York

Rawat GS, Chandawat RS, Sathyakumar S (1993) Conservation planning and management of protected areas in the Himalaya. Paper for the training seminar on Conservation of Biological Resources in the Himalaya. UNESCO., GBPIHED. Srinagar (Garhwal). pp 1–17

Rikhari HC, Chandra R, Singh SP (1989) Pattern of species distribution and community characters along a moisture gradient within an Oak zone of Kumaun Himalaya. Proc. Indian National Science Academy, pp 431–438

Rodgers WA, Panwar HS (1988) Planning a wildlife protected area network in India, vol 1. Wildlife Institute of India, Dehradun

Rodgers WA, Panwar HS, Mathur VB (2000) Wildlife Protected Area Network in India: a review. Wildlife Institute of India, Dehradun, India

Roy PS, Ranganath BK, Diwakar PG, Vohra TPS, Bhan SK, Singh IJ, Pandian VC (1991) Tropical forest type mapping and monitoring using remote sensing. Int J Remote Sens 12(11):2205–2225

Roy PS, Porwal MC, Singh S, Negi DS, Kumar K (1992) Grassland mapping using satellite remote sensing in alpine zone. Pilot project in Kinnaur District (H. P.). IIRS Project Report. IIRS, Dehradun

Roy PS, Ravan SA, Rajadhya N, Dass KK, Abhineet J, Singh S (1995) Habitat suitability analysis of *Nemorhaedus goral*—remote sensing and Geographical Information System approach. Curr Sci 69(8):685–691

Rutledge RD (1982) The method of bounded counts: when does it work? J Wildlife Manage 46(3):757–761

Saxena AK, Pandey P, Singh JS (1982) Biological spectrum and other structural functional attributes of the vegetation of Kumaun Himalaya. Vegetatio 49:111–119

Singh G, Kachroo P (1978) Plant community characteristics in Dachigam Sanctuary, Kashmir. Natraj Publications. Dehra Dun

Singh S (1999a) Assessment of flora and habitat diversity and collection of baseline data to monitor vegetation of Great Himalayan National Park Conservation Area. Final Report of Forestry Research and Extension Project. Great Himalayan National Park, Wildlife Institute of India. DehraDun

Singh SK (1999b) A study on the plant community composition and species diversity in Great Himalayan National Park, Western Himalaya. Unpublished Ph.D. thesis, Kumaon University, Nainital, India

Singh SK, Rawat GS (1999) Floral diversity and vegetation structure in Great Himalayan National Park, Western Himalaya. Final Report of Forestry Research and Extension Project. Great Himalayan National Park, Wildlife Institute of India. DehraDun

Singh SK, Rawat GS (2000) Flora of Great Himalayan National Park. Himalchal Pradesh. Bishen Singh Mehra Pal Singh. Dehra Dun. p 304

Yadava PS, Singh JS (1977) Grassland Vegetation: It's structure, function, utilization and management. Today and Tomorrows Printers and Publishers, New Delhi, 132p

Chapter 6
Vegetation Diversity of Kashmir and Its Sustainable Use

Javaid M. Dad

Abstract A unique biospheric unit, in terms of altitudinal, topographic, and edaphic heterogeneity, in northwestern Himalayas, the Kashmir Himalayan region offers vast array of habitats which are characterized by their own suite of species and physiognomic characteristics. While the floristic explorations in the region started nearly two centuries ago, but yet owing to the tough terrain and arduous topography, many of its areas still remain uncharted. The natural vegetation of the region exhibits a greater diversity and varies from rich marsh and aquatic vegetation in plains to lush evergreen conifers on slopes. Occupying the major proportion of its landscapes, its terrestrial vegetation ranges from domesticated plant systems in valley floor to forests and grasslands in mountains. From fulfilling basic human subsistence requirements to economic well-being, this rich floristic diversity also contributes immensely to food and habitat needs of wild and domesticated animals. However, undue human interference via habitat fragmentation, urbanization, ill-planned economic development, overgrazing, and deforestation has affected this region badly and puts numerous species at edge. In this scenario, the sustainable utilization of natural resources assumes greater significance and while authorities have taken certain conservation measures but yet the situation is disappointing. The present paper while providing a snapshot of various conservation threats concludes by listing few conservation measures.

Keywords Kashmir · Himalayas · Flora · Uses

6.1 Introduction

Located in northwestern extremity of Himalayan biodiversity hotspot between 32°10′ and 37°10′N latitudes and 72°30′ and 80°30′E longitudes, Jammu and Kashmir fall politically within Indian union. The area is a scenic collage of charming landscape that includes mountains, valleys, streams, rivers, forests,

J. M. Dad (✉)
Central Institute of Temperate Horticulture, Srinagar, Jammu and Kashmir, India

D. Egamberdieva, M. Öztürk (eds.), *Vegetation of Central Asia and Environs*,
https://doi.org/10.1007/978-3-319-99728-5_6

Fig. 6.1 Map showing the location of Kashmir Himalayas

meadowlands, and other ecological elements. For this reason, the region is aptly called as heaven on earth. Owing to its natural resource profusion, it is also described as the "biomass state of India." The area enjoys strategic importance as it shares its borders with China (east), Pakistan (west), and Afghanistan and Russia (north) (Fig. 6.1). The state has a total area of about 222,236 km^2 (including 120,849 km^2 under the occupation of Pakistan and China). Agriculture is the mainstay of economy with more than 75% of population being directly or indirectly dependent on this sector (Anonymous 2010).

The state is divided into three provinces, viz., Jammu, Kashmir, and Ladakh, which owing to the vast contrasts in topography and climate differ markedly from one another in physiography, culture, and vegetation. But despite these differences, these provinces appear as a unified natural system in their own rights and within their confinement contain a vast and extremely rich physiographic, ethnic, and cultural diversity. Climatologically, Ladakh experiences cold desert-like conditions and with very severe winters and scorching summers and receives little precipitation. At Jammu, climate is subtropical at lower altitudes and moist temperate at higher reaches. On contrary, marked by well-defined seasonality, the Kashmir valley experiences a predominantly four season's temperate climate: winter (Dec. to Feb.), spring (March to May), summer (June to Aug.), and autumn (Sep. to Nov.). These climatic differences reflect in differing vegetation types across three provinces. On a whole, Ladakh offers a short-growing season with extreme climate and chiefly supports xerophytic vegetation. The latest systematic floristic survey of Ladakh estimates the vascular plant taxa at 1180 (Klimeš and Dickore 2006). The

Table 6.1 Synoptic view of the faunal diversity of Jammu and Kashmir

Group	No. of species		%	No. of endemic species	
	India	J and K		India	J and K
Mammals	372	73	20	30	1
Birds	1175	358	30	42	–
Reptiles	399	68	17	132	1
Amphibians	181	14	8	112	1
Butterflies	1500	158	11	–	–
Total	3627	571	16		

Source: Anonymous (2010)

subtropical Jammu grows a mix of dry and deciduous vegetation. Several floristic studies of Jammu, besides presenting floristic accounts (Sharma and Kachroo 1981), have concluded that the province is floristically very rich (Bhellum and Magotra 2012). In contrast, the vegetation of Kashmir displays tremendous variability. From distinctive hydrophytic formations that spread across beds and banks of streams and canals, lakes, and wetlands through various domesticated plant systems in main valley to terrestrial vegetation formations across forests, grasslands, and pasturelands, the natural vegetation of Kashmir is far from uniform. This rich vegetation is the life support of almost whole populace and has a definite bearing on not only on their daily lives but it also contributes to the food and habitat needs of both wild and domesticated animals and thereby supports the amazing faunal diversity of the region (Table 6.1).

6.2 Floristic Features of Kashmir Himalaya

(a) **General Overview**

Located at the junction of Holarctic and Paleotropical floristic realms, Kashmir Himalaya is floristically very rich. Due to little ice age, the original tropical Kashmir Himalayan flora is believed to have been transformed into subtropical-temperate-alpine flora (Puri 1943). Subsequently, floristic elements from both and far off places got added to it via both natural and human dispersal mechanisms and made this flora into an admixture of various phytogeographical regions. This mesmerizing floristic richness attracted everyone from a layman to naturalist, and starting about two centuries back from first documented botanical collections of William Moorcroft to present time, various workers explored this floristic richness. These explorations provided a significant insight into various floristic features, and based on these extant records, an estimate for the angiosperms, gymnosperms, bryophytes, and pteridophytes of Kashmir Himalaya today stands at 2312 species. This translates into about 12% of India's total angiosperm flora, and considering that area-wise the region represents

Table 6.2 Floristic diversity of various taxonomic groups in Kashmir

			Distribution		
No	Taxonomic group		Species	Genera	Families
1	Bryophytes		210	95	41
2	Pteridophytes		90	29	12
3	Gymnosperms		12	08	04
4	Angiosperms	Monocotyledons	410	162	25
		Dicotyledons	1590	548	107
	Total		2312	842	189

Source: Dar and Naqshi (2002)

only 0.48% of the total land area of India is significantly huge. These species are grouped under 842 genera and 189 families (Table 6.2).

As is provided in Table 6.2, lower plants like bryophytes and pteridophytes are poorly represented in Kashmir Himalaya. This is presumably because compared to higher plant groups, only few efforts have been directed to record and document these lower plant groups. In Kashmir Himalaya, the largest genera among liverworts are *Plagiochila*, *Porella*, *Marchantia*, and *Riccia*, while *Brachythecium*, *Bryum*, *Orthotrichum*, *Funaria*, and *Grimmia* constitute largest genera (Dar and Khuroo 2013). Among gymnosperms, the chief genera include *Abies*, *Taxus*, *Ephedra*, and *Juniperus*. These gymnosperms are integral to our forest ecosystems and are of tremendous immense socioeconomic and ecological importance. With over 2000 species (410 monocotyledons and 1590 dicotyledons), the angiosperms are the most speciose group of this region. Ara et al. (1995) reported that, distributed over 120 genera and 60 families, the overall arboreal flora of this region is represented by 295 species with Rosaceae and Caprifoliaceae being two dominant arboreal families, while *Cotoneaster*, *Lonicera*, *Rubus*, *Berberis*, *Salix*, *Spiraea*, *Acer*, *Clematis*, *Prunus*, and *Rosa* are the genera with larger number of arboreal species. However, undue human interference over the years has affected these gymnosperms greatly, and subsequently various arboreal species have become endangered. These among others include *Ulmus wallichiana*, *Morus nigra*, *Alnus nitida*, and *Corylus jacquemontii*.

Based on the available floristic information, the first ten dominant angiosperm families of Kashmir include Asteraceae, Poaceae, Brassicaceae, Rosaceae, Lamiaceae, Fabaceae, Cyperaceae, Scrophulariaceae, Ranunculaceae, and Apaceae, while *Carex*, *Taraxacum*, *Polygonum*, *Potentilla*, *Artemisia*, *Nepeta*, *Ranunculus*, *Veronica*, *Gentiana*, and *Pedicularis* are ten largest angiosperm genera of Kashmir valley (Table 6.3).

(b) **Phytogeographical analysis**

The phytogeographical analysis of the floristic wealth of Kashmir Himalaya suggests that throughout the geomorphological and geological history, it has undergone numerous changes. The recurrent glaciations and severe climatic and biotic forces heavily disturbed the original vegetation. This phenomenon

Table 6.3 The first ten large families and genera of angiosperms in Kashmir

No	Family	Number of Species	Genera	Genus	Species
1	Asteraceae	260	72	*Carex*	36
2	Poaceae	160	73	*Taraxacum*	35
3	Brassicaceae	115	44	*Polygonum*	25
4	Rosaceae	98	32	*Potentilla*	22
5	Lamiaceae	88	32	*Artemisia*	20
6	Fabaceae	84	34	*Nepeta*	20
7	Cyperaceae	81	09	*Ranunculus*	19
8	Scrophulariaceae	77	19	*Veronica*	19
9	Ranunculaceae	70	17	*Gentiana*	18
10	Apiaceae	68	37	*Pedicularis*	18

Source: Dar and Khuroo (2013)

led to disappearance of many broad-leaved species and their replacement by coniferous types, such as deodar (Inayatullah and Ticku 1964), and made the present-day vegetation quite different than what it was originally. Particularly during the glacial phase of Pleistocene, the typical Himalayan or the Sino-Japanese tropical cover first changed to subtropical vegetal and then finally to temperate types. With regard to Kashmir Himalaya, the extant flora is Holarctic with elements belonging to two subkingdoms, Boreal and Tethyan, the former being represented by Circumboreal, Eurasian, and Eastern Asiatic regions and later by Mediterranean and Irano-Turanian regions. It has the floristic representation of various phytogeographical regions like Circumboreal, European-North Asiatic, Eastern Asiatic (Sino-Japanese), Mediterranean, Western Asiatic, and Central Asiatic. Among the major phytochoria, the Central Asiatic influence on Kashmir Himalayan flora is more dominant over that of Sino-Japanese and European/Mediterranean. Furthermore, as opined by Dhar and Kachroo 1983, the Kashmir Himalayan flora generally tends to lean more toward Central and Northeast Asia than in opposite direction. These authors have also observed that among neighboring floristic units, its affinity is greatest with Northern Pakistan and Western Himalaya but progressively decreases with Afghanistan, Eastern Himalaya, Tibet, and Iran.

(c) **Alien Flora**

A recent inventorization reported a total of 571 plant species, belonging to 352 genera in 104 families, as alien in the Kashmir Himalaya (Khuroo et al. 2007). With dicotyledons contributing the maximum, i.e., 425 alien plant species, these aliens constitute about 29% of total flora of the region. The highest number of alien plant species belongs to families of Poaceae (60 species), Asteraceae (54), and Brassicaceae (30). Some notorious alien invasive species include *Anthemis cotula*, *Xanthium strumarium*, *X. spinosum*, *Galinsoga parviflora*, *Cannabis sativa*, *Conyza canadensis*, *C. bonariensis*, *Salvinia natans*, *Ceratophyllum demersum*, and *Potamogeton nodosus*.

(d) **Endemism**

Despite an admixture of various floristic regions, the Kashmir Himalayan flora contains various species which are restricted in their distribution and do not stretch beyond this region. These species are called as endemics. While, Dhar and Kachroo (1983) had reported that the percentage of endemic taxa in dicots is over 30%, but a recent inventorization has recorded that about 153 angiosperm plants are endemic to Kashmir Himalayan region (Dar et al. (2012). Considering the geographical area of the region, this is highly significant because as against the 186, 609, and 757 km^2, respectively, in the Himalaya, India, and Ladakh, the Kashmir Himalaya has one endemic taxon per 104 km^2. While a number of endemic species are reported high for Asteraceae (Table 6.4), the plant genera of *Berberis*, *Jaeschkea*, *Impatiens*, *Taraxacum*, *Scorzonera*, *Doronicum*, *Lagotis*, *Saussurea*, *Pedicularis*, and *Nepeta* display maximum incidence of endemism (Dar and Nasreen 2003).

(e) **Medicinal and Aromatic Flora**

The Kashmir Himalayan flora has good representation medicinal plants. It is widely believed that being a rich repository of medicinal plants, the Kashmir valley has provided opportunity for sustaining the Unani system of medicine during onslaught of western system. More recently, Dar et al. (2007) revealed that about 17% of the flora in Kashmir Himalaya has known or potential medicinal value. Various authors like Dad and Khan (2013) have studied the distribution of medicinal plants in Kashmir Himalaya and documented their traditional uses across the region. While these plants have a sporadic distribution, these grow in abundance in areas like Gurez and Tilail valley, Lolab valley, Pir Panjal, Gulmarg, Sonamarg, Khilanmarg, Pahalgam, and Yusmarg. Most of these medicinal and aromatic plant species (MAPS) belong to the dicotyledons (Table 6.5), the Asteraceae having the highest number of MAPS, followed by Ranunculaceae, Lamiaceae, Apiaceae, and Scrophulariaceae (Table 6.6; Fig. 6.2).

Despite these explorations, very little attention has been paid to study the ethnobotanical and ethnomedicinal uses of available medicinal herbs in general and traditional knowledge and practices developed by the migratory and tribal societies at higher altitudes in particular. It is for these reasons that the region has been recognized as floristically unexplored. Moreover, the addition of species notably

Table 6.4 Families with larger number of endemic species in Kashmir

No	Family	Number of endemic species
1	Asteraceae	21
2	Brassicaceae	17
3	Rosaceae	11
4	Fabaceae	9
5	Scrophulariaceae	9
6	Gentianaceae	8

Source: Dar and Khuroo (2013)

Table 6.5 A numerical summary of the medicinal and aromatic flora in Kashmir

	Number of		
Plant groups	Families	Genera	Species
Dicotyledons	79	239	400
Monocotyledons	13	28	37
Gymnosperms	4	4	7
Pteridophytes	5	5	6
Total	101	276	450

Source: Dar and Khuroo (2013)

Table 6.6 First ten families with larger number of medicinal plants in Kashmir

No.	Name of family	Total number of species	Number of MAPS
1.	Asteraceae	260	55
2.	Ranunculaceae	70	33
3.	Lamiaceae	88	31
4.	Apiaceae	68	19
5.	Scrophulariaceae	77	17
6.	Brassicaceae	115	16
7.	Papilionaceae	84	16
8.	Rosaceae	98	16
9.	Polygonaceae	52	15
10.	Gentianaceae	50	11

Source: Dar and Khuroo (2013)

Arabis nova, *Thlaspi coclearioides*, *G. venusta*, and *Primula involucrata* exclusively from remote areas of the valley (Dad and Khan 2012) reveals that the floral inventory of the area is not yet complete. Systematic surveys of the high-altitude vegetation that can overcome the climatic and other logistic problems and record the seasonal variation in the occurrence of various early and late season plants are likely to add many more species.

6.3 Brief Overview of Vegetation of Kashmir

Owing to unmatched variability in geological strata, altitude, climate, land use, topography, edaphic features, and meticulous geographical settings, the natural vegetation of Kashmir exhibits a great diversity and shows distinctive vegetation types. From rich marsh and aquatic vegetation in plains to lush evergreen conifers on gentle slopes at high altitudes, the vegetation of Kashmir displays a pronounced zonation. In fact the vegetation of Kashmir is a mosaic of hydrophyte formations, dwarf scrubs, meadows, turf, open scree, and rock vegetation types, with different life forms and plant species corresponding to a definite habitat type across different

Fig. 6.2 Photographs showing economical use of plant diversity in Kashmir Himalayas

elevation belts. On the basis of altitude, two broad elevation zones have been recognized in Kashmir Himalaya. Till 1800 m, occurs the valley zone which includes the main valley basin. A striking geological characteristic of this zone is the "Karewas," which are plateau-like tablelands of lacustrine origin formed during Pleistocene Ice Age. Beyond 1800 m occurs the montane zone which comprises the hillocks and mountains surrounding the main valley and side valleys, including the forests and meadows (Table 6.7; Fig. 6.3).

Each habitat has its own suite of vegetation and physiognomic characteristics which differentiate it from others. In fact, there are many unique vegetation types in various ecoclimatic zones of Kashmir Himalaya, and to give a detailed account of all those types is well beyond the scope of present article. Instead, to serve as a baseline, a broad account of the natural vegetation under three separate headings of forest, grassland, and aquatic is provided, and after discussing the threats faced by

Table 6.7 A brief outline of habitat diversity in the Kashmir Himalaya

Valley zone (1300–1800 m)		Montane zone (>1800 m)
Artificial habitats	Natural habitats	Montane streams and nullas
Habitation sites	Riverine sites	Montane slopes
Plantation sites	Valley lakes	Montane lakes
Cultivated fields	Wetlands	Forestlands
Orchards	Springs and canals	Temperate (1600–2700 m)
Mud-/Masonry walls	Karewa mounds	Subalpine (2700–3500 m)
Roadsides/Railway tracks	Grasslands	Alpine (>3500 m)
	Wastelands	Meadowlands
	Graveyards	Subalpine Alpine
		Glacier sites
		Alpine glacial lakes
		Alpine rocky and scree slopes

Source: Dar and Christensen (1999)

Fig. 6.3 Synoptic view of grassland, forest, and aquatic vegetation of Kashmir Himalaya

them, few suggestive measures are listed for better management of this natural vegetation.

1. **Forest Vegetation**
   Forests are vital for existence of humankind on earth. Being mountainous, the state is endowed with rich forest resources, which cover about 20,230 km^2 of state area. Of this, the highest forest cover (about 51% of its geographical area) is in the Kashmir region. These forests, besides being central in preserving the fragile mountain ecosystem, also serve as catchments for important water resources of entire Himalayan region. From lush green alpine pastures to evergreen conifers on gentle slopes of middle and greater Himalayas, and from foothill scrubs to deciduous forests at Pir Panjal Range, the forest vegetation of

Table 6.8 Major forest types and species of Kashmir Himalaya

No	Forest type	% Area	Forest species	% Age
1	Subtropical dry deciduous	4.02	Deodar	6.13
2	Subtropical pine	13.50	Fir	22.77
3	Subtropical evergreen	3.40	Kail	13.64
4	Himalayan moist temperate	44.06	Chir	Neg.
5	Himalayan dry temperate	6.57	Others	57.46
6	Dry/moist alpine	28.25		
Total		100.00		100.00

Source: Anonymous (2010)

Kashmir is highly heterogeneous, exhibits tremendous species diversity, and displays a remarkable variation in terms of species composition and distribution on both altitudinal and latitudinal gradient. On a whole, six broad vegetation types, namely, subtropical dry evergreen, subtropical pine, Himalayan moist temperate, Himalayan dry temperate, subtropical pine, and subalpine and alpine forests, occur in state of Jammu and Kashmir (Table 6.8). At lower slopes of Siwaliks, the subtropical rain forests with Teak, Sal, Shisham, and Pipal dominate. However, on a latitudinal scale, as one moves northward, temperate vegetation predominates.

At lower submontane altitudes (up to 1800 m), broad leaf forest formations mostly by poplar, chinar, maple, and willow occur. Higher up, at altitudes between 1800 and 2600 m asl, the vegetation comprises usually the conifers, such as blue pine (*Pinus wallichiana*), silver fir (*Abies pindrow*), Himalayan deodar (*Cedrus deodara*), the Himalayan yew (*Taxus wallichiana*), and spruce (*Picea smithiana*), with some associated broadleaved trees and shrubs like *Padus cornuta* and *Acer caesium*. In this elevational gradient, the forest formations intermix and appear as mixed stands, but nevertheless pure vegetation stands do occur also. Moving upward (3200 m) in subalpine zone, silver fir assumes dominance at lower elevations, while natural stands of birch (*Betula utilis*) occur above 3200 m asl and usually form the tree line. Beyond this altitude, the quality of forests in terms of density and quality degrades, and bushland and alpine scrub vegetation—comprising mainly the species of *Juniperus*, *Rhododendron*, *Salix*, *Lonicera*, and *Cotoneaster*—appears. Species-wise, forest vegetation of Kashmir Himalaya is chiefly coniferous (82.57%), with few species like fir (*Abies* spps. −22.77% acreage), kail (*Pinus* spps. −13.64% acreage), and deodar (*Cedrus* spps. 6.13%) being most important. The other important plant species found in these forests include *Pinus roxburghii*, *Dalbergia sissoo*, *Pinus gerardiana*, *Juglans regia*, *Populus ciliata*, *Fraxinus floribunda*, *Quercus* spp., and *Parrotia jacquemontiana.*

In terms of their extent, distribution, and potential, these forests contribute significantly toward meeting the needs of local population for timber, firewood, fodder, and other forest produce and contribute significantly toward state economy. However, as with other ecosystems, they are faced with numerous challenges. Conversion of these forests into non-forest uses like agriculture and horticulture, uncontrolled fuel wood collection by locals, large-scale commercial extraction of timber, encroachments, and illicit tree felling are few important challenges. The infrastructural deficiencies and trained manpower shortcomings have further aggravated these challenges and pushed nearly half our forests into the category of open forests.

2. **Grassland Vegetation**
   Grasslands and meadows are a common feature of Kashmir Himalaya with about 5.23% geographical area of state being under alpine grasslands, permanent pastures, and other grazing lands (Anonymous 2010). Locally called as Margs or Bahks, these ecosystems serve as potential summer pastures for low-lying populace wherein people along with their herds move from one area to another along the traditionally assigned routes and treks with well-defined and fixed haltages enroute—a process called as transhumance. Representing an interlocking system which is functionally and integrally incorporated in the socioeconomic life of nomads, this movement starts early spring when the ethnic tribes of Chopans, Gujjars, and Bakerwals move across Pir Panjal and Kashmir valley to high-altitude pastures in Greater Himalayas and stay there till summer ending. Afterward, they start moving back to lower foothills and stay in winter habitats on southern valleys and plains of Jammu. Besides this cultural significance, these grasslands possess a rich plant wealth. On a whole, these grasslands are rich in native grasses and legumes with a high fodder value. The major forage genera exhibiting forage biodiversity are represented by Poaceae, Cyperaceae, and Juncaceae with *Poa annua*, *P. bulbosa*, *P. pratense*, *Chrysopogon gryllus*, *Stipa sibirica*, *Phleum himalaicum*, *P. alpinum*, *Carex nivalis*, *Fimbristyllis dichotoma*, *Juncus thomsonii*, *J. articulates*, and *Dactylis glomerata* being important species. In addition, *Agrostis*, *Agropyron*, *Dactylis*, *Elymus*, *Festuca*, *Lolium*, and *Stipa* also offer a significant diversity, while *Astragalus*, *Lotus*, *Medicago*, and *Trifolium* constitute important legumes. More specifically, the grasslands which occur beyond tree line are endowed with characteristic herbaceous elements, such as species of *Aconitum*, *Aquilegia*, *Gentiana*, *Iris*, *Pedicularis*, *Potentilla*, *Primula*, *Ranunculus*, and *Saussurea*. The grassland vegetation also harbors numerous medicinal herbs which are extracted by both local and migratory communities for their own consumption as well as for sale. However, the indiscriminate and nonsystematic collection of important commercial drug-yielding plants like *Artemisia absinthium*, *Atropa acuminate*, *Dioscorea deltoidea*, *Digitalis purpurea*, *Fumaria indica*, *Origanum vulgare*, *Podophyllum hexandrum*, *Sausurea costus*, *Arnebia benthamia, and Gentiana kurroo* has led to a severe pressure on their availability in their wild habitats.
3. **Aquatic Vegetation**
   Kashmir Himalaya abounds in a vast array of freshwater bodies like rivers, lakes, wetlands, and streams which besides being of great ecological, aesthetic, socioeconomic, and cultural significance are also an important repository of plant and animal biodiversity. The important river ecosystems of Kashmir include Jhelum and its various tributaries, while the world famous Dal, Anchar, Wular, Gangabal,

Tarsar, Marsar, and Manasbal constitute the important lake ecosystems. Of many, Malangapora, Narkora, Tullamulla, Shalabogh, Kranchu, Haigam, Malgam, Nowgam, Mirgund, and Hokarsar are the important wetland ecosystems of Kashmir. Although all these ecosystems are highly rated for their plant and animal resources, but for plant resources, lakes and wetlands are particularly important. While the macrophytic vegetation seems to be least influenced by altitude, most of the macrophytic vegetation found in lacustrine ecosystems of Kashmir grows at altitudes below 3000 m. At altitudes above 4000 m, aquatic habitats are generally devoid of any vegetation. The important factors governing the occurrence and growth of various macrophytes in these lakes are water depth, nutrient status, and water turbidity. Depending on where the plant species grow, the three prominent vegetation zones present in these ecosystems are emergents, rooted floating leaf types and submerged. Generally, lakes with lower water turbidity are reported conducive for submerged forms while as floating forms dominate in turbid waters. Among emergents, *Myriophyllum verticillatum*, *Juncus articulates*, *Sagittaria sagittifolia*, *Equisetum debile*, *E. diffusum*, *Potentilla reptans terristrial*, *Epilobium hirsutum*, *Scirpus palustris*, *Phragmites australis*, *Typha angustata*, *Sparganum ramosum*, and *Eleocharis palustris* are the important species that dominate along lake and wetland littorals in Kashmir, while *Salvinia natans*, *Spirodela polyrhiza*, *Nymphea alba*, *N. stellata*, *Nelumbium nucifera*, *Trapa natans*, *Ranunculus aquatalis T. bispinosa*, *Hydrocharis morus-ranae*, *Nymphoides peltatum*, *Potamegeton natans*, *P. lucens*, and *P. crispus* constitute the floating leaf species, and *Hydrilla verticillata*, *Potamogeton crispus*, *P. lucens*, *P. pucillus*, *Najas graninae*, *Myriophyllum spicatum*, and *Ceratophyllum demursum* are important submerged plant species (Pandit 1992).

Besides the ecological significance, these aquatic habitats are of special value to surrounding populace and provide them manifold resources (Table 6.9).

However, years of mismanagement, reduction of plant cover in catchment areas, lake encroachment, higher silt load through feeding channels and agricultural runoff, and overfishing have put tremendous pressures on the aquatic habitats of Kashmir and affected the distribution of many important species like *Chara fragilis*, *Nitella acuminate*, *N. dispersa*, *Nitellopsis obtusa*, *Eurale ferox*, and *Potamageton perfoliatus*.

6.4 Brief Overview of the Utilization of Plant Resources in Kashmir Himalaya

Mankind's continuation on earth is dependent on numerous natural resources, among which plants play an important role, supplying food, clothing, timber, fuel, medicine, and the like. To study the utilization of plant resources across Kashmir Himalaya, various workers have conducted studies across Kashmir Himalaya and provided information on wide-ranging uses of plants across valley. A brief account

Table 6.9 Name, usage, and part used of few aquatic plants in Kashmir Himalaya

No	Name of Plant species	Part utilized	Usage
1.	*Nelumbium nucifera*	Rhizomes and seeds	Human food
	Nymphea alba	Petioles and fruit	
	N. candida	Petioles and fruit	
	N. tetragonia	Petioles and fruit	
2.	*Phragmites australis*	Shoots	Cattle feed (fodder)
	Echinochola crusgalli		
	Scirpus lacustris		
	Nymphoides peltata	Leaves	
	Hydrocharis morsusranae		
3.	*Ceratophyllum demursum*	Plant as a whole	Green manure, compost, mulch etc.
	Myriophyllum spicatum		
	Hydrilla verticillata		
	Najas graminea		
	Salvinia natans		
4.	*Typha angustata*	Rhizomes	Medicinal uses like diarrhea, dysentery, vomiting, skin diseases, small pox.
	Acorus calamus	Rhizomes, shoots	
	Nelumbium nucifera	Rhizomes	
	Spirodela polyrhiza	Whole plant	
5	*Typha angustata*	Leaves	Thatching and roofing material
	Phragmites australis	Culms	
	Sparganum erectum	Leaves	
6	*Nelumbium nucifera*	Plant as a whole	Ornamental and indoor aquaria
	Potamogeton lucens		
	Ceratophyllum demursum		
	Najas graminea		
	Utricularia flexuosa		

Source: Pandit (1992)

of those uses is provided, with an understanding that those uses would highlight the intimate relationship that exists between plants and human beings and provides encouragement for preservation of traditional knowledge and will help in the conservation of the documented species.

1. **Wild Foods**: Historically, majority of plant species were being collected and used as wild foods in Kashmir, but owing to changing food habits and reduced

plant availability, their usage has fallen drastically. However, still these plants form an important component of dietary habits and among others include *Trapa natans* (*used by most people during winter months*), *Morus nigra*, *Ficus palmata*, *Berberis lyceum*, *Frageria nubicola*, and *Pyrus pashia.*

2. **Vegetable (Wild and Cultivated)**: Wild plants have always played a crucial role in meeting nutritional needs in general and in remote rural areas in particular. Plants like *Coronopus didymus* and *Eremurus persicus* and species of *Gagea*, *Lactuca*, *Silene*, and *Ranunculus* have been used as vegetables in Kashmir, since ages. Various other plants used as leafy vegetables at higher elevations include *Centaurea iberica*, *Taraxacum officinale*, *Dipsacus inermis*, *Rheum webbianum*, *Eremurus himaliacus*, *Malva neglecta*, *Capsella bursa-pastoris*, and *Medicago sativa*.
3. **Forages**: A rich diversity of plant wealth is found in Kashmir which serves as important forage resources. While grasses, sedges, and legumes are found distributed across montane, subalpine, and alpine altitudes, the distribution of these forage species in alpine habitats is of particular significance because it serves as a resource base for the migrating animal population during the summer months. Among others, the chief forage supplying plants found in Kashmir valley include *Chrysopogon gryllus*, *Themeda anthera*, *Trifolium repens*, *Trifolium pretense*, *Poa annua*, *P. bulbosa*, *Phleum alpinum*, *Lotus corniculatus*, *Robina pseudoacacia*, *Aesculus indica*, and *Morus alba.*
4. **Wood**: Numerous plants harvested from the forests and plantations are being used as a source of wood and timber by people across Kashmir valley. The highly valued timber species include *Cedrus deodara*, *Pinus wallichiana*, *Abies pindrow*, and *Picea smithiana.* The best quality timber is obtained from *C. deodara a*nd is much in demand for houseboats and bridges as it is impervious to water. Historically, old shrines, mosques, and temples in Kashmir are all made of deodar. The other important species valued for making furniture, walking sticks, cricket bats, and other allied tools include *Populus nigra*, *Juglans regia*, *Platanus orientalus*, *Celtis australis*, *Betula utilis*, and *Salix alba.*
5. **Medicine**: The medicinal plants have traditionally occupied a central position in the sociocultural, spiritual, and medicinal arena of people of Kashmir. In fact, various ethnic tribes have learned and practiced the medicinal usage of plants that grow in their proximity since ages. The important drug-yielding plants of Kashmir include *Aconitum heterophyllum*, *Acorus calamus*, *Arnebia benthamii*, *Artemisia absinthium*, *Bergenia ciliate*, *Euphorbia hispida*, *Viola indica*, *Podophyllum hexandrum*, *Picrorhiza kurroa*, *Iris hookeriana*, *Dioscorea deltoidea*, *Ephedra gerardiana*, *Ajuga bracteosa*, *Saussurea costus*, *Digitalis purpurea*, and *Rheum webbianum.* However, anthropogenic disturbance, indiscriminate and nonsystematic collection, and illicit trade have led to considerable population depletion of these plants, rendering them very rare and threatened.
6. **Arts and Crafts**: To meet their artistic ability and craftsmanship, the people of Kashmir have been utilizing plant resources, since ages. From making floorings, footwear, and furniture to boxes, baskets, buckets, and other household items,

numerous plants are being used meticulously by people of Kashmir. The chief plants of this category include *Typha angustata*, *Parrotiopsis jacquemontiana*, *Cotoneaster nummularia*, *Salix purpurea*, and *Indigofera heterantha*. Besides contributing heavily to the economy of these artists, these plants are also an inseparable part of the Kashmiri culture and tradition.

7. **Perfumery**: Various plants growing across Kashmir form an indispensable part of perfume industry and are valued high due to the presence of various aromatic oils, alkaloids, and various other compounds. Of particular importance is *Saussurea lappa*—an alpine plant, whose rhizomes are used both as incense and medicine. Similarly, the male flowers of *Salix aegyptiaca* and leaves of *Skimmia anquetilia* are pleasantly scented and thus used commonly in perfume industry. The other important plants of this group include *Pedicularis brevifolia*, *Origanum normale*, *Morina longifolia*, and *Cyperus rotundus*.
8. In addition to the above major uses, the people of Kashmir have been utilizing the plant resources in numerous other ways which includes plants used in:

    (a) **Incense and Religious and Social Rites**: Irrespective of the religion, both Hindu and Muslims in valley use plants for performing religious and social rites and averting the evil eye. These among others include *Cotula anthemoides*, *Prunella vulgaris*, *Peganum harmala*, *Crocus sativa*, *Mentha royleana*, and *Jurinea ceratocarpa*. Though, of late their use in urban centers has dwindled and these plants are now used sporadically, in far-off rural areas, these are used quite often.
    (b) **Spices, Condiments, and Flavorings**: The important plants of this group include *Crocus sativa*, *Bunium persicum*, *Thymus linearis*, and *Foeniculum vulgare*. While *C. sativa* (saffron) of Kashmir is world famous and is used as condiment, medicine, flavoring, and coloring agent, the seeds of *B. persicum* and *F. vulgare* are used for the preparation of the food delicacies, flavoring soups, meat dishes, sauces, and pastries.
    (c) **Fibers**: The important fiber-yielding plants of Kashmir valley include *Abutilon theophrasti*, *Cannabis sativa*, *Linum usitatissimum*, and *Ulmus villosa*. On a local level, these plants are being used mostly for making ropes and other allied things.
    (d) **Dyes**: Similar to fiber yielding plants, various plants growing in Kashmir valley have been used for dyeing since ages and among other include *Ulmus villosa*, *Rubia cordifolia*, *Leonurus cardiaca*, *Geranium nepalense*, *Lycopus europaeus*, *Polygonum amphibium*, and *Datisca cannabina*.
    (e) **Narcotics**: The common narcotic hallucinogenic plants of Kashmir include *Nicotiana tabacum*, *Phytolacca acinosa*, *Datura stramonium*, *Cannabis sativa*, and *Atropa acuminata*. Illegally, these plants are cultivated and yield various hallucinogenic compounds like tobacco and bhang.

From the above, it is well evident that from basic subsistence requirements to economic well-being and religious sanctity to cultural ethos, the vegetation formations of Kashmir are all but central, and thus we must use them judiciously and in a sustainable way. Unfortunately, undue human interference has put these formations under tremendous pressure and brought them on the brink of extinction.

6.5 Major Conservation Threats

Exponential rise in human population; urbanization; industrialization; habitat destruction and fragmentation; overgrazing; unrestricted tourist flow; deforestation; ecologically insensitive economic development and adoption of unsustainable agricultural practices; shrinkage of forests, grasslands, wetlands, and lakes; nutrient loss in agricultural and other terrestrial ecosystems and a nutrient buildup (eutrophication) in aquatic ecosystems; continuing deforestation and illegal collection of various rare and threatened medicinal herbs; biological invasions; and overexploitation of plant biodiversity across Kashmir Himalaya are some of the measure threats across Kashmir Himalaya. These factors have brought dramatic changes in physical environment of Kashmir Himalaya and threatened the existence of various species. Dar and Naqshi (2002) have found that about 355 plant species (282 dicotyledons and 73 monocotyledons) in Kashmir Himalaya are threatened with various degrees of extinction (Table 6.10).

The endemic plants are generally prone to threat because of their rarity and ecosystem/habitat specialization. It is estimated that ca. 40% of endemic plant species in Kashmir Himalaya are endangered (Dhar and Kachroo 1983). A few important threatened plants include *Aconitum heterophyllum*, *Picrorhiza kurroa*, *Fritillaria roylei*, *Rheum webbianum*, *Saussurea costus*, *Dactylorhiza hatagirea*, *Aconitum deinorrhizum*, *A. kashmiricun*, *Astragalus anomalus*, *A. bakeri*, *Artemisia dolichocephala*, *Androsace aizoon*, *A.mucronifolia*, *Atropa acuminate*, *Allium auriculatum*, *Berberis kashmiriana*, *Carex kashmirensis*, *Corydalis cashmeriana*, *C. crassissima*, *Cremanthodium amicoides*, *Cypripedium cordigerum*, *Delphinium uncinatum*, *Dioscorea deltiodea*, *Fritillaria roylei*, *Lavetra kashmiriana*, *Lactuca benthamii*, *Inula racemosa*, *Meconopsis latifolia*, *Podophyllum hexandrum*, *Papaver himalaicum*, *Potentilla kashmirica*, *Picrorhiza kurroa*, *Puccinellia kashmiriana*, *Primula clarkei*, *Trollius acaulis*, *Saussurea atkinsoni*, and *Stipa chitralensis*. Similarly, in aquatic ecosystems, the manyfold increase in agricultural activity and reduction in

Table 6.10 Plant species under different threat categories in the Kashmir Himalaya

No	Threat category	Number of species
1.	Endangered	40
2.	Vulnerable	50
3.	Rare	155
4.	Indeterminate	110
	Total	355

Source: Dar and Naqshi (2002)

plant cover in catchment area, encroachment on the lake area for human settlements and other allied things, higher silt load through feeding channels and agricultural runoff, unrestricted tourist flow, and greater recreation demands together with the added construction activities have also resulted in the deterioration of water qualities and thereby reduction in the available plant resource.

6.6 Management Options

Protecting the rich and dynamic habitats of Kashmir Himalaya and conserving their rich biodiversity on a long-term basis are formidable challenges before policymakers. The present conservation scenario of Kashmir Himalaya is dismal and disappointing, and despite series of conservation measures taken by government like enacting laws and declaring areas as protected (Table 6.11), no proper conservation and management plan is available for the region, on a whole.

As a starting point, floral explorations and systematic documentations and inventorization in remote, inaccessible, and underexploited areas; compilation of a vegetation database across ecosystems to study the spread and shrinkage dynamics; continuous monitoring across various ecosystems; widespread auto-ecological studies on rare, endemic, and threatened (RET) plants like *Aconitum heterophyllum*, *Picrorhiza kurroa*, *Fritillaria roylei*, *Rheum webbianum*, *Saussurea costus*, and

Table 6.11 Protected Area Network of Jammu and Kashmir

Status	Jammu	Kashmir	Ladakh	No	Area (km^2)
NP	Kishtwar	Dachigam, City Forest Kazinag	Hemis	05	4014.07
WS	Ramnagar, Nandni, Jasrota, Surinsar Mansar, Trikuta	Overa-Aru, Hirpora, Gulmarg Thajwas, Limber, Lachipora Daksum	Changthang, Karakoram	14	10,479.72
CR	Sudhmahadev, Jawahar Tunnel, Thein, Bahu	Khiram, Panyar, Khrew Khanagund, Ajas, Achabal Shikargah, Khonmoh, Brain-Nishat, Naganari, Khimber, Chatergul, Zaloora-Harwan	Sabu, Boodh Karbu, Kanji	20	676.75
WR	Gharana, Pargwal, Kukarian, Naga, Sangral-AsaChak	Hokesar, Shallabugh, Hygam Malgam, Mirgund, Chatlam, Kranchoo	Tsomoriri, Norrichai, PangongTso, Hanley Masrhes	15	142.48
Total				54	15,313.02

Source: Anonymous (2010)

Note: NP, WS, CR, and WR designate national park, wildlife sanctuary, conservation reserve, and wildlife reserve, respectively

Dactylorhiza hatagirea aimed at assigning them authentic threat status based on the latest IUCN regional guidelines; rehabilitation of these RET plants through ex situ means followed by their subsequent reintroduction into suitable habitat types; efforts for the long-term conservation of local land races and cultivars; documentation of fading traditional knowledge particularly among the ethnic tribes of Gujjars and Bakarwals; and involvement of local communities in conservation programs and plans together with other legal and administrative measures are some of the options which could be used for better utilization of our plant resources. Specifically in forest areas, demarcation of forest land, regeneration of native species in open forests, strengthening the infrastructure and manpower, and controlling overgrazing are few important measures in this direction.

Acknowledgment Author is highly thankful to Dr. A. R. Naqshi, ex. Director, Centre for Plant Taxonomy, University of Kashmir for his assistance and encouragement.

References

Anonymous (2010) Digest of Forest Statistics, 2008-09. Jammu and Kashmir Forest Department. Government of Jammu and Kashmir, Srinagar, Kashmir

Ara S, Naqshi AR, Baba MY (1995) Indigenous and exotic trees and shrubs of Kashmir Valley. Indian J For (8):233–272

Bhellum B, Magotra R (2012) A catalogue of flowering plants of Doda, Kishtwar and Ramban districts (Kashmir Himalayas). Bishen Singh Mahendra Pal Singh Publishers, Dehradun, India

Dad JM, Khan AB (2012) Filling gaps on the distribution of *Arabis nova* Vill, *Thlaspi cocleariodes* Hook.f and Thomas, *Gentiana venusta* Wall. ex Griseb and *Primula involucrata* Wall.ex Dubey in Gurez valley, Kashmir, India. Checklist 8(6):1319–1320

Dad JM, Khan AB (2013) Threatened Medicinal Plants of Kashmir valley, India. Med Plants 5(1):1–5

Dar GH, Christensen KI (1999) Habitat diversity and zonality of vegetation in Sind Valley, Kashmir Himalaya. Nat Biosphere 4(1 & 2):49–71

Dar GH, Naqshi AR (2002) Threatened flowering plants of Kashmir Himalaya—a checklist. Orient Sci 6(1):23–53

Dar GH, Nasreen A (2003) Endemic angiosperms of Kashmir: assessment and conservation. National seminar on recent advances in plant science research, 12–14 Oct, 2003, Department of Botany, University of Kashmir, Srinagar, p 63

Dar GH, Khuroo AA, Khan ZS, Dar AR (2007) Medicinal flora of the Kashmir Himalaya: a taxonomic perspective. J Himalayan Ecol Sustain Dev (2):13–20

Dar GH, Khuroo AA, Nasreen A (2012) Endemism in the angiosperm flora of Kashmir Valley, India: stocktaking. In Mukherjee SK, Maiti GG (eds), Proceedings of International Seminar on Multidisciplinary Approaches in Angiosperm Systematics, Kolkata

Dar GH, Khuroo AA (2013) Floristic diversity in the Kashmir Himalaya: progress, problems and prospects. Sains Malaysiana 42(10):1377–1386

Dhar U, Kachroo P (1983) Alpine Flora of Kashmir Himalayas. Scientific Publishers, Jodhpur, India

Inayatullah M, Ticku BL (1964) A preliminary study of the forest typology of Jammu and Kashmir. Indian Forester 90(6):332–347

Khuroo AA, Irfan R, Zafar R, Dar GH, Wafai BA (2007) The alien flora of Kashmir Himalaya. Biol Invasions 9(3):269–292

Klimeš L, Dickore WB (2006) Flora of Ladakh (Jammu & Kashmir, India). A preliminary check-list published at http://www.butbn.cas.cz/klimes

Pandit AK (1992) Macrophytes as components of Dal lake ecosystem. In: Mishra SR, Saksens DN (eds) Aquatic ecology. Ashish Publishing House, New Delhi, India

Puri GS (1943) The occurrence of *Woodfordia fruticosa* (L.) S. Kurz in the Karewa deposits of Kashmir with remarks on changes of altitude and climate during the Pleistocene. J Indian Bot Society (22):125–131

Sharma BM, Kachroo P (1981) Flora of Jammu and Plants of Neighborhood. Bishen Singh Mahendra Pal Singh, Dehradun, India

Chapter 7
Medicinal Plants of Tajikistan

Farukh Sharopov and William N. Setzer

Abstract Many people in Tajikistan rely on medicinal plants as their traditional form of medicine to prevent and cure health disorders. Medicinal plants, in particular, have played an important role for the local people. In this review, we present a summary of the phytochemical contents of medicinal plants from Tajikistan and their uses in traditional medicine.

Keywords Medicinal plants · Tajikistan · Phytochemical contents

Abbreviations

5-LOX	5-Lipoxygenase
ABTS	2,2′-Azino-bis(3-ethylbenzothiazoline-6-sulfonic acid) radical cation scavenging assay
Caco-2	Human colorectal adenocarcinoma cancer
CAE	Caffeic acid equivalent
CCRF-CEM	Human T-cell lymphoblastic leukemia cancer
CEM/ADR5000	Adriamycin-resistant leukemia cancer
DPPH	2,2-Diphenyl-1-picrylhydrazyl radical scavenging assay
DW	Dried weight
FRAP	Ferric reducing antioxidant power assay
HeLa	Human cervical cancer
MBC	Minimum bactericidal concentration
MCF-7	Human breast adenocarcinoma cancer
ME	Methanol extract

F. Sharopov (✉)
Department of Pharmaceutical Technology, Avicenna Tajik State Medical University, Dushanbe, Tajikistan

W. N. Setzer
Department of Chemistry, University of Alabama in Huntsville, Huntsville, AL, USA
e-mail: wsetzer@chemistry.uah.edu

D. Egamberdieva, M. Öztürk (eds.), *Vegetation of Central Asia and Environs*,
https://doi.org/10.1007/978-3-319-99728-5_7

MIC	Minimum inhibitory concentration
MRSA	Methicillin-resistant *Staphylococcus aureus*
PTP	The enzyme of protein-tyrosine phosphatase
QE	Quercetin equivalent
TP	Total phenolic content
TF	Total flavonoid content
WE	Water extract

7.1 Introduction

Historically, the territory of present-day Tajikistan was part of an ancient network of the great Silk Road (around 120 BCE–1450 CE), as well as at the intersection centers of trade, culture, immigration, and military roads. Here occurred the exchange of various scientific and cultural information. The natural conditions of Tajikistan influence the biotransformation and accumulation of plant secondary metabolites. Tajikistan is diverse in terms of conditions for the development of the plants. Climate, high altitude, mountainous soil and minerals, and relatively large number of sunny days per year are essential for plant growth, synthesis, and accumulation of the biologically active secondary metabolites. There is a continuous interest to identify new sources of biologically active substances for use in practical medicine (Sharopov et al. 2015c). The medicinal plants are a great source of secondary plant metabolites. Interest in medicinal plants has been gradually increasing (Ozturk et al. 2011a, b, 2014, 2015, 2016a, b, 2017). Medical practice has shown that drugs with herbal origin are often safer compared to synthetic drugs. About 25% of the drugs in use today are obtained from medicinal plants, or they contain active ingredients of plant origin (Harvey 2008; Sahoo et al. 2010). Plants generally contain complex mixtures of biologically active secondary metabolites with multiple target effects and often low toxicity (Sharopov 2015). The plants produce a wide diversity of biologically active compounds such as alkaloids, glucosinolates, cyanogenic glucosides, flavonoids, tannins, coumarins, lignans, terpenoids, saponins, organic acids, and many others. They are also used as raw materials for the production of nutraceutical and pharmaceutical products.

7.2 Geography, Climate, and Vegetation

Tajikistan is a mountainous, landlocked country in Central Asia with a population of 8.5 million. It has borders with Afghanistan in the south, Uzbekistan in the west, Kyrgyzstan in the north, and China in the east, spreading over an area of 143,100 km^2 and lying between 67° to 75° E longitude and 36° to 41° N latitude. The country elevation starts from Beshai Palangon Nature Reserve (250–300 m above sea level) and ends with Ismoil Somoni Peak at 7495 m.

Repeats

Both geography and topography of the country present a challenging opportunity to pursue research in a broad spectrum of biological and ecological problems. It is part of the Western Pamir-Alai Mountains, one of the main formative processing centers in the plant kingdom. Vertical zone is expressed clearly in a relatively small area, from the deserts of the subtropical type to the snow-capped peaks of the Pamirs. It is very rich in endemic plant species and vegetation types because of its unique natural and climatic conditions and considerable biogeographical isolation by high ridges of the Karakorum, Hindu Kush, Kunlun, and Tian Shan Mountains (Nowak et al. 2016; Sharopov 2015; Sharopov et al. 2015c; Tayjanov et al. 2017).

High mountains dominate the country with about 50% more than 3000 m above sea level, and the relief differentiation has resulted in the formation of numerous micro- and macrohabitats. There are over 900 rivers in Tajikistan longer than 10 km. The high-mountain ecosystems of Tajikistan are regarded as biodiversity hotspots with around 4550 species of higher plants and nearly 30% endemism (Squires and Safarov 2013). The reasons for a high degree of biodiversity and endemism are the presence of high mountain ranges that serve as barriers towards migration of plants and animals. Additionally, the country is characterized by a low percentage of cloud cover, large temperature differences (absolute minimum of −63 °C in the Pamir and maximum of 48 °C in Panji Poyon), low humidity, and low precipitation (Sharopov et al. 2015c).

More than 20 types of ecosystems have been identified in Tajikistan, which can be conditionally grouped into 12 dominating natural and anthropogenic ecosystems: (1) nival-glacial, (2) high-mountain desert, (3) high-mountain meadow-steppe, (4) mountain coniferous forest, (5) mountain mesophilic forest, (6) mountain xerophytic light forest, (7) mountain savannoide, (8) foothill semidesert, (9) wetlands and water, (10) agroecosystem, (11) urban, and (12) ruderal-degraded ecosystems (Fig. 7.1) (National Biodiversity and Biosafety Center of the Republic of Tajikistan 2017; Novikov and Safarov 2002; Rahmonov et al. 2013; Vanselow 2011).

Nival-glacial ecosystem occupies the highlands of the country, a significant part of the Eastern and Western Pamirs. In the cold rocky glacial conditions of this ecosystem, approximately 16–17 species of flowering plants, including *Melandrium apetalum, Draba altaica, Astragalus nivalis,* and *Saussurea glacialis*, are distributed (National Biodiversity and Biosafety Center of the Republic of Tajikistan 2017; Novikov and Safarov 2002).

High-mountain desert ecosystem occupies vast areas of the Eastern and Western Pamirs, fragmentarily found in the basin of the Zeravshan River. The vegetation cover is dominated by the species of *Ceratoides krascheninnikovia, Artemisia pamirica, A. korshinskyi, Ajania tibetica, Stipa glareosa, Oxytropis immersa, Acantholimon diaspensioides*, and *A. pamiricum* (National Biodiversity and Biosafety Center of the Republic of Tajikistan 2017; Novikov and Safarov 2002).

High-mountain meadow-steppe ecosystem is fragmentary and found on all mountain ranges of Tajikistan, with great ecological significance. Main species in this ecosystem are *Festuca altaica, F. pamirica, Stipa kirghisorum, Poa alpina, Carex melanantha, C. stenocarpa, Cobresia stenocarpa, Oxytropis savellanica,* and

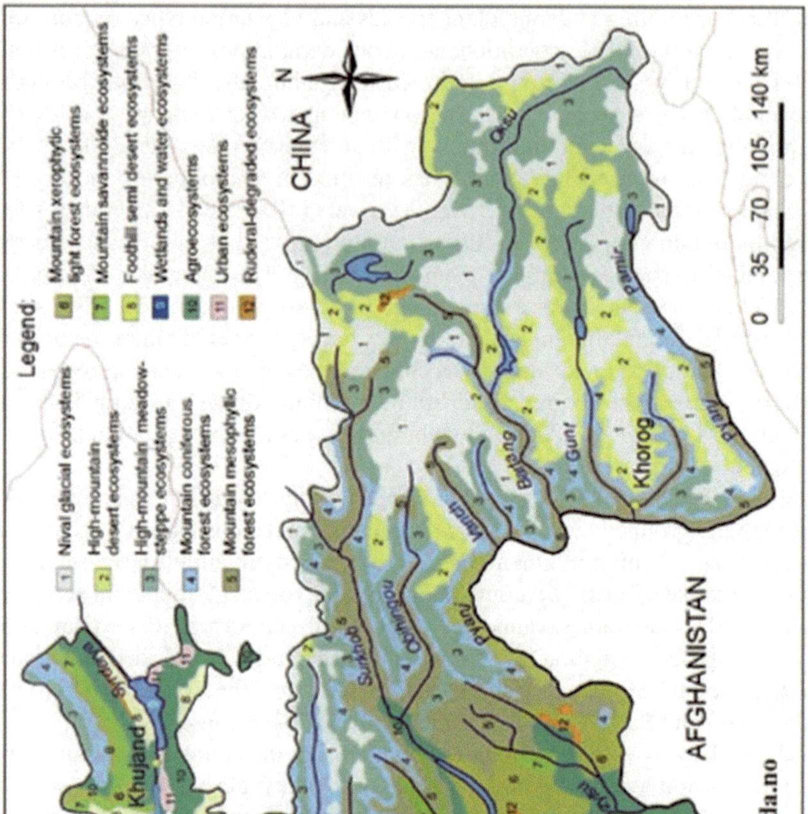

Fig. 7.1 Twelve dominating natural and anthropogenic ecosystems in Tajikistan (Novikov and Safarov 2002)

Thymus seravshanicus (National Biodiversity and Biosafety Center of the Republic of Tajikistan 2017; Novikov and Safarov 2002).

Mountain coniferous forest ecosystem makes up about 50% of the total area of the country's forest cover. Distributed in northern Tajikistan, small fragments of these are found in Central, Southwest Tajikistan and in the mountains of the Western Pamirs. Juniper forests and woodlands are mainly represented by four species: *Juniperus seravschanica, J. turkestanica, J. semiglobosa*, and *J. sibirica* (National Biodiversity and Biosafety Center of the Republic of Tajikistan 2017; Novikov and Safarov 2002).

Mountain mesophilic forest ecosystem is represented by maple-walnut, willow-poplar-birch forests with sparse mesophytic bushes. In the composition of these forests, a significant number of rare endemic plant species are distributed. Most valuable plants are broad-leaved mesophilic relict forests: *Juglans regia* and *Acer turkestanicum*—widespread in Central Tajikistan. Large areas of small-leaved forests of *Betula tianschanica* occur in the basin of the Zeravshan River. Most valuable maple-walnut forests are located in Sarikhosor, Childukhtaron, and Dashtijum. A significant number of rare and endangered species are found in the composition of these forests; notable among these are *Ungernia victoris, Ostrovskaia magnifica, Cousinia darwasica, Cousinia leptocampyla, Iskandera hissarica,* and *Stipa jagnobica*. In the forest plant communities, there is a significant number of wild relatives of fruits, including apple (*Malus*), pear (*Pyrus*), cherry plum (*Prunus*), hawthorn (*Crataegus*), and barberry (*Berberis*) (National Biodiversity and Biosafety Center of the Republic of Tajikistan 2017; Novikov and Safarov 2002).

Mountain xerophytic light forest ecosystems occupy vast territories of southern and western Tajikistan, and small fragments are also found in northern Tajikistan. They include *Pistacia vera, Acer regelii, Celtis caucasica, Ephedra equisetina, E. intermedia*, and *Calophaca grandiflora*. Moreover, *Hordeum spontaneum, Vicia tenuifolia, Amygdalus bucharica, Diospyros lotus, Ziziphus jujuba, Punica granatum,* and *Vitis vinifera* are found as a part of this ecosystem (National Biodiversity and Biosafety Center of the Republic of Tajikistan 2017; Novikov and Safarov 2002).

Mountain savannoide ecosystems are widespread in southern and northern Tajikistan. The dominant species here are *Hordeum bulbosum, Carex pachystylis, Ferula kokanica, F. kuhistanica,* and *Phlomis bucharica* (National Biodiversity and Biosafety Center of the Republic of Tajikistan 2017; Novikov and Safarov 2002).

Foothill semidesert ecosystems occupy the high terraces of the valley part of the lower reaches of the major rivers—Panj, Vakhsh, Kafirnigan, Syr Darya, and Zeravshan. About 30,000 ha of Beshai Palangon Nature Reserve in southern Tajikistan belong to this zone. The main dominants of the vegetation of this ecosystem are *Haloxylon persicum, Calligonum litvinovii, Salsola richteri, Artemisia tenuisecta, Hammada leptoclada, Carex physodes, Halostachys belangeriana*, and *Halocharis hispida* (National Biodiversity and Biosafety Center of the Republic of Tajikistan 2017; Novikov and Safarov 2002).

Wetlands and water ecosystems. Main dominants of the vegetation cover here are *Populus pruinosa, Elaeagnus angustifolia, Lycium dasystemum, Typha* *angustifolia, Imperata cylindrica, Phragmites communis, Saccharum spontaneum, Tamarix hispida*, and *Juncus articulatus* (National Biodiversity and Biosafety Center of the Republic of Tajikistan 2017; Novikov and Safarov 2002).

Agroecosystems are located in all natural zones, from the torrid foothills (300 m above sea level) to the high-mountain deserts of the Eastern Pamirs (3000–3500 m above sea level). A significant part of varieties of crops, local species of wild relatives of cereals, legumes, industrial plants, vegetables, and melons as well as fodder plants grow on the open ground (National Biodiversity and Biosafety Center of the Republic of Tajikistan 2017; Novikov and Safarov 2002).

Urban ecosystems cover the cities of Dushanbe, Kurgan-Tyube, Tursunzade, Khujand, and Kulyab and other large settlements and industrial enterprises (National Biodiversity and Biosafety Center of the Republic of Tajikistan 2017; Novikov and Safarov 2002).

Ruderal-degraded ecosystems are found in all zones of human activity. The main dominant communities of ruderal ecosystems include representatives from the families Asteraceae, Poaceae, Polygonaceae, Ranunculaceae, Clusiaceae, and often Lamiaceae (National Biodiversity and Biosafety Center of the Republic of Tajikistan 2017; Novikov and Safarov 2002).

7.3 Historical Aspects of the Medicinal Plants in Tajikistan

What is now present-day Tajikistan was part of an ancient network of the great Silk Road (around 120 BCE–1450 CE), which served as an intersection of trade, culture, immigration, and military conquests between East and West. The Tajik civilization was born at the crossroads of this Silk Road.

One of the ancient sources of Persian-Tajik medicine was the holy book of the Zoroastrians, *Avesta* (seven to six centuries BC). In the *Avesta* the names of more than 100 species of medicinal plants are given. According to the *Avesta*, medicinal plants were used as sedative, analgesic, antidote, emollient, restorative, and antiseptic (Hojimatov 1989b). Many herbs have been introduced in this book for various ailments. Some herbal extracts, infusions, decoctions, abortifacients, and disinfectants are also well described in the second section (*Vendidad)* and the fifth section (*Yashts)* of the *Avesta* (Adhami et al. 2007; Hojimatov 1989b). The *Vendidad* contains an indication of the first world leader of *Trita*, medical knowledge, treatment, medicinal plants, diseases, and medicines.

In the development of Persian-Tajik medicine, an important role was played by the Academy of Gundeshapur (226–651), founded by Sassanid King Shapur I. Gundeshapur scientists Barzuya, Jurjis ibn Jibril ibn Bukhtiyeshu, Jibril ibn Bukhtiyeshu ibn Jurjis, Mosawayh, Shapur ibn Sahl, Hunayn, Sahorbuht, Muhammad al-Harith ibn Kalada, Ganga al-Hindi, and others made important contributions in the development of pharmacology, pharmacognosy, pharmacy, and toxicology (Hojimatov 1989b).

The high level of the development of Persian-Tajik medicine was reached in Transoxiana (Mawarannahr) and Khorasan and matched with the culmination of the development of science and culture of the Samanid State (875–999). Ali ibn Sahl Rabban al-Tabari (circa 818–870), Abu Hanifa Dinavari (815–895), Abu Bakr Muhammad ibn Zakariyya al-Razi (Rhazes in Latin) (865–925), Abu Mansur Muwaffaq (second part of the tenth century), Abubakr Rabeq al Buchari (880–890), Abu Mansur Buchari (991), Abu Sahl Masehi (died in 1010), Miskawayh (940–1030), Abu Rayhan Biruni (973–1048), and Abu Ali ibn Sina (Avicenna in Latin) (980–1037) were well-known scholars in the history of Persian medicine and pharmacy. Most important of Persian-Tajik scholar works about medicinal herbs between the ninth and nineteenth centuries are presented on Table 7.1.

These ancient books contain many useful recipes based on medicinal plants for many simple and complex medical treatments. The work of Abu Ali ibn Sina was very versatile. In his book *the Canon of Medicine,* he describes 811 simple medicines, 612 being of plant origin. This number includes more than 150 species belonging to 112 genera and 55 families from Tajikistan (Rasulova and Yunusov 1980).

From the ancient times until today, natural vegetation of Tajikistan has attracted the attention of many geobotanists. Modern botanical research in Tajikistan began with the Russian researcher, Alexander Leman, who visited Tajikistan 180 years ago (1838–1841). He traveled a huge distance from the Urals to Bukhara and had the opportunity to travel through Zarafshan Valley (Margolina 1941). Leman collected valuable botanical material, which was processed and published by Prof. A.A. Bunge. A recent discovery by Leman is from Zeravshan which deal with three new types of *Astragalus* tragacanth (Margolina 1941). Geobotanists A.P. Fedchenko and O.A. Fedchenko have provided great contributions to the scientific investigations of the flora of Central Asia, especially Tajikistan (Pamir) (Margolina 1941; Romanyuk 2017). While traveling through Central Asia (1868–1871), A.P. Fedchenko collected material on the flora, fauna, geography, and ethnography of the region. In his honor, the expedition (1878) headed by V.F. Oshanin discovered the longest glacier in the world outside the polar region (77 km long), named after A.P. Fedchenko (Romanyuk 2017). The botanical collections of O.A. Fedchenko were exceptionally large and were processed by N.N. Kaufman, E.L. Regel, A.A. Bunge, S.A. Schmalgauzen, and others (Margolina 1941).

The German explorer W.R. Rickmers, as member of the Austro-German Pamirs Expedition (1913) and the Soviet-German Alai-Pamir Expedition (1928), had explored and mapped a large area of the unexplored core of the Pamirs (Romanyuk 2017).

N.I. Vavilov, as an organizer of biological sciences in the former Soviet Union, visited Tajikistan during Russian Empire in 1916; he repeatedly visited during the Soviet Union time with large scientific expeditions (Glazko and Bautin 2012) and published his work under the title of *Cultural Flora of Tajikistan in its Past and Future* in 1934. He wrote that in spite of relatively limited space of Tajikistan, it is one of the most interesting areas in the former Soviet Union. According to him

Table 7.1 Most important Persian-Tajik scholar works on medicinal herbs between the ninth and nineteenth centuries

Author	Books on medicinal plants or herbal remedies	Time period
Ali ibn Sahl Rabban al-Tabari	*Firdous-al-Hikmah* (*The Paradise of Wisdom*)	850
Abu Hanifa Dinavari	*Kitab-an-Nabot* (*Book of Plants*)	815–895
Abu Bakr Muhammad ibn Zakariyya al-Razi	*Al-Hawi-fi-Tibb* (Container (encyclopedia) of medicine) consist about 26 books	865–925
Abu Mansur Muwaffaq	*Kitab-al-Abniya–an-Haqoiq-al-Adviya* (The book of the foundations about the true properties of medicines), common name "Abu Mansur Pharmacopeia"	967–976
Abubakr Rabeq al Buchari	*Hidayat-ul-Mutaqallimin-fi-Tibb* (Textbook for students of medicine)	900–920
Abu Mansur Buchari	*Risola dar iloji amrozi sudur* (Treatise on the treatment of breast diseases) *Majmuai kabir dar adviyai mufrada* (A large collection of simple medicines)	970–980
Abu Sahl Masehi	"One hundred questions on the art of medicine" and "General Medicine"	1000
Miskawayh	"A book about drinks"; "A book about food"	1020–1030
Abu Rayhan Biruni	"*Kitab al-Saydala fi al-Tibb*" (Book on the Pharmacopoeia of Medicine)	1020–1045
Abu Ali ibn Sina	*al-Qanun fi al-Ṭibb* (*The Canon of Medicine*)	1000–1035
Ismail Jurjani	*Zakhirai Khorazmshokhi* (Treasury of Khorazmshakhi)	1113–1137
Mansur ibn Muhammad ibn Yusuf	*Kifoyai Mansuri* (Mansur abbreviated description)	1432
Yusuf Muhammad	*Jomeq –ul-Favoid or "Tibbi Yusufi*" (Yusuf Medicine)	1500
Subhonquli Bahadurkhan	*Ehyo-ut-Tibb* (Revival of Medicine)	1680–1702
Muhammad Husayn al-Oqili	*Karobodini kabir* (The Great Pharmacopoeia), *Makhzan-ul-Adwiya* (Treasury of medicines)	1777
Sultan Giyosuddin	*Kitob-us-Sinoat* (The book of art (healing))	XIV–XV
Bakhva ibni Havoskhon	*Maqdan-ush-Shifo* (The Source of Healing)	1489–1517
Muhammad Akbar Arzoni	*Muzharraboti Akbar* (Tests of Akbar)	XVII
Sulton Ali Tabibi Khurosoni	*Dastur-ul-Iloj* (Guide to Healing)	1527
Muhammad Aizamkhon	*Muhiti Aqzam* (The Great Ocean) or (Ocean of Wisdom)	1860–1865

the natural flora of Tajikistan is extremely rich with species, and there are at least 4000 species of flowering plants. N.I. Vavilov noted that there are great opportunities to search for new plant species and forms (fruits, rubber, essential oil-bearing plants, medicinal herbs, industrial plants, etc.) in Tajikistan (Glazko and Bautin 2012).

In 1927, Monteverde and Hammerman published their work on "Turkestans Collection of Medicinal Products of the Museum Main Botanical Garden" and characterized 284 medical preparations, obtained from 250 plant species (Monteverde and Hammerman 1927), but most plants were identified by their Tajik common names (Hojimatov 1989b). It is widely understood, however, that the flora of the Central Asia (Pamir) needs further studies (Nowak et al. 2016).

During the Soviet Union, Russian and Tajik botanists intensively studied the flora and vegetation of the country. The results of their work were published in ten volumes titled as *Flora of the Tajikistan SSR (Soviet Socialist Republic)*. These volumes contain interesting information on the medicinal characteristics of many plant species from Tajikistan. Special contributions have been made by the well-known researchers like B.A. Fedchenko, A.P. Fedchenko, V.L. Komarov, P.N. Ovchinnikov, V.I. Zapryagaeva, M.I. Ismailov, M.R. Rasulov, S.Y. Yunusov, and many others (Hojimatov 1989b).

According to the flora of the Tajikistan covering the Russian period, there are 4513 vascular plant species belonging to 116 families. Taxonomically richest families are Asteraceae (655 species), Fabaceae (520), Poaceae (325), Brassicaceae (248), and Lamiaceae (196) (Nowak and Nobis 2010).

In 1989, Hojimatov published his book with the title *Wild Medicinal Plants of Tajikistan*, where he described about 158 medicinal plant species of Tajikistan. Another valuable work on the Tajikistan medicinal plants was the dissertation of Prof. Y.J. Sadikov under the title "Biologically Active Substances of the Wild Medicinal Plants of Tajikistan." He carried out phytochemical screening of 304 wild medicinal plants of Tajikistan for the presence of alkaloids, coumarins, and essential oils. He and his coworkers isolated and identified 144 alkaloids (4 were new), 30 coumarin compounds, and 30 components of essential oils (Sadikov 2003).

Many people in Tajikistan rely on medicinal plants as their traditional form of medicine to prevent and cure health disorders. Medicinal plants have played an important role for the local people. Nearly 1500 plant species are used in the folk medicine in the country according to preliminary estimates, but only a small number of these are used in modern medicine (Hojimatov 1989b).

In this chapter we have attempted to describe the ethnopharmacology and phytochemistry of some medicinal plants from Tajikistan, with a focus on biologically active compounds (alkaloids, essential oils, polyphenols). This chapter will provide information of interest to a wide audience of international readers. Most of the data presented are taken from the original published literature written in either Russian or Tajik languages and are generally unavailable to the most readers.

7.4 Plant Secondary Metabolites

7.4.1 Alkaloid-Bearing Plants

7.4.1.1 *Anabasis turkestanica* (Korovin & Iljin) and *Anabasis aphylla* L.: Amaranthaceae

Aerial parts of *A. turkestanica* have been collected from the central part of Tajikistan (Kurgan-Tyube, 900 m) and seeds of *A. aphylla* from the northern part of Tajikistan (Isfara, 1400 m). The alkaloid contents of *A. turkestanica* and *A. aphylla* are 0.46% and 2.0%, respectively. Anabasine, anabasamine, and lupinine have been identified as major alkaloids in *A. turkestanica* and anabasine, lupinine, aphyllidine, and aphylline in *A. aphylla* (Sadikov 2003) (Fig. 7.2).

7.4.1.2 *Astragalus quisqualis* Bunge: Fabaceae

Genus *Astragalus* is the largest in the Fabaceae family. The alkaloid content in *A. quisqualis* is around 0.78% of the air-dry weight. Smirnovine and methylcytosine have been identified as major alkaloids in *A. quisqualis* (Sadikov 2003).

7.4.1.3 Genus *Berberis* (Zilol, Zirk (Tajik)): Berberidaceae

There are nearly 500 species, distributed mainly in East and Central Asia. Seven *Berberis* species grow in Tajikistan. It has a valuable fruit (berry) and medicinal plant (Zapryagaeva 1975), and it is also a decorative shrub. In traditional medicine of Central Asian inhabitants, extracts from the roots and stems of barberry were used in the treatment of various inflammatory cases like skin (eczema, ringworm, erysipelas, leprosy), stomach, liver, eyes, and mouth (stomatitis) diseases (Begenov et al. 2014). Main alkaloids of the bark and roots of *Berberis* growing in the flora of Tajikistan are recorded as berberine, berbamine, berbamunine, isoboldine, isocorydine, isotetrandrine, magnoflorine, oxyacanthine, palmatine, jatrorrhizine, and columbamine (Table 7.2). Berberine has choleretic properties, enhances uterine contractions, and lowers blood pressure. An application of various parts of barberry and its alkaloids is used for the treatment of malignancies.

The quantitative compositions of the fruits of some barberry species such as *B. vulgaris* L., *B. oblonga* (Regel.) C.K. Schneid., *B. heteropoda* Schrenk, *B. heterobotrys* E.L. Wolf., *B. amurensis* Rupr., *B. regeliana* (Regel) Koehne ex C.K. Schneid., and *B.* × *ottawensis* C.K. Schneid growing in the flora of Tajikistan have been studied. Authors emphasize that fruits of barberry contain isoquinoline alkaloids (0.01–0.63%): berberine, palmatine, jatrorrhizine, berbamine, and hydroxyacanthine (Yusufbekov et al. 1985). Sadikov and coworkers have synthesized a number

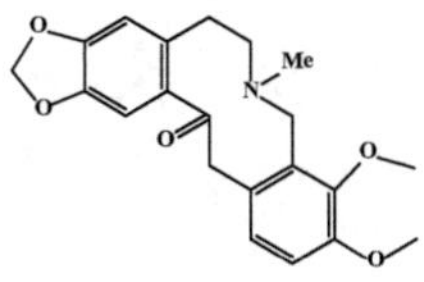

allocryptopine

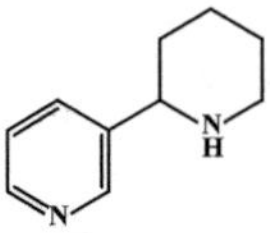

anabasine

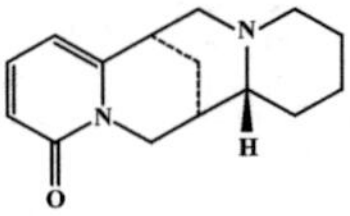

anagyrine

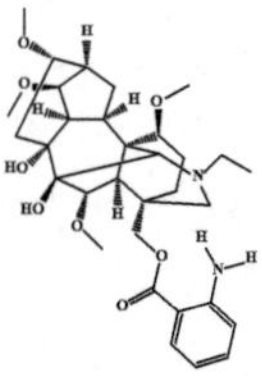

**anthranoyllyc
octonine**

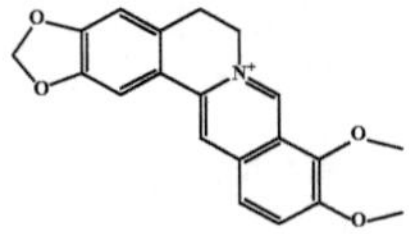

berberine

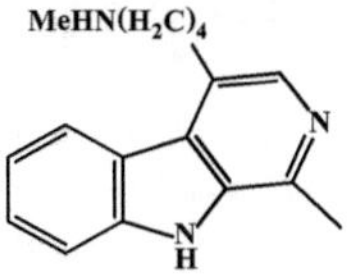

brevicarine

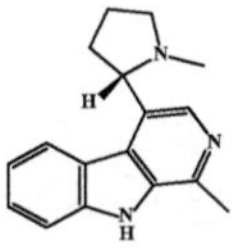

brevicolline

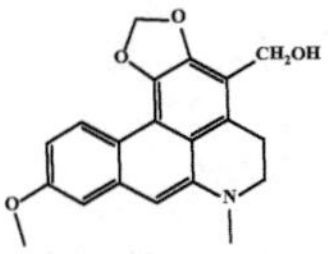

cabudine

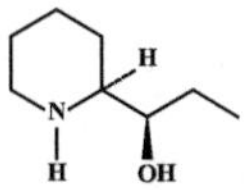

conhydrine

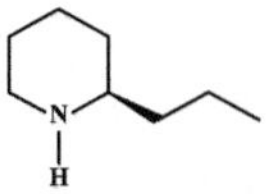

coniine

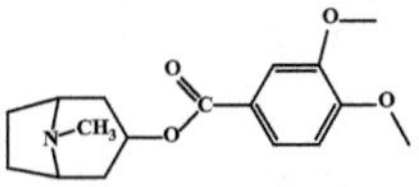

convolamine

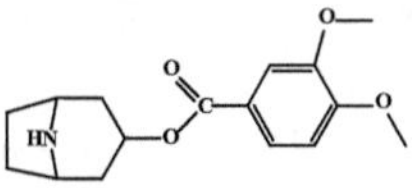

convolvine

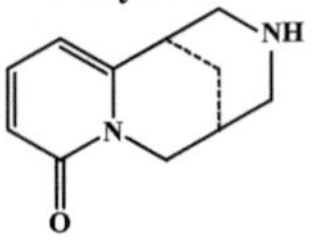

cytisine

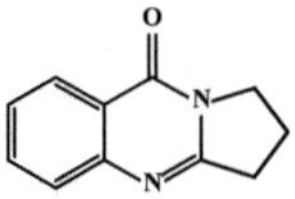

**deoxyvasicinon
e**

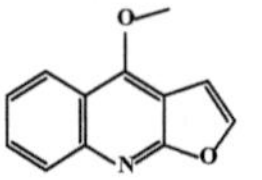

dictamnine

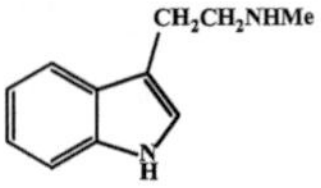

dipterine

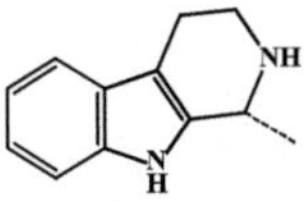

eleagnine

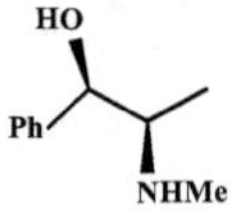

ephedrine

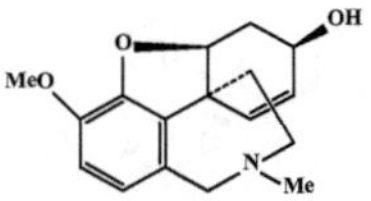

galantamine

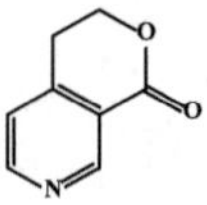

gentianadine

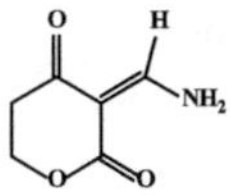

gentianaine

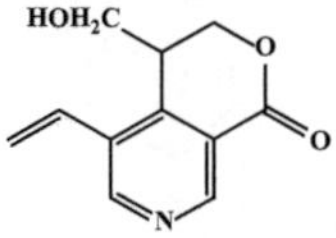

gentianamine

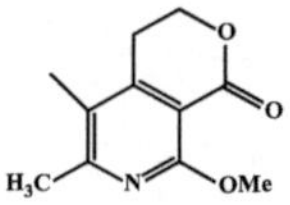

gentiananine

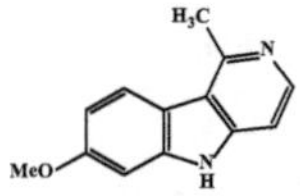

harmine

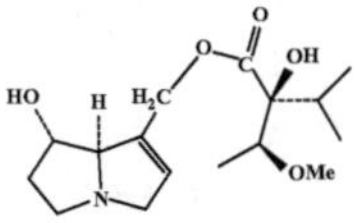

heliotrine

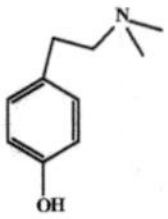

hordenine

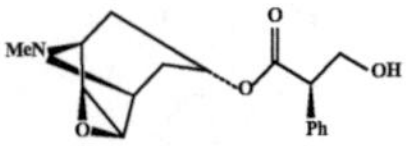

**hyoscine
(scopolamine)**

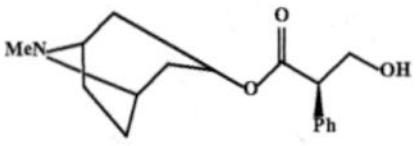

hyoscyamine

Fig. 7.2 Structures of the main alkaloids of the medicinal plants of Tajikistan

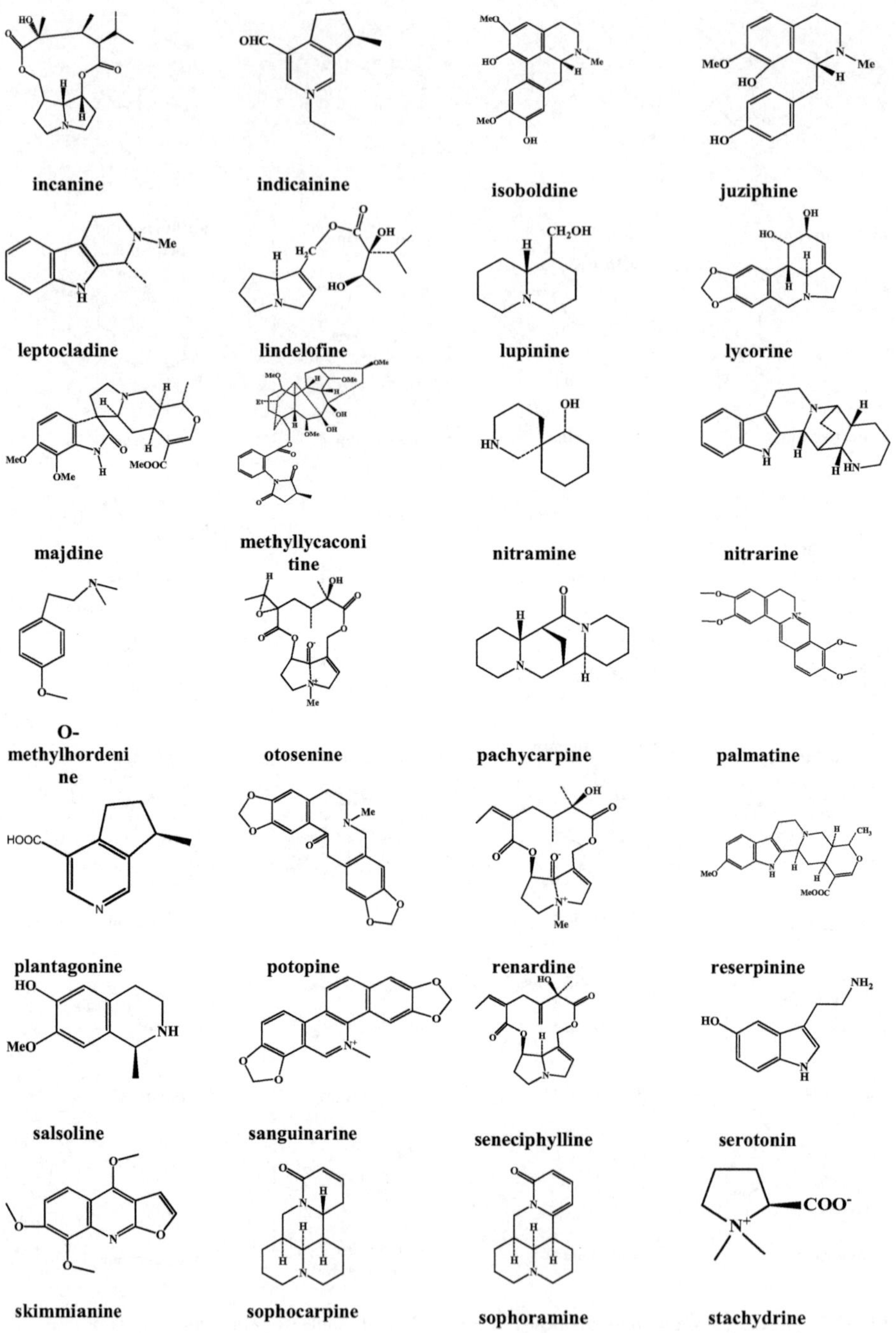

Fig. 7.2 (continued)

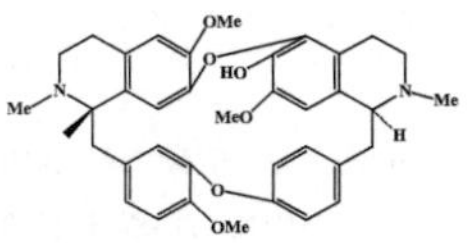

talmine

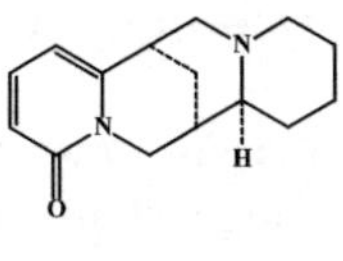

thermopsine

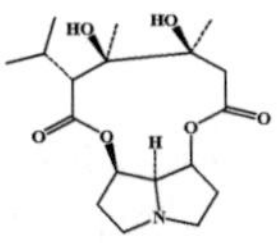

trichodesmine

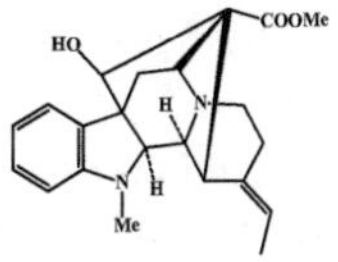

vincamajine

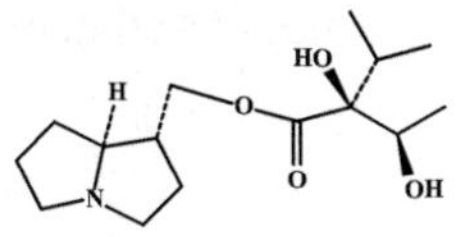

viridiflorine

Fig. 7.2 (continued)

Table 7.2 Alkaloids of *Berberis* growing in the flora of Tajikistan

Species	Distribution	Alkaloids
B. iliensis Popov	South of Tajikistan	Berberine, berbamine, oxyacanthine, jatrorrhizine
B. heterobotrys E.L. Wolf	Central, South and East parts of Tajikistan	Berberine, berbamine, oxyacanthine, palmatine
B. heteropoda Schrenk	Gissaro-Darvaz floristic region	Berberine, berbamine, oxyacanthine, palmatine jatrorrhizine, columbamine (Davidyans and Sadikov 1963)
B. integerrima Bunge	All part of Tajikistan, mainly Gissaro-Darvaz floristic region, West Pamir	Berberine, berbamine, berbamunine, isoboldine, isocorydine, isotetrandrine, magnoflorine, oxyacanthine, palmatine
B. nummularia Bunge	Pamir	Oxyacanthine, berberine

of derivatives of berberine with physiologically active acids (nicotinate, ascorbate, phenoxymethyl penicillate, pascate, acetylsalicylate, gallate) for increasing their biological activity (Davidyans and Sadikov 1963; Sadikov 2003).

7.4.1.4 *Capparis spinosa* L. (Kavar (Tajik)): Capparaceae

Capparis spinosa L. is a widely distributed shrub. The local population uses the root and fruits of *C. spinosa* as an anthelmintic, also for treating jaundice and paralysis (Dadobaeva 1972). The alkaloid content in the cortex of the roots of *C. spinosa* is around 0.47–0.91%. Main alkaloid of *C. spinosa* is the pyrrolidine alkaloid stachydrine (84–87%) (Sadikov 2003).

7.4.1.5 *Conium maculatum* L. (Tutakali (Tajik)): Apiaceae

The alkaloid content is around 0.12–1.44% of the air-dry weight of the *C. maculatum* (Hojimatov and Bobokhojaeva 1976). The highest alkaloid content has been observed during the flowering and the beginning of fruiting period (April to May) (Sadikov 2003). The alkaloids coniine (43–52%), conicine (20–27%), and conhydrine (15–21%) have been isolated and identified from the alkaloid fraction of *C. maculatum* (Sadikov and Hojimatov 1983).

7.4.1.6 *Convolvulus subhirsutus* (Regel & Schmalh.): Convolvulaceae

The alkaloid content of the aerial parts of *C. subhirsutus* collected from Varzob Region of Tajikistan (2000 m) lies around 0.43%. Convolvine and convolamine have been identified as the major alkaloids of *C. subhirsutus* (Sadikov 2003).

7.4.1.7 *Cyperus longus* L. (Salom aleykumi daroz (Tajik)): Cyperaceae

Cyperus longus L. is a perennial plant. In folk medicine, rhizomes are used as diaphoretic (sudorific) preparation. The alkaloid content of the aerial and underground parts of *C. longus* growing in Tajikistan lies around 0.095% and 0.044%, respectively (Sadikov and Begavatov 1990). The main components of the alkaloid composition of *C. longus* are brevicolline (61–84%) and brevicarine (16–39%) (Sadikov 2003).

Table 7.3 The content and identified alkaloids in *Corydalis* species growing in Tajikistan (% of the dry weight of raw materials)

Name of the plant	Organ	Amount of alkaloid	Alkaloids
C. nudicaulis Regel	Aboveground part	1.2	Sanguinarine, protopine, allocryptopine
C. macrocentra Regel	Fruit	2.1	Sanguinarine, corydaline, bulbocapnine
C. popovii Nevski ex Popov	Fruit	2.6	Isoboldine, protopine
C. ledebouriana Kar. et Kir.	Fruit	1.7	Corydaline, ledeboridine
C. sewerzowii Regel	Fruit	1.0	Bicuculline, corgoine, isocorydine
C. fimbrillifera Korch.	Fruit	1.3	Corlumine, corydaline, protopine
C. pseudostricta M. Pop.	Tubers	0.9	Bicuculline, hydrastine
C. stricta Stephan ex Fisch.	Tubers		Sanguinarine, bicuculline, hydrastine, corlumine, protopine

7.4.1.8 Genus *Corydalis*. (Havoboronak (Tajik)): Fumariaceae

Corydalis ledebouriana Kar. & Kir is widely distributed in Tajikistan. In Tajik traditional medicine, it is recommended as dormitive and calming and also for obesity and female diseases. *C. sewerzowii* Regel is used for bruises, fractures, eczema, and wounds (Hojimatov 1989b). *Corydalis* species contain a significant amount of alkaloids. The highest is observed in *C. ledebouriana* (2.6%) (Table 7.3). Fourteen individual alkaloids have been identified in the tubers of *C. ledebouriana*, growing in Tajikistan. *C. ledebouriana* can serve as an additional source of sanguinarine for medical practice (Sadikov 2003). The alkaloids isolated from *Corydalis* have an isoquinoline structure. These types of compounds generally exhibit bactericidal activity and are useful in the treatment of skin diseases, ulcers, suppurative inflammation, purulent wounds, and sinusitis (Sadikov 2003).

7.4.1.9 *Datura stramonium* L. (Bangi devona (Tajik)): Solanaceae

It is a very poisonous plant. In traditional medicine, leaves of *D. stramonium* were used as a soothing and analgesic remedy (Dadobaeva 1972). The content of alkaloids lies around 0.08–0.98% of the air-dry weight. The following alkaloids have been recorded: hyoscyamine (42–67%), hyoscine (13–25%), and atropine (0.01–0.14%) (Sadikov et al. 1978).

7.4.1.10 *Delphinium confusum* Popov: Ranunculaceae

Delphinium confusum Popov is a perennial herb growing at an altitude of 2400–2600 m above sea level and an endemic plant of the Pamir-Alai. The alkaloid content is around 0.43% of the air-dry weight of the *D. confusum*. Methyllycaconitine and anthranoyllycoctonine have been identified as major alkaloids of *D. confusum* (Narzullaev et al. 1976, 1981).

7.4.1.11 *Delphinium ternatum* Huth (Isparak (Tajik)): Ranunculaceae

Delphinium ternatum is a perennial herbaceous plant 60–120 cm tall that grows at altitudes of 1800–2400 m above sea level and is an endemic of the Pamir-Alai. In Tajik folk medicine, decoctions from the aerial parts of some species of *Delphinium* have been used in neoplasms and liver diseases and as an antihelminthic agent. The main component of the alkaloid composition of the roots of *D. ternatum* is methyllycaconitine (Narzullaev et al. 1976, 1981).

7.4.1.12 *Dictamnus tadshikorum* Vved.: Rutaceae

It belongs to Rutaceae family. The alkaloid content is around 0.56% of the roots of *D. tadshikorum* collected from Vahdat Region of Tajikistan (1450 m). Skimmianine, dictamnine, fagarine, and evoxine have been identified as major alkaloids of this species (Sadikov 2003).

7.4.1.13 *Elaeagnus angustifolia* L.: Elaeagnaceae

Elaeagnus angustifolia L. is a member of family Elaeagnaceae. The alkaloid content is around 0.95% in the roots collected from Dushanbe, Tajikistan (800 m). Eleagnine and trahydroharmol have been identified as major alkaloids (Sadikov 2003).

7.4.1.14 Genus *Ephedra* L.: Ephedraceae

There are six species of *Ephedra* in the flora of Tajikistan. They possess high content of alkaloids. Ephedrine is identified as the main alkaloid. The alkaloids and in particular ephedrine content in various types of *Ephedra* species growing in Tajikistan are shown in Table 7.4 (Sadikov 2003). The largest content of alkaloids is observed in *E. equisetina* and *E. valida.* They can serve as a source of ephedrine. Ephedrine is used in the group of adrenomimetic drugs (Dadobaeva 1972).

The *Ephedra* group of alkaloids is a small, important group of low molecular weight compounds with strong physiological activities (Buckingham et al. 2010).

7.4.1.15 Genus *Eremurus*: Asphodelaceae

Eremurus is a genus of perennial herbaceous plants of the family Asphodelaceae. The alkaloid contents of ten species of *Eremurus* growing in Tajikistan have been investigated. The contents of alkaloids and the main alkaloids of the *Eremurus* species (Sadikov 2003) are summarized in Table 7.5.

Table 7.4 Alkaloid content and ephedrine in the *Ephedra* species growing in Tajikistan

Species	Alkaloid content, %	Ephedrine, %
Ephedra equisetina Bunge	2.5	1.15
Ephedra procera C.A. Mey.	1.52	0.85
Ephedra intermedia Schrenk ex C.A. Mey.	1.35	0.88
Ephedra valida V.V. Nikitin	1.8	0.93
Ephedra gerardiana Wall. ex C.A. Mey.	0.78	0.57
Ephedra lomatolepis Schrenk	0.59	0.39

Table 7.5 The content of alkaloids in the *Eremurus* species growing in Tajikistan

Вид	Alkaloid content, %	Hordenine, % of the total alkaloid content	*O*-Methylhordenine, % of the total alkaloid content
Eremurus stenophyllus (Boiss. & Buhse) Baker	0.06	90	–
Eremurus comosus O. Fedtsch.	0.95	98	0.9
Eremurus brachystemon Vved.	0.57	84	15
Eremurus × borissianus Kamelin	0.07	93	3
Eremurus tadshikorum Vved	0.082	85	14
Eremurus olgae Regel	0.03	62	20
Eremurus aitchisonii Baker	0.44	83	13
Eremurus robustus Regel	0.08	81	18
E. hissaricus Vved.	0.06	73	–
Eremurus regelii Vved.	0.09	80	10

The highest alkaloid content has been reported in *E. comosus* and *E. brachystemon* (Table 7.5). The highest hordenine and *O*-methylhordenine content is found in *E. comosus* (98%) and *E. olgae*, respectively (Sadikov 2003).

7.4.1.16 *Gentiana cruciata* L.: Gentianaceae

Gentiana cruciata L. is a herbaceous perennial flowering plant. The alkaloid content is around 1.68% in the aerial parts collected from Darvoz Region of Tajikistan (3000 m). Gentiananine, gentianaine, gentianamine, and gentianadine have been identified as the major alkaloids of *G. cruciata* (Sadikov 2003).

7.4.1.17 *Glaucium fimbrilligerum* Boiss. (Kuknori jari (Tajik)): Papaveraceae

The alkaloid content in the aerial parts of *G. fimbrilligerum* collected from Varzob Region of Tajikistan (2100 m) lies around 0.78%. Protopine, allocryptopine, chelerythrine, sanguinarine, and codeine have been identified as the major alkaloids (Sadikov 2003).

7.4.1.18 *Hammada wakhanica* (Paulsen) Iljin: Chenopodiaceae

Aerial parts of *H. wakhanica* have been collected from Vanj Region of Tajikistan (1700 m). The alkaloid content is 1.12%. Leptocladine and dipterine are the major alkaloid components in *H. wakhanica* (Sadikov 2003).

7.4.1.19 *Haplophyllum acutifolium* (DC.) G. Don.: Rutaceae

The main alkaloids of this species growing in Tajikistan are skimmianine (41–48.5%) and evoxine (31–39%). It can serve as an additional source of skimmianine (Sadikov 2003; Sadikov and Hojimatov 1988). Also, skimmianine is found as a major alkaloid in *Dictamnus tadshikorum* (Sadikov and Hojimatov 1988).

7.4.1.20 *Heliotropium lasiocarpum* Fisch. & C.A. Mey. (Geliotrop opushennoplodniy Khazarrang (Tajik)): Boraginaceae

This species contains a significant amount of alkaloids (Yunusov and Sidyakin 1950). The highest amount of alkaloids in Tajikistan has been observed in the flowering period (0.63%) (Sadikov and Hojimatov 1977). The primary component of the total alkaloid fraction is heliotrine (90–93%) (Sadikov 2003). Heliotrine is toxic and causes chronic liver poisoning in animals. The compound also shows ganglion-blocking activity and antineoplastic properties (Buckingham et al. 2010).

7.4.1.21 *Hippophae rhamnoides* L. (Angat (Tajik)): Elaeagnaceae

The content of alkaloids is around 0.95% in the annual stems collected from Pamir, Tajikistan (2900 m). Serotonin and harmine have been identified as the major alkaloids (Sadikov 2003). It is an important resource in Tajikistan. Fruits contain up to 9% oil, which has epithelizing, wound healing, and analgesic properties. The oil contains fatty acids and carotenoids (40–360 mg/100 g oil) (Gachechiladze et al. 1981; Korzinikov et al. 1984).

7.4.1.22 *Lindelofia macrostyla* (Bunge) Popov: Boraginaceae

Lindelofia macrostyla (Bunge) Popov is a perennial herbaceous shrub and 50–100 cm tall. The plants in this genus are alkaloid-bearing (Kiyamidinova et al. 1967). Total alkaloid contents of *L. macrostyla* growing in Tajikistan are 1.2–1.5% in leaves and 2.1–2.5% in fruits. Main alkaloids of *L. macrostyla* are lindelofine (62–63%), viridiflorine (23–26%), and echinatine (9–10%) (Sadikov 2003).

7.4.1.23 *Lindelofia olgae* (Regel & Smirn.) Brand: Boraginaceae

The alkaloid content is 2.8% in the aerial parts collected from Karotegin Valley of Tajikistan (2800 m). Viridiflorine and lindelofine have been identified as the major alkaloids (Sadikov 2003).

7.4.1.24 *Mahonia aquifolium* (Pursh) Nutt.: Berberidaceae

The alkaloid composition of this species growing in the botanical garden of the Academy of Sciences of Tajikistan has been studied. The alkaloid composition is similar to the isoquinoline (protoberberine) alkaloids of barberry (berberine, palmatine, jatrorrhizine, columbamine) (Khusainova and Sadikov 1986).

7.4.1.25 *Nitraria schoberi* L.: Nitrariaceae

The content of alkaloids is around 0.62% in the underground part (bulbs) collected from northern Tajikistan (1000 m). Nitrarine, nitramine, and isonitrarine have been identified as the major alkaloids (Sadikov 2003).

7.4.1.26 *Papaver pavoninum* Schrenk (Lolaarusak (Tajik)): Papaveraceae

Papaver pavoninum is an annual plant growing at an altitude of 800–3000 m above sea level. The plant has large reserves of raw materials in Tajikistan. The content of alkaloids (% of dry weight) during flowering-fruiting varies in leaves (0.41%), stems (0.18%), flowers (0.06%), green fruits (0.44%), mature fruit (0.27%), and roots (0.05%). Main alkaloids in the plants growing in Tajikistan are protopine (71–82%) and α-allocryptopine (16–23%) (Sadikov and Hojimatov 1985).

7.4.1.27 *Pedicularis peduncularis* Popov: Orobanchaceae

In Tajik folk medicine, the decoction from aboveground part of *P. peduncularis* is used as a hemostatic for uterine bleeding, and its dried flowers are used as diuretic. Baths from fresh plants are taken to treat various skin diseases. Pyrimidine alkaloids—plantagonine, indicainine, plantagonin, indicin, peducularine, and *N*-methylcytisine—have been isolated and identified from the aboveground parts of *P. peduncularis* (Sadikov 2003).

7.4.1.28 *Pedicularis sarawschanica* Regel: Orobanchaceae

Alkaloid content in the fruits collected from Varzob Region of Tajikistan (3200 m) is around 0.91%. Plantagonine and peducularine have been identified as the major alkaloids.

7.4.1.29 *Peganum harmala* L. (Ispand (Tajik)): Nitrariaceae

Peganum harmala is a perennial alkaloid-rich, highly poisonous plant of Nitrariaceae (Zygophyllaceae). In Tajik folk medicine, it is used to treat Parkinson's disease and encephalitis (Dadobaeva 1972). The main alkaloids in the plants growing in

Tajikistan are harmine (10–75%) and deoxyvasicinone (5–70%). The highest content of harmine has been observed in green fruits and roots; deoxyvasicinone was the main alkaloid of the aerial part (Sadikov 2003).

7.4.1.30 *Petilium eduardii* Vved.: Liliaceae

The content of alkaloids is around 1.26% as determined in the aerial parts of the plant collected from Karotegin Valley of Tajikistan (3200 m). Imperialine, peimisine, and edpetiline have been identified as the major alkaloids (Sadikov 2003).

7.4.1.31 *Salsola pestifer* A. Nelson (Bodgelak (Tajik)): Chenopodiaceae

Salsola pestifer is an annual herbaceous plant growing at an altitude of 350–3000 m above sea level (Kinzikaeva 1968). In Tajik folk medicine, the decoction and juice are used to treat skin diseases. The content of alkaloids in leaves, stems, seeds, and plant roots studied on air-dry weight basis reveals a range of 0.01–0.51%. Highest content of alkaloids in the plant is observed in seeds (0.51%) and the lowest in roots (0.01%). D-salsoline and L-salsolidine are the main components of the alkaloid composition (Sadikov 2003).

7.4.1.32 *Senecio renardii* C. Winkl.: Asteraceae

This is a flowering plant with alkaloid contents of 0.98% in the samples collected from Darvoz Region of Tajikistan (2900 m). Renardine, seneciphylline, and otosenine have been identified as the major alkaloids (Sadikov 2003).

7.4.1.33 *Spartium junceum* L.: Fabaceae

This flowering plant grows up to 3 m tall, with stalks and twigs cultivated as a decorative plant in Tajikistan. The plant contains considerable amount of alkaloids (0.56–0.88% of the dry weight). Main alkaloids are cytisine, *N*-methylcytisine, and anagyrine. The highest accumulation of alkaloids in the plants is observed during the flowering period (May and June) (Sadikov and Hojimatov 1981).

7.4.1.34 *Thalictrum* Genus: Ranunculaceae

Thalictrum is a genus of 120–200 species of herbaceous perennial flowering plants. Aerial parts of *T. collinum* Wallr. have been collected from Varzob Region of Tajikistan (2400 m) and the roots of *T. kuhistanicum* Ovcz. & Kochk. during the flowering period from Vanj Region of Tajikistan (2500 m). The contents of the alkaloids of *T. collinum* and *T. kuhistanicum* have been determined as 1.23% and 1.13%, respectively. Talmine has been identified as the major alkaloid in both species (Kurbanov et al. 1988; Sadikov 2003).

7.4.1.35 *Thalictrum isopyroides* C.A. Mey.: Ranunculaceae

Thalictrum isopyroides is used in traditional medicine for fever and colitis and as spasmolytic. For the first time, new alkaloid cabudine is isolated from *T. isopyroides* growing in Tajikistan (Kurbanov et al. 1975).

7.4.1.36 *Thermopsis dolichocarpa* V. Nikitin (Mastak (Tajik)): Fabaceae

Thermopsis dolichocarpa is a perennial herbaceous plant that is 20–80 cm tall. It is distributed in the foothills at an altitude of 960–2800 m above sea level in Tajikistan. The plant is used as an expectorant. Its seeds are used as a source of cytisine. The amount of alkaloids in the plants growing in Tajikistan is around 2.37%, from which cytisine (66–68%), *N*-methylcytisine (6.03%), thermopsine (20–21%), and pachycarpine (5–10%) have been isolated and identified (Sadikov and Hojimatov 1990). There are large reserves of this plant in Tajikistan. This plant can be used as an additional source of cytisine production for medical application (Sadikov 2003).

7.4.1.37 *Trichodesma incanum* (Bunge) A. DC. (Kampirchapon (Tajik)): Boraginaceae

In Tajik folk medicine, this flowering plant is used for the treatment of furuncles and scabies. A decoction of the leaves and stems is used for the treatment of various types of external tumors (Dadobaeva 1972). Dynamics of the accumulation of alkaloids has been studied in various phases of development. Highest content of alkaloids has been observed in the phase of early vegetation (0.97%). Trichodesmine (53–66%) and incanine (1–20%) are the main components of alkaloid content of *Trichodesma incanum* (Sadikov 2003).

7.4.1.38 *Ungernia tadschicorum* Vved. ex Artjush. (Amonkara (Tajik)): Amaryllidaceae

The plant is used in the treatment of stomach ulcers, furuncles, and skin cancers in Tajik folk medicine (Hojimatov 1989b). Highest content of alkaloids has been recorded in the roots (1.75%) and fruits (1.3%) and lowest in stems (0.32%). Lycorine, galantamine, hippeastrine, pancratine, tanzettine, ungerine, and hordenine alkaloids have been isolated from *U. tadschicorum* (Sadikov 2003). Galantamine hydrobromide and lycorine hydrobromide are used in scientific medicine.

Galantamine is used for the treatment of mild to moderate Alzheimer's disease and various other memory impairments, in particular those of vascular origin (Wilcock et al. 2003). Lycorine has been seen to have promising biological and pharmacological activities such as antibacterial, antiviral, or anti-inflammatory effects and may have anticancer properties (Jahn et al. 2012).

7.4.1.39 *Ungernia victoris* Vved. (Amonkara (Tajik)): Amaryllidaceae

Ungernia victoris is a perennial bulbous plant. It grows in the Gissaro-Darvaz floristic region at an altitude of 1600–2400 m above sea level in Tajikistan (Vvedenskiy 1963). In Tajik folk medicine, the baked bulbs are used as a wound-healing remedy. The bulbs possess four alkaloids identified as galantamine, lycorine, narwedine, and hordenine (Sadikov 2003).

7.4.1.40 *Verbascum songaricum* Schrenk ex Fisch. & C.A. Mey. (Dumi govak (Tajik)): Scrophulariaceae

Verbascum songaricum is an alkaloid-rich biennial, flowering plant (Ziyaev et al. 1971). The content of alkaloids (in % of the dry weight) is as follows: in leaves (0.18%), stems (0.09%), buds (0.21%), flowers (0.68%), green fruits (0.26%), mature fruit (0.22%), and roots (0.07%). The main alkaloids of found are anabasine (62–66%) and plantagonine (29–36%) (Sadikov et al. 1984; Sadikov 2003).

7.4.1.41 *Vexibia alopecuroides* (L.) Yakovlev: Scrophulariaceae

Total alkaloid content of the seeds growing in Kharangon (1900 m above sea level) in Tajikistan is 1.43%. Sophoramine and sophocarpine have been identified as major components of the alkaloids (Sadikov 2003).

7.4.1.42 *Vexibia pachycarpa* (Synonym, *Sophora pachycarpa* Schrenk ex C.A. Mey) (Schrenk ex C.A. Mey.) Yakovlev (Talkhak, Kharbuya (Tajik)): Scrophulariaceae

Vexibia pachycarpa is a perennial herbaceous plant. In Tajik folk medicine, this plant reported to stimulate breathing and acts as a ganglion blocking agent (Dadobaeva 1972). However, it is considered as a poisonous plant, quarantine weed, and a strong insecticidal agent. The content of alkaloids is around 0.76–1.93% of the air-dry weight. Main alkaloids of the alkaloid are pachycarpine (44–46%), sophocarpine (30–31%), matrine (22–23%), and cytisine (0.02–0.09%) (Sadikov et al. 1985b). In medicine, pachycarpine hydroiodide is used as a ganglion blocker, mainly for spasms of peripheral vessels, as well as for induction of labor activity.

7.4.1.43 *Vinca major* L.: Apocynaceae

Vinca major is an evergreen creeper with stems that are 36–124 cm long and is cultivated in the Dushanbe Botanical Garden and green plantations of Dushanbe as a soil cover plant (Rasulova 1975). The alkaloid content has been observed as follows: in leaves (1.33%), stems (1.02%), flowers (0.06%), and roots (0.89%) (Sadikov et al. 1985a). The main alkaloids in the aerial parts in the flowering phase are reserpinine (28%), majdine (22.5%), vincamajine (19%), akuammine (14%), and majorine (10%) (Sadikov 2003).

7.4.1.44 *Ziziphus jujuba* Mill.: Rhamnaceae

Alkaloid content is around 1.26% in the plants collected from Shurobod Region of Tajikistan (2200 m). Isoboldine and juziphine have been identified as the major alkaloids (Sadikov 2003).

7.4.2 Phenolic-Rich Plants

7.4.2.1 *Hypericum scabrum* L. (Choykahak (Tajik)): Hypericaceae

The species belonging to the genus *Hypericum* are medicinal plants known since ancient times as healing herbs because of their medicinal value. In Tajik traditional medicine, *H. scabrum* is also used in place of *H. perforatum* L. (Sharopov et al. 2010). Eight pure phenolic compounds—3-8″-bisapigenin, quercetin, quercetin-3-*O*-α-L-arabinofuranoside, quercetin-3-*O*-α-L-rhamnoside, quercetin-3-*O*-β-D-glucopyranoside, quercetin-3-*O*-β-D-galactopyranoside, (−)-epicatechin, and (+)-catechin—were identified from the aerial part of *H. scabrum*. Total polyphenolic compounds and total flavonoid contents were determined in the extract of *H. scabrum* as 107 and 23 mg/g of the dried extract, respectively. Quercetin, quercetin-3-*O*-β-D-glucopyranoside, and quercetin-3-*O*-β-D-galactopyranoside showed very strong DPPH free radical scavenging activity. Seventy percent ethanol extract of *H. scabrum* gave best results for protein-tyrosine phosphatase 1B (PTP-1B) inhibition assay with IC_{50} of 1.57 μM (Jiang et al. 2015). The structures of the major phenolic compounds from medicinal plants of Tajikistan are shown in Fig. 7.3.

7.4.2.2 *Geranium collinum* Stephan ex Willd.: Geraniaceae

Geranium collinum is a perennial plant used for the treatment of rheumatism, gout, dysentery, and external and internal bleeding, as well as in the treatment of skin wounds, eczema, scabies, tenosynovitis, and pruritus (Numonov et al. 2017). This plant has been proven to be effective for curing pneumonia, catarrhal symptoms of

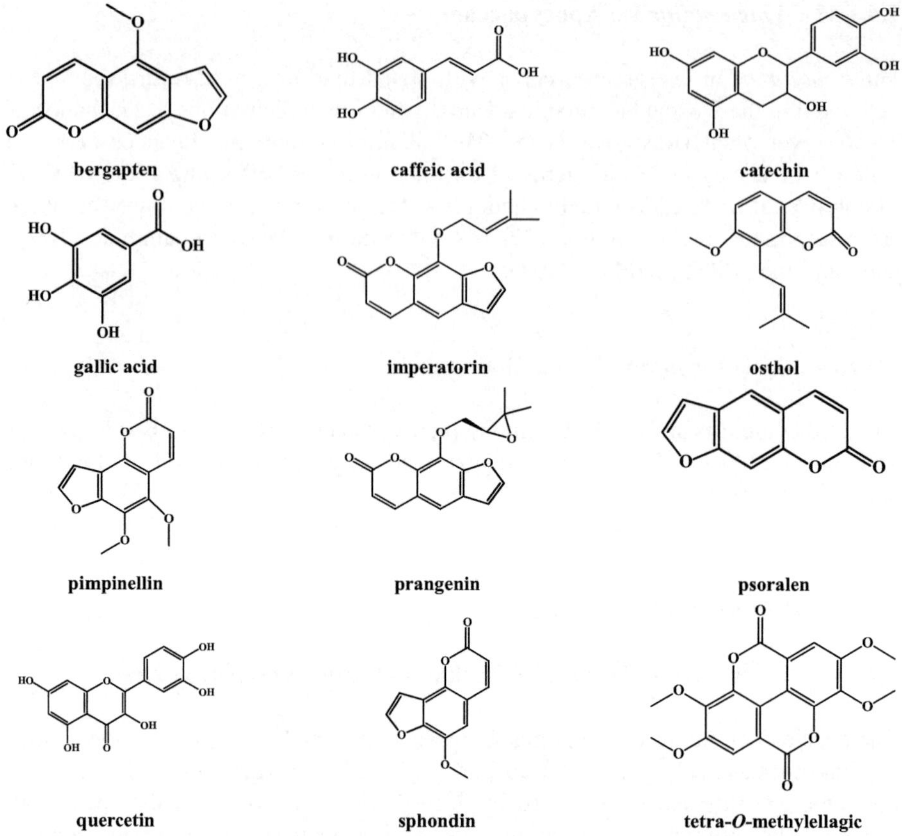

Fig. 7.3 Structure of main phenolic compounds of the medicinal plants of Tajikistan

the stomach and intestines, as well as certain gynecological diseases (Numonov et al. 2017). A 50% aqueous-ethanolic extract of the roots of this plant possesses potent antidiabetic activity, with IC_{50} values of 0.10 and 0.09 μg/mL for the enzymes protein-tyrosine phosphatase 1B (PTP-1B) and α-glucosidase, respectively. Phytochemical investigations on the 50% aqueous-ethanolic extract of *G. collinum* led to the isolation of ten pure compounds identified as 3,3′,4,4′-tetra-*O*-methylellagic acid, 3,3′-di-*O*-methylellagic acid, quercetin, caffeic acid, (+)-catechin, (−)-epicatechin, (−)-epigallocatechin, gallic acid, β-sitosterol-3-*O*-β-D-glucopyranoside, and corilagin. Three isolated compounds exhibited strong inhibitory activity against PTP-1B, with IC_{50} values below 0.9 μg/mL, more effective than the PTP-1B-enzyme inhibitor (3-(3,5-dibromo-4-hydroxy-benzoyl)-2-ethylbenzofuran-6-sulfonic acid-(4-(thiazol-2-ylsulfamyl)-phenyl)-amide) (1.46 μg/mL) (Numonov et al. 2017).

7.4.3 *Coumarins of Some Species of the Apiaceae (Umbelliferae) Family*

Hojimatov and coauthors have published data on the composition of coumarins of 38 species of the Apiaceae. High contents of coumarins have been found in the roots (as much as 6.5%) and fruits (up to 4.2%). Some species of *Prangos* Lindl., *Ferula* L., *Heracleum* L., *Pachypleurum* Hoff., *Conioselinum* Fisch., *Libanotis* L., and *Seseli* L. are reported to be rich in coumarin contents (Hojimatov et al. 1979).

7.4.3.1 *Prangos sarawschanica* Korovin (Yugan (Tajik)): Apiaceae

Prangos sarawschanica is a perennial large herbaceous plant, with large reserves in Tajikistan. This plant is used in Tajik traditional medicine for treating vitiligo. Representatives of the *Prangos* genus contain furocoumarins with antitumor activity (Zetlin and Nikonov 1965). Three coumarins, osthol, prangenin, and imperatorin, have been isolated and identified from the roots of *P. sarawschanica* densely distributed in Tajikistan (Sadikov 2003).

7.4.3.2 *Prangos fedtschenkoi* Korovin (Yugan (Tajik)): Apiaceae

Osthol, prangenin, and imperatorin have been isolated from the fruits of this species in Tajikistan (Sadikov 2003). It has also a large source in Tajikistan; fruits can serve as a source of imperatorin for the treatment of vitiligo.

7.4.3.3 *Heracleum lehmannianum* Bunge (Boldirgon (Tajik)): Umbelliferae

Heracleum lehmannianum is a biannual, strongly poisonous, and fragrant plant that is used for treating heart disease and rheumatism in the Tajik traditional medicine (Dadobaeva 1972). The highest number of coumarins is found in the roots (0.5%). Nine furocoumarins—pimpinellin, isopimpinellin, bergapten, isobergene, psoralen, sphondin, scopoletin, umbelliferone, and angelicin—have been isolated from the roots of this species (Sadikov 2003). In addition, 28 components of the essential oil have been identified in *H. lehmannianum* growing in Tajikistan. The main components of this oil are limonene (12%), hexyl butyrate (28%), octyl acetate (20%), and ethyl caprylate (21%) (Sadikov 1990).

7.4.4 Essential Oil-Bearing Plants

7.4.4.1 *Achillea filipendulina* Lam.: Asteraceae

This plant has been used since ancient times in traditional herbal medicines for a variety of ailments. Decoctions have been used to treat sciatica, gout, arthritis, congestion, cardiovascular diseases, gastrointestinal disturbances, and malaria, as well as a diuretic, anthelmintic, and purgative (Saeidnia et al. 2011). Externally, the plant has been used to treat scabies and wounds.

Essential oils in this species from two different sites in south-central Tajikistan have been studied following hydrodistillation and analysis by gas chromatography-mass spectrometry. The major components of the oil are santolina alcohol (43.6–46.3%), 1,8-cineole (8.8–11.4%), borneol (5.3–6.0%), isoborneol (4.8–5.4%), and *cis*-chrysanthenyl acetate (6.5–9.3%) (Sharopov and Setzer 2010) (Fig. 7.4). Antioxidant, antimicrobial, cytotoxic, and anti-inflammatory activities of plant extracts are summarized in Tables 7.6, 7.7, 7.8, 7.9, 7.10, 7.11, and 7.12.

According to DPPH and ABTS analysis, the volatile oil shows antioxidant activity with IC_{50} values of 4.83 and 2.01 mg/mL, respectively. FRAP value was 214.2 μM Fe (II)/mg oil. Antioxidant activity of the methanol and water extract of *A. filipendulina* plants from Tajikistan are much higher compared to its oil. IC_{50} values of DPPH and ABTS of methanol and water extracts vary between 4.2–15.6 and 88–23.6 μg/mL, respectively. FRAP values range from 627 to 708 μM Fe (II)/mg sample.

The cytotoxicity of the oil has been tested against HeLa, Caco-2, MCF-7, CCRF-EM, and CEM/ADR5000 cancer cell lines: IC_{50} values were 209.9 μg/mL for HeLa, 551.5 μg/mL for Caco-2, 223.7 μg/mL for MCF-7, 29.9 μg/mL for CCRF-CEM, and 45.1 μg/mL for CEM/ADR5000 cell lines. Antimicrobial activity of essential oil of this plant has been tested against *Escherichia coli (*ATCC 25922) and methicillin-resistant *Staphylococcus aureus* (MRSA) (NTCT 10442). MIC and MBC ranged between 5 and 10 mg/mL. The essential oil inhibits soybean 5-lipoxygenase (5-LOX) with an IC_{50} value of 221.3 μg/mL (Sharopov 2015).

7.4.4.2 *Anethum graveolens* L. (Dill): Apiaceae

Anethum graveolens has been widely used for flavoring foods and beverages due to its pleasant spicy aroma. It has been extensively used as a traditional herbal medicine throughout Europe, Asia, and America (Sharopov et al. 2013a). In Tajik traditional medicine, its leaves are used mainly for digestive and diuretic disorders (Nuraliev 1989). Major components of the essential oil of the plants growing in Tajikistan are carvone (51.7%), *trans*-dihydrocarvone (14.7%), dill ether (13.2%), α-phellandrene (8.1%), and limonene (6.9%). Its cytotoxicity has been tested against HeLa (human cervical cancer), Caco-2 (human colorectal adenocarcinoma), and MCF-7 (human breast adenocarcinoma) cancer cell lines: IC_{50} values were 93 μg/mL for HeLa,

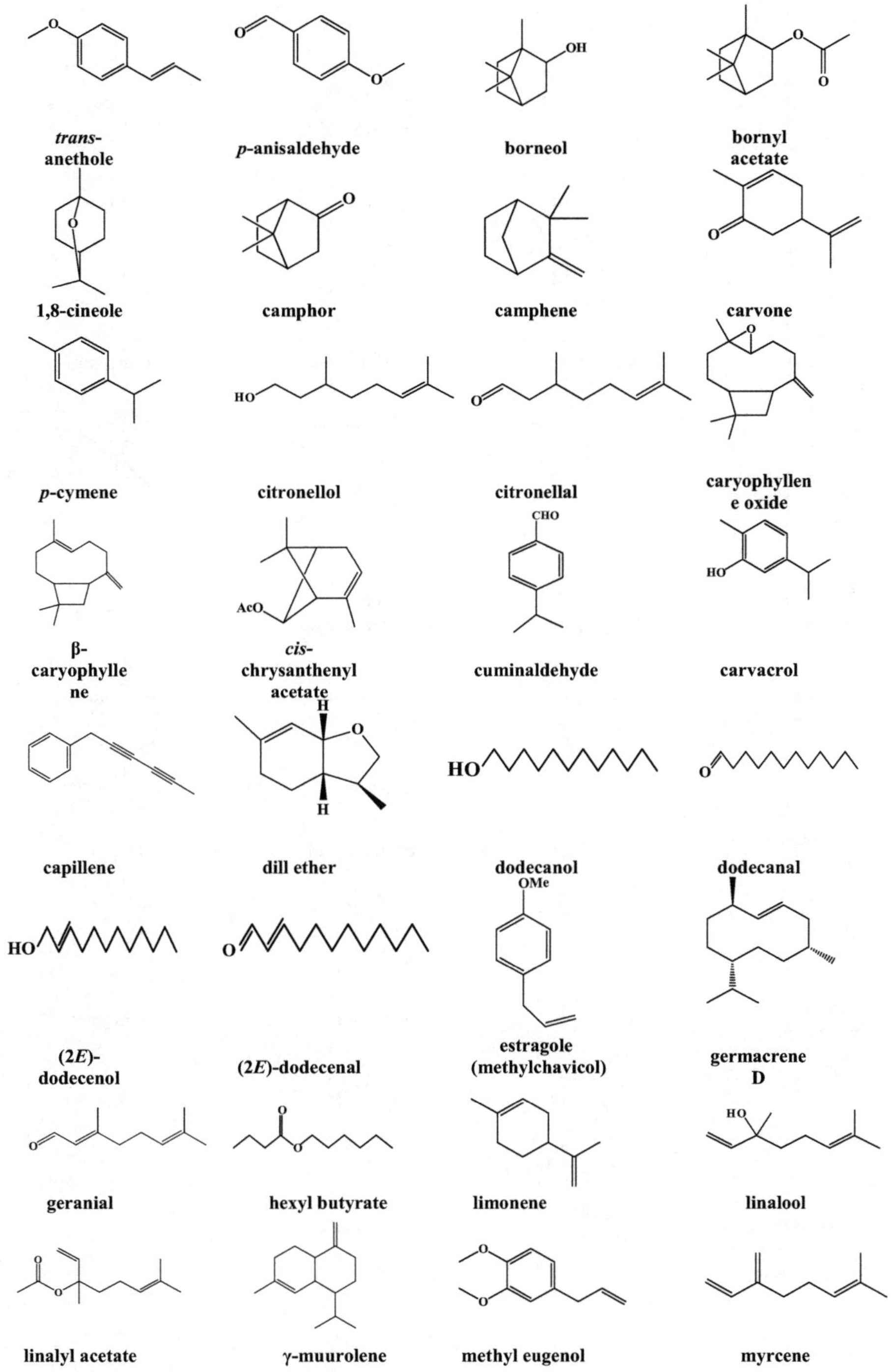

Fig. 7.4 Structure of the major compounds of the essential oils of Tajikistan plants

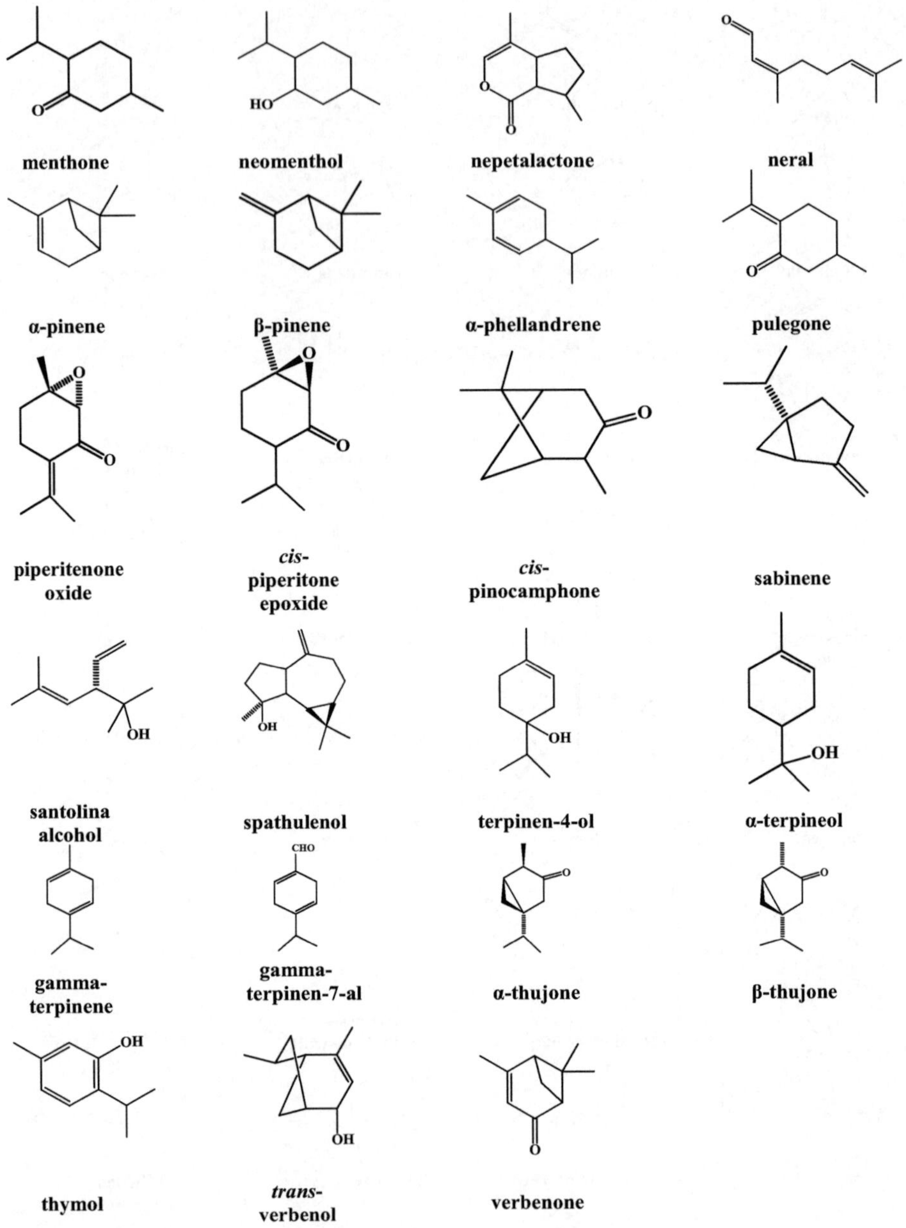

Fig. 7.4 (continued)

Table 7.6 Antioxidant activity of the essential oils (Sharopov 2015)

Species	DPPH, IC_{50}, mg/mL	ABTS, IC_{50}, mg/mL	FRAP, μM Fe (II)/mg sample	Degree of linoleic acid peroxidation, %; Csam. = 0.9 mg/mL
Achillea filipendulina	4.83	2.01	214.2	0.5
Anethum graveolens	4.98	4.12	47.9	0.4
Artemisia absinthium	1.35	0.87	338.9	33.3
Artemisia rutifolia	7.91	0.25	74.2	17.6
Artemisia scoparia	2.55	0.28	43.1	18.5
Ferula clematidifolia[a]	15.7	1.55	124.5	n.d.
Ferula clematidifolia[b]	14.72	0.45	434.75	n.d.
Foeniculum vulgare	15.6	10.9	193.5	n.d.
Galagania fragrantissima[c]	8.34	11.68	208.33	n.d.
Galagania fragrantissima[a]	8.21	6.46	315.83	n.d.
Galagania fragrantissima[d]	8.13	4.74	67.2	n.d.
Hypericum perforatum	3.71	0.48	98.25	n.d.
Hypericum scabrum	6.69	5.67	22.5	n.d.
Hyssopus seravschanicus	4.9	1.39	53.8	54.2
Mentha longifolia	2.31	0.67	76.9	30.6
Ocimum basilicum	5.94	7.98	51.6	n.d.
Origanum tyttanthum	0.28	0.12	699.2	79.7
Polychrysum tadshikorum	5.14	6.7	239.17	n.d.
Salvia sclarea[e]	12.5	5.03	54	28.5
Salvia sclarea [c]	n.a.	0.3217	277.25	n.d.
Salvia sclarea[a]	14.5	7.9	121.42	n.d.
Tanacetum parthenium	4.82	0.96	n.d.	37.1
Ziziphora clinopodioides	5.12	0.79	66.9	15.3
Ascorbic acid	0.002	0.005	1899.5	85.1

n.a. not active, *n.d.* not determined
[a]From leaves
[b]From root
[c]From flowers
[d]From stem
[e]From all part

Table 7.7 Total phenolic and flavonoid content of the plant extracts from Tajikistan (Sharopov 2015)

Species	TP, mg CAE in 100 g DW		TF, mg QE in 100 g DW	
	ME	WE	ME	WE
Achillea filipendulina	2661.7	1713.9	46.5	114.6
Anethum graveolens	846.8	369.3	117.9	6.5
Artemisia absinthium	1657	976.9	121.8	104.6
Artemisia rutifolia	1611.3	637.1	599.7	249.1
Artemisia scoparia	355.3	408.2	312.2	36.5
Ferula clematidifolia	1443.4	699.3	533.6	60.5
Galagania fragrantissima	2924.1	2368.3	733.3	258.7
Hypericum perforatum	559.5	800.6	462.3	98
Hypericum scabrum	758.3	1530.1	371.4	159.4
Melissa officinalis	129.1	70.4	177.1	3.5
Mentha longifolia	274.5	761.6	231.2	16.7
Ocimum basilicum	794.2	1108	649.5	25.7
Origanum tyttanthum	1589.3	1353.9	59.2	78.3
Polychrysum tadshikorum	1831.5	731.1	640.6	143.7
Salvia sclarea	612.2	1478.6	18.2	399.8
Tanacetum parthenium	1413.5	1524.1	759.7	98.4
Tanacetum vulgare	1387.3	423.9	655	168.3
Ziziphora clinopodioides	1508.4	822.8	825.2	108.8

216 μg/mL for Caco-2, and 67 μg/mL for MCF-7 cells. Its essential oil has proven active in the brine shrimp lethality test (LC_{50}, 15.9 ± 2.3 μg/mL) but showed only marginal antimicrobial activity against *Escherichia coli* (MIC, 625 μg/mL) (Sharopov et al. 2013a).

7.4.4.3 *Artemisia absinthium* L.: Asteraceae

Artemisia absinthium is traditionally used in Tajikistan and is especially involved in anthelmintic, digestive, antifungal, and antibacterial, but also balsamic, diuretic, and emmenagogue activities. The extract (~2 teaspoons chopped herbs in a glass of boiled water, the daily dose) is utilized to treat hyperacidity, gastric colic, gastritis, flatulence, and conditions of the liver and gallbladder. Chronic, large doses have been reported to upset the nervous system (Sharopov et al. 2012c). The major components of essential oil of *A. absinthium* from Tajikistan are myrcene (8.6–22.7%), *cis*-chrysanthenyl acetate (7.7–17.9%), a dihydrochamazulene isomer (5.5–11.6%), germacrene D (2.4–8.0%), thujone (0.4–7.3%), linalool acetate (trace-7.0%), α-phellandrene (1.0–5.3%), and linalool (5.3–7.0%) (Sharopov et al. 2012c). Its oil was screened in antioxidant [DPPH and ABTS radical scavenging, FRAP (ferric reducing antioxidant power)] assays and cytotoxic (HeLa, Caco-2, MCF-7, CCRF-CEM, and CEM/ADR5000 cell lines) and lipoxygenase inhibitory activities.

Table 7.8 Antioxidant activity of the methanol and water extracts of plants from Tajikistan (Sharopov 2015)

	DPPH, µg/mL		ABTS, µg/mL	FRAP, mM $FeSO_4$/1 g extract		
Species	ME	WE	ME	WE	ME	WE
Achillea filipendulina	15.6	4.2	23.6	8.8	627.2	708.3
Anethum graveolens	30.2	132.6	40.3	52.5	1119.6	579
Artemisia absinthium	20.1	45.8	34.1	34	460.2	541.8
Artemisia rutifolia	12.7	21.5	16.6	15.8	580.4	564.1
Artemisia scoparia	100.5	143.3	87.4	212.3	380.1	380
Caffeic acid	1.66		2.01		2393.2	
Ferula clematidifolia	10.2	55.4	3.8	64.6	976.1	550
Galagania fragrantissima	4.1	24.8	4	16.6	1217.9	1082
Hypericum perforatum	52.7	75.6	62.5	71.3	783.9	577.7
Hypericum scabrum	26.1	25.8	74.9	62.5	768.9	650.5
Melissa officinalis	212.3	n.a.	n.a.	250.3	245.2	73.2
Mentha longifolia	285.6	134.7	n.a.	54.9	250.7	268.2
Ocimum basilicum	100.6	63.1	123.4	56.1	542.9	475.9
Origanum tyttanthum	9.5	31. 3	23.6	34.8	744.5	565.5
Polychrysum tadshikorum	12.7	44.5	32.6	41.4	485	505.8
Salvia sclarea	39.2	23.4	84.8	15. 8	542	584.5
Tanacetum parthenium	18.9	34.8	16	32.7	569.3	558.7
Tanacetum vulgare	23.8	73.6	31.3	62.7	514.1	417.7
Ziziphora clinopodioides	20.5	354.6	36.2	63	611.6	520.8

n.a. not active

Antioxidant, antimicrobial, cytotoxic, and anti-inflammatory activities of plant extracts are summarized in Tables 7.6, 7.7, 7.8, 7.9, 7.10, 7.11, and 7.12. In the DPPH and ABTS radical scavenging assays, *A. absinthium* oil shows antiradical activity with IC_{50} values of 1.35 and 0.87 mg/mL, respectively. FRAP value was 338.9 µM Fe (II)/mg sample. The cytotoxic activities of the oil against HeLa, Caco-2, MCF-7, CCRF-CEM, and CEM/ADR5000 cell lines show IC_{50} values of 95.6, 190.2, 125.9, 60.34, and 85.2 g/mL, respectively. Antimicrobial activity has also been investigated on *E. coli* ATCC 25922 and MRSA NCTC 10442 microorganisms. The essential oil has exhibited activity against MRSA with MIC and MBC as 20 mg/mL. It has also exhibited remarkable lipoxygenase inhibitory activities with an IC_{50} of 57 µg/mL (Sharopov 2015; Sharopov et al. 2015b).

7.4.4.4 *Artemisia rutifolia* Stephan ex Spreng.: Asteraceae

Artemisia rutifolia is used as a tonic, febrifuge, and anthelmintic agent in traditional medicine. A powder of the plant mixed with honey is useful against worms. A tea prepared from the dried and chopped herb is drunk to treat asthma and weakness of the heart and also as anti-inflammatory, diuretic, and anthelmintic agent. In Tajikistan

Table 7.9 Cytotoxicity, IC_{50} (μg/mL) of essential oils (Sharopov 2015)

Species	HeLa	Caco-2	MCF-7	CCRF-CEM	CEM/ADR 5000
Achillea filipendulina	209.9	551.5	223.7	29.9	45.1
Anethum graveolens	93.2	215.8	65.6	16.3	27.5
Artemisia absinthium	95.6	190.2	125.9	60.34	85.2
Artemisia rutifolia	295.2	478.6	456.2	387.2	560.6
Artemisia scoparia	78.3	179.1	73.2	151.7	n.d.
Ferula clematidifolia	n.d.	n.d.	n.d	56.3	142.5
Foeniculum vulgare	206.7	74.8	58.6	32.3	165.5
Galagania fragrantissima	48.3	73.3	51.6	9.26	75.5
Hypericum perforatum	64.2	65.9	55.9	41.9	124.8
Hypericum scabrum	n.d.	n.d.	n.d.	158.9	280.9
Hyssopus seravschanicus	218.1	423.0	206.5	110.8	127.7
Mentha longifolia	42.3	62.8	58	12.1	25.4
Ocimum basilicum	n.d.	813.9	n.d.	798.4	1038.9
Origanum tyttanthum	43.3	78.1	54.6	7.46	n.d.
Polychrysum tadshikorum	n.d.	n.d.	n.d.	77.4	137.9
Salvia officinalis	n.d.	n.d.	n.d.	65.8	70.7
Salvia sclarea	176.6	210.8	146.4	354	n.d
Tanacetum parthenium	n.d.	285.7	n.d.	82.6	163.2
Tanacetum vulgare	n.d.	n.d.	n.d.	22.5	92.5
Ziziphora clinopodioides	164.4	407.2	134.9	70.95	73.6
Doxorubicin	4.5	8.1	3.3	2.3	5.2

n.d. not determined

the plants belong to the thujone-rich chemotype, in contrast to the cineole/camphor chemotype found in Mongolia, and is dominated by α-thujone (21–37%) and β-thujone (36–47%) as well as 1,8-cineole (3–12%) and germacrene D (2–3%) (Sharopov and Setzer 2011c). Antioxidant, antimicrobial, cytotoxic, and anti-inflammatory activities of plant extracts are summarized in Tables 7.6, 7.7, 7.8, 7.9, 7.10, 7.11, and 7.12. Bioactivity of *A. rutifolia* is most likely due to the thujones, present in its essential oil. However, thujone has psychotropic effects, acting on the γ-aminobutyric acid-gated chloride channel, a member of the superfamily of ligand-gated ion channel receptors. In addition to the essential oil, this plant is rich in guaianolide, germacranolide, and eudesmanolide sesquiterpenoids (Sharopov and Setzer 2011c).

7.4.4.5 *Artemisia scoparia* Waldst & Kit.: Asteraceae

In the folk medicine of Tajikistan, decoctions and infusions from the tops of the shoots of this plant are used to treat kidney disorders, as well as diaphoretic, diuretic, and anthelmintic. Decoctions are also considered useful against epilepsy, rheumatism, fever, and inflammation of the respiratory tract (Hojimatov 1989a).

Table 7.10 Cytotoxicity of the methanol plant extracts from Tajikistan (Sharopov 2015)

Species	IC_{50} µg/mL	
	CCRF CEM	ADR-5000
Achillea filipendulina	35.5	25.8
Anethum graveolens	205.3	201.4
Artemisia absinthium	52.3	133.1
Artemisia rutifolia	86.3	131.6
Artemisia scoparia	215.6	134.5
Ferula clematidifolia	73.85	99.6
Galagania fragrantissima	144.3	125.4
Hypericum perforatum	62.5	225.4
Hypericum scabrum	57.7	67.5
Melissa officinalis	41.8	120.8
Mentha longifolia	94.0	205.24
Ocimum basilicum	62.8	310.1
Origanum tyttanthum	158.4	167.9
Polychrysum tadshikorum	10.2	32.5
Salvia sclarea	80.3	148.7
Tanacetum parthenium	7.3	n.d.
Tanacetum vulgare	36	107.3
Ziziphora clinopodioides	151.2	175.9
Doxorubicin	2.3	5.2

n.d. not determined

Table 7.11 Antimicrobial activity of essential oils (Sharopov 2015; Sharopov et al. 2015b)

Species	*E. coli* ATCC 25922		MRSA NTCT 10442	
	MIC (mg/mL)	MBC (mg/mL)	MIC (mg/mL)	MBC (mg/mL)
Achillea filipendulina	5	10	5	5
Artemisia absinthium	>20	>20	>20	>20
Artemisia rutifolia	10	20	5	20
Artemisia scoparia	2.5	5	1.25	2.5
Ferula clematidifolia	>20	>20	>20	>20
Ferula foetida	>20	>20	>20	>20
Galagania fragrantissima	>20	>20	0.04	0.08
Hypericum perforatum	5	5	1.25	2.5
Hypericum scabrum	>20	>20	>20	>20
Hyssopus seravschanicus	10	10	5	10
Mentha longifolia	5	10	10	20
Ocimum basilicum	>20	>20	>20	>20
Origanum tyttanthum	0.31	0.31	0.62	1.25
Salvia sclarea	>20	>20	>20	>20
Tanacetum vulgare	>20	>20	20	20
Ziziphora clinopodioides	5	5	10	10
Ampicillin	0.003	0.006	0.008	0.016

Table 7.12 Anti-inflammatory activity of essential oils (Sharopov 2015; Sharopov et al. 2015b)

Species	5-Lipoxygenase inhibition, IC_{50}, µg/mL
Achillea filipendulina	221.3
Anethum graveolens	33.47
Artemisia absinthium	56.6
Artemisia rutifolia	75.6
Artemisia scoparia	184.3
Galagania fragrantissima	7.34
Hyssopus seravschanicus	100.7
Mentha longifolia	28.14
Origanum tyttanthum	14.78
Salvia sclarea	Not active
Tanacetum parthenium	21.6
Ziziphora clinopodioides	33.12
Sodium linoleate (negative control)	–

According to Kurbanov (1992) and Nazarov et al. (2002), the aerial parts of *A. scoparia* are useful as an expectorant. Its essential oils from different geographical locations exhibit a great variability, but those from Tajikistan are dominated by the diacetylenes 1-phenyl-2,4-pentadiyne (34%) and capillene (5%), as well as β-pinene (21%), α-pinene (5%), methyl eugenol (6%), myrcene (5%), limonene (5%), and (*E*)-β-ocimene (4%) (Sharopov and Setzer 2011b). Antioxidant, antimicrobial, cytotoxic, and anti-inflammatory activities of plant extracts are summarized in Tables 7.6, 7.7, 7.8, 7.9, 7.10, 7.11, and 7.12.

7.4.4.6 *Bunium persicum* (Boiss.) B. Festch. (Zira (Tajik)): Apiaceae

It is distributed in the mountainous regions of Tajikistan, Afghanistan, Iran, Pakistan, and the western part of Northern India (Kashmir, Punjab) and bears fruit that is an extremely popular seasoning material for meat dishes in Central Asia. In Tajikistan traditional medicine, the plant has been recorded to show several therapeutic effects on digestive and urinary tract disorders and is used for chronic diseases of the stomach (chronic gastritis), intestines (colitis), and liver (jaundice) as well as chronic cholangitis, swelling, and kidney stones (Dadobaeva 1972). Composition of the essential oil from its fruits has been studied by GLC, IR, NMR, and mass spectral methods. The essential oils of the fruit from Tajikistan are mainly composed of monoterpene hydrocarbons and oxygenated monoterpenoids. Main components are *p*-cymene (5.3–19.15%) and cumene aldehyde (36.0–40.66%) (Kurbanov and Sadikov 1978; Sharopov et al. 2015a), together with γ-terpinen-7-al (15.0%) and α-terpinen-7-al (13.1%), followed by γ-terpinene (10.9%) and β-pinene (9.1%) (Sharopov et al. 2015a).

7.4.4.7 *Coriandrum sativum* L. (Coriander): Apiaceae

This annual herb has been cultivated as a spice almost worldwide for centuries. In traditional medicine, it is used mainly for the treatment of appetite deficit and digestive problems. The fruits are used since ancient times for the treatment of wounds and burns. It is an important spice crop, with antioxidant properties due to its active phenolic acid compounds, including caffeic and chlorogenic acids. Its essential oil serves as counter irritant in painful joints, rheumatism, and menstrual disorders (van Wyk et al. 2015). Chemical composition of the essential oil from the plants grown in Tajikistan shows that it is dominated by aliphatic aldehydes and alcohols such as (2*E*)-dodecenal (16.5%), decanol (14.9%), decanal (11.3%), tetradecanol (9.2%), (2*E*)-decene-1-ol (7.4%), (8*Z*)-undecenal (6.2%), and dodecanal (4.4%). This essential oil shows cytotoxic activity against Caco-2, CCRF-CEM, and CEM/ADR 5000 tumor cell lines with IC_{50} values of 86.8, 16.5, and 38.5 μg/mL, respectively. The hemolytic activity (IC_{50}) of the essential oil against red blood cells (RBC) has been reported as 2.3 mg/mL (Sharopov et al. 2017).

7.4.4.8 *Ferula clematidifolia* Koso-Pol.: Apiaceae

Ferula clematidifolia is a herbaceous perennial plant, up to 1.5 m tall, with yellow flowers mainly distributed in Tajikistan and Afghanistan. It is used for the treatment of flu, fever, and colds in children. Three crystalline compounds (giferolide, malaphyll, and malaphyllin) have been isolated from an ethanolic extract of the roots (Sagitdinova et al. 1990). The essential oils in the leaves and root of this plant during its flowering season in Yovon Region of Tajikistan have revealed that it contains a total of 42 compounds representing 99.8% of the total oil composition. The major components of the oil are β-pinene (1.6–36.9%), myrcene (3.9–34.3%), limonene (1.0–30.1%), α-pinene (2.5–29.4%), sabinene (8.1–16.5%), and β-phellandrene (0.3–7.0%) (Sharopov et al. 2016a). The antioxidant activity of the essential oil has been evaluated by DPPH, ABTS, and FRAP assays. Antioxidant, antimicrobial, cytotoxic, and anti-inflammatory activities of plant extracts are listed in Tables 7.6, 7.7, 7.8, 7.9, 7.10, 7.11, and 7.12. The antioxidant activity values lie around 14.7–15.7 mg/mL for DPPH, 0.4–1.5 mg/mL for ABTS, and 124.5–434.7 μM Fe (II)/mg for FRAP. Its cytotoxicity on HeLa, Caco-2, and MCF-7 cancer cell lines has revealed that IC_{50} values were 93 μg/mL for HeLa, 216 μg/mL for Caco-2, and 67 μg/mL for MCF-7 cells. Cytotoxicity of its oil has been tested against CCRF-CEM (human T-cell lymphoblastic leukemia) and CEM/ADR5000 (adriamycin-resistant leukemia) cell lines using the MTT assay: IC_{50} values are reported as 56.3 μg/mL for CCRF-CEM and 142.5 μg/mL for CEM/ADR5000 cells (Sharopov 2015).

7.4.4.9 *Foeniculum vulgare* Mill. (Fennel): Apiaceae

Foeniculum vulgare is widely used for flavoring foods and beverages due to its pleasant spicy aroma. In traditional medicine, the plant and its essential oil have been extensively used as carminative, digestive, galactogogue, and diuretic agent and to treat respiratory and gastrointestinal disorders. In Tajik traditional medicine, plant extracts are commonly used for digestion and kidney problem (Williams 2010). The chemical composition of the essential oil from the plants grown in Tajikistan has been investigated, and it is dominated by *trans*-anethole, *p*-anisaldehyde, and α-ethyl-*p*-methoxy-benzyl alcohol (unpublished work). The essential oil exhibits moderate cytotoxicity in leukemia cells (CCRF-CEM) (Sharopov 2015).

7.4.4.10 *Galagania fragrantissima* Lipsky: Apiaceae

Galagania fragrantissima is well distributed in Afghanistan, Kyrgyzstan, Uzbekistan, and Tajikistan. Leaves and young shoots are used as a spice for soups and other dishes. Main constituents of the essential oil are aliphatic aldehydes and alcohols such as (2*E*)-dodecenal (83.6%), (2*E*)-dodecenol (7.8%), (2*E*)-tetradecenal (3.4%), and dodecanal (2.3%). Antioxidant, antimicrobial, cytotoxic, and anti-inflammatory activities of plant extracts are presented in Tables 7.6, 7.7, 7.8, 7.9, 7.10, 7.11, and 7.12. The cytotoxicity of the oil has been tested against HeLa, Caco-2, and MCF-7 cancer cell lines. IC_{50} values are reported as 0.206 mg/mL for HeLa, 0.074 mg/mL for Caco-2, and 0.058 mg/mL for MCF-7 cell lines. The cytotoxicity is most likely due to the major component (2*E*)-dodecenal—it is very electrophilic and can react with a variety of nucleophiles, such as amino groups either from proteins or DNA. Other aldehydes are also reactive and can form Schiff bases with free amino groups. Alkylation of amino groups of DNA bases could potentially lead to mutations (Sharopov 2015; Sharopov et al. 2013b).

7.4.4.11 *Hypericum perforatum* L. and *Hypericum scabrum* L.: Hypericaceae

The essential oils obtained by hydrodistillation from the aerial parts of the plants growing wild in Tajikistan using GC-MS have revealed 66 compounds. Those identified in the essential oils of *H. perforatum* include germacrene D (13.7%), α-pinene (5.1%), β-caryophyllene (4.7%), *n*-dodecanol (4.5%), caryophyllene oxide (4.2%), bicyclogermacrene (3.8%), and spathulenol (3.4%), as the main constituents. Twenty-six components have been identified in the oil of *H. scabrum* with α-pinene (44.8%), spathulenol (7.1%), verbenone (6.0%), *trans*-verbenol (3.9%), and γ-muurolene (3.5%) as the most abundant components (Sharopov et al. 2010). Antioxidant, antimicrobial, cytotoxic, and anti-inflammatory activities of plant extracts are listed in Tables 7.6, 7.7, 7.8, 7.9, 7.10, 7.11, and 7.12.

7.4.4.12 *Hyssopus seravschanicus* Pazij (Tajik name ushnondoru): Lamiaceae

Hyssopus seravschanicus is a perennial, branched, semishrub found growing wild in the northwestern part of Tajikistan. The hyssop plant is an important culinary and medicinal plant located mostly in Central Asia, South Europe, and North Africa. It is rich in essential oils, flavonoids, tannins, and marrubiin and has been used as a healing herb in the Turkish culture for centuries. It has also been used for its antispasmodic, stomachic, antifungal, and expectorant properties. The volatile compounds from the plants growing widely near the high ridge mountainous areas of Varzob, Northern Dushanbe, in Tajikistan at an altitude approximately 2500 m above sea level have been subjected to an extraction and analysis using gas chromatography-mass spectrometry (GC-MS). Eighty-seven chemical components of the essential oil have been reported, representing approximately 95.4% of the oil. Most abundant components are *cis*-pinocamphone (57.0–88.9%), β-pinene (0.4–6.0%), 1,8-cineole (1.8–3.6%), camphor (0.5–4.0%), and spathulenol (0.1–5.0%). This essential oil shows notable antimicrobial activity against *Bacillus cereus* and *Staphylococcus aureus* (Sharopov et al. 2012a; Sharopov 2002).

7.4.4.13 *Melissa officinalis* L.: Lamiaceae

Lemon balm has been used in Europe and Asia to reduce stress and anxiety, promote sleep, improve appetite, and ease pain and discomfort from indigestion [Dudchenko et al. 1989]. Avicenna believed that lemon balm refreshes and strengthens the heart and helps in digestion and hiccup. He recommended lemon balm as a tonic and for the treatment of melancholia (Sino 1982). The plant is also used as an additive in food and for the production of many phytopharmaceutical preparations, fragrances, and cosmetics (Dudchenko et al. 1989). The main constituents of the essential oils of plants from Tajikistan are listed as geranial (43%), neral (31.5%), *trans*-anethole (12%), β-caryophyllene (4%), and citronellal (3%). The essential oil is cytotoxic to MCF-7 cells (IC_{50}, 0.062 mg/mL) and active in the brine shrimp lethality test (LC_{50}, 21.8 μg/mL) but has shown only marginal antimicrobial activity against *Bacillus cereus* (MIC, 313 μg/mL) and *Aspergillus niger* (MIC, 625 μg/mL) (Sharopov et al. 2013c). The biological activities can be attributed to the monoterpenoid aldehydes which can form Schiff's bases with free amino groups of peptides and proteins, thus changing their biological properties (Eisenman et al. 2013). Citral, a mixture of geranial and neral, has shown in vitro cytotoxic activity (Adams and Taylor 2010). Lemon balm leaves are rich in the antioxidant rosmarinic acid which has antiviral properties and was used to treat herpes infections (Eisenman et al. 2013).

7.4.4.14 *Mentha longifolia* (L.) Huds.: Lamiaceae

The mints, *Mentha* species, are widely distributed in Eurasia, Australia, and South and North Africa. Various species of *Mentha* have been used as folk remedies for the treatment of bronchitis, flatulence, anorexia, ulcerative colitis, and liver complaints, due to their anti-inflammatory, carminative, antiemetic, diaphoretic, antispasmodic, analgesic, stimulant, emmenagogue, and anticatarrhal activities (Ozturk and Gork 1978a, b, 1979a, b, c, d; Sharopov et al. 2012b).

M. longifolia collected from three different sites in south-central Tajikistan has been analyzed to determine essential oil constituency. Essential oils have been extracted by hydrodistillation of the plants and subsequently analyzed by gas chromatography-mass spectrometry. A total of 82 compounds have been identified, representing 84.5–99.0% of total oil composition. Although qualitatively similar, the Tajikistan samples did show quantitative differences. Major components and their percentage of the oil are *cis*-piperitone epoxide (7.8–77.6%), piperitenone oxide (1.5–49.1%), carvone (0.0–21.5%), pulegone (0.3–5.4%), menthone (0.0–16.6%), thymol (1.5–4.2%), β-thujone (0.2–3.2%), carvacrol (0.0–2.7%), and β-caryophyllene (0.9–2.5%) (Sharopov et al. 2012b). Antioxidant, antimicrobial, cytotoxic, and anti-inflammatory activities of plant extracts are summarized in Tables 7.6, 7.7, 7.8, 7.9, 7.10, 7.11, and 7.12.

7.4.4.15 *Nepeta nuda* L.: Lamiaceae

Nepeta nuda L. (syn. *Nepeta pannonica* L.) is a herbaceous perennial medicinal aromatic plant, widely distributed in Central Asia. An infusion of this herb has been traditionally used to treat asthenia and syphilis (Eisenman et al. 2013). In the essential oil of *N. nuda* growing in Tajikistan, most abundant compounds are reported as 1,8-cineole (24.6%), 4a-α,7-β,7a-α-nepetalactone (21.0%), germacrene D (13.5%), and β-caryophyllene (12.7%). The potential antioxidant activity of these essential oils has been studied by using DPPH, ABTS, and FRAP assays. The essential oil shows weak in vitro free radical scavenging activity (Mamadalieva et al. 2017).

7.4.4.16 *Ocimum basilicum* L.: Lamiaceae

Ocimum basilicum, also called sweet basil, is a half-hardy annual plant, native to the tropical and subtropical regions of Asia. In Tajikistan, it is popularly cultivated and frequently used as a flavoring agent in food but is also well known for its use in folk medicine. The leaves and flowering tops are used as a remedy for inflammatory diseases of the upper respiratory tract (bronchitis, laryngitis, pharyngitis, etc.), chronic gastritis, enterocolitis, and food poisoning. Moreover, hot basil tea is consumed to treat nausea, flatulence, and dysentery (Hojimatov 1989b; Sharopov et al. 2016c). All 63 compounds have been identified in the essential oil of *O. basilicum* from Tajikistan, representing 99.9% of the total oil composition. Major components of the essential oil are linalool (47.2%), estragole (31.7%), pulegone (4.8%), τ-cadinol (2.1%), 1-phenylpenta-2,4-diyne (1.6%), and camphor (1.0%). The presence of the diacetylene, 1-phenylpenta-2,4-diyne, is unusual; this compound has never been previously reported in *O. basilicum* oils (Sharopov et al. 2016c). Antioxidant, antimicrobial, cytotoxic, and anti-inflammatory activities of plant extracts are summarized in Tables 7.6, 7.7, 7.8, 7.9, 7.10, 7.11, and 7.12.

7.4.4.17 *Origanum tyttanthum* Gontsch. (Synonym: *Origanum vulgare* L. subsp. *gracile* (K. Koch) Ietsw.,): Lamiaceae

Origanum tyttanthum, a perennial herb that is up to 60 cm tall and with ovate dark green leaves, about 2 cm long, is the most widely distributed medicinal plant in Central Asia. In Tajikistan, it blooms from June to September with small purple-red or red-violet flowers, covering a total area of over 140,000 ha, yielding a total stock of air-dry raw materials of 6490 tons, and is thus a natural resource with industrial value (Nuraliev 2008). As a medicinal plant, it has been used as an expectorant, carminative, diaphoretic, stimulant, stomachic, and tonic, as well as a gargle for laryngitis, stomatitis, and sore throat (Hojimatov 1989b; Nuraliev 1989; Sakhobiddinov 1948).

O. tyttanthum has been collected from two different sites in south-central Tajikistan, essential oils obtained by hydrodistillation and analyzed by gas chromatography-mass spectrometry. A total of 52 compounds have been identified representing 99.0–100% of total oil composition. Major components of its oil are carvacrol (34.3–59.2%), thymol (10.8–46.4%), *p*-cymene (0.7–7.3%), β-thujone (1.9–4.1%), piperitenone oxide (0.1–3.8%), γ-terpinene (0.3–3.5%), *cis*-piperitone epoxide (0.8–3.3%), carvacrol acetate (0.4–2.4%), menthone (0.6–2.1%), and borneol (1.0–2.3%) (Sharopov et al. 2011). Antioxidant, antimicrobial, cytotoxic, and anti-inflammatory activities of plant extracts are shown in Tables 7.6, 7.7, 7.8, 7.9, 7.10, 7.11, and 7.12.

7.4.4.18 *Pelargonium graveolens* L. Her. *ex* Ait.: Geraniaceae

Pelargonium graveolens is an important, high-value perennial, aromatic shrub. In Tajikistan, geranium is cultivated on an industrial scale, and approximately 3.5 tons of geranium oil is produced each year. The essential oil is strongly rose scented and extensively used in the perfumery and cosmetic industries. Geranium oil has also become important for skin care because it is good in opening skin pores and cleaning oily complexions. This oil has been used in reducing pain due to post-herpetic neuralgia as well as treating dysentery, hemorrhoids, inflammation, and heavy menstrual flows. The main constituents of the essential oil from the aerial parts growing in Tajikistan are citronellol (37.5%), geraniol (6.0%), caryophyllene oxide (3.7%), menthone (3.1%), linalool (3.0%), β-bourbonene (2.7%), *iso*-menthone

(2.1%), and geranyl formate (2.0%). The perfumery value (ratio of citronellol to geraniol) of the essential oil from Tajikistan lies around 6.25, noting that this oil contains higher amounts of citronellol and lower amounts of geraniol (Sharopov et al. 2014b).

7.4.4.19 *Polychrysum tadshikorum* Kovalevsk.: Asteraceae

This plant has several synonyms in the literature including *Cancrinia tadshikorum* (Kudrj.) Tzvelev, *Chrysanthemum myriocephalum* Rech., and *Tanacetum tadshikorum* Kudrj. It is an endemic plant of Central Asia (Tajikistan and Afghanistan). The herb is used traditionally against toothache and to reduce fever by the Tajik people. A total of 72 compounds have been identified representing 97.01% of the total oil composition. The major components of the oil are terpinen-4-ol (14.5%), sabinene (13.0%), *p*-cymene (7.4%), linalool (5.4%), and γ-terpinene (4.4%), which are typical secondary metabolites of the Asteraceae (Sharopov et al. 2016b). Antioxidant, antimicrobial, cytotoxic, and anti-inflammatory activities of plant extracts are summarized in Tables 7.6, 7.7, 7.8, 7.9, 7.10, 7.11, and 7.12.

7.4.4.20 *Salvia sclarea* L.: Lamiaceae

The essential oil from the aerial parts growing wild in Tajikistan has been obtained by hydrodistillation and analyzed by gas chromatography-mass spectrometry. A total of 59 compounds have been identified representing 94.2% of total oil composition. The major components of the essential oil are linalyl acetate (39.2%), linalool (12.5%), germacrene D (11.4%), α-terpineol (5.5%), geranyl acetate (3.5%), and β-caryophyllene (2.4%). The chemical composition, the large concentrations of linalool and linalyl acetate, and a cluster analysis based on principal components have revealed that Tajik *S. sclarea* oil is comparable to commercial *S. sclarea* oils (Sharopov and Setzer 2012). Antioxidant, antimicrobial, cytotoxic, and anti-inflammatory activities of plant extracts are shown in Tables 7.6, 7.7, 7.8, 7.9, 7.10, 7.11, and 7.12.

7.4.4.21 *Tanacetum parthenium* L. Schultz-Bip. (Feverfew): Asteraceae

This species has more than ten synonyms commonly used in the literature including *Chrysanthemum parthenium* and *Pyrethrum parthenium*. It is native to Eurasia and cultivated widely at a global scale. The plant has traditionally been used for reducing fever, women's ailments, inflammatory conditions, psoriasis, toothache, insect bites, rheumatism, asthma, and stomachache (Pareek et al. 2011). It has been increasingly evaluated for the treatment of migraine. The major components of the essential oil growing wild in Tajikistan are camphor (69.7–94.0%), camphene

(1.7–12.2%), and bornyl acetate (4.2–8.7%). Antioxidant, antimicrobial, cytotoxic, and anti-inflammatory activities of plant extracts are shown in Tables 7.6, 7.7, 7.8, 7.9, 7.10, 7.11, and 7.12. According to DPPH and ABTS assays, the volatile oil has an antioxidant activity with IC_{50} values of 4.82 and 0.96 mg/mL, respectively. Lipid peroxidation has been inhibited by 37.1% by 1.125 mg/mL oil. The cytotoxicity of the oil has been tested against HeLa, CCRF-CEM, and CEM/ADR5000 cancer cell lines: IC_{50} values were 158.6 μg/mL for HeLa, 69.5 μg/mL for CCRF-CEM, and 83.9 μg/mL for CEM/ADR5000 cell lines. The essential oil of *T. parthenium* inhibits soybean 5-lipoxygenase (5-LOX) with an IC_{50} value of 21.6 mg/mL indicating a low anti-inflammatory activity (Sharopov et al. 2014a).

7.4.4.22 *Ziziphora clinopodioides* Lam.: Lamiaceae

The essential oils from the aerial flowering parts in this species, collected during 2 different years, were obtained by hydrodistillation and analyzed by gas chromatography-mass spectrometry. Forty-five components representing 100 and 94.7% of the total oil have been identified. The main constituents of the essential oils are pulegone (72.8 and 35.0%), neomenthol (6.5 and 23.1%), menthone (6.2 and 13.3%), *p*-menth-3-en-8-ol (1.7 and 3.5%), piperitenone (2.6 and 1.1%), and piperitone (0.7 and 1.2%) (Sharopov and Setzer 2011a). Antioxidant, antimicrobial, cytotoxic, and anti-inflammatory activities of plant extracts are presented in Tables 7.6, 7.7, 7.8, 7.9, 7.10, 7.11, and 7.12.

7.4.4.23 *Ziziphora pamiroalaica* Lam.: Lamiaceae

This plant is used in Tajikistan for the treatment of the diseases of cardiovascular system and gastrointestinal tract. The dominant components of the essential oil of the plants growing in Tajikistan (Pamir) are pulegone (52.7–60.4%), thymol (7.5–11.1%), isomenthone (7.1–10.6%), and menthone (3.9–6.3%) (Akobirshoeva and Olennikov 2017).

7.4.5 Other Biologically Active Compounds

7.4.5.1 *Artemisia annua* L. (Yavshoni yaksola (Tajik)): Asteraceae

In folk medicine, aqueous extract (decoction) of herbs is used to treat scabies and rheumatism (Dadobaeva 1972). Greatest content of sesquiterpene lactones is observed in the seeds (0.04–0.09%). Two sesquiterpene lactones, artemisinin and arteannuin B, and the tropane alkaloid scopolamine have been found in the *A. annua* in Tajikistan (Sadikov 2003).

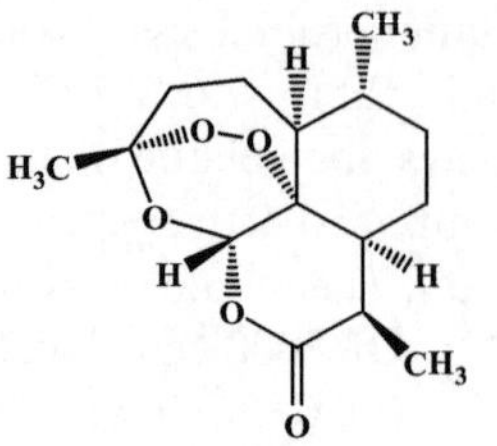

Artemisinin

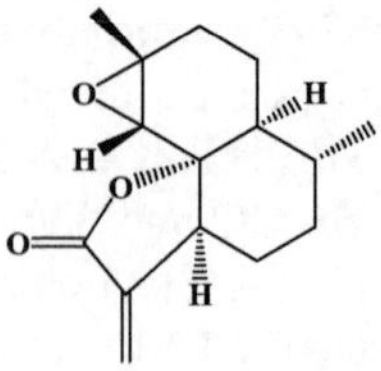

Arteannuin B

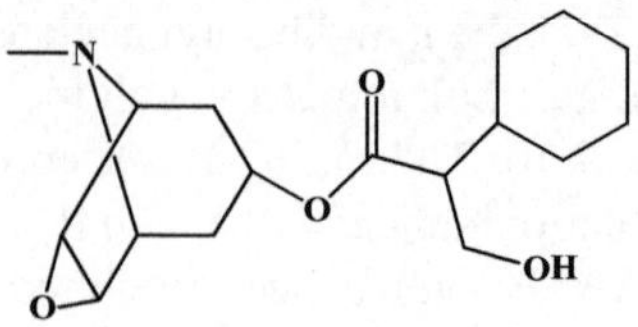

Scopolamine

7.4.5.2 *Helianthus tuberosus* L. (Noki zamini (Tajik)): Asteraceae

Helianthus tuberosus, a perennial herb, is the best known inulin-bearing plant and widely used in many food products, including diabetic food. It is commonly named as Jerusalem artichoke, topinambur, sunroot, earth apple, or erdbirne. The qualitative and quantitative analysis of amino acids and carbohydrates has been carried out on the plants growing in Tajikistan. It is rich in inulin carbohydrate (Saforzada et al. 2014).

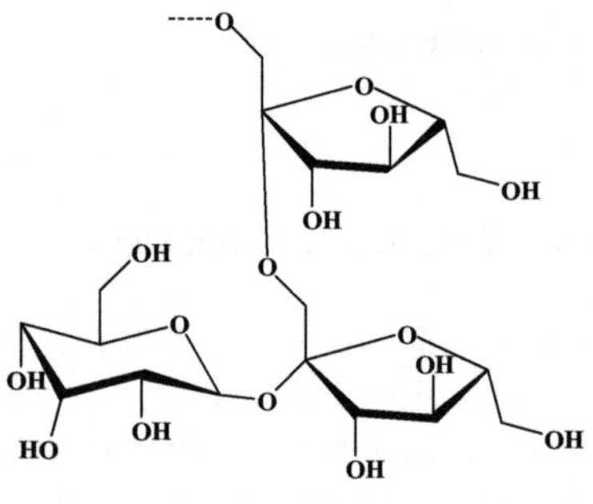

Inulin structural fragment

7.5 Conclusions

A perusal of the studies conducted on the medicinal plants of Tajikistan has revealed that alkaloid-containing species mainly belong to Berberidaceae, Papaveraceae, and Fabaceae (Leguminosae). The alkaloid contents vary between 1.1 and 7% in Berberidaceae, 1.2 and 1.9% in Papaveraceae, and 0.035 and 0.95% in Fabaceae.

Most of the essential oil-bearing plant species belong to the Apiaceae, Asteraceae, and Lamiaceae.

We hope this review will give impetus in the future practical applications of biologically active secondary metabolites from medicinal plants of Tajikistan in pharmaceutical, food, medicine, cosmetic, perfumery, agriculture, and other industries.

Most of the data summarized here has been published in Russian and Tajik languages and is not generally available to the mainstream researchers in other countries. It will prove helpful to a wider audience of international workers.

References

Adams TB, Taylor SV (2010) Safety evaluation of essential oils: a constituent-based approach. In: Baser KHC, Buchbauer G (eds) Handbook of essential oils: science, technology, and applications. CRC Press, Boca Raton, p 994

Adhami HR, Mesgarpour B, Farsam H (2007) Herbal medicine in Iran. Herb Gram 74:35–43

Akobirshoeva A, Olennikov DN (2017) Chemical composition of the essential oil of *Ziziphora pamiroalaica* Lam. growing in Tajikistan. Chem Plant Resour 1:51–58

Begenov A et al (2014) Assessment of the current status of populations of Kazakh rare plants (*Berberis iliensis* M. Pop.). World Appl Sci J 30:105–109

Buckingham J, Baggaley KH, Roberts AD, Szabo LF (2010) Dictionary of alkaloids, 2nd edn. CRC Press, Taylor & Francis, Boca Raton

Dadobaeva O (1972) Dictionary of scientific and local names of medicinal plants of north of Tajikistan. Irfon, Dushanbe

Davidyans SB, Sadikov YJ (1963) Preparation of berberine salts on the base of physiological active acids. Reports of the Tajikistan Academy of Sciences, pp 36–38

Dudchenko LG, Kozyakov AS, Krivenko VV (1989) Priyno-aromaticheskie i pryano-vkusovie rasteniya. Naukova dumka, Kiev

Eisenman SW, Zaurov DE, Struwe L (2013) Medicinal plants of Central Asia: Uzbekistan and Kyrgyzstan. Springer, New York

Gachechiladze ND, Korzinikov YS, Glazunova EM, Yusufbekov CY, Bondar BB, Krimskaya NB, Potapova KM (1981) Biochemical and morphological charakteristic form of *Hippophae rhamnoides* L., growing in west of Pamir. Plant Resour 17:37–42

Glazko VI, Bautin VM (2012) N.I. Vavilov as Organizator of Sciences. Reports of the Timiryazev Agricultural Academy, pp 151–181

Harvey AL (2008) Natural products in drug discovery. Drug Discov Today 13:894–901

Hojimatov M (1989a) Dikorastushie lekarstvennie rasteniya Tadjikistana. Tadj. Sovet. Ensclopedii, Dushanbe

Hojimatov M (1989b) Wild medicinal plants of Tajikistan. Tadj. Sovet. Ensclopedii, Dushanbe

Hojimatov M, Bobokhojaeva S (1976) Composition and contents of biological active compounds of the *Conium maculatum* L. Reports of Academy of Sciences of the Republic of Tajikistan, vol 63, pp 124–127

Hojimatov M, Ermatov N, Sadikov YJ, Kurbanov M, Nuraliev YN (1979) Composition of coumarins in the some of species of Apiaceae family. Reports of Academy of Sciences of the Republic of Tajikistan, vol 75, p 64

Jahn S, Seiwert B, Kretzing S, Abraham G, Regenthal R, Karst U (2012) Metabolic studies of the *Amaryllidaceous* alkaloids galantamine and lycorine based on electrochemical simulation in addition to in *vivo* and in *vitro* models. Anal Chim Acta 756:60–72

Jiang L, Numonov S, Bobakulov K, Qureshi MN, Zhao H, Aisa HA (2015) Phytochemical profiling and evaluation of pharmacological activities of *Hypericum scabrum* L. Molecules 20:11257–11271. https://doi.org/10.3390/molecules200611257

Khusainova KS, Sadikov YJ (1986) Alkaloids of *Mahonia aquifolium* L. Reports of the Tajikistan Academy of Sciences, pp 64–69

Kinzikaeva GK (1968) The Genius of Salsola, vol 3. Flora of Tajikistan SSR. Nauka, Leningrad
Kiyamidinova F, Akramov ST, Yunusov SY (1967) Alkaloids of *Lindelofia stylosa*. Chem Nat Compd 3:411–413
Korzinikov YS, Golovatiy AI, Guryanov AF (1984) Morphological and biochemical importance of *Hippophae rhamnoides* L. in Tajikistan. Plant Resour 20:19–25
Kurbanov B (1992) Lekarstvennie rasteniya – pomoshnik cheloveka. Irfon, Dushanbe 24
Kurbanov M, Sadikov YJ (1978) Composition of the essential oil of the seeds of *Binium persicum*. Reports of Academy of Sciences of the Republic of Tajikistan, vol 21, pp 33–36
Kurbanov M, Khusainova KS, Hojimatov M, Vezen AE, Haydarov KH, Burichenko VK (1975) Kabudin - new alkaloids of *Thalictrum isopyroides* C. A. Mey Reports of Academy of Sciences of the Republic of Tajikistan, vol 18, pp 20–22
Kurbanov M, Nuraliev YN, Hojimatov M, Isobaev MJ (1988) Alkaloids of *Thalictrum kuhistanicum* Ovcz. et. Koczkl. Plant Resour 24:125–128
Mamadalieva NZ, Sharopov F, Satyal P, Wink M (2017) Chemical composition of the essential oils of some Central Asian Nepeta species (Lamiaceae) by GLC-MS. Nat Prod Commun 12:1891–1893
Margolina DL (1941) Flora and vegetation of Tajikistan (Bibliography). Academy of Sciences USSR, Moscow
Monteverde NA, Hammerman AF (1927) Turkestans collection of medicinal products of the Museum of Main Boanical Garten, Main Botanical Garten Publication 26(4):291–358. [in russian]
Narzullaev AS, Sadikov YJ, Hojimatov M (1976) Alkaloids of *Delphinium ternatum* и *Delphinium confusum*. Reports of Academy of Sciences of the Republic of Tajikistan, pp 87–88
Narzullaev AS, Matveev BM, Sobirov SS (1981) Alkaloids os some species of Delphinum L. Plant Resour 17:45–49
National Biodiversity and Biosafety Center of the Republic of Tajikistan (2017)
Nazarov MN, Nazarov NM, Kholov AH, Isupov SJ, Sabzaev AR (2002) Rukovodstvo po sboru i sushke lekarstvennich rasteniy Tadjikistana. Medpress-Inform, Moscow
Novikov V, Safarov N (2002) Tajikistan 2002. State of the environment report: ecosystems. Research Laboratory for Nature Protection (Tajikistan) in working partnership with UNEP/GRID-Arendal (Norway). http://www.enrin.grida.no
Nowak A, Nobis M (2010) Tentative list of endemic vascular plants of the Zeravshan MTS in Tajikistan: distribution, habitat preferences and conservation status of species. Biodivers Res Conserv 19:65–80
Nowak A, Nowak S, Nobis M, Nobis A (2016) Vegetation of screes of the montane and colline zones in the Pamir-Alai Mountains in Tajikistan (Middle Asia). Tuexenia 36:223–248
Numonov S et al (2017) Evaluation of the antidiabetic activity and chemical composition of *Geranium collinum* root extracts — computational and experimental investigations. Molecules 22:983–1002. https://doi.org/10.3390/molecules22060983
Nuraliev YN (1989) Medicinal Plants. Maorif, Dushanbe
Nuraliev YN (2008) Phytotherapy in tajik traditional medicine and its perspective for modern medicine. In: Nuraliev YN (ed) The problems of phitotherapy and phitofarmacology. Irfon, Dushanbe, p 206
Ozturk M, Gork G (1978a) Biotic and abiotic factors effecting the distribution of *Mentha* species in West Anatolia. Plant 2:155–165
Ozturk M, Gork G (1978b) Studies on the chorology and economical evaluation of *Mentha* species in West Anatolia. Ege Univ Sci Fac J II(4):339–356
Ozturk M, Gork G (1979a) Ecological factors effecting the distribution and plasticity of *Mentha* species in West Anatolia. Plant 6:39–51
Ozturk M, Gork G (1979b) Studies on the morphology and taxonomy of *Mentha* species in West Anatolia. Ege Univ Sci Fac J III:43–55
Ozturk M, Gork G (1979c) Ecology of *Mentha pulegium*. Ege Univ Sci Fac J III:57–72
Ozturk M, Gork G (1979d) Edaphic relations of *Mentha* species in West Anatolia. Ege Univ Sci Fac J III:95–110

Ozturk M, Gucel S, Altundag E, Celik S (2011a) Turkish Mediterranean medicinal plants in the face of climate change, Chapter 5. In: Ahmad A et al (eds) Medicinal plants in changing environment. Capital Publishing, New Delhi, pp 50–71

Ozturk M, Gucel S, Altundag E, Mert T, Gork C, Gork G, Akcicek E (2011b) An overview of the medicinal plants of Turkey, Chap 8. In: Singh R (ed) Genetic resources,chromosome engineering & crop improvement: medicinal plants, vol 6. CRC Press, LLC,Taylor & Francis, Boca Raton, pp 181–206

Ozturk M, Altay V, Gucel S, Guvensen A (2014) Halophytes in the East Mediterranean-their medicinal and other economical values. In: Khan et al (eds) Sabkha ecosystems, vol IV Cash crop halophytes biodiversity & conservation. Tasks for Vegetation Science Series. Springer, New York, pp 247–272

Ozturk M, Altay V, Gucel S (2015) Herbals from the dry lands of Southeast Anatolia. In: Khan et al (eds) Halophytes for food security in dry lands. Academic Press, Elsevier, Amsterdam, pp 1–30

Ozturk M, Altay V, Altundag E, Gucel S (2016a) Halophytic plant diversity of unique habitats in Turkey: salt mine caves of Çankırı and Iğdır. In: Khan et al (eds) Halophytes for food security in dry lands. Academic Press, Elsevier, New York, pp 291–315

Ozturk M, Altay V, Gucel S, Altundağı E (2016b) Plant diversity of the drylands in Southeast Anatolia-Turkey: role in human health and food security. In: Ansari, Gill (eds) Plant biodiversity: monitoring, assessment and conservation. CABI, Oxfordshire, p 66

Ozturk, M., Altay, V., Gönenç, T.M. 2017. High altitude medicinal and aromatic plants from theEast Mediterranean in the era of climate change. In: Bhojraj et al. (eds) International training workshop on 'drug discovery from herbs-approaches and applications'. NAM S & T Centre, Daya Publishing House, New Delhi, pp 327–367

Pareek A, Suthar M, Rathore GS, Bansal V (2011) Feverfew (*Tanacetum parthenium* L.): a systematic review. Pharmacogn Rev 5:103–110

Rahmonov O, Majgier L, Andrejczuk W, Banaszek J, Karkosz D, Parusel T, Szymczyk A (2013) Landscape diversity and biodiversity of Fann Mountains (Tajikistan). Ekologia 32:388–395

Rasulova MR (1975) The genus of Vinca L., vol 5. Flora of Tajikistan SSR. Nauka, Leningrad

Rasulova MR, Yunusov SY (1980) Medicinal plants of the flora of Tajikistan, given in the "Canon" by Abuali ibn Sino. Reports of the Tajikistan Academy of Sciences, vol 80, pp 55–62

Romanyuk M (2017) Pamirs. Historical Photos from the Pamir Archive.

Sadikov YJ (1990) Chemical copmostion of *Heracleum lehmannianum.* Reports of the Academy of Sciences of the Republic of Tajikistan, vol 34, pp 174–177

Sadikov YJ (2003) Biologically active substances of wild medicinal plants of Tajikistan. Institute of Chemistry of the Academy of Sciences of the Republic of Tajikistan

Sadikov YJ, Begavatov YM (1990) *Cyperus longus* L. Report of Academy of Sciences of the Republic of Tajikistan, pp 31–35

Sadikov YJ, Hojimatov M (1977) The dynamic of alkaloid accumulation of *Heliotropium losiocarpum,* Reports of the Academy of Sciences of the Republic of Tajikistan, vol 20, pp 43–45

Sadikov YJ, Hojimatov M (1981) Alkaloids of *Spartium junceum.* Reports of Academy of Sciences of the Republic of Tajikistan, vol 24, pp 289–292

Sadikov YJ, Hojimatov M (1983) Alkaloids of *Conium maculatum.* Reports of Academy of Sciences of the Republic of Tajikistan, vol 26, pp 511–515

Sadikov YJ, Hojimatov M (1985) Alkaloids of *Papaver pavoninum.* Reports of the Tajikistan Academy of Sciences, pp 73–75

Sadikov YJ, Hojimatov M (1988) Alkaloids of *Haplophyllum acutifolium* (DC) G. Don. Fil. Plant Resour 24:77–81

Sadikov YJ, Hojimatov M (1990) Alkaloids of *Thermopsis dolichocarpa.* Reports of Academy of Sciences of the Republic of Tajikistan, pp 65–67

Sadikov YJ, Hojimatov M, Hafizov H, Begavatov YM (1978) Alkaloids of *Datura stramonium* L. Reports of the Academy of Sciences of the Republic of Tajikistan, vol 21, pp 34–37

Sadikov YJ, Hojimatov M, Degtyarev VA (1984) Alkaloids of *Verbasum songoricum* Schrank. Reports of the Academy of Sciences of the Republic of Tajikistan, vol 27, pp 452–456

Sadikov YJ, Hojimatov M, Degtyarev V (1985a) Alkaloids of *Vinca major* L. Report of Academy of Sciences of the Republic of Tajikistan, vol 23, pp 289–292

Sadikov YJ, Hojimatov M, Degtyarev VA (1985b) Alkaloids of *Vexibia pachycarpa* from Tajikistan. Reports of Academy of Sciences of the Republic of Tajikistan, vol 28, pp 161–164

Saeidnia S, Gohari AR, Dezfuli NM, Kuiuchi F (2011) A review on phytochemistry and medicinal properties of the genus *Achillea*. Daru 19:173–187

Saforzada RS, Khalifaev DR, Popov DM, Sharopov FS (2014) Amino acid composition of *Jerusalem artichoke* plant cultivated in Tajikistan. Drug Dev Registr 3:134–137

Sagitdinova GV, Saidkhodzhaev AI, Malikov VM, Pimenov MG, Melibaev S (1990) Sesquiterpene lactones of *Ferula clematidifolia* and *Ligularia alpigena*. Chem Nat Compd 53:553–555

Sahoo N, Manchikanti P, Dey S (2010) Herbal drugs: standards and regulation. Fitoterapia 81:462–471

Sakhobiddinov SS (1948) Wild Medicinal Plants of Central Asia. Gosizdat UzSSR, Tashkent

Sharopov FS (2002) Essential oil of *Hyssopus seravshanicus* and its sorption on bentonite clays. Chemistry Institute of the Tajikistan Academy of Sciences

Sharopov FS (2015) Phytochemistry and bioactivities of selected plant species with volatile secondary metabolites. Ruperto-Carola University of Heidelberg

Sharopov FS, Setzer WN (2010) Composition of the essential oil of *Achillea filipendulina* Lam. from Tajikistan. Pharm Chem 2:134–138

Sharopov FS, Setzer WN (2011a) Chemical diversity of *Ziziphora clinopodioides*: Composition of the essential oil of *Z. clinopodioides* from Tajikistan. Nat Prod Commun 6:695–698

Sharopov FS, Setzer WN (2011b) The essential oil of *Artemisia scoparia* from Tajikistan is dominated by phenyldiacetylenes. Nat Prod Commun 6:119–122

Sharopov FS, Setzer WN (2011c) Thujone-rich essential oils of *Artemisia rutifolia* Stephan ex Spreng. growing wild in Tajikistan. J Essent Oil Bearing Plants 14:136–139

Sharopov FS, Setzer WN (2012) The essential oil of *Salvia sclarea* L. from Tajikistan. Rec Nat Prod 6:75–79

Sharopov FS, Gulmurodov IS, Setzer WN (2010) Essential oil composition of *Hypericum perforatum* L. and *Hypericum scabrum* L. growing wild in Tajikistan. J Chem Pharm Res 2:284–290

Sharopov FS, Kukaniev MA, Setzer WN (2011) Composition of the essential oil of *Origanum tyttanthum* from Tajikistan. Nat Prod Commun 6:1719–1722

Sharopov F, Kukaniev M, Thompson RM, Satyal P, Setzer WN (2012a) Composition and antimicrobial activity of the essential oil of *Hyssopus seravschanicus* growing wild in Tajikistan. Pharm Chem 4:961–966

Sharopov FS, Sulaimonova VA, Setzer WN (2012b) Essential oil composition of *Mentha longifolia* from wild populations growing in Tajikistan. J Med Active Plants 1:74–84

Sharopov FS, Sulaimonova VA, Setzer WN (2012c) Composition of the essential oil of *Artemisia absinthium* from Tajikistan. Rec Nat Prod 6:127–134

Sharopov FS, Wink M, Gulmurodov IS, Zhang H, Setzer WN (2013a) Composition and bioactivity of the essential oil of *Anethum graveolens* from Tajikistan. Int J Med Arom Plants 3:125–130

Sharopov FS, Wink M, Khalifaev DR, Zhang H, Dosoky NS, Setzer WN (2013b) Chemical composition and antiproliferative activity of the essential oil of *Galagania fragrantissima* Lipsky (Apiaceae). Am J Essent Oils Nat Prod 1:11–13

Sharopov FS, Wink M, Khalifaev DR, Zhang H, Dosoky NS, Setzer WN (2013c) Composition and bioactivity of the essential oil of *Melissa officinalis* L. growing wild in Tajikistan. Int J Tradit Nat Med 2:86–96

Sharopov FS, Setzer WN, Isupov SJ, Wink M (2014a) Composition and bioactivity of the essential oil of *Tanacetum parthenium* from a wild population growing in Tajikistan. Am J Essent Oil Nat Prod 2:13–16

Sharopov FS, Zhang H, Setzer WN (2014b) Composition of geranium (*Pelargonium graveolens*) essential oil from Tajikistan. Am J Essent Oils Nat Prod 2:13–16

Sharopov F, Kukaniev M, Zhang H, Setzer WN (2015a) Essential oil constituents of Zira (*Bunium persicum* [Boiss.] B. Fedtsch.) from Tajikistan. Am J Essent Oils Nat Prod 2:24–27

Sharopov FS, Braun MS, Isupov S, Wink M (2015b) Antimicrobial, antioxidant and anti-inflammatory activities of essential oils of the selected aromatic plants from Tajikistan. Foods 4:645–653. https://doi.org/10.3390/foods4040645

Sharopov FS, Zhang H, Wink M, Setzer WN (2015c) Tajik aromatic medicinal plants. Medicines 2:28–46

Sharopov FS, Satyal P, Wink M (2016a) Composition of the essential oil of *Ferula clematidifolia*. Chem Nat Compd 52:518–519

Sharopov FS, Satyal P, Wink M (2016b) Composition of the essential oil of *Polychrysum tadshikorum*. Chem Nat Compd 52:523–524

Sharopov FS, Wink M, Kukaniev MA, Setzer WN, Satyal P, Ali N (2016c) The essential oil compositions of *Ocimum basilicum* from three different regions: Nepal, Tajikistan, and Yemen. Chem Biodivers 13:241–248

Sharopov F, Valiev A, Satyal P, Setzer WN, Wink M (2017) Chemical composition and anti-proliferative activity of the essential oil of *Coriandrum sativum* L. Am J Essent Oils Nat Prod 5:11–15

Sino AAI (1982) The Canon of Medicine (Канон врачебной науки), vol 2. Academy of Sciences UzSSR, Tashkent

Squires VR, Safarov N (2013) High altitude ecosystems and biodiversity of Tajikistan: conservation and management. High-altitude rangelands and their interfaces in the Hindu Kush Himalayas. International Centre for Integrated Mountain Development, Nepal

Tayjanov K, Mamadalieva NZ, Wink M (2017) Diversity of the mountain flora of Central Asia with emphasis on alkaloid-producing plants. Diversity 9:1–33

Vanselow KA (2011) The High-Mountain Pastures of the Eastern Pamirs (Tajikistan). An Evaluation of the Ecological Basis and the Pasture Potential. Universität Erlangen, Nürnberg

Vvedenskiy AI (1963) The Genus Ungernia, vol 2. Flora of Tajikistan SSR Nauka, Leningrad

Wilcock G, Howe I, Coles H, Lilienfeld S, Truyen L, Zhu Y, Bullock R (2003) A long-term comparison of galantamine and donepezil in the treatment of Alzheimers disease. Drugs Aging 20:777–789

Williams K (2010) Medicinal plants in Tajikistan: an alternative livelihood option. Int Soc Hortic Sci 954:109–116

van Wyk BE, Wink C, Wink M (2015) Handbuch der Arzneipflanzen - Ein Bildatlas. Wissenschaftliche Verlagsgesellschaft, Stuttgart

Yunusov SY, Sidyakin GP (1950) About alkaloids of seed of *Heliotropium losiocarpum*. Report of Academy of Sciences of the Republic of Uzbekistan, pp 3–6

Yusufbekov CY, Khusainova KS, Sadikov YJ, Aknazarov OA, Poryadina TV (1985) Berberis alkaloids. Reports of the Academy of Sciences of the Republic of Tajikistan, vol 28, pp 712–713

Zapryagaeva VI (1975) The Genus of Berberis L., vol 4. Flora of Tajikistan SSR. Nauka, Leningrad

Zetlin AL, Nikonov GK (1965) Antitumor activity of natural coumarins. Plant Resour 1:507–511

Ziyaev R, Abdusamatov A, Yunusov SY (1971) Alkaloids of *Verbascum songoricum* Sch. Chem Nat Compd:853–854

Chapter 8
Medicinal Plants of Uzbekistan and Their Traditional Uses

Dilfuza Egamberdieva and Dilfuza Jabborova

Abstract Uzbekistan is known throughout the world for its rich history of herbal plants and their use in traditional medicine. Uzbekistan significantly surpasses other regions of the world in terms of the endemism and species richness found in this country. These plants are a source for biologically active compounds that are used in pharmacological tests, with a high probability of new drug discovery. Pharmacological studies have confirmed that plant extracts and individual compounds have various biological activities such as hepatoprotective, dermatological, antimicrobial, antiviral, anti-ulcer, antioxidant, neuroprotective, and antiinflammatory properties. In the present review, an attempt was made to provide an up-to-date report on the diversity of medicinal plants widely used in Uzbekistan and their pharmacological properties. The present report can be useful for the continuous investigation of the Uzbek with the aim of obtaining new biologically active compounds.

Keywords Medicinal plants · Uzbekistan · Phytochemical · Pharmacological

8.1 Introduction

Today there is much interest in the cultivation of medicinal and aromatic plants to meet the great demand of biologically active compounds used by the food and pharmaceutical industries and in healthcare. Plant-based medicines have been used in traditional knowledge practices worldwide for treating various human diseases (Alves and Rosa 2007). Historically, scientist and practioners from many countries and continents have enriched our knowledge with their expertise in traditional medicine (Che et al. 2017). The remedies within natural resources include forest, mountain, grassland, and desert plants that are widely used, as in age-old tradition, to treat

D. Egamberdieva (✉)
Faculty of Biology, National University of Uzbekistan, Tashkent, Uzbekistan

D. Jabborova
Institute of Genetics and Plant Experimental Biology, Academy of Sciences of Uzbekistan, Tashkent, Uzbekistan

D. Egamberdieva, M. Öztürk (eds.), *Vegetation of Central Asia and Environs*,
https://doi.org/10.1007/978-3-319-99728-5_8

diverse ailments and diseases including gastrointestinal symptoms, cardiovascular diseases, skin disorders, and respiratory and urinary problems (McChesney et al. 2007). Traditional medicine is especially popular in the developing countries of the world because of the high cost of synthetic medicinal drugs (Abu-Irmaileh and Afifi 2000). The chemical substances in plants are vital to assess the value of medicinal plants (Edeoga et al. 2005; Egamberdieva et al. 2017).

The Republic of Uzbekistan is located at the center of the Eurasian continent, and the northern part is occupied by a desert, Kyzil-Kum. The Pamir-Alai and Tyan-Shan mountains are situated in the southern and southeastern regions. Various landscape features including deserts, high mountain ranges, wide steppes, and wetlands are found here. Uzbekistan is a globally and regionally important area and has an ancient research base of herbal medicine. *The Book of Healing*, introduced by Avicenna (considered to be "the father of modern medicine"), is an example of a classic consolidated work (Abu Ali Bin Sina (Avicenna) 1956; UNESCO 1993). This book contains descriptions of practices mainly for to prevent disease and to treat various ailments through the use of herbal plants and their components. The book contains 811 chapters, of which 550 are about herbal plants and their uses (Buranova 2015); notably 40% of these plants are found in Uzbekistan. Among 1000 species described in *The Book of Healing*, 20 species are known as food plants and are still consumed today by major populations.

The flora of Uzbekistan includes more than 4500 species of vascular plants, and about 20% of these have been used to treat various ailments (Mamedov et al. 2004). About 70% of Uzbek households use traditional remedies in their daily lives. The phytochemical composition and biological activity of these plants are still not known fully, and they may contain pharmaceutically important constituents. The plants commonly grown in Uzbekistan include families such as Apiaceae, Asteraceae, Amaryllidaceae, Amaranthaceae, Ranunculaceae, Rosaceae, Lamiaceae, Fabaceae, Alliaceae, Caryophyllaceae, Malvaceae, Brassicaceae, Capparidaceae, Cesalpinaceae, Crassulaceae, Euphorbiaceae, Boraginaceae, Geraniaceae, Gentianaceae, Hypericaceae, Araceae, Asparagaceae, Poaceae, Polygonaceae, Ranunculaceae, Rutaceae, Saxifragaceae, Zygophyllaceae, Berberidaceae, Biebersteiniacea, Campanulaceae, Orchidaceae, Solanaceae, Dipsacaceae, Ephedraceae, Onagraceae, Equisetaceae, Geraniaceae, Euphorbiaceae, Rubiaceae, Papaveraceae, Elaeagnaceae, Juglandaceae, and Scrophulariaceae (Table 8.1). However, about 100 plant species belong to the Lamiaceae family and are considered to be native plants of Uzbekistan.

Plant species such as *Apium graveolens* L., *Asparagus officinalis* L., *Cannabis sativa* L., *Citrullus colocynthis* L., *Ferula assafoetida* L., *Juglans regia* L., *Pyrus malus* L., *Malva silvestris* L., *Melilotus officinalis* L., *Mellissa officinalis* L., *Punica granatum* L., *Sesamum indicum* L., and *Trachyspermum ammi* L. were described in the book by Avicenna and are still used today in traditional medicine (Buranova 2015). Some of these plants are endemic and are found in the mountains, others in deserts, grasslands, and forests. In the Chatkal Biosphere Reserve alone more than 40 species of rare and endemic plants were found, and they are included in the *Red Data Book of Uzbekistan* (2009). A nature reserve that was established in 1947 is

Table 8.1 Summary of reported pharmaceutical properties of medicinal plants from Uzbekistan

Plant species	Pharmacological properties	Phytochemical constituents	References
Achillea filipendulina Lam. (Asteraceae)	Gastric diseases, hemorrhoids, an abortifacient	Essential oil, alkaloid traces, asparagine, amino acids, nitrogen-containing substances	Eisenman et al. (2013)
Ungernia victoris Vved (Amaryllidaceae)	Bronchitis, ulcers, poliomyelitis, neurological diseases	Alkaloids, galanthamine, narwedine, lycorine, ungerine, ungeridine, hippeastrine, haemanthidine, tazettine, nortazettine, hordenine	Mamedov and Craker (2001), Berkov et al. (2009), Khodjimatov et al. (1995)
Aconitum talassicum Popov (Ranunculaceae)	Rheumatism and malaria	Talatizamine, talatizine, talatizidine, isotalatizine, condelphine	Eisenman et al. (2013)
Acroptilon repens (Asteraceae)	Malaria, epilepsy, other diseases	Saponins, tannins, lycoalkaloids, essential oils, sesquiterpene lactones	Eisenman et al. (2013)
Agrimonia asiatica Juz. (Rosaceae)	Gastrointestinal diseases, as an astringent, rheumatism, intestinal infections, fever, edema, hemorrhoids	Ursolic acid, tannins, flavonol glycosides, B vitamins, saponins, trace alkaloids	Eisenman et al. (2013)
Anethum graveolens (Apiaceae)	Antimicrobial activity	Essential oil (carvone, limonene, *cis*-dihydrocarvone, diplaniol, 1,2-diethoxyethane)	Yili et al. (2009)
Prangos pabularia (Apiaceae)	Antibacterial and antiinflammatory activity	Coumarins, terpenoids and glycosides, gamma-pyrone (pabularin A, B, C)	Tada et al. (2002)
Ajuga turkestanica (Lamiaceae)	Weight deficiency, reduced hair growth, ulcers, burns	Oxysteroid compounds and phytoecdysones (turkesterone, ecdysterone, cyasterone)	Eisenman et al. (2013)
Alhagi pseudalhagi (Fabaceae)	Flavonoids catechin, epigallocatechin, gallocatechin, leucodelphinidin, quercetin, rutin	Colitis, gastritis, stomach ulcers, dysentery, cervical erosion	Eisenman et al. (2013)
Ferula pallida (Apiaceae)	Antioxidant, skin wounds	Sesquiterpene phenylpropanoid derivatives, pallidones, and sesquiterpene chromone derivatives	Kogure et al. (2004), Su et al. (2000)
Ferula varia (Schrenk) Trautv. (Apiaceae)	Fever, intestinal parasitosis	Eudesmane, guaiane- and one new germacrane-type sesquiterpene lactone glucosides, sesquiterpene lactone glucosides	Kurimoto et al. (2012) Sakhobiddinov (1995)

(continued)

Table 8.1 (continued)

Plant species	Pharmacological properties	Phytochemical constituents	References
Allium karataviense Regel (Alliaceae)	Lung diseases and shortness of breath	Oils, volatile organic compounds, flavonol glycosides, phenols, vitamins, ascorbic acid, mineral salts, microelements, saponins and sapogenins	Eisenman et al. (2013)
Artemisia balchanorum Krasch. (Asteraceae)	Cytotoxic activity	1,8-Cineole, α-thujone, camphor, β-thujone, *cis*-2,7-dimethyl-4-octen-2,7-diol	Baser et al. (1997)
Allium suvorovii Regel (Alliaceae)	Hemoptysis, tuberculosis, skin diseases, especially eczema and psoriasis	Carbohydrate stachyose	Eisenman et al. (2013)
Arctium lappa L. (Asteraceae)	Rheumatism, eczema, ulcers, skin eruptions, wounds	Polysaccharide inulin, protein, essential oil, carotinoids, flavonoids and tanning agents	Azizov et al. (2012)
Allochrusa gypsophiloides (Regel) Schischk. (Caryophyllaceae)	Bronchitis, gastrointestinal, skin, venereal diseases, spleen, liver, kidney diseases	Saponins, polysaccharides	Eisenman et al. (2013)
Helichrysum maracandicum M. Pop. ex Kirp. (Asteraceae)	Antiproliferative activity, gallbladder disorders	Chalcone glycoside, isosalipurposide, naringenin chalcone	Yagura et al. (2008)
Helichrysum sp. (Asteraceae)	Antimicrobial, antiinflammatory, antioxidative, cytotoxic activities	Flavonoids, phloroglucinol derivatives and diterpenes	Suzgec et al. (2005), Sala et al. (2002), Bigović et al. (2011)
Rhaponticum integrifolium C. Winkl. (Asteraceae)	Loss of appetite, weakening of the reparative functions of the organism	Ecdysteroids (ecdysterone, integristerone A, dehydromakisterone A)	Aripov (1995)
Althaea nudiflora Lindl. (Malvaceae)	Diarrhea, sialorrhea, tumors	Vitamin C, lipids, kaempferol glycosides	Eisenman et al. (2013)
Serratula sogdiana Bunge (Asteraceae)	Heal wounds, liver diseases	Ecdysteroids (ecdysterone, viticosterone E, viticosterone E, sogdysterone)	Saatov et al. (1999)

Artemisia baldshuanica Krasch. et Zarp. (Asteraceae)	Antiplasmodial activity	Essential oils (α- and β-pinenes, thujol alcohol, thujyl acetate, and thujyl valerate), hydrocarbons 5,5-dimethyl-1-ethyl-1,3-cyclopentadiene, tetratriacontane, nonacosane, the monoterpenoids 1,8-cineol, α-thujone, β-thujone, camphor, sesquiterpenoid chrysanthenone, sesquiterpene lactone ambrosine	Goryaev et al. (1962), Mukhamatkhanova et al. (2011)
Silybum marianum (Asteraceae)	Hepatitis, gastrointestinal disorders, antiinflammatory, anti-oxidant, anti-metastatic activity	Lipids (parafinic, olefinic and isoprenoid hydrocarbons, fatty acid esters with triterpenols and sterols, triacylglycerins, triterpenols, sterols, chlorophylls, pheophytins, triterpene acids, chlorophylls, pheophytins)	Mirzaeva et al. (2011)
Althaea officinalis L. (Malvaceae)	Sore throat, hepatitis, cystitis, prostate tumors, chronic prostatitis	Sucrose, betaine, flavonoids, coumarins, phenolic acids, fatty oil, vitamin C, carotene, phospholipids and pectin	Eisenman et al. (2013)
Crambe kotschyana Boiss. (Brassicaceae)	Upper respiratory tract congestion	Alkaloids (goitrin and goitridin), coumarins, vitamin C, and carotene, goitridin, triterpenol, sterol, triacylglycerols, isoprenes, triterpenes, sterols	Okhunov et al. (2011), Bakker et al. (2003)
Capparis spinosa L. (Capparidaceae)	Healing wounds, asthma, gastrointestinal diseases, hepatitis	Glycolipids (monogalactosyldiacylglycerides, digalactosyldiacylglycerides, sterolglycosides and their esters and phospholipid (phosphatidylinositols, phosphatidylethanolamines, and phosphatidylcholines)	Yili et al. (2009), Yuldasheva et al. (2008)
Amaranthus retro flexus L. (Amaranthaceae)	Colitis, intestinal colic, hemoptysis, menstrual, hemorrhoid hemorrhages, antiprotist and antibacterial	Fatty acids (miristic, palmitic, stearic, linoleic, linolenic acid)	Eisenman et al. (2013)

(continued)

Table 8.1 (continued)

Plant species	Pharmacological properties	Phytochemical constituents	References
Silene wallichiana Klotzch (Caryophyllaceae)	Anabolic adaptogenic, tonic antioxidant, antimicrobial	Ecdysteroids (viticosterone E 22-*O*-benzoate, 2-deoxy-α-ecdysone 22-*O*-benzoate, viticosterone E, 2-deoxy-α-ecdysone, 2-deoxyecdysterone, ecclysterone 22-*O*-benzoate, ecdysterone)	Saatov et al. (1999), Mamadalieva et al. (2000)
Silene brahuica Boiss. (Caryophyllaceae)	Stress-protective effect	Ecdysteroids (viticosterone E, polypodine B, ecdysterone, integristerone A, sileneoside A, α-ecdysone 22-sulfate)	Saatov et al. (1999)
Gleditsia triacanthos L. (Cesalpinaceae)	Spastic colitis, chronic cholecystitis, stomach ulcers, bronchial asthma	Neutral lipids (triacylglycerides, carotinoids, fatty acid esters, free fatty acids, sterols, diacylglycerides), glycolipids (mono- and digalactosyldiacylglycerides, sterolglycosides, sterolglycoside esters), phospholipids (phosphatidylcholines, phosphatidylethanolamines, phosphatidylinosites), alkaloids, flavonic pigments	Rakhmanberdyeva et al. (2002)
Rhodiola litvinovii (Crassulaceae)	Treating tiredness, neurotic ailments, adaptogenic	*p*-Tyrosol, salidroside, gallic acid, epigallocatechin-3-*O*-gallate)	Melikuziev et al. (2013)
Anagallis arvensis L. (Myrsinaceae; formerly Primulaceae)	Tuberculosis, gynecological disorders, rabies	Glucoside cyclamine, saponoids, flavonoids (kaempferol, quercetin), phenylcarbonic acids (caffeic, ferulic, etc.), anthocyans and fatty oil	Eisenman et al. (2013)
Euphorbia ferganensis B. Fedtsch. (Euphorbiaceae)	Dysentery, diarrhea, viral infections including AIDS	Polyphenols (gallic acid, quercitrin, quercetin, kaempferol-3-*O*-glucoside, 3-*O*-galloyl-4,6-hexahydroxydiphenoyl-β-D-glucose, geraniin, 1-*O*-galloyl-4,6-trihydroxy-3,4,3-1-trimethoxydiphenoyl-β-D-glucose, 2,3-di-*O*-galloyl-β-D-glucose, and 1,2,3-tri-*O*-galloyl-β-D-glucose)	Kim et al. (2003), Abdulladzhanova et al. (2001)

Glycyrrhiza glabra (Fabaceae)	Antiviral, antiinflammatory, antioxidant, bronchial asthma	Cresol, guaiacol, guaiacylethane, guaiacylpropan-1-ol	Mamedov and Craker (2001), Sdykov and Abduazimov (2000)
Amorpha fruticosa L. (Fabaceae)	Antibacterial, dysentery, nerve disorders, atherosclerosis	Glycosides, aglycons, amorphin	Khodjimatov et al. (1995)
Anchusa azurea Mill. (Boraginaceae)	Neurasthenia, asthma, laxative, febrifuge	Alkanin, anchusin, anchusa acid, resins, stearidonic acid	Eisenman et al. (2013)
Astragalus unifoliolatus Bunge (Fabaceae)	Cytotoxic activity, immune stimulant	Triterpene glycosides (oleanolic acid, cyclounifolioside A)	Kucherbaev et al. (2002)
Ammothamnus lehmannii Bunge (Fabaceae)	Against gastric ulcer disease	Flavanoids (ammothamnidin, lehmannin, iuteolin, cynaroside, quercetin, and isoquercetin)	Bakirov et al. (1987), Syrov et al. (2010)
Thermopsis altherniflora Rgl. et Schmalh. (Fabaceae)	Hypolipidemic, antisclerotic activity	Flavanoids (flavones and isoflavones, formononetin, chrysoeriol, apigenin, luteolin, thermopsoside, cynaroside)	Satimov et al. (1998)
Vicia subvillosa L. (Fabaceae)	Ascites, paralysis, epilepsy, flu	Flavanoids (apigenin, luteolin, quercetin, cinaroside)	Yuldashev (2005)
Artemisia absinthium L. (Asteraceae)	Vermifuge, dyspepsia, loss of appetite, insomnia, diseases of the liver, stomach, spleen, gall bladder, fever, hemorrhoids, malaria	Lactones (absinthin, anabsinthin and artabasin), flavonoids (artemetin), tannins, organic acids, vitamin C and carotene	Eisenman et al. (2013)
Geranium saxatile Kar. (Geraniaceae)	Antihypoxic properties for acute normobaric and hemic hypoxia	Phenolic compounds (gallic acid, ellagic acid, quercetin-3-*O*-glucoside)	Siddikov et al. (2011)
Gentiana olivieri (Gentianaceae)	Treating wounds, diarrhea, common cold, ease of digestion	Glycosides (secoiridoid glycosides, olivierosides, gentiopicroside, sweroside, 69-*O*-β-D-glucosylgentiopicroside, swertiapunimarin, eustomoside, eustomorusside, septemfidoside) alkaloids, iridoid and secoiridoid glucosides	Mamedov et al. (2004), Rakhmatullaev et al. (1969), Ersöz et al. (1991), Honda (1999)

(continued)

Table 8.1 (continued)

Plant species	Pharmacological properties	Phytochemical constituents	References
Artemisia dracunculus L. (Asteraceae)	Edema, scurvy, dyspepsia, to improve appetite	Essential oils, carotene, vitamin C and alkaloid traces	Eisenman et al. (2013)
Hypericum perforatum (Hypericaceae)	Antiinflammatory, antimicrobial, healing of kidney and heart diseases, diarrhea	Essential oils (β-caryophyllene, caryophyllene oxide, spathulenol, α-pinene)	Baser et al. 2002a, b
Artemisia leucodes Schrenk (Asteraceae)	Angioprotective action, myocarditis, artherosclerosis	Essential oils, lactones, anhydroaustricine, parishin B, parishin C, artelin, and artelein	Eisenman et al. (2013)
Artemisia annua L. (Asteraceae)	Rheumatism, skin diseases (scabies, abscesses, bacterial and fungal diseases, etc.)	Actones (artemisinin and arteannuin), coumarins (scopoletin), essential oil, tannins, alkaloids, resins, sugars, and vitamin	Eisenman et al. (2013)
Ajuga turkestanica (Rgl.) Brig. (Lamiaceae)	Anabolic activity and cholagogic action	Ecdysterone, cyasterone, ajugalactone, ajugasterone, α-ecdysone, and turkesterone, turkesterone, ecdysterone, phytoecdysteroids, irrdoids and their glycosides	Mamatkhanov et al. (1998), Abdukadirov et al. (2005); Mamadalieva et al. (2014)
Artemisia scoparia Waldst. & Kit. (Asteraceae)	Respiratory disease, rheumatism, radiculitis, irregularities in the menstruation cycle	Organic acids (citric, malic, oxalic, acetic, propionic, and valerianic), tannins, essential oils	Eisenman et al. (2013)
Perovskia scrophularifolia Bunge (Lamiaceae)	Dermatitis; human intestinal parasites, antioxidant activity	Anthocyanin, coumarins essential oils, rosmanol, epirosmanol, carnosol, diosmetin, eugenol β-D-glucopyranoside, salidroside, uridine	Nuriddinov et al. (1997), Asilbekova et al. (2000), Takeda et al. (2007), Nakatani and Inatani (1984), Masuda (2004)
Arischrada korolkowii Regel et Schmalh. Pobed. (Lamiaceae)	Antimicrobial activity	Essential oil (1,8-cineole, camphor, β-caryophyllene, bornyl acetate, caryophyllene oxide, borneol)	Baser et al. (2002a, b)
Artemisia vulgaris L. (Asteraceae)	Nervous diseases, epilepsy, neurasthenia, anticonvulsant, tuberculosis, and to increase the appetite	Essential oils (contains cineol, thujone, borneol, and aldehydes), flavonoids, alkaloids, carotene, and ascorbic acid	Eisenman et al. (2013)
Hyssopus seravschanicus (Lamiaceae)	Antiinflammatory, astringent, rheumatism, gastrointestinal and lung disease	Essential oil (sabinene, β-pinene, myrcene, *p*-cymene, 1,8-cineole, pinocamphone, carvacrol)	Dzhumaev et al. (1990)

Arum korolkowii Regel (Araceae)		Saponins, alkaloids, lipids, carotenoid, lycopene	Eisenman et al. (2013)
Leonurus turkestanicus	Stomachaches, heart, disease, hypertension, hysteria, epilepsy, tachycardia, gastrointestinal, and female diseases	Sterols, flavonoids, iridoids	Mamadalieva et al. (2014)
Scutellaria ramosissima Popov (Sershoh ko'kamaron)	Antioxidant activity, allergies, chorea, nervous tension, high blood pressure		Mamadalieva et al. (2014)
Origanum tyttanthum Gontsch (Lamiaceae)	Antimicrobial, hypocholesterolemic, hypolipidemic activity, healing tuberculosis and against human intestinal parasites	Essential oil (thymol, carvacrol, ursolic and oleanolic acids, 15-methylhexadecanoic acid, coumarins, thymol β-D-glucoside, naringenin, eryodictiol, rosmarinic acid), phenolic glucoside 4-*O*-β-D-glucopyranosylbenzyl-3′-hydroxyl-4′-methoxybenzoate	Baser et al. (2002a, b), Asilbekova et al. (2000), Passreiter et al. (2004), Rivero-Cruz et al. (2004), Petersen and Simmonds (2003), Takeda et al. (2008), Tkachenko et al. (1999), Nuraliyev and Zubaidova (1994)
Scutellaria ramosissima (Lamiaceae)	Epilepsy, allergy, various inflammations, nervous disorders, hypertension, cytotoxic and antimicrobial activity	Flavanoids (scutellarin, chrysin, apigenin, apigenin-7-*O*-glucoside, cynaroside and pinocembrine), lipids (glycolipids, neutral lipids, phospholipids)	Mamadalieva et al. (2001), Yuldasheva et al. (2014), Parajuli et al. (2009)
Asparagus persicus Baker (Asparagaceae)	Numerous diseases	Alkaloids, essential oils, vitamins, asparagine, saponins, steroid sapogenins and related substances	Eisenman et al. (2013)
Althaea nudiflora, *A. armeniaca* (Malvaceae)	Pneumonia, kidney ailments, antihemorrhagic agents	Hydrocarbons and esters of sterols and triterpenes, triacylglycerines, free fatty acids, triterpenols, diacylglycerines, sterols	Sagdullaev et al. (2001)
Phlomis bucharica	Stimulant, tonic, wound healing		Mamadalieva et al. (2014)
Berberis integerrima Bunge (Berberidaceae)	Kidney stones, lung tuberculosis, chest pains, headaches	Alkaloids (including berberine, columbamine, jatrorrhizine and oxyacanthine) and organic acids	Eisenman et al. (2013)

(continued)

Table 8.1 (continued)

Plant species	Pharmacological properties	Phytochemical constituents	References
Arundo donax L. (Poaceae)	Diaphoretic, diuretic, emollient, treatment of dropsy	Alkaloids (deoxyvasicinone, arundine, ardine, donaxarine, donaxanine, donaxaridine, donine, arundinine, arundamine, arundacine, arundarine, arundavine, arundafine, bufotenine)	Khuzhaev et al. (2004)
Rheum maximowiczii (Polygonaceae)	Stomach disorders, antioxidant	Phenyl butanoids, stilbene dimers, rhododendrol, epirhododendrin, lindleyin, torachrysone, resveratrol, and physcion	Kogure et al. (2004), Shikishima et al. (2001)
Astragalus sieversianus Pall. (Fabaceae)	Sedative, antibacterial, antiinflammatory	Saponins, alkaloids (especially smirnovine), coumarins, tannins, flavonoids, vitamins C, E, and P, and carotene	Eisenman et al. (2013)
Aconitum talassicum Popov (Ranunculaceae)	Rheumatism and malaria	Alkaloid (talatizamine, talatizine, isotalatizine, condelphine), talassicumines, A, B, and C, talatisamine, isotalatisidine, talatisine	Nishanov et al. (1991)
Crataegus turkestanica (Rosaceae)	Nervous and cardiovascular diseases	Lipids (phosphatidylinositol, phosphatidylcholine, phosphatidylglycerol, phosphatidic acid, phosphafidylethanolamin)	Gazizov and Glushenkova (1995)
Atraphaxis spinosa L. (Polygonaceae)	Fever	Alkaloids and tannins	Eisenman et al. (2013)
Haplophyllum leptomerum Lincz. et Vved. (Rutaceae)	Cytotoxic activity	Alkaloids (skimmianine, γ-fagarine, *N*-methyl-2-phenylquinolin-4-one, leptomerine, acutine, and 2-heptylquinolin-4-one, dictamnine)	Akhmedzhanova et al. (2010)
Bergenia hissarica Boriss. (Saxifragaceae)	Astringent, hemostatic, antiphlogistic agents in diseases of the gastrointestinal tract	Phenolic compounds (anthraquinones: aloe-emodin, physeion, aloe-emodin 8-*O*-β-glucoside, chrysophanein, emodin 1-*O*-β-D-glucopyranoside)	Yuldasheva et al. (2013)
Nitraria schoberi L. (Zygophyllaceae)	Antispasmodic, antineuropathic, anti-arrhythmic agent	Alkaloids (schoberine, nitraraine, nitraramine, dehydroschoberine, sibiridine, vasicinone, peganol)	Tulyaganov and Kozimova (2005)

Biebersteinia multifida DC. (Biebersteiniacea)	Hemostatic for postnatal bleeding, gastric diseases	Tannins, carbohydrates, saponins, alkaloids, polysaccharides	Eisenman et al. (2013)
Campanula glomerata L. (Campanulaceae)	Sore throat, hoarseness, headache	Alkaloids, vitamin C	Eisenman et al. (2013)
Capparis spinosa L. (Capparidaceae)	Analgesic, vermifuge, asthma, gastrointestinal diseases, hemorrhoids, toothaches	Rutin, quercetin, vitamin C, saponins, pigments, glycosides, flavonoids, thioglycosides, alkaloids (stachydrine), sugars, coumarins, other substances	Eisenman et al. (2013)
Carum carvi L. (Apiaceae)	Sedative, expectorant, diuretic, included in a preparation used as a carminative, laxative, sedative, and to increase appetite	Essential oil, fatty oil, tannins, flavonoids quercetin, camphorol, isorhamnetin, and polyenes	Eisenman et al. (2013)
Convolvules fruticosus		Alkaloid (cuscohygrine)	Mirzaev et al. (2016)
Clematis orientalis L. (Ranunculaceae)	Tuberculosis, sedative, analgesic, diuretic, diaphoretic, to treat cystitis and as an antiinflammatory to treat rheumatism, gout, and chronic gonorrheas	Alkaloid clematine, green resin with melissic acid, myricyl alcohol, and caulosapogenin glycoside	Eisenman et al. (2013)
Cnicus benedictus L. (Asteraceae)	Jaundice, liver diseases, hypochondria, respiratory tract catarrh, intermittent fever, gastrointestinal atonia, gout, ulcers, kidney diseases, urination disorders, and indigestion	Cnicin (a sesquiterpene lactone), resin, mucilage, sterins, tannins, essential oils, vitamin C, aromatic aldehydes (cinnamaldehyde, benzaldehyde, cuminaldehyde), and monoterpenes (citronellol, fenchone, *p*-cymene)	Khalmatov et al. (1984)
Codonopsis clematidea (Schrenk) C.B. Clarke (Campanulaceae)	Hepatitis and cholecystitis	Polysaccharides, alkaloids (codonopsin, codonopsinin)	Rakhimov et al. (2003), Yunusov (1974), Tashkhodzhaev et al. (2004)
Conium maculatum L. (Apiaceae)	Epilepsy, pertussis, migraine headaches, cancer, uterine fibroids	Alkaloids (conhydrine, pseudoconhydrine, g-coniceine, and methyl-coniine), essential oil, vitamin C, carotene, and caffeic acid	Khalmatov (1964), Lopez et al. (1999)

(continued)

Table 8.1 (continued)

Plant species	Pharmacological properties	Phytochemical constituents	References
Convolvulus arvensis L. (Convolvulaceae)	Asthma, lung disease, chest pains, liver and spleen diseases	Flavonoids (quercetin and kaempferol), caffeic acid, carotene and vitamin C	Khalmatov (1964)
Convolvulus subhirsutus Regel & Schmalh. (Convolvulaceae)	Asthma and lung tuberculosis	Alkaloids	Khalmatov et al. (1984), Yunusov (1974)
Crambe kotschyana Boiss. (Brassicaceae)	Gastric diseases	Sugar, oils	Khalmatov (1964)
Crataegus altaica (Loudon) Lange (Rosaceae)	Rheumatic heart disease, cardiosclerosis, stenocardia, vegetative neurosis	Flavonoids, saponins, tannins, polysaccharides, fatty oil, and phenolcarbonic acids	Khodzhimatov (1989)
Crataegus songarica K. Koch. (Rosaceae)	Rheumatism	Vitamin C, carotene, tannins, and leucoanthocyanidin	Kuchin (1955), Petrova (1972)
Dactylorhiza umbrosa (Kar. & Kir.) Nevski (Orchidaceae)	Gastritis, enterocolitis, and other gastrointestinal diseases	Alkaloids and saponins	Khodzhimatov (1989)
Datura stramonium L. (Solanaceae)	Fevers, neuralgia, rheumatism, radiculitis	Alkaloids, scopolamine, tannins, essential oils, and carotene	Khalmatov et al. (1984), Khodzhimatov (1989)
Daucus carota L. (Apiaceae)	Liver and kidney diseases, chronic cholestasis	Flavonoids, coumarins, anthocyanins, vitamins, pantothenic acid, anthocyanidin, essential oils, sugars	Gammerman et al. (1990)
Delphinium confusum Popov (Ranunculaceae)	Liver and kidney diseases	Alkaloids	Khalmatov et al. (1984), Altimishev (1991)
Delphinium semibarbatum Bienert. (Ranunculaceae)	Eczema and scabies	Flavonoids	Khalmatov (1964), Khalmatov et al. (1984)
Descurainia sophia (L.) Webb ex Prantl (Brassicaceae)	Antihelmintic, diuretic	Linolenic, linoleic, arachic, erucic acids	Bekker et al. (2005)
Dianthus superbus L. (Caryophyllaceae)	Various uterine diseases	Phytoecdysteroids	Khalmatov (1964), Saatov et al. (1990)

Dictamnus angustifolius G. Donfil. ex Sweet (Rutaceae)	Rheumatism, bleeding, itching, jaundice, chronic hepatitis, skin	Limonoids and coumarins	Wu et al. (1999)
Dipsacus dipsacoides (Kar. & Kir.) Botsch. (Dipsacaceae)	Rheumatism, ulcers, and stomach cancer	Alkaloids, flavonoids	Alimbaeva and Goncharova (1971), Mukhamedziev and Alimbaeva (1969)
Dipsacus laciniatus L. (Dipsacaceae)	Rheumatism	Iridoids, alkaloids, flavonoids	Alimbaeva et al. (1986), Kocsis et al. (1993)
Dodartia orientalis L. (Phrymaceae, formerly Scrophulariaceae)	Syphilis	Alkaloids, saponins	Khalmatov (1964)
Eminium regelii Vved. (Araceae)	Stomachaches, abdominal pain	Carotinoids, lipids	Sezik et al. (2004), Chernenko et al. (2005)
Ephedra equisetina Bunge (Ephedraceae)	Bronchial asthma, rheumatism, malaria, altitude sickness, fever	Alkaloids, tannins	Mamedov and Craker (2001), Khalmatov (1964), Khalmatov et al. (1984)
Epilobium hirsutum L. (Onagraceae)	Hemostatic, astringent	Flavonoids, alkaloids, coumarins	Vandisheva et al. (1977), Plant Resources of the USSR 1987
Equisetum arvense L. (Equisetaceae)	Kidney and bladder diseases, edema, rheumatism	Flavonoids, saponins, alkaloids	Kurochkin (1998), Khalmatov et al. (1984), Oh et al. (2004)
Eremurus regelii Vved. (Asphodelaceae, formerly Liliaceae)	Pyoderma	Polysaccharides	Mamedov et al. (2004), Yuldasheva et al. (1993)
Erodium cicutarium (L.) L'Her. ex Aiton (Geraniaceae)	Astringent, antiinflammatory, hemostatic	Geraniin, didehydrogeraniin, corilagin, rutin, hyperin, quercetin, isoquercitrin, kaempferol, myricetin, polyphenolic acids	Zavrazhanov et al. (1977), Fecka and Cisowski (2005)
Eryngium biebersteinianum Nevski (Apiaceae)	Edema, scrofula, gonorrhea, headaches, heart pain, diaphoretic, diuretic, anemia	Oils, saponins, tannids	Khalmatov (1964), Khalmatov and Kosimov (1994)
Erysimum diffusum Ehrh. (Brassicaceae)	Heart weakness, tachycardia, hypertension	Fatty oil	Khodzhimatov (1989), Khalmatov et al. (1984)

(continued)

Table 8.1 (continued)

Plant species	Pharmacological properties	Phytochemical constituents	References
Euphorbia jaxartica Prokh. (Euphorbiaceae)	Fungal skin diseases and scabies	Caoutchouc, resins, euphorbin	Khalmatov (1964), Pavlov (1947)
Euphorbia rapulum Kar. & Kir. (Euphorbiaceae)	Tuberculosis	Resins, caoutchouc	Khalmatov (1964)
Ferula foetida (Bunge) Regel (Apiaceae)	Diabetes, hypolipidemic activity, anticonvulsant, vermifuge, nervous diseases	Disulfide, hexenyl-disulfides, paraoxycoumarins, asaresinotannol, asaresinol, asaresin A, farnesferol A, B, C	Al-Awadi and Shoukry (1988), Kurmukov and Akhmedkhodzhaeva (1994)
Ferula kuhistanica Korovin (Apiaceae)	Tumors	Anisic and angelic acids, and umbelliferone	Khalmatov and Kosimov (1994), Khalmatov (1964)
Ferula moschata (Reinsch.) Koso-Pol. (Apiaceae)	Anti-HIV activity	Coumarins, sesquiterpene lactones	Zhou et al. (2000)
Fumaria vaillantii Loisel. (Fumariaceae) (Papaveraceae)	Jaundice, headache, fever, gonorrhea, uterine bleeding, erysipelas, rashes, pimples	Phospholipids	Khalmatov et al. (1984), Gazizov and Glushenkova (1997)
Galium verum L. (Rubiaceae)	Analgesic, sedative, kidney diseases	Flavonoids, tannins	Akopov (1990), Zhao et al. (2008)
Gentiana olivieri Griseb. (Gentianaceae)	Rheumatic pain, chest pains	Alkaloids (gentianine, gentiananine, gentianaine, gentianadine, gentio flavine, gentiotibetine, oliverine, oliveridine, oliveramine)	Khalmatov (1964)
Geranium collinum Steph. ex Willd. (Geraniaceae)	Malignant tumors, broken bones, fever	Tannins, flavonoids	Amirov (1974), Chumbalov et al. (1970)
Geum rivale L. (Rosaceae)	Headaches, insomnia, eye diseases, rheumatism, hemorrhoids		Krilov (1972)
Glaucium fimbrilligerum Boiss. (Papaveraceae)	Emetic and soporific effects, strong laxative	Alkaloids, fatty oil	Khodzhimatov (1989), Karimova et al. (1980)
Gleditsia triacanthos L. (Fabaceae)	Spastic colitis, chronic cholecystitis, stomach ulcers, bronchial asthma	Carbohydrates, lipids, fatty acids	Rakhmanberdyeva et al. (2002), Khalmatov et al. (1984)

Glycyrrhiza glabra L. (Fabaceae)	Chest pains, lung, gastritis, fever	Flavonoids	Mashkovskii (1984)
Glycyrrhiza uralensis Fisch. ex DC. (Fabaceae)	Pneumonia, bronchitis, asthma, ulcers	Flavonoids	Khodzhimatov (1989), Nakanishi et al. (1985)
Haplophyllum acutifolium (DC.) G. Don. f. (Rutaceae)	Various skin diseases	Kusunokinin, β-sitosterol, oleanolic acid, cholesterol ahexadecanoic acid	Khalmatov (1964), Ali et al. (2001)
Helichrysum maracandicum Popov ex Kirp. (Asteraceae)	Hemorrhoidal bleeding, common cold remedy, ascaridosis	Flavonoids, glycosides, coumarins, sterins, essential oil, fatty acids	Khodzhimatov (1989). Baimukhamedov and Komissarenko (1990), Yagura et al. (2008)
Herniaria glabra L. (Caryophyllaceae)	Syphilis, pulmonary	Coumarins, flavonoids, triterpene	Khalmatov et al. (1984), Khodzhimatov (1989), Schröder et al. (1993)
Hibiscus trionum L. (Malvaceae)	Catarrh in the upper respiratory tract, a diuretic	Fatty acids	Khalmatov (1964), Hu et al. (2006)
Hippophae rhamnoides L. (Elaeagnaceae)	Rheumatis, a laxative	Folic acid, sugars, organic acids, quercetin, flavonoids	Khodzhimatov (1989), Khalmatov et al. (1984), Yue et al. (2004)
Hyoscyamus niger L. (Solanaceae)	Toothaches	Alkaloids, flavonoids	Khalmatov et al. (1984), Gammerman et al. (1990)
Hypericum perforatum L. (Hypericaceae)	Kidney diseases, heart diseases, diarrhea, hemoptysis	Terpenes, tannins, alkaloids	Khalmatov et al. (1984), Nahrstedt and Butterweck (1997)
Hypericum scabrum L. (Hypericaceae)	Liver, heart, stomach, intestinal and bladder diseases	Flavonoids	Khodzhimatov (1989), Tanaka et al. (2004)
Hyssopus seravschanicus (Dub.) Pazij (Lamiaceae)	Stomatitis, bronchial asthma, acute respiratory infections	Steroids, flavonoids, triterpenoids, fatty oil (palmitic, stearic, oleinic, linoleic, and linolenic acids)	Zotov (1975); Dzhumaev (1980)
Impatiens parviflora DC. (Balsaminaceae)	Hemostatic and various uterine diseases	Parinaric acid	Khalmatov (1964), Tsevegsuren et al. (1998)
Inula britannica L. (Asteraceae)	Cystitis, diabetes, jaundice, respiratory catarrh, bone tuberculosis, rheumatism, hemorrhoids, antiinflammatory and astringent remedy	Vitamin C	Khodzhimatov (1989), Han et al. (2001)

(continued)

Table 8.1 (continued)

Plant species	Pharmacological properties	Phytochemical constituents	References
Inula grandis Schrenk ex Fisch. & C.A. Mey. (Asteraceae)	Brucellosis, tuberculosis, gastrointestinal diseases, as a vermifuge	Sequiterpene lactones	Khalmatov and Kosimov (1994), Su et al. (2000)
Juglans regia L. (Juglandaceae)	Skin diseases, venereal diseases, tuberculosis	Tannins, pigments	Akopov (1981), Li et al. (2007)
Juniperus seravschanica Kom. (Cupressaceae)	Rheumatism	Diterpenes, sesquiterpenes	Khodzhimatov (1989), Okasaka et al. (2006)
Zygophyllum oxianum Boriss. (Zygophyllaceae)	Rheumatism, wounds, carbuncles	Tritepenoid glycosides, cincholic acid	Seredin and Sokolov (1969), Feng et al. (2007)
Ziziphora tenuior L. (Lamiaceae)	Diarrhea, children's colitis, neurasthenia	Essential oils	Khalmatov (1964)
Ziziphora pedicellata Pazij & Vved. (Lamiaceae)	Cardiotonic, decreased arterial pressure, increased diuresis	Terpenes	Khalmatov (1964)
Ziziphora bungeana Juz. (Lamiaceae)	Gastritisis, frequent vomiting, meteorism	Pulegone, isomenthone, 1,8-cineole, piperitenone	Sonboli et al. (2006)
Vinca erecta Regel & Schmalh. (Apocynaceae)	Diarrhea, gastrointestinal disorders, headaches and dizziness, astringent and heal wounds	Alkaloids	Khalmatov et al. (1984), Khodzhimatov (1989), Yagudaev et al. (1983)
Verbascum songaricum Schrenk (Scrophulariaceae)	Stomach, intestinal catarrh, gallbladder	Saponins	Seredin and Sokolov (1969), Khalmatov (1964)
Tussilago farfara L. (Asteraceae)	Tuberculosis	Rutin, arnidiol, faradiol, taraxanthin, stigmasterin, cytosterin, phytosterins	Maznev (2004), Khodzhimatov (1989)
A. karataviense Regel	Vitamin deficiency		Khojimatov et al. (2015)
A. praemixtum Vved.	Vitamin deficiency		Khojimatov et al. (2015)
Bunium persicum L.	Stomach diseases, spice		Khojimatov et al. (2015)
Berberis integerima Bunge	Liver and kidney diseases		Khojimatov et al. (2015)
Morus alba L.	Diabetes		Khojimatov et al. (2015)

situated in Ugam-Chatkal National Park, within the Chatkal mountain range of the West Tien-Shan Mountains, and almost all types of vegetation of the Central Asian mountains grow within that region.

Among 4500 plant species grown in Uzbekistan, about 200 species are traditionally used in daily life for foods and nutrition (Zakirov 1989). Several plant species belonging to the family Alliaceae are used in salads, such as *Allium barsczewskii* Lipsky, *A. majus* Vved., *A. praemixtum* Vved., *A. suworowii* Regel, *Capsella bursa-pastoris* (L.) Medic, *Rheum maximoviczi* A. Losinsk, and *Urtica dioica* L. (Khojimatov et al. 2015). *Bunium persicum* L. is commonly used as a spice, and several plant species are consumed as fruits, such as *Elaegnus angustifolia* L., *Morus alba* L., *Punica granatum* L., and *Crataegus pontica* C. Koch. (Khojimatov et al. 2015).

8.2 Phytochemical and Pharmacological Properties

Medicinal plants contain a wide range of essential oils rich in phenolic compounds and a wide array of other biologically active compounds that are traditionally used as remedies for the treatment of various diseases (Palombo and Semple 2001; Van Wyk and Wink 2004). Among them, alkaloids, tannins, phenolic compounds, coumarins, flavonoids, saponins, terpenes, and ecdysteroids are widely used in the pharmaceutical industry (Tulyaganov and Kozimova 2005; Gapparov et al. 2007; Mamadalieva et al. 2011). According to Aripov (1995), about 300 plant species contain alkaloids, such as quinoline, isoquinoline, indole, pyrrolizidine, amaryllis, steroids, quinazolone, pyrrolidine, and diterpenes.

An endemic plant of Uzbekistan, *Crataegus turkestanica*, is commonly used for the treatment of cardiovascular diseases (Yu et al. 1995). *Ferula pallida* and *F. penninervis* were reported to contain new antioxidants with a unique mechanism of action (Kogure et al. 2004). Sakhobiddinov (1995) reported the common use of *Ferula varia* roots to heal intestinal parasitosis. The plant species *Euphorbia ferganensis* contains about 9.1% polyphenols (Abdulladzhanova et al. 2001), *Rhodiola litvinovii* contains *p*-tyrosol, salidroside, gallic acid, and epigallocatechin-3-*O*-gallate (Melikuziev et al. 2013), and *Geranium saxatile* contains gallic acid, ellagic acid, and quercetin-3-*O*-glucoside (Siddikov et al. 2011).

The flavonoids of *Thermopsis altherniflora* such as flavones, isoflavones, chrysoeriol, apigenin, luteolin, and cynaroside inhibited the development of hyperlipidemia induced in experimental animals by injections of Triton WR-1339 (Faizieva et al. 2003). The plant *Scutellaria ramosissima*, used in traditional medicine, is reported to have cytotoxic activity for several cancer cell lines and is also used in the treatment of various inflammations and nervous disorders (Mamadalieva et al. 2011). *Helichrysum maracandicum* was effective on SENCAR mouse skin transformed (SST) cells (Yagura et al. (2008). *Geranium saxatile*, widely grown in Uzbekistan, showed antihypoxic activity for acute normobaric hypoxia (Siddikov et al. 2011).

The endemic plant *Ungernia victoris* is distributed in different ecosystems of Uzbekistan and contains various alkaloids, such as galacturonic anhydride (Malikova and Rakhmanberdyeva (2013), which is isolated and identified from the leaves of *U. victoris*. The alkaloids, namely liriodenine, lysicamine, *O*-methylmoschatoline, and lanuginosine, are found in *Liriodendron tulipifera* L. (Ziyaev et al. 1987). Other plant species such as *Nitraria schoberi* contain alkaloids such as nitraraine, dehydroschoberine, sibiridine, vasicinone, and peganol (Tulyaganov and Kozimova 2005). Ecdysterone, cyasterone, ajugalactone, ajugasterone, α-ecdysone, and turkesterone were observed in *Ajuga turkestanica* (Abdukadirov et al. 2005).

Essential oils are also important phytochemical constituents of plant with wide biological activity. *Hypericum scabrum* and *H. perforatum* contain essential oils, namely, β-caryophyllene, caryophyllene oxide, spathulenol, and α-pinene, whereas *H. scabrum* contain α-pinene, spathulenol, *p*-cymene, acetophenone, and carvacrol (Baser et al. 2002a, b).

The flavonoids, pulegone, sterols, and essential oils extracted from *Ziziphora* species showed antibacterial and antifungal activities (Tada et al. 2002). *Hypericum perforatum* contains tannins, flavonoids, xanthones, hyperforin, and essential oils (Dall'Agnol et al. 2003). These compounds showed various biological activities including sedative, antihelmintic, analgesic, antibacterial, and antiinflammatory activity. *Silene wallichiana* is considered as a source for ecdysteroid compounds, and several were found in the plant including 20-hydroxyecdysone-22-benzoate, 2-deoxy-20-hydroxyecdysone, 2-deoxyecdysone, 22-benzoate-2-deoxy-20-hydroxyecdysone, and 25-acetate-2-deoxy-20-hydroxyecdysone (Mamadalieva et al. 2000). The coumarins, vitamin C, and β-carotene were found in the endemic plant of Uzbekistan *Crambe kotschyana* Boiss (Nikonov et al. 1965). The coumarins, terpenoids, and glycosides from the roots of *Prangos pabularia* (Apiaceae) were reported by Tada et al. (2002): these compounds showed antibacterial activity.

Many plant species contain lipids, such as glycolipids, neutral lipids, and phospholipids found in *Scutellaria ramosissima* (Yuldasheva et al. 2014); paraffinic, olefinic, and isoprenoid hydrocarbons, triterpenols, sterols, and palmitic and linoleic acids found in *Silybum marianum* (Mirzaeva et al. 2011); and triacylglycerines, triterpenols, diacylglycerines, and sterols in *Althaea nudiflora* and *A. armeniaca* (Sagdullaev et al. 2001).

The plant glycosides are known to have wide biological properties. However, there are few studies on the glycosides of the medicinal plants of Uzbekistan. The plant species *Gentiana olivieri* and *Origanum tyttanthum* were found to synthesize secoiridoid glycosides, olivierosides, gentiopicroside, sweroside, 69-*O*-*β*-D-glucosylgentiopicroside, swertiapunimarin, eustomoside, eustomorusside, and septemfidoside (Takeda et al. 1999, 2008). Kurimoto et al. (2012) reported eudesmane, guaiane sesquiterpene lactone glucosides, in *Ferula varia* (Schrenk).

The phytochemical constituents of plants showed various biological activities including antimicrobial, anabolic, antiinflammatory, antioxidative, and cytotoxic activity. For the treatment of cancer, several plants belonging to the genera *Ajuga*, *Scutellaria*, and *Stachys* have been reported (Konoshima et al. 2000; Shang et al. 2010). Azizov et al. (2012) reported the plant species *Arctium lappa*, which is commonly used in Uzbekistan for treatment of skin disease, eczema, and ulcers, as well as kidney ailments. The hepatoprotective and antioxidant activities of *Silybum marianum* were also reported by Minakhmetov et al. (2001). In an earlier report Nuraliyev and Zubaidova (1994) observed the antimicrobial, hypocholesterolemic, and hypolipidemic activity of *Origanum tyttanthum*. Mamadalieva et al. (2013) observed the antibacterial and antiproliferative properties of phytoecdysteroids (viticosterone E, 20-hydroxyecdysone-22-benzoate, 2-deoxy-20-hydroxyecdysone, 2-deoxyecdysone, 20-hydroxyecdysone, and integristerone A) from *Silene wallichiana*. Among human pathogenic bacteria tested, *Pseudomonas aeruginosa* and *Staphylococcus aureus* were inhibited strongly by plant-derived compounds. *Ferula pallid* (Apiaceae), a plant native to Uzbekistan, contains pallidones and sesquiterpene chromone derivatives (Su et al. 2000).

Antimicrobial activity of *Hypericum perforatum* has been reported against *Escherichia coli*, *Klebsiella pneumoniae*, and *Pseudomonas aeruginosa* (Egamberdieva et al. 2013). The inhibition of cell growth against HeLa, HepG-2, and MCF-7 cells by flavonoids of *Scutellaria immaculata* and *S. ramosissima* was reported by Mirzaeva et al. (2011). Akhmedzhanova (2010) reported the cytotoxic activity of the flavonoid dictamnine derived from *Haplophyllum leptomerum* against the human cancer cell lines HeLa and HCT-116.

In an earlier study, Barnaulov (1982) studied the anti-ulcer activity of flavonoids such as luteolin, quercetin, 4-methoxyscutarellin, and herbacetin. *Ajuga turkestanica* is an endemic plant of Uzbekistan widely studied for its biological activity, such as ecdysteroids and iridoids that are reported to have anabolic activity and cholagogic action (Soatov et al. 1994). The pharmacological activity of medicinal plants is presented in Table 8.1. *Anethum graveolens* is known for its antimicrobial activity against human pathogens such as *Candida albicans* and *Staphylococcus aureus* (Yili et al. 2009). The antibacterial activity of plant extracts derived from *Silene walchiana*, *S. viridiflora*, and *S. brahuica* was observed (Mamadalieva et al. 2008).

Yuldashev et al. (1993) characterized, in their study, the *Bergenia hissarica* plant as an astringent, antiphlogistic, and hemostatic agent used in gastrointestinal tract ailments.

The most widely distributed plant in the salt-affected sites of Uzbekistan is liquorice (*Glycyrrhiza glabra*) (Egamberdieva and Mamedov 2015). In traditional medicine, licorice roots have been used in treating chest and lung diseases, pneumonia, bronchitis, arthritis, kidney diseases, heart diseases, gastric ulcer, coughs, swellings, low blood pressure, allergies, catarrhs of the upper respiratory tract, liver toxicity, pancreatic disorders, and certain viral infections (Armanini et al. 2002).

8.3 Conclusion

In this chapter, the herbal plants of Uzbekistan are seen as very diverse and containing various biological active compounds. The plants commonly grown in Uzbekistan include various families, and the majority of the plants belong to the families

Apiaceae, Amaranthaceae, Ranunculaceae, Rosaceae, Lamiaceae, Fabaceae, Alliaceae, Malvaceae, Brassicaceae, Geraniaceae, Asparagaceae, Poaceae, Polygonaceae, Ranunculaceae, Zygophyllaceae, Orchidaceae, Solanaceae, Dipsacaceae, Ephedraceae, Rubiaceae, Papaveraceae, Elaeagnaceae, Juglandaceae, and Scrophulariaceae. Pharmacological studies have confirmed that plant extracts and individual compounds have various biological activities such as hepatoprotective, dermatological, antimicrobial, antiviral, anti-ulcer, antioxidant, neuroprotective, and antiinflammatory properties. However, the phytochemical composition and biological activity of these plants are still not known fully, and possibly contain pharmaceutically important constituents. Thus, more investigation is required, and these plants could be a promising source of novel drugs and potentially useful new pharmaceuticals. The present report can be useful for the continued investigation of the Uzbek with the aim of obtaining new biologically active compounds.

References

Abdukadirov I, Yakubova M, Nuriddinov K, Mamatkhanov A, Turakhozhaev M (2005) Ecdysterone and turkesterone in *Ajuga turkestanica* determined by HPLC. Chem Nat Compd 41(4):475–476

Abdulladzhanova N, Mavlyanov S, Dalimov D (2001) Phenolic compounds of *Euphorbia ferganensis* B. Fedtsch Chem Nat Comp 37(2):193–194

Abu Ali Bin Sina (Avicenna) (1956) Canon of medicine, vol 2. Academy of Sciences of Uzbekistan, Tashkent (in Russian)

Abu-Irmaileh BE, Afifi FU (2000) Treatment with medicinal plants in Jordan. Dirasat 27:53–74

Akhmedzhanova V, Angenot L, Shakirov R (2010) Alkaloids from *Haplophyllum leptomerum*. Chem Nat Compd 46(3):502–503

Akopov I (1981) Hemostatic plants. Meditsina, Tashkent (in Russian)

Akopov I (1990) The most important native medicinal plants and their uses. Meditsina, Tashkent (in Russian)

Al-Awadi F, Shoukry M (1988) The lipid lowering effect of an antidiabetic plant extract. Acta Diabetol 25(1):1–5

Ali M, Pervez M, Saleem M, Tareen R (2001) Haplophytin-A and B: the alkaloidal constituents of *Haplophyllum acutifolium*. Phytochemistry 57(8):1277–1280

Alimbaeva P, Goncharova A (1971) Dikorastushchiye lekarstvenniye rasteniya v Kirgizii (farmakolognosticheskiye issledovaniye) [Wild-growing medicinal plants of Kyrgyzstan (pharmagonostic studies)]. Kyrgyzstan, Frunze (in Russian)

Alimbaeva P, Mukhamedziev M, Akimaliev A (1986) Lekarstvenniye rasteniya semeistva vorcyankovuikh florui Kirgizii (Medicinal plants of the family Dipsacaceae from the flora of Kyrgyzstan). Ilim, Frunze (in Russian)

Altimishev A (1991) Prirodniye tselebnuiye sredstva, izd. 2 (Natural medicinal remedies), 2nd edn. Kyrgyzstan, Bishkek (in Russian)

Alves RR, Rosa IM (2007) Biodiversity, traditional medicine and public health: where do they meet? J Ethnobiol Ethnomed 3:14

Amirov M (1974) A study of the connection between anti-tumor activity and the chemical structures of plant tannins. Zdravookhraneniye Kazakhstana 8:36–38 (in Russian)

Aripov KN (1995) The chemistry of plant substances in Uzbekistan. Chem Nat Compd 31:2–15

Armanini D, Fiore C, Mattarello MJ, Bielenberg J, Palermo M (2002) History of the endocrine effects of licorice. Exp Clin Endocrinol Diabetes 110:257–261

Asilbekova D, Glishenkova A, Azcan N, Ozek T, Baser K (2000) Lipids of *Origanum tyttanthum*. Chem Nat Compd 36:124–127

Azizov U, Khadzhieva A, Rakhimov D, Mezhlumyan L, Salikhov S (2012) Chemical composition of dry extract of *Arctium lappa* roots. Chem Nat Compd 47(6):1038–1039

Baimukhamedov M, Komissarenko N (1990) Coumarins of *Helichrysum maracandicum*. Chem Nat Compd 25(5):617

Bakirov E, Yusupova S, Sattikulov A, Vdovin A, Malikov V, Yagudaev M (1987) Flavonoids of *Ammothamnus lehmannii*. Structure of lehmannin and of ammothamnidin. Chem Nat Compd 23:429

Bakker E, Ghys A, Kester A, Vles J, Dubas J, Blanco C, Hornstra G (2003) Long-chain polyunsaturated fatty acids at birth and cognitive function at 7 y of age. Eur J Clin Nutr 57(1):89–95

Barnaulov OD, Manicheva OA, Zapesochnaya GG, Shelyuto VL, Glyzin VI (1982) Effect of some flavonoids on the ulcerogenic activity of reserpine in mice. Khim-Farm Zh 16(3):300–303

Baser K, Demircakmak B, Nuriddinov K, Nigumatullaev A, Aripov K (1997) Composition of the essential oil of *Origanum tyttanthum* Gontsch. from Uzbekistan. J Essent Oil Res 9:611–612

Baser KHC, Ozek T, Nuriddinov HR, Demirci AB (2002a) Essential oils of two *Hypericum* species from Uzbekistan. Chem Nat Compd 38(1):54–57

Baser K, Nuriddinov H, Ozek T, Demirci B, Azcan N, Nigmatullaev A (2002b) Essential oil of *Arischrada korolkowii* from the Chatkal Mountains of Uzbekistan. Chem Nat Compd 38(1):51–53

Bekker N, Ul'chenko N, Glushenkova A (2005) Lipids from *Descurainia sophia* seeds. Chem Nat Compd 41(3):346–347

Berkov S, Georgieva I, Kondakova V, Atanassov A, Viladomat F, Bastida J, Codina C (2009) Plant sources of galanthamine: phytochemical and biotechnological aspects. Biotechnol Biotec Eq 23(2):1170–1176

Bigović D, Savikin K, Janković T, Menković N, Zdunić G, Stanojković T, Djurić Z (2011) Antiradical and cytotoxic activity of different *Helichrysum plicatum* flower extracts. Nat Prod Commun 6(6):819–822

Buranova D (2015). The value of Avicenna's heritage in development of modern integrative medicine in Uzbekistan. Integr Med Res 4(4):220–224

Che CT, George V, Ijinu TP, Pushpangadan P, Andrae-Marobela K (2017) Traditional medicine. In: Delgoda R (ed) Pharmacognosy, fundamentals, applications and strategies. Academic Press, Boston, pp 15–30

Chernenko T, Gazizov F, Glushenkova A, Nigmatullaev A (2005) Lipids of *Eminium lehmannii* leaves and tubers. Chem Nat Compd 41(2):141–143

Chumbalov T, Mukhamed'yarova M, Chanysheva I (1970) Flavonoids of *Atraphaxis spinosa*. I. Chem Nat Compd 6(5):639

Dall'Agnol R, Ferraz A, Bernardi AR, Albring D, Nцr C, Sarmento L, Lamb L, Hass M, von Poser G, Schapoval EES (2003) Antimicrobial activity of some *Hypericum* species. Phytomedicine 10:511–516

Dzhumaev K (1980) The most prospective wild-growing aromatic plants of the Surkhan-Sherabad valley and their economic value. In: Aktualnuiye voprosy izucheniya I ispolzobaniya e fi rnikh masel (Proceedings of 3rd symposium "Important questions in the study of using essential oil bearing plants and essential oil"). Simferopol, 24–26 Sept 96–97 (in Russian)

Dzhumaev K, Zenkevich I, Tkachenko K, Tsiburskaya I (1990) Essential oil of the leaves of *Hyssopus seravschanicus* from south Uzbekistan. Chem Nat Compd 1:123–125

Edeoga HO, Okwu DE, Mbaebie BO (2005) Phytochemical constituents of some Nigerian medicinal plants. Afr J Biotechnol 4(7):685–688

Egamberdieva D, Mamedov N (2015) Liquorice: with potential use in phytoremediation of salt affected soils. In: Öztürk M, Ashraf M, Aksoy A, Ahmad MSA, Hakeem KR (eds) Plants, pollutants and remediation. Springer, Dordrecht. https://doi.org/10.1007/978-94-017-7194-8_13

Egamberdieva D, Mamadalieva N, Khodjimatov O, Tiezzi A (2013) Medicinal plants from Chatkal Biosphere Reserve used for folk medicine in Uzbekistan. Med Aromat Plant Sci Biotechnol 7(1):56–64

Egamberdieva D, Wirth S, Behrendt U, Parvaiz A, Berg G (2017) Antimicrobial activity of medicinal plants correlates with the proportion of antagonistic endophytes. Front Microbiol 8:199. https://doi.org/10.3389/fmicb.2017.00199

Eisenman S, Zaurov D, Struwe L (2013) Medicinal plants of Central Asia: Uzbekistan and Kyrgyzstan. Springer, New York, pp 1–340

Ersöz T, Çalis I, Carcia J, Chulia A (1991) Constituents of *Gentiana olivieri*. Fitoterapia 62:184–185

Faizieva SK, Khushbaktova ZA, Syrov VN (2003) A total flavonoid extract from *Thermopsis altherniflora:* development of the production technology, creation of a medicinal preparation, and evaluation of hypolipidemic activity. Pharm Chem J 37(4):42–44

Fecka I, Cisowski W (2005) Tannins and flavonoids from the *Erodium cicutarium* herb. Z Naturforsch B Chem Sci 60(5):555–560

Feng Y, Li H, Xu L, Yang S (2007) 27-Nor-triterpenoid glycosides from the barks of *Zygophyllum fabago* L. J Asian Nat Prod Res 9(6):505–510

Gammerman A, Kadaev G, Yatsenko-Hmelevsky A (1990) Lekarstvennuiye Rasteniya (Medicinal plants). Vuishaya shkola, Moscow (in Russian)

Gapparov AM, Razzakov NA, Aripova SF (2007) Alkaloids of *Convolvulus subhirsutus* from Uzbekistan. Chem Nat Compd 43(3):291–292

Gazizov F, Glushenkova A (1995) Phospholipids of the pericarp of *Crataegus turkestanica*. Chem Nat Compd 31(4):456–466

Gazizov F, Glushenkova A (1997) Phospholipids of the seeds of *Fumaria vaillantii*. Chem Nat Compd 33(6):627–629

Goryaev M, Bazalitskaya V, Polyakov P (1962) Chemical composition of sages. Izd Akad Nauk Kaz SSR, Alma-Ata

Han J, Lee B, Kim Y, Yoon J, Jin H, Hong S, Lee H, Lee K, Lee H (2001) Ergolide, sesquiterpene lactone from *Inula britannica*, inhibits inducible nitric oxide synthase and cyclo-oxygenase-2 expression in RAW 264.7 macrophages through the inactivation of NF-KB. Br J Pharmacol 133(4):503–512

Honda G (1999) Report on traditional medicine of Turkic people. Kyoto University, Kyoto, pp 53–74

Hu K, Hou X, Gong Y, Abbas A (2006) Analysis of physical-chemical properties and fatty acid compositions of *Hibiscus trionum* L. seed oil. Food Sci 27(11):455–457 (in Chinese)

Karimova S, Israilov I, Yunusov M, Yunusov S (1980) Alkaloids of *Glaucium fimbrilligerum*. Chem Nat Compd 16(2):177–180

Khalmatov K (1964) Dikorastushchiye Lekarstvenniye Rasteniya Uzbekistana (Wild-growing medicinal plants of Uzbekistan). Meditsina, Tashkent (in Russian)

Khalmatov K, Kosimov A (1994) Dorivor Usimliklar (Malummotnoma) [Medicinal plants (references)]. Ibn Sino, Tashkent (in Uzbek)

Khalmatov K, Kharlamov I, Alimbayeva P, Karriev M, Khaetov I (1984) Osnovnuiye Lekarstvennuiye Rasteniya Srednei Azii (The main medicinal plants of Central Asia). Meditsina, Tashkent (in Russian)

Khodjimatov K, Aprasidi G, Khodjimatov A (1995) Wild-growing medicinal plants of Central Asia. Abu Ali ibn Sino, Tashkent, p 112 (in Russian)

Khodzhimatov M (1989) Dikorastushchiye Lekarstvennuiye Rasteniya Tadzhikistana (Wild-growing medicinal plants of Tadjikistan). Glavnaya nauchnaya redaktsiya Tadzhikskoi Sovetckoi entsiklopedii, Dushanbe (in Russian)

Khojimatov O, Abdiniyazova G, Pak V (2015) Some wild growing plants in traditional foods of Uzbekistan. J Ethn Foods 2:25–28

Khuzhaev V, Zhalolov I, Turgunov K, Tashkhodzhaev B, Levkovich M, Aripova S, Shashkov A (2004) Alkaloids from *Arundo donax*. XVI. Structure of the new dimeric indole alkaloid arundavine. Chem Nat Compd 40:261–265

Kim D, Chun O, Kim Y, Moon H, Lee C (2003) Quantification of polyphenolics and their antioxidant capacity in fresh plums. J Agric Food Chem 51(22):6509–6515

Kocsis A, Szabo L, Podanyi B (1993) New bis-iridoids from *Dipsacus laciniatus*. J Nat Prod 56(9):1486–1499

Kogure K, Yamauchi I, Tokumura A, Kondou K, Tanaka N, Takaishi Y (2004) Novel antioxidants isolated from plants of the genera *Ferula*, *Inula*, *Prangos* and *Rheum* collected in Uzbekistan. Phytomedicine 11:645–665

Konoshima T, Takasaki M, Tokuda H, Nishino H (2000) Cancer chemopreventive activity of an iridoid glycoside, 8-acetylharpagide, from *Ajuga decumbens*. Cancer Lett 157:87–92

Krilov G (1972) Travui Zhizni i Ikh Iskateli (Herbs for life and their seekers). Zapad-Sibirskoye Knezhnoye Izdatelstvo, Novosibirsk (in Russian)

Kucherbaev K, Uteniyazov K, Kachala V, Saatov Z, Shashkov A (2002) Triterpene glycosides from plants of the *Astragalus* genus. Structure of cyclounifolioside *a* from *Astragalus unifoliolatus*. Chem Nat Compd 38(2):175–178

Kuchin N (1955) About using an infusion of the flowers of *Crataegus songarica* for the treatment of patients with rheumatism. Synopsis of dissertation for the degree of Ph.D. in Medical Sciences, Alma Ata (in Russian)

Kurimoto S, Suzuki K, Okasaka M, Kashiwada Y, Kodzhimatov O, Takaishi Y (2012) New sesquiterpene lactone glucosides from the roots of *Ferula varia*. Phytochem Lett 5:729–733

Kurmukov A, Akhmedkhodzhaeva K (1994) Estrogennuiye lekarstvennuiye preparatui iz rastenii roda *Ferula* (Estrogenic medicinal preparations of plants in the genus *Ferula*). Ibn Sino, Tashkent (in Russian)

Kurochkin E (1998) Lekarstvennuiye Rasteniya (Medicinal plants). Parus, Samara (in Russian)

Li L, Tsao R, Yang R, Kramer J, Hernandez M (2007) Fatty acid profiles, tocopherol contents, and antioxidant activities of heartnut (*Juglans ailanthifolia* var. *cordiformis*) and Persian walnut (*Juglans regia* L.). J Agric Food Chem 55(4):1164–1169

Lopez T, Cid M, Bianchini M (1999) Biochemistry of hemlock (*Conium maculatum* L.) alkaloids and their acute and chronic toxicity in livestock. A review. Toxicon 37(6):841–865

Malikova MK, Rakhmanberdyeva RK (2013) Pectinic substances from leaves of *Ungernia victoris*. Chem Nat Compd 49(4):596–598

Mamadalieva NZ, Ramazanov NS, Dinan L, Saatov Z (2000) Phytoecdysteroids of plants of the *Silene* genus. 2-Deoxyecdysterone-3-*O*-benzoate from *Silene wallichiana*. Chem Nat Compd 36:513–515

Mamadalieva N, El-Readi M, Hermann F, Tahrani A, Hamoud R, Azimova S (2001) Flavonoids in *Scutellaria immaculata* and *S. ramosissima* and their biological activities. J Pharm Pharmacol 63(10):1346–1357

Mamadalieva N, Syrov V, Egamberdieva D, Edwards Jones V (2008) Phytoecdysteroids from *Silene viridiflora* and their actoprotective, adaptogenic and antimicrobial activity. FEBS J 275(1):457

Mamadalieva N Z, El-Readi M Z, Hermann F, Tahrani A, Hamoud R, Azimova S S, Egamberdieva D, Wink M, (2011). Flavonoids in Scutellaria immaculata and S. ramosissima and their biological activities. J Pharm Pharm 63(10):1346–1357

Mamadalieva N, Egamberdieva D, Tiezzi A (2013) In vitro biological activities of the components from *Silene wallichiana*. Med Aromatic Plant Sci Biotechnol 7(1):1–6

Mamadalieva N, Mamedov N, Craker L, Tiezzi A (2014) Ethnobotanical uses and cytotoxic activities of native plants from the Lamiaceae family in Uzbekistan. Acta Hortic 1030:61–70

Mamatkhanov A, Yakubova M, Syrov V (1998) Isolation of turkesterone from the epigeal part of *Ajuga turkestanica* and its anabolic activity. Chem Nat Compd 34:150–154

Mamedov N, Craker L (2001) Medicinal plants used for treatment of bronchial asthma in Russia and Central Asia. J Herbs Spices Med Plants 8(2/3):91–117

Mamedov N, Gardner Z, Craker L (2004) Medicinal plants used in Russia and Central Asia for the treatment of selected skin conditions. J Herbs Spices Med Plants 11:191–222

Mashkovskii M (1984) Lekarstvennuiye Sredstva: Posobiye po Farmakoterapii Dlya Vrachei, 9 izd (Medicinal remedies: manual on pharmacotherapy for doctors), 9th edn. Meditsina, Moscow (in Russian)

Masuda T (2004) Antioxidant mechanism of the potent diterpenoid antioxidants found in sage and rosemary. Foods & Food Ingredients J Jpn 209:858–865

Masuda T (2004) Antioxidant mechanism of the potent diterpenoid antioxidants found in sage and rosemary. Food Food Ingredients J Jpn 209:858–865

Maznev N (2004) Bolshaya Entsiklopediya Narodnoi Meditsinui (Big encyclopedia of traditional medicines). Martin, Moscow (in Russian)

McChesney JD, Venkataraman SK, Henri JT (2007) Plant natural products: back to the future or into extinction? Phytochemistry 68:2015–2022

Melikuziev F, Nishanbaev S, Turgunov K, Bobakulov K (2013) Phenolic compounds from roots of *Rhodiola litvinovii*. Chem Nat Compd 49(2):349–350

Minakhmetov RA, Onuchak LA, Kurkin VA, Avdeeva EV, Volotsueva AV (2001) Analysis of flavonoids in *Silybum marianum* fruit by HPLC. Chem Nat Compd 37(4):318–321

Mirzaev Y, Okhunov I, Bobaculov K, Abdullaev N, Khujaev V, Aripova S (2016) Compounds of the plant *Convolvules fruticosus* and pharmacological activity of the major alkaloid-cuscohydrine. Uzbek Biol J 2:3–5

Mirzaeva N, Chernenko T, Gusakova S, Sagdullaev S (2011) Lipids and flavolignans from *Silybum marianum*. Chem Nat Compd 47:355

Mukhamatkhanova R, Shamyanov I, Zakirov S, Tashkhodzhaev B, Levkovich M, Abdullaev N (2011) A new melampolide and other secondary metabolites from *Artemisia baldshuanica*. Chem Nat Compd 47(5):726–730

Mukhamedziev M, Alimbaeva P (1969) Triterpene glycosides of *Dipsacus azureus*. Chem Nat Compd 5(5):376–377

Nahrstedt A, Butterweck V (1997) Biologically active and other chemical constituents of the herb of *Hypericum perforatum* L. Pharmacopsychiatry 30:129–134

Nakanishi T, Inada A, Kambayashi K, Yoneda K (1985) Flavonoid glycosides of the roots of *Glycyrrhiza uralensis*. Phytochemistry 24(2):339–341

Nakatani N, Inatani R (1984) Two antioxidative diterpenes from rosemary (*Rosmarinus officinalis* L.) and a revised structure for rosmanol. Agric Biol Chem 48:2081–2085

Nikonov GK, Ermatov IE, Davidyants SB (1965) Coumarin and its derivatives in plant of Tadzhikistan. Dokl Akad Nauk Tadzh SSR 8(10):22

Nishanov A, Sultankhodzhaev N, Yunosov M, Yusupova I, Tashkhozhaev B (1991) Alkaloids of *Aconitum talassicum*. Chem Nat Compd 27(1):82–86

Nuraliyev Y, Zubaidova T (1994) Hypocholesterinemic and hypolipidemic effects of ether oil of *Origanum tyttanthum* Gontsch. Khim Farm Zh 28:63–64

Nuriddinov K, Khodzimatov K, Aripov K, Ozek T, Demirchakmak B, Basher K (1997) Essential oil of *Perovskia scrophularifolia*. Chem Nat Compd 33:299–300

Oh H, Kim D, Cho J, Kim Y (2004) Hepatoprotective and free radical scavenging activities of phenolic petrosins and flavonoids isolated from *Equisetum arvense*. J Ethnopharmacol 95(2/3):421–424

Okasaka M, Takaishi Y, Kashiwada Y, Kodzhimatov O, Ashurmetov O, Lin A, Consentino L, Lee K (2006) Terpenoids from *Juniperus polycarpus* var. *seravschanica*. Phytochemistry 67(24):2635–2640

Okhunov I, Levkovich M, Abdullaev N, Khuzhaev V, Aripov S (2011) Alkaloids from *Crambe kotschyana* endemic to Uzbekistan. Chem Nat Compd 47(3):487–489

Palombo EA, Semple SJ (2001) Antibacterial activity of traditional Australian medicinal plants. J Ethnopharmacol 77:151–157

Parajuli P, Joshee N, Rimando A, Mittal S, Yadav A (2009) In vitro antitumor mechanisms of various *Scutellaria* extracts and constituent flavonoids. Planta Med 75(1):41–48

Passreiter C, Wilson J, Andersen R, Isman M (2004) Metabolism of thymol andtrans-anethole in larvae of *Spodoptera litura* and *Trichoplusia* Ni (Lipidoptera: Noctuidae). J Agric Food Chem 52:2549–2551

Pavlov N (1947) Rastitelnoye Suiryo Kazakhstana (Rasteniya; ikh Veshchestva i Ispolzovaniye) [Plant raw material of Kazakhstan (Plants: their substances and uses)]. Izdatelstvo Akademii Nauk SSSR, Moscow (in Russian)

Petersen M, Simmonds M (2003) Rosmarinic acid. Phytochemistry 62:121–125

Petrova VP (1972) Catechins and leucoanthocyanins from the fruits of wild hawthorn species (*Crataegus* L.). Ukr Botanichnii Zh 29(2):144–147 (in Ukrainian)

Plant Resources of the USSR (1987) Fam. Hydrangeaceae – Haloragaceae, vol 3. Nauka, Leningrad (in Russian)

Rakhimov D, Salimova E, Zakirova R (2003) Polysaccharides of plant tissue culture. III. Polysaccharides of *Codonopsis clematidea* callus culture. Chem Nat Compd 39(1):46–47

Rakhmanberdyeva R, Talipova M, Gazizov F, Rakhimov D (2002) Carbohydrates and lipids of *Gleditsia triacanthos* seeds. Chem Nat Compd 38(1):24–26

Rakhmatullaev T, Akramov S, Yunusov Y (1969) The alkaloids of *Gentiana olivieri*. Chem Nat Compd 5:531–532

Rivero-Cruz J, Chai H, Kardono L, Setyowati F, Afriatini J, Riswan S, Farnsworth N, Cordell G, Pezzuto J, Swanson S, Kinghorn A (2004) Cytotoxic constituents of the twigs and leaves of *Aglaia rubiginosa*. J Nat Prod 67:343–347

Saatov Z, Gorovits M, Abubakirov N (1990) Phytoecdysteroids of the plant *Dianthus hoeltzeri*. Chem Nat Compd 26(6):717–718

Saatov Z, Agzamkhodzhaeva D, Syrov N (1999) Distribution of phytoecdysteroids in plants of Uzbekistan and the possibility of using drugs based on them in neurological practice. Chem Nat Compd 35:186–191

Sagdullaev B, Shakhidoyatov R, Khodzhaeva M, Chernenko T, Turakhozhaev M, Abduazimova M (2001) Lipids and carbohydrates from *Althaea nudiflora* and *A. armeniaca* roots. Chem Nat Compd 37(3):208–212

Sakhobiddinov S (1995) Wild medicinal plants of Uzbekistan. Gosizdat UzSSR, Tashkent (in Russian)

Sala A, Recio M, Giner R, Manez S, Tournier H, Schinella G, Rios J (2002) Anti-inflammatory and antioxidant properties of *Helichrysum italicum*. J Pharm Pharmacol 54(3):365–371

Satimov G, Mamatkhanov A, Kotenko L, Turakhodzhaev M, Madrakhimov S, Faizieva S, Khushbaktova Z, Syrov V (2003) A Total Flavonoid Extract from Thermopsis altherniflora: Development of the Production Technology, Creation of a Medicinal Preparation, and Evaluation of Hypolipidemic Activity. Pharmaceutical Chemistry Journal 37(4):203–206

Satimov G, Mamatkhanov A, Kotenko L, Turakhodzhaev M, Madrakhimov S, Mamatkhanov A, Yakubova M, Syrov V (1998) Isolation of turkesterone from the epigeal part of *Ajuga turkestanica* and its anabolic activity. Chem Nat Compd 34:150–154

Schröder H, Schubert-Zsilavecz M, Reznicek G, Cart J, Jurenitsch J, Haslinger E (1993) A triterpene saponin from *Herniaria glabra*. Phytochemistry 34(6):1609–1613

Sdykov T, Abduazimov K (2000) Lignins of *Allochruza paniculata* and *Glycyrrhiza glabra*. Chem Nat Compd 38(3):288–290

Seredin R, Sokolov S (1969) Lekarstvennuiye Rasteniya i Ikh Primeneniye, izd 2 (Medicinal plants and their uses), 2nd edn. Stavropolskoye knizhnoye izdatelstvo, Stavropol (in Russian)

Sezik E, Yesilada E, Shadidoyatov H, Kulivey Z, Nigmatullaev A, Aripov H, Takaishi Y, Takeda Y, Honda G (2004) Folk medicine in Uzbekistan. I. Toshkent, Djizzax, and Samarqand provinces. J Ethnopharmacol 92(2/3):197–207

Shang X, He X, He X, Li M, Zhang R, Fan P, Zhang Q, Jia Z (2010) The genus *Scutellaria*: an ethnopharmacological and phytochemical review. J Ethnopharmacol 128:279–313

Shikishima Y, Takaishi Y, Honda G, Ito M, Takeda Y, Kodzhimatov O, Ashurmetov O (2001) Phenylbutanoids and stilbene derivatives of *Rheum maximowiczii*. Phytochemistry 56:377–381

Siddikov D, Nishanbaev S, Narbutaeva D, Vdovin A, Abdullaev N (2011) Polyphenols from *Geranium saxatile*. Chem Nat Compd 47(4):630–631

Soatov Z, Syrov VN, Mamatkhanov AU, Abubakirov NK (1994) Phytoecdysteroids of *Ajuga* plants and their biological-activity: distribution and chemical structures of isolated compounds. Chem Nat Compd 152–160

Sonboli A, Mirjalili M, Hadian J, Ebrahimi S, Yousefzadi M (2006) Antibacterial activity and composition of the essential oil of *Ziziphora clinopodioides* subsp. *bungeana* (Juz.) Rech. f. from Iran. Z Naturforsch 61(9/10):677–680

Su B, Takaishi Y, Honda G, Itoh M, Takeda Y, Kodzhimatov O, Ashurmetov O (2000) Sesquiterpene phenylpropanoid and sesquiterpene chromone derivatives from *Ferula pallida*. J Nat Prod 63:520–522

Suzgec S, Mericli A, Houghton P, Cubukcu B (2005) Flavonoids of *Helichrysum compactum* and their antioxidant and antibacterial activity. Fitoterapia 76:269–272

Syrov V, Yuldashev M, Tursunova N, Khushbaktova Z (2010) Medicinal plants isolation, structural studies, and antiulcer activity of lehmanin and ammothamnidin from *Ammothamnus lehmannii*. Pharm Chem J 44(2):16–18

Tada Y, Shikishima Y, Takaishi Y, Shibata H (2002) Coumarins and gamma pyrone derivatives from *Prangos pabularia*. Phytochemistry 59:649–654

Takeda Y, Masuda T, Honda G, Takaishi Y, Ito M, Ashurmetov OA, Khodzhimatov OK, Otsuka H (1999) Secoiridoid glycosides from *Gentiana olivieri*. Chem Pharm Bull 47(9):1338–1340

Takeda Y, Hayashi T, Masuda T, Honda G, Takaishi Y, Ito M, Otsuka H, Matsunami K, Khodzhimatov O, Ashurmetov O (2007) Chemical constituents of an Uzbek medicinal plant, *Perovskia scrophularifolia*. J Nat Med 61:84–85

Takeda Y, Tomonari M, Arimoto S, Masuda T, Otsuka H, Matsunami K, Honda G, Ito M, Takaishi Y, Kiuchi F, Khodzhimatov OK, Ashurmetov OA (2008) A new phenolic glucoside from an Uzbek medicinal plant, *Origanum tyttanthum*. J Nat Med 62:71–74

Tanaka N, Takaishi Y, Shikishima Y, Nakanishi Y, Bastow K, Lee K, Honda G, Ito M, Takeda Y, Kodzhimatov O, Ashurmetov O (2004) Prenylated benzophenones and xanthones from *Hypericum scabrum*. J Nat Prod 67(11):1870–1875

Tashkhodzhaev B, Aripova S, Turgunov K, Abdilalimov O (2004) Stereochemistry of the alkaloids codonopsine and codonopsinine. Chem Nat Compd 40(6):618–619

Tkachenko K, Kazarinova N, Muzychenko L, Shurgaya A, Pavlova O, Savonova N (1999) Antibiotic properties of essential oil of some plant species. Rastitel'nye Resursy 35:11–24

Tsevegsuren N, Christie W, Lösel D (1998) *Tanacetum* (*Chrysanthemum*) *corymbosum* seed oil: a rich source of a novel conjugated acetylenic acid. Lipids 33(7):723–727

Tulyaganov T, Kozimova N (2005) Alkaloids from *Nitraria schoberi*. *O*-Acetylnitraraine. Chem Nat Compd 41(5):578–579

UNESCO: International Bureau of Education, vol. XXIII, no. 1/2, 1993, p. 53–69

Van Wyk B, Wink W (2004) Medicinal plants of the world. Timber Press, Portland

Vandisheva B, Yusupova A, Alimbaeva P, Nuralieva Z (1977) Lekarstvennuiye Rasteniya Kirgizii i Perspektivui Ikh Ispolzovaniya (Medicinal plants of Kyrgyzstan and perspectives on their uses). Ilim, Frunze (in Russian)

Wu T, Li C, Leu Y, Hu C (1999) Limonoids and alkaloids of the root bark of *Dictamnus angustifolius*. Phytochemistry 50(3):509–512

Yagudaev M, Khalmirzaev M, Yunusov S (1983) NMR study of alkaloids. V. ^{13}C-NMR spectra and reconsideration of the structures of 11- and 10-hydroxypleiocarpamines and the structures of new alkaloids from *Vinca erecta*. Chem Nat Compd 19(4):454–458

Yagura T, Motomiya T, Ito M, Honda G, Iida A, Kiuchi F, Tokuda H, Nishino H (2008) Anticarcinogenic compounds in the Uzbek medicinal plant, *Helichrysum maracandicum*. J Nat Med 62:174–178

Yili A, Aisa H, Maksimov V, Veshkurova O, Salikhov S (2009) Chemical composition and antimicrobial activity of essential oil from seeds of *Anethum graveolens* growing in Uzbekistan. Chem Nat Compd 45(2):280–281

Yu F, Gazizov A, Glushenkova I (1995) Phospholipids of the pericarp of *Crataegus turkestanica*. Chem Nat Compd 31(4):465–466

Yue M, Jiang T, Shi Y (2004) Fast determination of flavonoids in *Hippophae rhamnoides* and its medicinal preparation by capillary zone electrophoresis using dimethyl-β-cyclodextrin as modifier. Talanta 62(4):695–699

Yuldashev M (2005) Flavonoids from the aerial part of *Vicia subvillosa*. Chem Nat Compd 41(1):99–100

Yuldashev MP, Batirov IK, Malikov VM (1993) Anthraquinones of *Bergenia hissarica*. Chem Nat Compd 29(4):543–544

Yuldasheva N, Rakhimov D, Turakhozhaev M (1993) Polysaccharides of *Eremurus*. XXVI. Study of the structure of the pectin from the leaves of *Eremurus regelii*. Chem Nat Compd 29(2):154–156

Yuldasheva N, Ul'chenko N, Glushenkova A (2008) Lipids of *Capparis spinosa* seeds. Chem Nat Compd 44(5):637–638

Yuldasheva N, Ulchenko N, Glushenkova A (2013) Lipids from fruits of *Solenanthus turkestanicus* and *Onosma irrigans*. Chem Nat Compd 49(4):599–602

Yuldasheva N, Ul'chenko N, Mamadalieva N, Glushenkova A, Ovidi E, Triggiani D, Tiezzi A (2014) Lipids from the aerial part of *Scutellaria ramosissima*. Chem Nat Compd 50(1):68–71

Yunusov S (1974) Alkaloidy izd. 2 (Alkaloids), 2nd edn. FAN, Tashkent (in Russian)

Zakirov KZ (1989) Classification of Central Asian vegetation. FAN, Tashkent

Zavrazhanov V, Kitaeva R, Khmelev K (1977) Lekarstvennuiye Rasteniya Tsentralnogo Chernozemya (Medicinal plants of Central Chernozem). Izdatelstvo Voronezh Inst, Voronezh (in Russian)

Zhao C, Shao J, Li X, Kang X, Zhang Y, Meng D, Li N (2008) Flavonoids from *Galium verum* L. J Asian Nat Prod Res 10(7):611–615

Zhou P, Takaishi Y, Duan H, Chen B, Honda G, Itoh M, Takeda Y, Kodzhimatov O, Lee K (2000) Coumarins and bicoumarin from *Ferula sumbul*: anti-HIV activity and inhibition of cytokine release. Phytochemistry 53(6):689–697

Ziyaev R, Abdusamatov A, Yunusov SY (1987) Alkaloids of *Liriodendron tulipifera*. Chem Nat Compd 5:628–638

Zotov E (1975) Phytochemical study of *Hyssopus seravshanicus* and *ferganensis* and production of medicinal forms. Synopsis of dissertation for the degree of Ph.D. in pharmacology. Frunze-Tartu (in Russian)

Chapter 9
Current State and Prospects for Studies on the Diversity of Medicinal Flora in Kazakhstan

Nadezhda G. Gemejiyeva and Lyudmila M. Grudzinskaya

Abstract A large territory and diverse soil and climatic conditions are responsible for the rich genetic resources of Kazakhstan. Approximate 6000 species of vascular plants from 160 families are distributed in the country. It is home to hundreds of species of economically valuable plants, including those used for medicinal purpose, as animal forage and in the food industry. Medicinal plants can be evaluated as raw material in different industries in Kazakhstan. These are an extremely valuable natural resource. Therefore, an appraisal of the current state and potential for studying the diversity of medicinal flora in Kazakhstan is extremely important.

Analysis of species diversity and the potential for their use in conventional and alternative medicine has confirmed that Kazakhstan's medicinal flora accounts for a quarter (26%) of all species of vascular plants of Kazakhstan and numbers 1406 species belonging to 612 genera and 134 families. Out of these 230 species from 71 families are used in conventional medicine. Raw material reserves have been registered for 141 (10%) resource species of medicinal flora in Kazakhstan, which belong to 100 genera and 47 families. Approximately 82 species are used in conventional medicine, whereas 59 species are widely used in experimental and alternative medicine. Greatest species diversity is concentrated in the mountain ecosystems of Kazakhstan. Reserves of commercial value in the North and West Tien Shan ranges create no less than 25 of the 141 resource species of medicinal flora. In the Kazakhstan Altai ranges, at least 10 of the 141 resource species of medicinal flora are recognized as having commercial value. Country exports raw materials for the following valuable medicinal plants: *Allochrusa gypsophiloides*, *Cistanche salsa*, *Ferula foetida*, *Glycyrrhiza glabra*, and *G. uralensis*. Approximately 700 species (49%) of medicinal plants have been crop tested, while experimental data on the productivity of raw materials and changes in the data have been generated for 250 species. Sixty-five species are classed as rare and endangered, and their cultivation is the priority for introduction and agricultural technology in this particular area. Current state of medicinal plant resources in Kazakhstan requires the realization of

N. G. Gemejiyeva (✉) · L. M. Grudzinskaya
Ministry of Education and Science, Institute of Botany and Phytointroduction, Almaty, Kazakhstan

D. Egamberdieva, M. Öztürk (eds.), *Vegetation of Central Asia and Environs*,
https://doi.org/10.1007/978-3-319-99728-5_9

a regional and nationwide long-term program for a comprehensive study and development of medicinal plant resources. There is great need for the creation of a collection of useful plants within the Kazakhstan flora; development of technology for introducing the cultivation of species of local flora is used intensively and with a limited realm to create a sustainable raw materials base for the domestic pharmaceutical industry. We also need to continue ethnobotanical research to highlight new sources of medicinal raw materials based on indigenous species. Contemporary systematic resource research into the medicinal flora in Kazakhstan will provide a scientific basis for its preservation and sustainable use.

Keywords Kazakhstan · Medicinal plants · Raw material reserves

9.1 Introduction

Maintaining and ensuring the rational use of our planet's plant resources is now a global problem. Climate change is threatening the preservation of natural vegetation and the human environment (Proskuryakov 2012). According to the *Convention on Biological Diversity* (1992), two thirds of all plants in the world are facing a threat of extinction as a consequence of human impact and climate change (Ozturk et al. 2010, 2016). Given the huge potential of floral diversity on one hand, and degradation of the ecosystems due to human impact and climate change, on the other hand, the Republic of Kazakhstan has developed a National Strategy and Action Plan on Conservation and Sustainable Use of Biological Diversity (Kokshetau 1999). The "strategy for preserving biodiversity" at regional and global levels should focus on assessing the current state of biological resources, creating a systemized list of all plants whose numbers are growing and those facing the threat of extinction, including rare, endemic, and economically important species to preserve them as in situ as well as ex situ (Ryabushkina et al. 2016). Immense territory and varying soil and climatic conditions of Kazakhstan are the key factors responsible for the rich genetic resources of the country. Plants are an important part of the biodiversity and determine the global stability of ecosystems. Their sustainable use and preservation too are an integral part of natural resources and play an important role in the human activity.

Flora of Kazakhstan has approximately 6000 species of vascular plants belonging to 160 families (Abdulina 1999; Baitenov 2001). This rich diversity has an immense potential as a source of medicines and represents a science-intensive and competitive product that is in growing demand globally. Hundreds of these plant species in Kazakhstan have an economic value, for instance, as medicines (1400 species), forage for livestock (1028 species), and in the food chain (about 500 species) (Grudzinskaya and Tazhkulova 2008). Medicinal plants, specifically pharmacopeia species, can be used as raw materials for herbal medicine production in Kazakhstan. These are an extremely valuable and irreplaceable natural resource. Therefore, an appraisal of the current state and potential for studying the diversity of medicinal flora in the country is extremely important. The production of healthcare products from medicinal plants is directly linked to supplying plant raw materials to the pharmaceutical industries.

Our aim here is to analyze the current state of and potential for studying the diversity of Kazakhstan's medicinal flora, in particular the plants used in scientific and alternative medicine in the country. For this purpose common resource-based information (Methodology for determination of medicinal plant reserves 1986), geobotanical studies (Bykov 1957; Korchagin 1964; Ponyatovskaya 1964), and introductory studies (Methodology of surveys during introduction of medicinal plants 1984; Recommendations for the Study of the Ontogenesis of Introduced Plants in the Botanical Gardens of the USSR 1990) have been evaluated. This analysis resource will serve as a scientific basis for the balanced use and preservation of plant resources, primarily medicinally valuable plants which have the potential for use as raw material in Kazakhstan's medical, pharmaceutical, and food industries. A sustainable raw material base will make the production of contemporary dietary supplements (DS), herbal preparations, and effective domestic plant-based preparations made from natural renewable raw materials will prove more appropriate.

9.2 Kazakhstan Geographical Features

Kazakhstan is located in the center of the Eurasian continent. This is the significant factor in the formation of a number of ecological systems. The country experiences a continental climate with insufficient humidity over large areas. It is the ninth largest country in the world, located in Central Asia, lying between the Caspian Sea, the Lower Volga Delta, the Lower Ural Mountains, Siberia, and China (Fig. 9.1).

Over 25% of Kazakhstan's territory is steppe, 50% is desert and semidesert, with the remaining 25% classed as mountains, seas, lakes, and rivers. Topography of the country rises from huge depressions below sea level to mountains at over 5000 m above sea level. From north to south, the country's territory is divided into natural and climatic zones, such as forest steppe, semidesert and desert areas, and then foothills and mountainous regions. The soil cover is also varied, with a large part of the forest steppe region covering chernozem. To the south the soil is either more dark chestnut, light chestnut, and brown desert. Soils of the mountain foothills are serozems. The deserts in the country are characterized by a severe continental climate, with low rainfall and high air temperatures in summer, severe cold in winter, and frequent winds, extremely dry air, and extreme temperature fluctuations during the summer months and even during a single day.

In the south and southeast of the country, sand approaches the mountains of Tien Shan system, which stretches for 2400 km and incorporates the Barlyk, Dzhungar Alatau, Ile Alatau, Talas Alatau, and Ketmen ranges. The highest mountain peak is Khan Tengri at 6992 m above sea level. To the east, Kazakhstan is flanked by the southern Altai. The huge glaciers and permanent snow on the high Tien Shan and Altai mountains provide irrigation water and water used to generate electricity,

Fig. 9.1 Location of the Republic of Kazakhstan

while the mountain slopes and valleys are covered with a rich and luscious vegetation, making excellent summer pastures (Zhailyau) (Republic of Kazakhstan 2006).

The altitudinal zonality of the varied mountain systems, which is defined by their biological diversity, endemism, and high ecological and economic value, is represented by desert, semidesert, steppe, forest steppe and forest areas, mountain meadows, and a snow belt. Due to the variations in the soil, moisture, and orography, there are also many intrazonal types of (floodplain forest and meadow, swamps, solonchak) vegetation.

Kazakhstan is home to extensive water bodies and a number of rivers. It has a 2340-km-long border to the west and southwest along the Caspian Sea—the largest lake in the world. The Aral Sea to the east of the Caspian Sea is located in the sandy desert. The latter covered an area of 65,000 km^2 before 1960 but has now shrunk to less than 10% of the previous volume. The southern sandy area of Central Kazakhstan is home to another vast lake—Balkhash—which currently covers an area of 18.2 thousand km^2. Other major lakes include Zaisan (in the east), Alakol (in the southeast), and Tengiz (in the center of the country). There are nearly 7000 natural lakes whose total surface area exceeds 45,000 km^2. Major rivers are Irtysh, Ishim, Ural, Syr Darya, Ile, Shu, and Tobol (Natural Conditions and Resources in Kazakhstan. Source: http://www.ecololocate.ru/locats-141-1.html). Therefore, country's climatic, geological soil and orographic conditions manifest themselves in the taxonomic and ecological diversity of plant species and have created the uniqueness and originality in Kazakhstan flora.

9.3 Medicinal Flora

The plant diversity in the country has a great potential as a source of raw materials to generate plant-based preparations with a wide range of fast-acting pharmaceutical and therapeutic properties. They do not possess cumulative properties and have fewer adverse side effects. The growing global trend of research into the chemistry of natural compounds has been a factor in the constant increase in the number of medicines available. In recent years, Kazakhstan, under the management of National Science Academy, has used wild plant species to create new original plant-based preparations such as anticancer drug *Arglabin*, the hepatoprotector *Salsokollin*, the anti-atherosclerotic preparation *Aterolid*, the antiparasitic preparation *Sausalin*, the adaptogenic *Ekdifit*, and others (International Scientific and Production Holding Company Phytochemistry 2018).

A great raw material base will secure the production of effective domestic plant-based products, contemporary DS, and herbal remedies made from natural and renewable raw materials, which will help as many people as possible.

Analysis of species diversity and the potential for the use in the country as medicinal plants in conventional and alternative medicine has helped much. Kazakhstan is currently home to at least 1406 species of medicinal plants (MP) grouped under 612 genera, belonging to 134 families (Annotated List of Medicinal plants of Kazakhstan 2014) (Fig. 9.2).

Domestic conventional medicine (CM) makes use of only 230 (16.4%) species of medicinal plants from 161 genera, 71 families, of which 29 species are registered in the State Pharmacopeia of the Kazakhstan (SPRK) (2009) and 92 species are registered in the State Register of Medicines of the Republic (2013). In taxonomic terms, 7 families are most common accounting for 10 and 30 medicinal species, such as Lamiaceae (10), Ranunculaceae (11), Apiaceae (13), Fabaceae (15), Polygonaceae (17), Rosaceae (24), and Asteraceae (30). At the level of genera leading families are Asteraceae (22), Rosaceae (13), Apiaceae (11), and Fabaceae (10).

Clinical experimental studies (CE) have tested 262 (18.6%) species from 60 families, 153 genera. The most common are families incorporating 12–41 medicinal

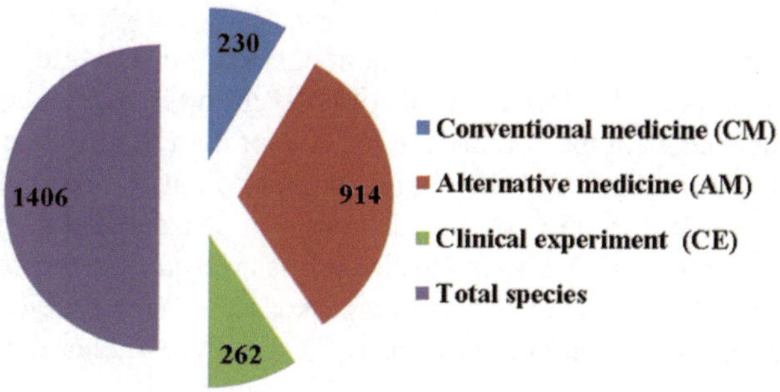

Fig. 9.2 Prospects for the use of medicinal plants in Kazakhstan in conventional, alternative, and experimental medicine

species, such as Apiaceae (12), Lamiaceae (13), Poaceae (14), Scrophulariaceae (17), Ranunculaceae (22), Rosaceae (30), and Asteraceae (41). On the basis of species, leading families are Asteraceae (16), Ranunculaceae (15), Poaceae (12), and Rosaceae (11).

Plants used in alternative medicine (AM) include over 900 (65%) species (Fig. 9.2). This group combines medicinal species used in alternative medicine not only in Mongolia, China, Tibet, and Buryatia but in Central and East Asia as well. Alternative medicine practiced in Kazakhstan has been using the species from such genera as *Aconitum*, *Ferula*, *Artemisia*, *Polygonum*, *Chenopodium*, *Mentha*, *Fritillaria*, *Eminium*, *Angelica*, *Rhodiola*, *Stemmacantha*, and few others.

Studying the medicinal plants used in alternative medicine in Kazakhstan will help identify previously unknown economic and valuable information on the species and expand the range of valuable domestic plant-based preparations with a wide spectrum of pharmacological activity and medicinal activity. Species names and plant family characteristics have been taken from Cherepanov (1995).

9.4 Taxonomical Aspects

In all floral reports, the number of vascular plants is roughly reported as 6000 species, varying between 5100 species (Illustrated Identifier of Kazakhstan Plants 1969, 1972) and 6000 species (Abdulina 1999; Baitenov 2001). Our information is based on the later data from Abdulina (1999), which records 5658 species belonging to 1067 genera and 159 families of vascular plants in Kazakhstan. On the whole, the medicinal flora accounts for a quarter (26%) of all species of vascular plants in the country. But, these species include 57% of the total genera and in 85% of the families known in the country. The ratio of the number of species, genera, and families of medicinal plants to the total number of species, genera, and families of vascular plants in Kazakhstan is shown in Fig. 9.3.

The vast majority of known medicinal species (98.1) belong to the angiosperms. The specific nature of the country's geography, climate, and soil explains the extremely low share of ferns (0.8%), gymnosperms (0.9%), and horsetails (0.2%) (Fig. 9.4).

Of all flowering plants, only 11% are monocotyledonous, while 89% are dicotyledonous. This explains the dominant role of latter group in the medicinal flora. The medicinal plant species in the country are known for their diverse life forms, majority or 65% being perennials and 16% being annual herbaceous plants. Fifteen leading families account for 65% of the diversity of Kazakhstan's medicinal flora (Table 9.1). The highest numbers in species terms are found in the following families: Asteraceae (196 species, 72 genera), Rosaceae (89 species, 27 genera), Fabaceae (78 species, 27 genera), Lamiaceae (77 species, 30 genera), Ranunculaceae (73 species, 20 genera), Apiaceae (68 species, 38 genera), Brassicaceae (62 species, 33 genera), Scrophulariaceae (45 species, 14 genera), Polygonaceae (42 species, 11 genera), Caryophyllaceae (41 species, 21 genera), Chenopodiaceae (39 species, 21

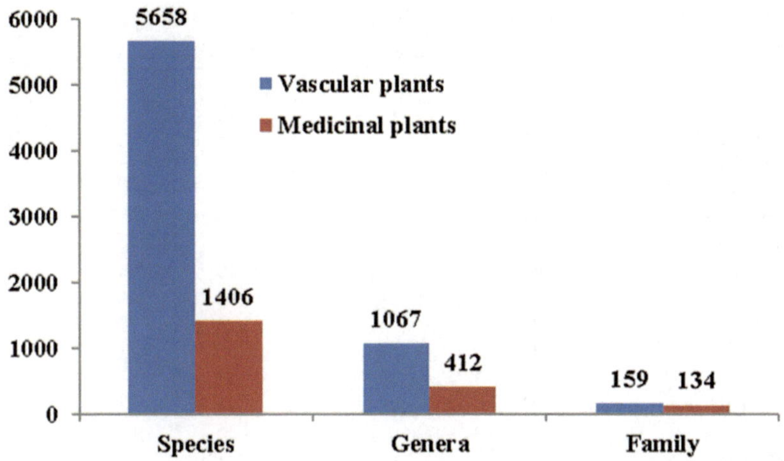

Fig. 9.3 Ratio of the number of species, genera, and families of medicinal plants to total number of species, genera, and families of vascular plants of Kazakhstan flora

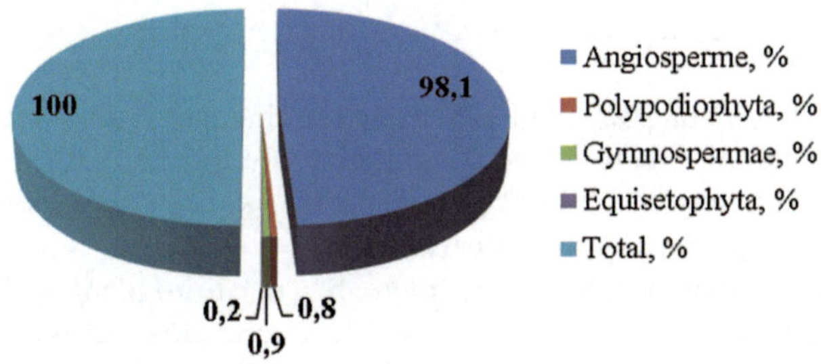

Fig. 9.4 Ratio of medicinal flora species by vascular plant groups

genera), Poaceae (35 species, 29 genera), Boraginaceae (30 species, 20 genera), Euphorbiaceae (21 species, 3 genera), Crassulaceae (19 species, 6 genera). The largest genera are *Artemisia* L. (42 species), *Potentilla* L. (25 species), and *Euphorbia* L. (18 species).

9.5 Ecogeographical Analysis

Representatives of Kazakhstan's medicinal flora flourish under most varied ecological conditions and are made up predominantly of mesophilic and xerophilic ecomorphs, the ratio of which is extremely varying depending on the family. For example, in the Apiaceae family, xerophytes tend to be the most common; in the Brassicaceae, mesophytes is common; and in the Chenopodiaceae, expectedly the most common by a long way are xerophilic ecoforms. The Asteraceae,

Table 9.1 Number of medicinal plant species and genera in leading families

Family	Number of genera	Number of species
Asteraceae	72	196
Lamiaceae	30	77
Ranunculaceae	20	73
Brassicaceae	33	62
Scrophulariaceae	14	45
Chenopodiaceae	21	39
Boraginaceae	20	30
Crassulaceae	6	19
Rosaceae	27	89
Fabaceae	27	78
Apiaceae	38	68
Polygonaceae	11	42
Caryophyllaceae	21	41
Poaceae	29	35
Euphorbiaceae	3	21
Total number	372	915

where mesophytes and xerophytes prevail, psychrophytes, halophytes, psammophytes, and others, are sufficiently widespread. Hygrophilic and hydrophilic ecoforms are poorly represented among Kazakhstan's medicinal plants, accounting for less than 4–6%.

Majority of medicinal plants in Kazakhstan are characterized by relatively extensive geographical distribution. Most common are plants with Palearctic and Holarctic area, and only a small number of species have relatively confined Central Asian-Turan and Turan geographic distribution.

In Apiaceae and Asteraceae families, more than one third of all medicinal species are represented by plants with Palearctic and Holarctic areas. Common species are even better represented in the Brassicaceae, Caryophyllaceae, Fabaceae, Lamiaceae, and Ranunculaceae families. In the Polygonaceae and Rosaceae families, plant species with extended areas are better represented, close to 50%. At the same time, in the Chenopodiaceae and Liliaceae families, more than half of species are distributed in the Turan lowland.

Medicinal plants with local Kazakhstan distribution (endemics) are represented by a total of 19 species. These are *Allium pskemense* B. Fedtsch., *Artemisia cina* Berg ex Poljakov, *A. karatavica* Krasch. et Abolin ex Poljakov, *A. transiliensis* Poljakov, *Berberis iliensis* Popov, *Crataegus almaatensis* Pojark., *Daphne altaica* Pall., *Echinops albicaulis* Kar. et Kir., *Ephedra lomatolepis* Schrenk, *Erysimum croceum* Popov, *Ferula iliensis* Krasn. ex Korovin, *Iris alberti* Regel, *Lonicera iliensis* Pojark., *Oxytropis almaatensis* Bajt., *Ribes nigrum* L., *R. turbinatum* Pojark., *Thesium minkwitzianum* B. Fedtsch., *Thymus altaicus* Klokov et Des.-Shost., and *Tulipa greigii* Regel. This fact testifies in particular to the extreme lack of research in the domestic flora, because against the 776 endemics known in Kazakhstan (Gemedjieva et al. 2010), medical properties are known for only 0.02% of them.

9.6 Regional Distribution of Medicinal Plants in Kazakhstan

Analysis of the distribution of medicinal plants in Kazakhstan (Republic of Kazakhstan 2006) shows that the greatest number of medicinal plants grow in the south of the country (1037 species—71%). The medicinal species intensity is roughly 1.3 species per 1000 km^2. The second region in terms of total medicinal species is the east of the country where 933 species (64%) grow. The richness of medicinal species in the east of the country is approximately 3.5 species per 1000 km^2. Six hundred and fifty species (44%) of medicinal plants grow in the north of Kazakhstan, registering a species richness of 1.5. The central desert and steppe regions of the country are home to 562 medicinal species (38%), with richness of 0.9 per 1000 km^2. In the west of Kazakhstan, that number is 550 species (37%), with richness of 0.9 (Fig. 9.5).

The highlighted regions are extremely varying in terms of area and ecological and geographical elements, which is why the distribution of medicinal plants in specific regions differs. These differences are most felt in the south (5) and east (3) of the country where desert and steppe territory directly borders the country's high mountain systems. The mountain ecosystems in the south (from the Dzhungar Alatau to the Western Tien Shan) (Fig. 9.5—5a region) are home to 933 species of medicinal plants (64%), while species richness is 4.4 per 1000 km^2. The Kazakhstani Altai and Tarbagatai mountain ecosystems (Fig. 9.5—3a) are home to 914 species of medicinal plants (62%), while their richness is 6.1 species per 1000 km^2. Consequently, the majority of medicinal species are found on the edges of the mountain systems in the east and south of Kazakhstan.

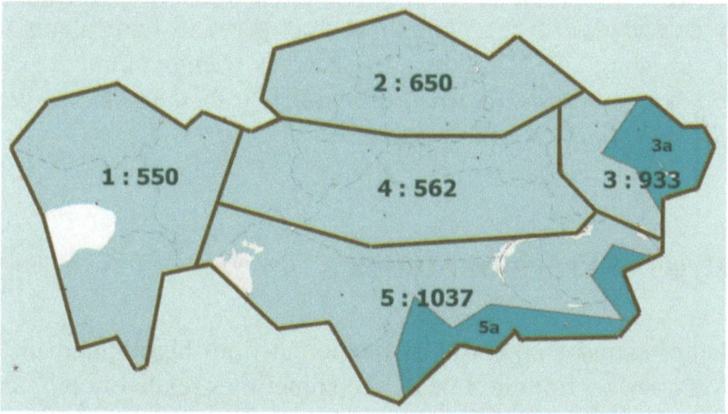

Fig. 9.5 Distribution of medicinal plant species in Kazakhstan. *1* western region, *2* northern region, *3* eastern region, *3a* mountainous territory in the eastern region, *4* central region, *5* southern region, and *5a* mountainous territory in the southern region

9.7 Analysis of Phytochemical Studies

Kazakhstan's medicinal plants incorporate the majority of known classes of biologically active elements phytochemically. Most common are the species containing flavonoids and their derivatives (60% of species), alkaloids (42%), organic and phenol acids (34%), vitamins (32%), tanning agents (29%), coumarins (25%), and other groups of bioactive substances.

Due to their phytochemical composition, medicinal plants in the country show extremely wide range of pharmacological effects. The medicinal flora incorporates plants whose preparations may be used to test for and treat allergies (*Stellaria dichotoma*, *Ephedra distachya*, *E. equisetina*, *Glycyrrhiza glabra*, *G. korshinskyi*, *G. uralensis*, *Vexibia alopecuroides*, *Agrimonia asiatica*, and others), cardiovascular illnesses (*Convallaria majalis*, *Hedysarum alpinum*, *Lathyrus pilosus*, *Ziziphora clinopodioides*, *Tulipa turkestanica*, *Colchicum luteum*, *Leptopyrum fumarioides*, *Pulsatilla campanella*, *Aconitum nemorum*, *Atragene sibirica*, certain *Adonis* species, and others), infectious diseases (*Artemisia lercheana*, *Euphorbia seguieriana*, *Cirsium esculentum*, *Crepis tectorum*, *Pulicaria salviifolia*, *Althaea cannabina*, *Convolvulus ammanii*, *Gentiana macrophylla*, *Fritillaria verticillata*), oncological (certain species of *Angelica*, *Amaranthus*, *Aegopodium alpestre*, *Sagittaria sagittifolia*, *Anthriscus sylvestris*, *Apium graveolens*, *Bupleurum bicaule*, *Castanea sativa*, *Cicuta virosa*, *Conium maculatum*, *Ferula tenuisecta*, *Phlojodicarpus villosus*, *Scandix pecten-veneris*), and other illnesses.

It is worth pointing out that there is a gap in the previous research studies on the chemical and therapeutic features of local Kazakhstani plant species. Of the nearly 800 endemics of Kazakhstan, phytochemical data is only known for a few, with 20 species being used medicinally. The list of medicinal plants in Kazakhstan could be significantly extended with the plants that may serve as equivalents to medicinal species not growing in Kazakhstan, such as *Gnaphalium kasachstanicum*, *Niedzwedzkia semiretschenskia*, *Zostera minor*, *Picea schrenkiana*, *Petasites spurius*, *Trapa kasachstanica*, and few others.

9.8 Analysis of Previous Studies

Analysis of the resource potential of medicinal flora highlighted in the previous research is very low in volume. Local, and sometimes relatively old, and even contemporary data on reserves and raw materials is studied in pharmacopeia and the species used in conventional medicine (Guidelines for working with Medicinal plants 1999; Kukenov 1999) available for 141 (10%) of species from 47 families (Table 9.2). Most common of these previous resources are the State Pharmacopeia of the Republic of Kazakhstan (2009) (Annotated List of Medicinal plants of Kazakhstan 2014).

Table 9.2 List of species of medicinal plants for which reserves of raw materials have been recorded in Kazakhstan (conventional medicine, CM; alternative medicine, AM; clinical experiment, CE; State Pharmacopeia of the Republic of Kazakhstan, SPRK)

Species	Family	Use
Achillea millefolium L.	Asteraceae Dumort.	CM, SPRK
Achillea nobilis L.	Asteraceae Dumort.	AM
Aconitum altaicum Steinb.	Ranunculaceae Juss.	AM
Aconitum anthoroideum DC.	Ranunculaceae Juss.	AM
Aconitum apetalum (Huth) B. Fedtsch.	Ranunculaceae Juss.	AM
Aconitum leucostomum Worosch.	Ranunculaceae Juss.	CM
Aconitum volubile Pall. ex Koelle	Ranunculaceae Juss.	CE
Acorus calamus L.	Araceae Juss.	CM
Adonis tianschanica (Adolf) Lipsch.	Ranunculaceae Juss.	CM
Agrimonia asiatica Juz.	Rosaceae Juss.	CE
Agropyron cristatum (L.) Beauv.	Poaceae Barnhart	AM
Ajania fastigiata (C. Winkl.) Poljak.	Asteraceae Dumort.	AM
Alhagi pseudalhagi (M. Bieb.) Fisch.	Fabaceae Lindl.	CM
Allium nutans L.	Alliaceae J. Agardh	AM
Allochrusa gypsophiloides (Regel) Schischk.	Caryophyllaceae Juss.	CM
Althaea officinalis L.	Malvaceae Juss.	CM
Amygdalus spinosissima Bunge	Rosaceae Juss.	AM
Anabasis aphylla L.	Chenopodiaceae Vent.	CM
Arctium tomentosum Mill.	Asteraceae Dumort.	CM
Armeniaca vulgaris Lam.	Rosaceae Juss.	CM
Artemisia leucodes Schrenk	Asteraceae Dumort.	CM, SPRK
Artemisia absinthium L.	Asteraceae Dumort.	CM, SPRK
Artemisia annua L.	Asteraceae Dumort.	CE
Artemisia cina Berg. ex Poljak.	Asteraceae Dumort.	CM
Artemisia dracunculus L.	Asteraceae Dumort.	CE
Artemisia glabella Kar. et Kir.	Asteraceae Dumort.	CM, SPRK
Artemisia rutifolia Steph. ex Spreng.	Asteraceae Dumort.	CE
Artemisia santolinifolia (Turcz. ex Pamp.) Krasch.	Asteraceae Dumort.	CE
Artemisia sieversiana Willd.	Asteraceae Dumort.	CE
Artemisia terrae-albae Krasch.	Asteraceae Dumort.	AM
Artemisia vulgaris L.	Asteraceae Dumort.	CM
Berberis sphaerocarpa Kar. et Kir.	Berberidaceae Juss.	AM
Bergenia crassifolia (L.) Fritsch	Saxifragaceae DC.	CM
Betula pendula Roth	Betulaceae S.F. Gray	CM, SPRK
Betula pubescens Ehrh.	Betulaceae S.F. Gray	CM, SPRK

(continued)

Table 9.2 (continued)

Species	Family	Use
Bidens tripartita L.	Asteraceae Dumort.	CM
Bupleurum longifolium L. (*B.aureum* Fisch.)	Apiaceae Lindl.	CM
Bupleurum multinerve DC.	Apiaceae Lindl.	CM
Capparis herbacea Willd.	Capparaceae Juss.	CM
Capsella bursa-pastoris (L.) Medik.	Brassicaceae Burnett	CM
Carum carvi L.	Apiaceae Lindl.	CM
Chamaenerion angustifolium (L.) Scop.	Onagraceae Juss.	CE
Chamomilla recutita (L.) Rauschert	Asteraceae Dumort.	CM, SPRK
Chartolepis intermedia Boiss.	Asteraceae Dumort.	CM
Chelidonium majus L.	Papaveraceae Juss.	CM
Cichorium intybus L.	Asteraceae Dumort.	CM
Cistanche salsa (C. A. Mey.) G. Beck	Orobanchaceae Vent.	CE
Conium maculatum L.	Apiaceae Lindl.	CM
Corydalis sewerzowii Regel	Fumariaceae DC.	AM
Crataegus almaatensis Pojark.	Rosaceae Juss.	AM
Crataegus korolkowii L. Henry	Rosaceae Juss.	CM, SPRK
Crataegus pontica C. Koch	Rosaceae Juss.	AM
Crataegus sanguinea Pall.	Rosaceae Juss.	CM
Crataegus songarica C. Koch	Rosaceae Juss.	AM
Crepis sibirica L.	Asteraceae Dumort.	AM
Daphne altaica Pall.	Thymelaeaceae Juss.	AM
Delphinium confusum M. Pop.	Ranunculaceae Juss.	CM
Delphinium dictyocarpum DC.	Ranunculaceae Juss.	CM
Delphinium elatum L.	Ranunculaceae Juss.	CM
Echinops albicaulis Kar. et Kir.	Asteraceae Dumort.	CE
Elaeagnus oxycarpa Schltdl.	Elaeagnaceae Juss.	AM
Ephedra equisetina Bunge	Ephedraceae Juss.	CM
Ephedra intermedia Schrenk	Ephedraceae Juss.	CM
Equisetum arvense L.	Equisetaceae Rich. ex DC.	CM
Eremurus regelii Vved.	Asphodelaceae Juss.	CE
Ferula foetida (Bunge) Regel	Apiaceae Lindl.	CM
Ferula soongarica Pall. ex Spreng.	Apiaceae Lindl.	AM
Filipendula ulmaria (L.) Maxim.	Rosaceae Juss.	CM
Frangula alnus Mill.	Rhamnaceae Juss.	CM
Fritillaria verticillata Willd.	Liliaceae Juss.	AM
Glycyrrhiza glabra L.	Fabaceae Lindl.	CM, SPRK
Glycyrrhiza korshinskyi Grig.	Fabaceae Lindl.	CM
Glycyrrhiza uralensis Fisch.	Fabaceae Lindl.	CM, SPRK

(continued)

Table 9.2 (continued)

Species	Family	Use
Hedysarum neglectum Ledeb.	Fabaceae Lindl.	CE
Helichrysum arenarium (L.) Moench	Asteraceae Dumort.	CM
Helichrysum maracandicum M. Pop. ex Kirp.	Asteraceae Dumort.	AM
Hippophae rhamnoides L.	Elaeagnaceae Juss.	CM
Hypericum perforatum L.	Hypericaceae Juss.	CM, SPRK
Hypericum scabrum L.	Hypericaceae Juss.	CE
Hyssopus ambiguus (Trautv.) Iljin	Lamiaceae Lindl.	AM
Inula helenium L.	Asteraceae Dumort.	CM
Inula macrophylla Kar. et Kir.	Asteraceae Dumort.	AM
Leonurus turkestanicus V. Krecz. et Kuprian.	Lamiaceae Lindl.	AM
Limonium gmelinii (Willd.) O. Kuntze	Limoniaceae Ser.	SPRK
Lonicera altaica Pall. ex DC.	Caprifoliaceae Juss.	AM
Malus sieversii (Ledeb.) M. Roem.	Rosaceae Juss.	CM
Melilotus officinalis (L.) Pall.	Fabaceae Lindl.	CM
Mentha longifolia (L.) L.	Lamiaceae Lindl.	CE
Nepeta pannonica L.	Lamiaceae Lindl.	AM
Nuphar lutea (L.) Smith	Nymphaeaceae Salisb.	CM
Origanum tyttanthum Gontsch.	Lamiaceae Lindl.	AM
Origanum vulgare L.	Lamiaceae Lindl.	CM, SPRK
Padus avium Mill.	Rosaceae Juss.	CM
Patrinia intermedia (Horn.) Roem. et Schult.	Valerianaceae Batsch	CM
Paeonia anomala L.	Paeoniaceae Rudolphi	CM, SPRK
Peganum harmala L.	Peganaceae (Engl.) Tieg. ex Takht.	CM
Pentaphylloides fruticosa (L.) O. Schwarz	Rosaceae Juss.	CE
Peucedanum morisonii Bess. ex Spreng.	Apiaceae Lindl.	CM
Phragmites australis (Cav.) Trin. ex Steud.	Poaceae Barnhart	CE
Pinus sylvestris L.	Pinaceae Lindl.	CM
Plantago major L.	Plantaginaceae Juss.	CM
Polemonium caeruleum L.	Polemoniaceae Juss.	CM
Polygonum coriarium Grig.	Polygonaceae Juss.	AM
Polygonum hydropiper L.	Polygonaceae Juss.	AM
Polygonum nitens (Fisch.et C.A. Mey.) V. Petrov ex Kom.	Polygonaceae Juss.	CM
Polygonum persicaria L.	Polygonaceae Juss.	CM
Polygonum scabrum Moench.	Polygonaceae Juss.	CM
Psoralea drupacea Bunge	Fabaceae Lindl.	CM
Pulsatilla patens (L.) Mill.	Ranunculaceae Juss.	CE
Rheum tataricum L. fil.	Polygonaceae Juss.	CM
Rhodiola rosea L.	Crassulaceae DC.	CM

(continued)

Table 9.2 (continued)

Species	Family	Use
Ribes nigrum L.	Grossulariaceae DC.	CM
Rosa acicularis Lindl.	Rosaceae Juss.	CM, SPRK
Rosa alberti Regel	Rosaceae Juss.	CM
Rosa majalis Herrm.	Rosaceae Juss.	CM, SPRK
Rubus idaeus L.	Rosaceae Juss.	CM
Rumex confertus Willd.	Polygonaceae Juss.	CM
Salvia deserta Schangin	Lamiaceae Lindl.	AM
Sanguisorba officinalis L.	Rosaceae Juss.	CM
Saussurea salsa (Pall.) Spreng.	Asteraceae Dumort.	AM
Serratula coronata L.	Asteraceae Dumort.	CM, SPRK
Sorbus sibirica Hedl.	Rosaceae Juss.	AM
Sorbus tianschanica Rupr.	Rosaceae Juss.	AM
Sphaerophysa salsula (Pall.) DC.	Fabaceae Lindl.	CM
Stemmacantha carthamoides (Willd.) Dittrich	Asteraceae Dumort.	CM
Syrenia siliculosa (M. Bieb.) Andrz.	Brassicaceae Burnett	AM
Tamarix ramosissima Ledeb.	Tamaricaceae Link	CE
Tanacetum vulgare L.	Asteraceae Dumort.	CM
Thermopsis alterniflora Regel et Schmalh.	Fabaceae Lindl.	CM
Thymus marschallianus Willd.	Lamiaceae Lindl.	CE
Tribulus terrestris L.	Zygophyllaceae R.Br.	CM
Tussilago farfara L.	Asteraceae Dumort.	CM
Urtica cannabina L.	Urticaceae Juss.	CE
Urtica dioica L.	Urticaceae Juss.	CM
Vaccinium myrtillus L.	Ericaceae Juss.	CM
Valeriana dubia Bunge	Valerianaceae Batsch	AM
Veratrum lobelianum Bernh.	Melanthiaceae Batsch	CM
Veratrum nigrum L.	Melanthiaceae Batsch	AM
Verbascum thapsus L.	Scrophulariaceae Juss.	CM
Vexibia pachycarpa Schrenk ex C. A. Mey.) Jakovl.	Fabaceae Lindl.	CM
Ziziphora clinopodioides Lam.	Lamiaceae Lindl.	CE

Out of 230 species belonging to 33 families, plants officially used in medicine and the reserves of raw materials have only been studied for 84 (36.5%), which is no less than 6% of the total number of medicinal plant species in the country. Reserves of raw materials have not been calculated for the remaining 146 species (63.5%). The vast majority of these plants grow in the wild and do not have commercial value, some are found locally or randomly, or are grown in the form of crops and used as raw materials. A part of the species is recognized as rare, and some are difficult to access as in the marshy areas and in mountainous tops. Some examples include ruderal plants growing alongside the roads, in populated areas, on fallow land, and so on (Gemejiyeva 2015).

Of those species used in conventional medicine, the ones with potential for subsequent resource research are *Alhagi kirghisorum*, *Rhamnus cathartica*, and *Filipendula vulgaris* species and the genera like *Polygonum*, *Salix*, *Potentilla*, and *Juniperus*.

The distribution of the quantity of medicinal plant species officially used and for which data on reserves and family data exists shows that of the 84 species, there is one species each from 28 families such as Araceae, Brassicaceae, Capparaceae, and so forth and only two species per family in Betulaceae and Ephedraceae. Of the remaining families, most common in species terms are Asteraceae (17 species), Rosaceae (11), Fabaceae (8), Polygonaceae (5), Apiaceae (6), and Ranunculaceae (5).

Analysis of previous resource works depicts that primarily; out of 29 pharmacopeia species from 11 families included in the second edition of the State Pharmacopeia of the Republic of Kazakhstan (2009), raw material reserves have been studied for 21 (72.4%) of species with the greatest species representation found in the Asteraceae family (6 species) (*Achillea millefolium*, *Artemisia absinthium*, *A. glabella*, *A. leucodes*, *Chamomilla recutita*, *Serratula coronata*) and the Rosaceae family (4 species) (*Crataegus korolkowii*, *C. sanguinea*, *Rosa acicularis*, *R. majalis*). A raw material base is found for two pharmacopeia species in the Fabaceae (*Glycyrrhiza glabra* and *G. uralensis*) and Betulaceae (*Betula pendula*, *B. pubescens*) families. In the remaining seven families (Hypericaceae, Lamiaceae, Limoniaceae, Malvaceae, Melanthiaceae, Nymphaeaceae, Paeoniaceae) raw material data was discovered for only one species. No data on raw materials is available for eight (or 27.6%) pharmacopeia species in Kazakhstan flora, such as *Ajania fruticulosa*, *Salsola collina*, *Thymus serpyllum*, *Althaea armeniaca*, and so forth, which is due to the lack of wild growth with commercial value.

A taxonomic analysis of the 141 species for which raw material data is available shows that the most studied in resource species terms are the families: Apiaceae (7 species), Polygonaceae (8 species), Lamiaceae (9 species), Fabaceae, Ranunculaceae (10 species each), Rosaceae (19 species), and Asteraceae (30 species). The greatest number of species studied in resource terms are (94 species) 66.7% of the total number of resource species focused on 7 families (14.9%) of 47. Two species have been studied in nine families (19.1%): Betulaceae, Hypericaceae, Melanthiaceae, Urticaceae, Brassicaceae, Ephedraceae, Elaeagnaceae, Poaceae, and Valerianaceae. The least studied in resource terms are 31 families (or 66%) of the 47 for which one species has been studied. Analysis of the 7 leading families for the given index shows that the leader in terms of genus (17) and species (30) studied was family Asteraceae¸ which represented 17 species officially using conventional medicine (including 6 official) and 13 species used in experimental and alternative medicine.

The extensively studied taxa representatives in resource terms are genus *Artemisia* (11 species), of which 5 species are officially used in medicine (*Artemisia absinthium*, *A. cina*, *A. glabella*, *A. leucodes*, *A. vulgaris*) and 6 (*Artemisia annua*, *A. dracunculus*, *A. rutifolia*, *A. santolinifolia*, *A. sieversiana*, *A. terrae-albae*) are used in alternative and experimental medicine. Furthermore, raw material numbers have been identified for two milfoil species (*Achillea millefolium*, *A. nobilis*) and two strawflower species (*Helichrysum arenarium*, *H. maracandicum*) used in conventional and experimental medicine. The raw material numbers for a large number of species used in conventional medicine have not been studied. These include *Ajania fruticulosa*, *Arctium lappa*, *Artemisia taurica*, *A. vulgaris*, *Centaurea cyanus*, *Chamomilla suaveolens*, *Cnicus benedictus*, *Echinops ritro*, *Silybum marianum*, *Solidago virgaurea*, and *Taraxacum officinale*. Out of the listed ones, except for *Artemisia vulgaris* and *Taraxacum officinale*, nearly all species grow under natural conditions randomly and are not available in commercial quantities.

Family Rosaceae embodies the second highest number of resource species, 19 of which are registered as raw materials species. Nearly 11 of the 23 officially recognized, 8 species are used in the alternative and experimental medicine. Raw materials for nearly 50% of the officially recognized species of the family have not been investigated, and those under review are *Amygdalus communis*, *Coluria geoides*, *Filipendula vulgaris*, *Fragaria vesca*, *Geum urbanum*, *Potentilla argentea*, *Potentilla erecta*, *Rosa corymbifera*, *R. fedtschenkoana*, and *R. laxa*. The following taxa have a potential in terms of the registration as reserves: *Rosa*, *Filipendula vulgaris*, *Amygdalus communis*, *Fragaria vesca*, and *Geum urbanum*.

Fabaceae family is third in terms of number of resource species. Out of its 10 resource species, raw materials have been identified for 2 pharmacopeia species and 7 of the 15 officially recognized species, and 1 species is used in the alternative and experimental medicine. Among officially recognized species, no study has been carried out on the raw materials situation of *Alhagi kirghisorum*, *Glycyrrhiza echinata*, *Hedysarum alpinum*, *Ononis arvensis*, and *Thermopsis alterniflora*. These species have a potential in terms of recording reserves of raw materials.

In the Ranunculaceae family, raw material status has been recorded for only 5 of the 11 officially recognized species. These five species are used in alternative and experimental medicine. Raw material reserves have not been studied for nearly 50% of officially recognized species such as *Aconitum soongaricum*, *Adonis vernalis*, *Thalictrum foetidum*, *T. isopyroides*, *T. minus*, and *Nigella damascena*. There is need for their cultivation as a crop. *Adonis vernalis* is a rare species, but *Aconitum soongaricum* is relatively extensively distributed, but major commercial volumes are lacking. Furthermore, populations of this species are prone to uncontrolled use by the local population, making use of its underground part as a medicinal raw material. It will be a good idea to enforce strict quotas of *A. soongaricum* raw materials for pharmaceutical industry needs together with the obligatory root development of plants in their natural habitats and creation of crop plantations. In vitro species studies have a potential by using biotechnology methods.

Only representative of genus *Aconitum* among the pharmaceutical species is *Aconitum leucostomum*, which has a raw material base. Resource investigations have revealed that the raw material base on the North Tien Shan and Kazakhstan Altai ranges is sufficient to proceed with commercial harvesting. The closest to this species is *Aconitum apetalum*, which has potential for cultivation in zongorin in commercial volumes due to the stable raw material base in the Dzhungar Alatau and Listvyaga (Altai) mountain ranges.

The Lamiaceae family has nine resource species. The raw materials have been identified for only one species *Origanum vulgare* and for eight species used in the alternative and experimental medicine. For officially recognized medicinal plants such as *Marrubium vulgare*, *Melissa officinalis*, *Phlomis pungens*, *Salvia aethiopis*, *S. sclarea*, *Stachys betoniciflora*, and *Thymus serpyllum*, no data exists on the raw materials. *Marrubium vulgare*, *Salvia aethiopis*, *S. sclarea*, and *Phlomis pungens* do have a potential for the study of reserves. Remaining species are often found less and no growth with commercial potential has been found.

In the Polygonaceae family, raw materials have been identified only for five officially recognized species, and two species are used in alternative and experimental medicine. For the 11 species used in Kazakhstan's conventional medicine, raw materials have not been studied in the 6 species of genus *Polygonum*. These are *Polygonum amphibium*, *P. aviculare*, *P. bistorta*, *P. minus*, *P. soongoricum*, and *P. viviparum*, as well as *Rheum compactum*. In addition four species of genus *Rumex* too have not been evaluated: *Rumex acetosella*, *R. marschallianus*, *R. rossicus*, and *R. thyrsiflorus*. However, natural growth of *Polygonum aviculare* and *P. bistorta* is sufficient to prepare cuttings. Remaining species are found less and lack commercial potential.

For the species of the family Apiaceae, raw materials have been identified for only five officially recognized species, and two species are used in alternative and experimental medicine. For the species used in official medicine in Kazakhstan, raw material resource has not been identified in *Daucus carota*, *Ferula sumbul*, *Oenanthe aquatica*, and *Pimpinella saxifraga*. The above species grow randomly in the wild, and it was not possible to locate growth at commercial level. The plantations are preferable for *Coriandrum sativum*, *Foeniculum vulgare*, and *Pastinaca sativa*.

Analysis of data on the distribution of raw materials of resource species in Kazakhstan has helped us to establish the most investigated stocks of the medicinal species. These are in the south (91 species or 64.5%) and east (59 species or 41.8%) of Kazakhstan. The raw materials status has been identified for 19 species (13.5%) in the north of Kazakhstan, while an insignificant share of medicinal species have been studied from the central part (7 species or 5%) and western (6 species or 4.3%) regions of the country (Fig. 9.6).

In Kazakhstan Altai ranges, commercial reserves account for at least 10 (7.1%) of the 141 resource species of the medicinal flora, *Aconitum leucostomum* leading the group, generating commercial growth in 6 mountain ranges. *Veratrum lobelianum* and *Achillea millefolium* possess reserves in five ranges. At least seven species generate commercial reserves in three of the ten Kazakhstan Altai ranges. For the remaining species, raw material bases have been recorded only in one or two ranges.

The facts cited above depict that a large part of resource species with a guaranteed raw materials base is focused in the North and South Tien Shan ranges located in the south and southeast of Kazakhstan (Gemejiyeva and Grudzinskaya 2016).

The mountain ranges with the maximum potential for generating medicinal plant harvestings are in the North and West Tien Shan (Dzhungar Alatau (27 species), Ile Alatau (26 species), Ketmen (14 species), Kyrghyz Alatau (14 species), Kungei Alatau (12 species), Karzhantau (12 species), Terskey Alatau (9 species), Boroldaitau

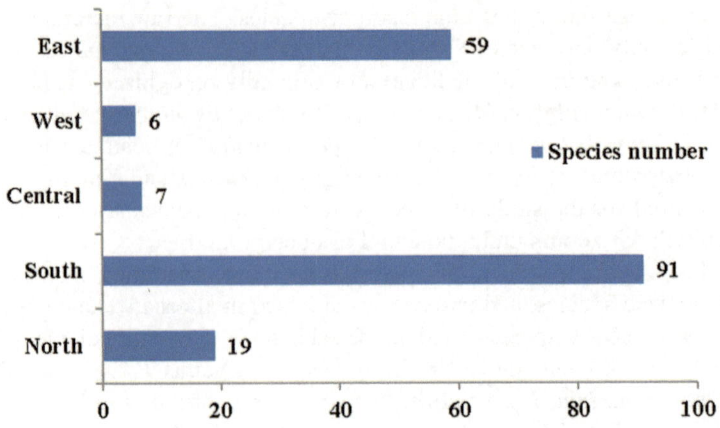

Fig. 9.6 Distribution of raw materials of resource species in Kazakhstan

(6 species)) and the Kazakhstani Altai ranges (West Tarbagatai (21 species), Kurshum (13 species), Kalbin (11 species), Sarymsakty (9 species), Asutau (9 species), Narym (9 species), Saur, Manrak (8 species), Altai Tarbagatai (7 species), South Altai (6 species) and Listvyaga (4 species)).

It is worth pointing out that the raw material data provided for 141 resource species covers a 20–25-year study period. For the remaining species, raw material data covers the period between 1961 to 1963 (*Anabasis aphylla*), 1968 (*Polygonum coriarium*, *Limonium gmelinii*), 1983 (*Sphaerophysa salsula*), and 1986 (*Leonurus turkestanicus*). The resource studies of above mentioned species have not been undertaken after those dates. Analysis of available raw material data shows that data received before 1990 generates lower numbers (31 species). The highest number is 70 species, which covers resource studies before 2000 and up to 2010. In the last 4 years, the number of species registering for reserves of raw materials has fluctuated insignificantly, 49 and 47 species, respectively.

This justifies the need to launch new resource studies in future, primarily on the species that are deemed economically viable and have values, such as *Glycyrrhiza glabra* and *G. uralensis*, *Cistanche salsa*, and *Ferula foetida*; wild relatives of crops like capers and rose hip; and a number of grasses with medicinal, nutritional, forage, and other benefits. These include *Artemisia leucodes*, *A. glabella*, *A. annua*, *Serratula coronata*, *Saussurea salsa*, and *Chartolepis*, and the species of genera *Aconitum*, *Delphinium*, and others.

9.9 Cultivation of Medicinal Plants

The position of introduction studies on the medicinal plants in Kazakhstan has been somewhat higher than resource studies. Roughly 700 species, roughly 49% of the total list of MP, have been crop tested. The core permanent introductory medicinal plant centers in Kazakhstan are botanical gardens in various ecological and climate zones. The country's main botanical garden (Almaty city) houses crop testing data for 528 species of medicinal plants (Plants of the Natural Flora of Kazakhstan in Introduction 1990; Guidelines for working with Medicinal plants 1999). The Altai Botanical Garden (Ridder town) has tested crops of 211 species of medicinal plants, including such rare and valuable species as *Cypripedium calceolus*, *C. guttatum*, *C. macranthon*, *Oxycoccus quadripetalus*, *Rhodiola algida*, and several others. The Zhezkazgan Botanical Garden (Zhezkazgan city) has tested cultivation of 96 species of medicinal plants of Kazakhstan, mainly drought-resistant plants in arid areas, such as *Peganum harmala*, the species of genera *Glycyrrhiza*, *Salsola*, as well as medicinal shrubs of the Asteraceae, Chenopodiaceae, Poaceae families, together with a number of rare medicinal plants, including *Niedzwedzkia semiretschenskia*. On the Mangyshlak peninsula, the Mangyshlak Experimental Botanical Garden (Aktau city) has tested a number of natural flora plants, such as *Armeniaca vulgaris*, *Crataegus ambigua*, *Juniperus communis*, *Malacocarpus crithmifolius*, and species of the genera *Lonicera*. Before 1990, the Karaganda Botanical Garden (Karaganda city) tested cropping of 200 species of medicinal plants, predominantly those indigenous to the steppe such as *Achillea, Allium*, *Artemisia*, *Campanula*, *Centaurea*, *Gypsophila*, *Potentilla*, *Sedum*, *Trifolium*, *Adonis volgensis*, *Artemisia cina*, *Chartolepis intermedia*, *Filipendula vulgaris*, *Limonium gmelinii*, *Patrinia intermedia*, *Valeriana rossica*, and others (Plants of the Natural Flora of Kazakhstan in Introduction 1990). In addition to the botanical gardens, medicinal plant collections are known in the Kazakh Scientific Agricultural Research Institute (Almaty Oblast), in the Kazakhstan Research and Development Institute (RDI) for the Potato and Vegetable Industry (Almaty Oblast), in the South-Kazakhstan RDI for the Agricultural Industry (South-Kazakhstan Oblast), in the Uralsk State Agricultural Testing Station (Uralsk city), in the Central Kazakhstan RDI for the Agricultural Industry (Karaganda Oblast), and in the Kazakh Institute of Agroforestry (Shuchinsk town) (Guidelines for working with Medicinal plants 1999).

Expanded introductory testing for identifying plant productivity is performed in the Almaty Botanical Garden, with data generated for more than 250 species of global flora, of which 70 are Kazakhstan species (Grudzinskaya and Tazhkulova 2008, 2013; Grudzinskaya 2009; Grudzinskaya and Arysbayeva 2011, 2014, 2016; Grudzinskaya et al. 2017). Extremely limited data is available on the commercial cultivation of Kazakhstan medicinal plants. At the end of the last century, roughly 14 species were grown under semicommercial conditions around the Ile Alatau foothill zone, while arid zones in Central Kazakhstan have used agricultural technology in experimental nurseries to develop another six pharmacopeia species. The yield of cultivated medicinal plants and the agricultural technology to grow these has been followed under commercial conditions covering 33 species.

Among the medicinal plants of Kazakhstan, special value is given to those species whose raw materials are used in pharmaceutical production. These include 170 species in Kazakhstan. These are primarily the medicinal plants most in demand, whose raw materials go directly, without processing, into the pharmacy network, including 27 pharmacopeia species. According to the State Register of Medicines of the Republic of Kazakhstan, domestic manufacturers currently use only 25 species to prepare infusions, extracts, and medical forms. Another 64 species of medicinal

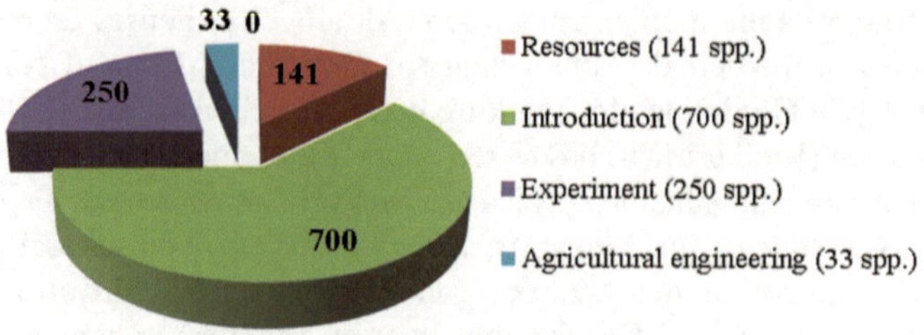

Fig. 9.7 Level of previous resource and introductory studies on Kazakhstan's medicinal plants

plants of Kazakhstan are used by overseas firms. The demand for plant raw materials in domestic pharmaceutical firms is currently met by using naturally growing populations or purchasing raw materials from overseas. Supplies of cultivated species of Kazakhstan are extremely small and sporadic (Fig. 9.7).

Out of 230 species of CM, 137 species (60%) have been introduced in the country (predominantly in botanical gardens). More detailed experimental research (identification of raw material productivity) has been done on 65 species (28%), while production testing is known for only 21 species of CM (9%). No crop testing data exists in Kazakhstan for over 90 species of CM. Twelve species of CM have been recognized as a priority for introductory research across Kazakhstan, including *Artemisia cina*, *Ephedra intermedia*, *Erysimum cheiranthoides*, *Hedysarum alpinum*, *Oxycoccus quadripetalus*, *Sphaerophysa salsula*, and *Thermopsis lanceolata*. In addition, practically all these species require new agricultural technology to help in their production at commercial scale.

9.10 Rare and Protected Medicinal Species

All medicinal species have direct practical applications; they are harvested from nature. Of the medicinal plants highlighted, 65 species have been recorded in the Red Data Book of Kazakhstan (2014), i.e., they cannot be heavily harvested from their natural habitats. The status of critically endangered (red book) species of medicinal plants varies. Category A includes those species facing the threat of extinction, such as *Artemisia cina* and *Ferula iliensis*. Category B includes rare species of which there are 37, including popular plants like *Allium microdictyon*, *Rheum altaicum*, *Stemmacantha carthamoides*, and *Rhodiola rosea*. Category C includes 25 widespread species of medicinal plants, which are rare in Kazakhstan, including *Adonis vernalis*, *Convallaria majalis*, *Rheum wittrockii*, and *Oxycoccus microcarpus*. This situation prioritizes the need to cultivate critically endangered species of Kazakhstan medicinal plants and places special emphasis on their introduction and agricultural development in this sector. The group of medicinal plants used in conventional medicine (CM) has specific value in terms of practical application. Of the 230 species of plants in this group, there is only 1 endemic species—*Artemisia cina*—and 16 critically endangered species: *Adiantum capillus-veneris*, *Adonis tianschanica*, *Adonis vernalis*, *Allochrusa gypsophiloides*, *Alnus glutinosa*, *Armeniaca vulgaris*, *Artemisia cina*, *Convallaria majalis*, *Huperzia selago*, *Ledum palustre*, *Malus sieversii*, *Orchis militaris*, *Quercus robur*, *Stemmacantha carthamoides*, *Rhodiola rosea*, and *Vitis vinifera*.

9.11 Conclusions

The medicinal flora in Kazakhstan includes 1,406 species belonging to 612 genera and 134 families of vascular plants, of which only 230 species are from 71 families, including 29 pharmacopeia species, used in conventional medicine. The raw material reserves have been registered for 141 resource species of medicinal flora in Kazakhstan. These cover 100 genera and 47 families, while 82 species are used in conventional medicine, with the remaining 59 species widely used in experimental and alternative medicine. More than 50% of the resource species (94) are concentrated in 7 leading families from a total of 47, including Apiaceae, Polygonaceae, Lamiaceae, Fabaceae, Ranunculaceae, Rosaceae, and Asteraceae, whereas 31 families are less well researched, with one species each. Most studied stocks of medicinal species are in the south and east of Kazakhstan. In the north of the country, raw materials have been identified for 13.5% of species, while figures for the central (5%) and western (4.3%) regions are insignificant.

The greatest species diversity (about 90 species) of medicinal flora is concentrated in Kazakhstan's mountain ecosystems. Raw material reserves of commercial value in the North and West Tien Shan ranges are responsible for no less than 25 of the 141 resource species of medicinal flora. In the Kazakhstan Altai ranges, at least 10 of the 141 resource species of medicinal flora are recognized as having a commercial value. The mountain ranges with the most potential for medical raw material cultivation are North and West Tien Shan, responsible for 6–27 species such as Dzhungar Alatau, Ile Alatau, Ketmen, Kyrghyz Alatau, Kungei Alatau, Karzhantau, Terskey Alatau, and Boroldaitau, and the Kazakhstani Altai ranges are responsible for 4–21 species such as West Tarbagatai, Kurshum, Kalbin, Sarymsakty, Asutau, Narym, Saur, Manrak, Altai Tarbagatai, South Altai, and Listvyaga. The following make up the raw materials base for pharmacopeia species and species used in conventional medicine: *Hypericum perforatum*, *Achillea millefolium*, *Origanum vulgare*, *Patrinia intermedia*, *Ziziphora clinopodioides*, *Hypericum scabrum*, *Mentha longifolia*, *Polygonum nitens*, and *Rumex tianschanicus*, with a commercial cultivation in four to seven ranges.

Previous introduction studies in Kazakhstan's medicinal plants are somewhat greater than resource research. Approximately 700 species (49%) of medicinal plants have been crop tested. Data on raw material yield levels and the specifics of agricultural technology in commercial and semicommercial conditions is available for 33 species. Experimental data on raw material productivity and annual fluctuations has been generated for 250 species. Of the 230 species of CM, 137 species

(60%) have been introduced in the country (predominantly in its botanical gardens). More detailed experimental research (identification of raw material productivity) has been performed for 65 species (28%), while production testing is known for only 21 species of CM (9%). Of the medicinal plants highlighted, 65 species have entered the Red Data Book of Kazakhstan (2014) and have special introduction and agricultural development value.

The current state of medicinal plant resources in Kazakhstan requires the realization of a regional and nationwide long-term program for a comprehensive study and development of medicinal plant resources, creation of a collection of useful plants within the Kazakhstan flora, development of technology for introducing the cultivation of species of local flora used intensively and with a limited data to create a sustainable raw materials base for the domestic pharmaceutical industry, and to continue ethnobotanical research to highlight new sources of medicinal raw materials based on indigenous species. Contemporary systematic resource research into the medicinal flora in Kazakhstan will provide a scientific basis for its preservation and sustainable use. Such studies for the whole country will help to identify potential harvesting and the subsequent development of regions, highlight those species with potential for commercial cultivation and requiring protection, and guarantee a scientific basis for their rational use, which will help best to transform Kazakhstan's natural riches into sustainable economic growth ("Kazakhstan-2050" Strategy. Source: http://online.zakon.kz/Document/?doc_id=31305418&mode=p).

Acknowledgment The materials and results used here cover studies supported by the project grant on "Screening of the Natural Flora of Kazakhstan for Plants with Medicinal Properties and the Development of Proposals to Improve their Use" (2012–2014). Some of the results have been published reference book "Annotated List of Medicinal plants of Kazakhstan" (2014).

References

"Kazakhstan-2050" Strategy (2012) II. Ten global challenges for the XXI century. Sixth challenge—the exhaustion of natural resources. Address of the President of the Republic of Kazakhstan, Leader of the Nation Nazarbayev NA to the People of Kazakhstan. Astana, 14 December. Source: http://online.zakon.kz/Document/?doc_id=31305418&mode=p

Abdulina SA (1999) List of vascular plants of Kazakhstan. In: Kamelin RV (eds). Almaty, 187 pp

Annotated List of Medicinal plants of Kazakhstan (2014) Reference book.In: Grudzinskaya LM, Gemejiyeva NG, Nelina NV, Karzhaubekova GG (eds). Almaty, 228 pp

Baitenov MS (2001) Flora of Kazakhstan in 2 volumes. Vol. 2, Flora lineal complex. Gylym, Almaty 280 pp

Bykov BA (1957) Geobotany. Nauka, Alma-Ata, pp 22–23

Cherepanov SK (1995) Vascular plants of Russia and adjacent states (the former USSR). Cambridge University Press, New York 516 pp

Convention on Biological Diversity (1992) United Nations, 30 pp

Gemedjieva N, Teixeira da Silva JA, Ryabushkina N (2010) Representation of endemics in floristic subprovinces of Kazakhstan. Asian Australas J Plant Sci Biotechnol 4:56–63

Gemejiyeva NG (2015) Analysis of the species and resource potential of Kazakhstan's medicinal flora. Problems of the botanist in South Siberia and Mongolia: collection of scientific articles

taken from materials of the XIV international science and practical conference (25–29 May 2015, Barnaul). AltGU, Barnaul, pp 173–181
Gemejiyeva NG, Grudzinskaya LM (2016) State and prospects for the development of Kazakhstani plant resources "conservation and sustainable use gene pool of plant world in Eurasia at the present stage" international scientific conference within "day of Kazakhstan" (September 3, 2016, EXPO-2016 Antalya, Turkey). Antalya, pp 48–50
Grudzinskaya LM (2009) Introductory analysis of plants in the Fabaceae Lindl. family. Botanical Research in Siberia and Kazakhstan, Issue 15. Kemerovo, pp 94–102
Grudzinskaya LM, Arysbayeva R (2011) Introductory evaluation of medicinal plants in the Asteraceae Dumort. family cultivated in the Almaty Botanical Garden. Botanical Research in Siberia and Kazakhstan, Issue 17. Kemerovo, pp 141–156
Grudzinskaya LM, Arysbayeva RB (2014) Quality of families of collection medicinal plants in the Lamiaceae Juss. family. Botanical Research in Siberia and Kazakhstan, Issue 20. Kemerovo, pp 77–85
Grudzinskaya LM, Arysbayeva RB (2016) Productivity of collection medicinal plants in the Ranunculaceae Juss. family. KAS RK News. Biological and Medical Series. Almaty, KAS RK, no. 2 (314), pp 96–103
Grudzinskaya LM, Tazhkulova N (2008) Introductory valuation of plants in the Apiaceae Lindl. family. Fundamental and applied botany problems at the start of the 21st century: all-Russian conference. Petrozavodsk, pp 217–220
Grudzinskaya LM, Tazhkulova N (2013) Introductory evaluation of medicinal plants in the Rosaceae Juss. family cultivated in the Almaty Botanical Garden. Botanical Research in Siberia and Kazakhstan, issue 19. Kemerovo, pp 123–130.
Grudzinskaya LM, Arysbayeva RB, Ramazanova MS, Sadakmende TA (2017) Storage periods for collection medicinal plant seeds. In: Kupriyanov AN (ed) Botanical research in Siberia and Kazakhstan: collection of research works, Issue 23. KREPO Irbis, Kemerovo, pp 96–105
Guidelines for working with Medicinal plants (1999) In: Beklemishev ND (ed) Government Enterprise Dari-Darmek, Almaty, 230 pp
Illustrated Identifier of Kazakhstan Plants (1969), vol 1. Alma-Ata, 645 pp
Illustrated Identifier of Kazakhstan Plants (1972), vol 2. Alma-Ata, 522 pp
International Scientific and Production Holding Company Phytochemistry (2018) Source: http://phyto.kz/ru/vedushchie-uchenye.html
Kokshetau (1999) National strategy and action plan on conservation and sustainable use of biological diversity in the Republic of Kazakhstan, Ministry of natural resources and protection of environment of the Republic of Kazakhstan, 112 pp
Korchagin AA (1964) Species (floral) composition of plant communities and methods for their study. Field Geobot 3:39–60
Kukenov MK (1999) Botanical resources of Kazakhstan. Gylym, Almaty 160 pp
Methodology for determination of medicinal plant reserves (1986) Moscow, 50 pp
Methodology of surveys during introduction of medicinal plants (1984) Moscow, 39 pp
Natural Conditions and Resources in Kazakhstan (2018) Source: http://www.ecololocate.ru/locats-141-1.html
Ozturk M, Aksoy A, Gucel S, Altundag E (2010) Biodiversity and conservation—a case study from Europe and Turkey. In: Bhatti R, Soomro SA (eds) Proceedings of 2nd international conference "biodiversity is our life". Shah Abdul Latif University, Khairpur, pp 1–12
Ozturk M, Altay V, Gucel S, Altundağı E (2016) Plant diversity of the drylands in Southeast Anatolia-Turkey: role in human health and food security. In: Ansari, Gill (eds) Plant biodiversity: monitoring, assessment and conservation. CABI, Oxfordshire 66 pp
Plants of the Natural Flora of Kazakhstan in Introduction (1990) Guide book. Gylym, Alma-Ata 287 pp
Ponyatovskaya VM (1964) Recording the abundance and the location of species in natural plant communities. Field Geobot 3:209–237
Proskuryakov MA (2012) Chronobiological analysis of plants under a changing climate. Almaty, 230 pp

Recommendations for the Study of the Ontogenesis of Introduced Plants in the Botanical Gardens of the USSR (1990) Kiev, pp 6–21

Red Book of Kazakhstan (2014) In: Baitulin IO, Sitpayeva GT (eds), vol 2, Astana, 452 pp

Republic of Kazakhstan (2006) Natural conditions and resources. Almaty, pp 287–289

Ryabushkina N, Abugalieva S, Turuspekov Y (2016) Problems Of Study And Conservation Of Flora Biodiversity In Kazakhstan. Eurasian J Appl Biotechnol. https://doi.org/10.11134/btp.3.2016.2

State Pharmacopeia of the Republic of Kazakhstan (2009) Astana, vol. 2, 790 pp

State Register of Medicines of the Republic of Kazakhstan (2013) List of medicines registered and deployed for use and production by the Ministry of Health of the Republic of Kazakhstan: adilet.zan.kz›docs/U950002655

Chapter 10
Medicinally Important Plants of Kazakhstan

Kanat N. Sarsenbayev

Abstract The flora of Kazakhstan includes about 6,400 species of higher vascular plants (Aralbai et al. 2017), approximately 5000 fungus, 486 lichens, more than 2000 algae, and about 500 bryophytes. The phytobiota's wealth in Kazakhstan is estimated at 14500-15500 taxa. Out of 6,400 species of higher vascular plants, about 730 species are endemic (Baitenov, 2001; Aralbai, 2011), representing 11-12% of the total species composition. There are more than 1,000 species with curative values (Aralbai, 2016). In addition, there are more than 500 species of introduced, cultivated, and imported plants. In Kazakhstan and Central Asia nearly 80% of the plant resources, in particular medicinal plants, were harvested during the soviet period. The plant diversity of Kazakhstan is an inexhaustible source of biologically active substances. However, knowledge about the biochemical composition of plants is poor; there is a lack of information on the regulation of biologically active compound synthesis, and an insufficient number of pharmaceutical profile productions. The main sources obtained include licorice, *Cistanche*, and *Ferula*. In earlier times during the USSR period, *Ephedra* was mainly collected from the Almaty region of Kazakhstan and a major part of the raw material was exported under the informal scheme, which means permitted harvesting of plants is significantly exceeds quantities. Currently nearly 10 species of plants are exported, and in the case of *Cistanche*, under another name through the Customs. This chapter presents information on some of these valuable plants of Kazakhstan, which are being harvested, exported from Kazakhstan, and resold by China to other countries. There is an urgent need for Kazakhstan to establish a business partnership with other countries on a mutual long-term basis and start sharing the supply of plant resources and processing of their products.

Keywords Plant resources · *Ephedra* · *Ferula* · *Cistanche* · Glycyrrhiza

---

K. N. Sarsenbayev (✉)
Biological Sciences, Eurasian National University named after N. Gumilyov,
Astana, Kazakhstan

D. Egamberdieva, M. Öztürk (eds.), *Vegetation of Central Asia and Environs*,
https://doi.org/10.1007/978-3-319-99728-5_10

10.1 Introduction

The present state of natural vegetation reveals that there is a significant change in species composition in natural cenoses, a reduction in the area occupied by medicinal plants, and a decrease in wild stocks of raw materials (Shtephan 2012). Biodiversity of flora and vegetation in the Republic of Kazakhstan is conserved through the creation of protected areas such as national natural reserves, national parks, and state natural reserves. The total area specially protected is currently 5678.7 thousand hectares (ha), which include 10 state nature reserves of 1612 thousand ha; 11 State National Parks found on 2249.3 thousand ha, and four state natural reserves spread over 1817.4 thousand ha (Nyssanbayev 2013). Of the 6000 species of plants distributed in Kazakhstan, 1 of 10 is reported as being endangered. More than 120 species are wild relatives of agri-crops, and 70 species from 29 genera have been recorded as forage crops. More than 120 species are wild relatives of agricultural plants, and more than 700 species of medicinal plants are mentioned in folk medicine; 263 plants have been investigated scientifically (Urazaliev et al. 2004; Shtephan 2012). The reason for this great diversity is the location of Kazakhstan at the junction of three genetic centers, as reported by Vavilov (1926). The diverse soil and climatic features provide habitats for the growth and development of a substantial number of species and very large populations of plants (Ivanov and Buhteeva 1981). The plant diversity of Kazakhstan, Siberia, and the Altai has been investigated by botanists in detail during the past two centuries. Agro-botanical expeditions have been undertaken continually for collecting source materials for the All-Russian Institute of Plants (AIP), named after I.N. Vavilov, from 1969 until 1979 (Vavilov 1965). Many more recent expeditions have been organized to investigate natural plant communities and to collect seeds of wild medicinal plant taxa (Kukenov 1996; Egeubaeva 2002) in Central Kazakhstan (Shaushekov 2004) and in Northern Kazakhstan (Shtefan 2001, 2004).

The best way to preserve natural plant communities is to ensure that their existence continues in their natural habitats. "In situ" conservation is possible in nature reserves, national parks, and plantation forests, but its disadvantages are that such preservation involves large areas. However, we can study plant reproduction and introduction to nature. Introduction of wild species with a large habitat and multiple forms has ecological flexibility and the variability as source material for breeding, in particular medicinal plants. On the other hand, "ex situ" adds to the former conservation practice. Researchers have restarted examining the natural vegetation of the northern, eastern, and central regions of Kazakhstan. Loss of biodiversity, in particular certain species of fodder and medicinal plants, is of great concern. Much degradation in the populations of medicinal plants and loss of taxa are taking place because of the uncontrolled harvesting of wild medicinal herbs.

The plant diversity of the country is represented by meadow–forest, meadow–steppe, and fescue–feather grass and mixed grass species. All these flourish on lands removed from agricultural use (Kurkin 1998). Nearly 80 species of medicinal plants grow in Northern Kazakhstan, mostly in the meadow–steppes: *Achillea millefolium*,

Taraxacum officinale, *Salvia stepposa*, *Artemisia glabella*, *Plantago lanceolata*, *Capsella bursa-pastoris*, *Tanacetum vulgare*, and *Equisetum arvense*. These taxa are widely used as medicinal plants. Surveys in Akmola, North Kazakhstan, and Kostanay have revealed that there is a lack of extensive natural phytogenesis because of widespread cultivation practices. The natural vegetation, is preserved in places occupied by birch and birch–aspen groves and forests, in land depressions, and in nonagricultural areas. At some sites we come across secondary mixed grass communities where some of the plant taxa serve as indicators of human impact (Kleptsova et al. 2001; Tkachenko 2006). *Phlojodicarpus sibiricus* and *Ziziphora bungeana* dominate the sagebrush–forb cenosis, found on the stony-gravelly soils on the hill slopes of the North Kazakhstan region. *Plantago major*, *Plantago media*, *Rumex confertus*, and *Galium boreale* prefer moist sites In the mixed grass meadows, among the scrublands in Kostanai region we sometimes come across *Echinops sphaerocephalus*, *Eryndium planum*, *Sanguisorba offficinalis*, and *Salvia nutans*. In the open birch forests, *Thymus marschallianus*, *Thymus serphyllum*, *Echinops ritro*, *Filipendula ulmaria*, *Phlomis tuberosa*, *Cichorium intybus*, *Bidens tripartite*, and *Helichrysum arenarium* are widespread in the secondary mixed grass cenoses. *Helichrysum arenarium* has a wide application in herbal medicine and is an excellent raw material for production of choleretic action drugs (Sokolov and Zamotaev 1991). The decrease in its populations has stressed the fact that its propagation area in nature is needed as such, so cultivation has been started. *Urtica dioica* and *Artemisia dracunculus* are found in man-made forests. The hummocky semi-desert area of the Karaganda region shows mainly xeromorphic taxa such as *Artemisia*, *Testuca sulcata*, *Stipa capillata*, and *Ceratocarpus arenarius*. On the slopes of hills we find *Silene multiflora*, *T. serpyllum*, and *Caragana frumex*. Around the bases of hills species composition is represented by mixed grass cenoses: *Bromus arvensis*, *Glycyrrhiza uralensis*, *Onobryshis arenaria*, *Xanthium strumarium*, *Thermopsis lanceolata*, *Salvia stepposa*, and *Ammophila arenaria*. *Glycyrrhiza uralensis* occupies nearly 50–60% of the total plant cover.

The steppe zone of the Pavlodar region is not rich; it is represented mainly by xerophytes such as *Caragana frutex*, *S. capillata*, and *Calamagrostis epigeios*. *Eryngium planum*, *Linaria vulgaris*, and *T. serpyllum* show insignificant distribution as medicinal plants. In East Kazakhstan steppes transition to the Altai foothills, an extension of the West Siberian Plain. The plant diversity found here includes *Salvia stipposa*, *Plantago major*, *Verbascum thapsus*, and *Echinops ritro* located among strips of pine forest. Some medicinal plants found here are *Helichrysum arenarium*, *Artemisia dracunculus*, and *Potentilla erecta*, but these are not of industrial value. Many medicinal plants are found on the Altai hills and foothills, growing sparsely in small clumps: *Origanum vulgare*, *Hypericum scabrum*, *Ziziphora bungeana*, *Phlojodicarpus sibiricus*, *Echinops sphaerocephalus*, and *A. millefolium*. *Glycyrrhiza uralensis* and *Glycyrrhiza korshinskyii* take a special place among the medicinal plants of Kazakhstan. Their main thickets of industrial value are located in the south, where they cover more than 30,000 ha, and in the north, where natural populations occur as separate arrays of 0.5–2.5 ha in Kostanay. *Adonis vernalis* does not have industrial value. The concentration of cardiac glycosides in this plant makes it indispensable in the treatment of cardiovascular disease (Sokolov and Zamotaev 1991).

10.2 Studies of Some Economically Important Plants

10.2.1 Ephedra

The industrially valuable *Ephedra equisetina* in Kazakhstan grows in Almaty and Zhambyl regions (Fig. 10.1), East Kazakhstan, and South Kazakhstan. The most important reserves are concentrated in the Almaty region, where it is possible to harvest more than 323.8 tons a year (Egeubaeva et al. 2006).

Genus *Ephedra* has a centuries-old history of its use as medicine, particularly as an adrenomimetic and bronchodilator (Egeubaeva et al. 2006). This perennial shrub, although flourishing well mainly in the Mediterranean floristic region, is also found in the mountains of Central Asia, Kashmir, and Kazakhstan as well as Western China (Ozturk et al. 1996a, b; Egeubaeva et al. 2006). There are 30 species of *Ephedra* reported in the literature: of these, 7 species and 2 hybrid forms are found in Kazakhstan. Of these, the alkaloid-producing plants are *E. distachya*, *E. equisetina*, *E. intermedia*, and *E. lomatolepis* (Sarsenbaev and Polymbetova 1986). Only 2 species are mountainous. These contain a significant alkaloid content (2–3%) compared to other species and are used to produce ephedrine. *Ephedra* species growing in Kyzylorda with typical East Mediterranean-type climatic features are

Fig. 10.1 General habitat of *Ephedra equisetina* from Almaty region

industrially valuable together with those in Almaty, as well as in Shymkent and Zhambyl states (*Ephedra equisetina*) in the Southern Kazakhstan. The reserves in Almaty region are industrially feasible, with a gross productive capacity of more than 350 tons per year. Some species have been recommended by local botanists as the main raw material for the production of ephedrine in the Shymkent Chemical and Pharmaceutical Plant. Its green branches contain up to 3.2% alkaloid content, and that of L-ephedrine on average is 48–70%. In *E. intermedia*, pseudoephedrine prevails, with a value of 70–95%. Many of the *Ephedra* taxa contain tannins, leucoanthocyanides, flavone dyes, and aromatic and other compounds (Gubanov and Sinitsin 1966; Collective Authors 1980; www.fito.nnov.ru).

A detailed study on the identification of new *Ephedra* thickets has been carried out in the country together with its mapping, as well as the assessment of the raw material resource potential in general. The study was related to the interests of the pharmaceutical and tanning industries. The main aim was to study *Ephedra* species of Kazakhstan to elucidate the optimal conditions for the accumulation of alkaloids and tannins in its green branches and at the same time to determine the raw material potential for its use in the production of ephedrine and tannins. The qualitative composition of tanning substances in *Ephedra* also has been thoroughly studied. The limited availability of *Ephedra* has contributed to the development of commercially available methods for the synthesis of ephedrine, which has a significant drawback on a cost basis. The preparation of a racemate was followed by separation into right- and left-rotating optical isomers. Despite success in the field of artificial synthesis, the quality of natural ephedrine was higher. Globally there is increasing demand in the use of ephedrine obtained from *Ephedra*. Data from pharmacological studies reveal that natural alkaloids have a higher quality of therapeutic effect with a longer duration. With a full-fledged plant, raw material, and sufficient reserves or with a successful plant culture, one or another alkaloid is undoubtedly more convenient and economically more profitable for production from the natural plant material (Collective Authors 1969, 1980; Egeubaeva et al. 2006).

10.2.1.1 Morphological Features and Distribution of Some *Ephedra* Species

Ephedra lomatolepis is a dioecious shrub, 0.5 m tall, with almost whorled branches and often with underground stem-like shoots; sprigs up to 2 mm thick, green, hard, medium interstitial up to 6 cm long; leaflets are 2–3, reduced to the vagina, up to 4 mm in length, green and grassy on the back, in the areas of adhesion as a light and narrow film; anthers rounded, in capitate inflorescences, up to 6 mm in length; flowers 4–8 in number, single or paired; the spike or single solitary, at the nodes or at the base of the branches, sometimes up to 5 cm long, 3–4 spikelets; bracts paired or whorled, free or almost free, broadly ovate, posteriorly grassy, marginally webbed and slightly gnarled or whole; tubule twisted, up to 1.5 mm in length; fruits ovoid or spherical, up to 6 mm in length; seeds flat-convex, up to 4 mm long, brown. Flowering in May, fruiting in July. Distributed in Betpakdalin, Muyun-Kum, and

Ephedra distachya *Ephedra monosperma*

Fig. 10.2 A general view of two *Ephedra* species from their habitat

Balkhash-Alakol natural zones (Gubanov and Sinitsin 1966; Collective Authors 1969, 1980; www.fito.nnov.ru).

Ephedra distachya (Kuzmichev grass, steppe raspberry) (Fig. 10.2) is a dense short branching shrub, up to 20 cm tall, with a creeping rhizome and a dark-gray bark; branches erect, yellowish-green, usually straight or slightly curved, longitudinally furrowed along the ribs and finely tuberous, up to 1 mm in diameter, with internodes up to 3 cm in length. Leaves reduced, small, up to 2 mm in length, membranous, obtuse, incised into two triangular lobes and adpressed together into small vaginas, which collapse as the branches grow (Gubanov and Sinitsin 1966; Collective Authors 1969, 1980; www.fito.nnov.ru).

The anther spikelets lie on short shoots, collectively looking like small oblong inflorescences, about 110 mm in length, bracts 2–2.5 mm long, oblong, leathery, sometimes slightly dyed; perianth consists of rounded leathery lobes; anthers are 8 in number. The fruiting spikelets are roundish-ovate, small, up to 6–7 mm long, 5–15 mm long, solitary or assembled into bundles, with 2–3 flowers, 3–4 bracts, of which the lower ones are one-third intertwined and on the edge narrow-webbed, while the innerside ones are half-jointed and surround the tube; tube on the edge with a tongue, straight, 1.5 mm long. Fruits are almost spherical, berry-shaped, dark red, 6–7 mm long; seeds are paired, ovate or ovate-oblong, 4.0–5.5 mm long, semicircular, dark brown, convex beneath. Flowering occurs in May and June, fruiting in July. Distributed in the Irtysh, Semipalatinsk hog, Kokchetav, Caspian, Mugodzhar, Turgai, Kazakh Uplands, Priaral, Kzyl-Orda, Betpakdala, Balkhash-Alakol natural zones (Gubanov and Sinitsin 1966; Collective Authors 1969, 1980; www.fito.nnov.ru).

Ephedra monosperma (Fig. 10.2) is a dioecious shrub, 15–25 cm tall, with a long knobby underground stalk on the surface of the soil ending in a bundle of ascending stiff branches; the branches are yellowish-green, about 1 mm wide, straight or curved, smooth or hardly scabrous, finely scabrous, with interstices up to 2.5 cm in length; leaves are reduced to membranous vaginas; anther spikelets back-ovate, located as doubles, anthers 6–8 on a column exhibited from hymen scales; fruiting spikelets on short downward curved shoots with 2–3 pairs widely oval, along the margins of narrowly membranous bracts, the inner of them are equal to each other, ½–¾ of their length fused; tubule serpentiformly curled, 1.5–2 mm long; fruits are berry-shaped, globose, red, 6–9 mm long, in the predominant part of single-seeded, sometimes double-seeded; seeds oval, brown, 4–6 mm long, convex on both sides, if they are 2, then flatly convex. Flowering in June, fruiting in August. In Kazakhstan it is found in Altai, Djungar Alatau, Western Tien Shan (Gubanov and Sinitsin 1966; www.fito.nnov.ru).

Ephedra strobilacea differs from other species in stickiness and dry fruits. The dioecious shrub is up to 2 m in height; young branches up to 4 mm in thickness, grayish-green; leaves are scaly, pointed, about 2 mm long, fused to one third or one half, fusion sites membranous, and on the back grassy, greenish; anthers lie in the nodes of young branches; flowers 2–3; tube up to 3 mm in length, straight; seeds 5–6 mm long, greenish or grayish in color, oblong or elliptical, slightly keeled on the back. Flowering in May, fruiting in July. It occurs in the Kzyl-Kum desert (Gubanov and Sinitsin 1966; Collective Authors 1969, 1980; www.fito.nnov.ru).

Ephedra fedtschenkoae is a shrub with an underground trunk, ending in a bundle of ascending yellowish-green branches, about 1.5 mm in length; 2–7 cm tall, straight or more often curved, interstitial 1–1.5 cm long; vagina of young branches greenish-yellow, leaves forming them in the upper half are loose, broad, gray, at the base brownish; on a very short stem, usually 4–5 flowering stalks, anthers 6–7 on short stalks or almost sessile on a column; fruiting on short stalks or sessile, single or paired, two-flowered, lower bracts about 2 mm long, flattened, wide-oval, upper one-third or half-fold at the edge narrow-webbed, reaching up to half the length of the internal part, of which usually two tubes are straight or slightly curved up to 2 mm in length, fruits are berry-shaped, red, 7 mm long. Seeds are plano-convex, dark brown, shiny, about 4 mm in length, one of them is usually less developed on the column. Flowering in July, fruiting in August. Found on Tien Shan (Gubanov and Sinitsin 1966; Collective Authors 1969, 1980; www.fito.nnov.ru).

Ephedra regeliana is a dioecious shrub, up to 3–8 cm tall; from the underground stem a bunch of yellowish-green rough branches grow, 1–1.5 flowering, anthers 6–7; fruiting spikelets generally two-flowered, 2 or single, with 2–3 pairs of bracts; bracts finely bordered, inner ones cover the flower and exceed it; tubules 2–2.5 mm long, slightly curved or straight; seeds 3.5–4.5 mm in length, black, shining. Flowering in June, fruiting in July. Found in Zailiysky, Kungei Alatau, Kyrgyz Alatau, Karatau, Western Tien Shan (Gubanov and Sinitsin 1966; Collective Authors 1969, 1980; www.fito.nnov.ru).

10.2.1.2 Ecological Characteristics of Some *Ephedra* Species

Ephedra lomatolepis grows in sandy habitats, in deserts, and is generally absent in the vicinity of stony rocks, which is a characteristic ecological difference from other *Ephedra* taxa in general. Antheral and fruit-bearing spikelets are attached to the twigs on fairly long stalks (3–10 mm). The branches are bordered, much more tender and thinner. This species can be easily distinguished according to its ecological isolation, presence of relatively long stalks in anthers and fruiting spikelets, and relatively soft, smooth, and thin light green twigs (Gubanov and Sinitsin 1966; Collective Authors 1969, 1980; www.fito.nnov.ru).

Ephedra distachya grows in the steppes, on plains, on the rubble slopes of steep mountains, on hills, and alongside the river valleys, but thickets occur in limited areas; grows on the slopes of hills, in the lower belt of mountains, on sandy massifs, among stones. As per its ecological characteristics, it is a xerophyte and oligotroph, adapted to dry habitats and soils poor in organic matter. Optionally calciphyllous, light-loving, prefers soils with a loose mechanical composition, poor dry stony, calcareous and sandy, as well as on sand and chalk outcrops. Distributed up to an altitude of 900 m (Gubanov and Sinitsin 1966; Collective Authors 1969, 1980; www.fito.nnov.ru).

Ephedra strobilacea is peculiar to the sand deserts.

Ephedra fedtschenkoae grows in steppes, plains, on gravelly slopes of steppe mountains, on hills, and alongside the river valleys, and as thickets in limited areas (Gubanov and Sinitsin 1966; Collective Authors 1969, 1980; www.fito.nnov.ru).

Ephedra regeliana grows on stony and gravelly slopes of the upper parts of the mountains. The morphological differences of individuals are very significant. The size of the bushes varies greatly; the thickness of the branches and the length of the segments and other indices also vary. The color of branches is mostly bright green, sometimes bluish, and rarely quite blue. The reason for its blue color is not clear (Gubanov and Sinitsin 1966; Collective Authors 1969, 1980; www.fito.nnov.ru).

Gubanov and Sinitsin (1966) have studied these taxa for a number of years. They surveyed and mapped the *Ephedra* taxa in the flora of Kazakhstan. The industrial stock of medicinal raw materials lies in the Dzungarian, Zailiysky, Kirghiz Alatau, and Ketmen ranges. In the mountains of Karatau it is estimated to produce 2000 tons. In the system of harvesting of plant raw materials, *Ephedra* occupies an important place, because harvesting significantly affects its profitability. It is the true national wealth of Kazakhstan, and the state of our own medical industry depends to a certain extent on the careful, diligent use of this taxon (Collective Authors 1969, 1980; www.fito.nnov.ru).

These authors also studied the bioecology of *Ephedra*, the methods of responsible harvesting, and the restoration of thickets. They proved the possibility of obtaining their crops by sowing seeds on the treated soil of foothills. Experiments on *E. equisetina* culturing showed that for 6–7 years of cultivation the plant height reaches 0.6–0.7 m, and the content of alkaloids in green branches is 1.7%. The restoration of thickets, and the rational use and safety of the populations of *Ephedra* is therefore of paramount importance (Sarsenbaev et al. 2013). The reserves of *Ephedra* in Kazakhstan are quite rich and require responsible use. In addition to its medicinal value, this plant has great ecological significance as a stabilizer, in particular, *E. equisetina* for mountain slopes, on stony and gravelly talus. Therefore, the rational use and conservation of *Ephedra* populations is of paramount importance (Epgeubaeva et al. 2006).

10.2.1.3 Biochemical Features of Some *Ephedra* Species

Of the nine taxa of *Ephedra* distributed in Kazakhstan, *E. equisetina* and *E. intermedia* are of practical use. These taxa accumulate a large amount of ephedrine. During the past few years interest in the collection of natural raw material from *Ephedra* has increased significantly because the dried and crushed sprigs of *Ephedra* are the main component of biologically active additives for weight loss. These additives based on ephedrine are produced in China and other Southeast Asian countries. They are sold in large quantities in the USA and Europe. *Ephedra* is also used as a tea as an energy-giving herbal drink. The latter is popular in the US among Asians. A person drinking *Ephedra* tea is said to be able to work actively for about 20 h a day (Collective Authors 1969). The marketing value of *Ephedra* is very large, and industries are approaching Kazakhstan for its purchase in large quantities. Information on the current reserves of *Ephedra* in Kazakhstan is not known. Most important is that not many data are available on the quality of raw material from its various populations. Studies have been undertaken on the population polymorphism and the biochemical features of the taxa from this genus in places of its mass growth.

A comparative analysis of the accumulation of alkaloids in six samples of *Ephedra* needles has shown the presence of ephedrine and pseudoephedrine alkaloids in all samples. Thin-layer chromatographic evaluation revealed that ephedrine prevails in samples from (a) Terekt canyon near Taldi Korgan city and (b) 61 km from the Terekt canyon near Taldi Korgan city; pseudoephedrine; although in much smaller quantities; predominated in samples from the Terekt canyon near Taldi Korgan city (2 km inside) and the mountains of Kzyl-tal (near the river). The presence of other alkaloids with Rf = 0.8 and Rf = 0.55 was noted in traces. The alkaloids revealed differences between the samples. The concentration of unidentified alkaloids was slightly less than pseudoephedrine, but higher than that of other samples. There are also differences in the alkaloid with Rf = 0.55; the concentration is more significant in some samples. In the samples collected 61 km from Terekt canyon near Taldi Korgan city, the presence of another unidentified alkaloid with Rf = 0.65 was noted with good amounts, but it was absent in other samples.

Spectrophotometric study has shown that ephedrine content in the samples varies from 2.16 to 4.6% in general, with maximum quantity found in the samples from Terekt canyon near Taldi Korgan city (4.6%) and Kzyl-tal Mountains (3.6%); whereas it is a minimum in the samples taken from Kzyl-tal Mountain near the dry river bed (2.16%). On an average basis, levels about 3% indicate the suitability of ephedra for industrial harvesting. The morphometric parameters in different populations have shown that we can use such indicators as the height of the plant, number of shoots, stem thickness, needle length, and weight of the aerial parts for collecting good samples. Most developed plants are from the Kzyl-Tal mountain range, being 107 cm tall; number of shoots is 5, stem thickness at the base of the shoot is 2.8 cm, needle length is 24.5 cm, and the weight of the aboveground parts of the bush is 2.4 kg. The plants of *Ephedra* located in the foothills near the village of Biljan are less productive, being only 48 cm tall with only 9 shoots, stem thickness at the base of the shoots is 0.8 cm, needle length just 23.5 cm, and the weight of the aboveground

Table 10.1 Morphometric parameters of the *Ephedra* samples from nine localities

No	Place of growth	Height of plant (cm)	Number of shoots (cm)	Thickness of stem at the base (cm)	Length of leaves (cm)	Above ground weight (kg)
1	Aul Bilzhan, (Sarkand region)	48 ± 3.9	9.0 ± 0.2	0.8 ± 0.2	23.5 ± 0.4	0.3 ± 0.1
2	Mountains Kyzyl-Tal (Sarkand region)	107 ± 2.3	15 ± 1.3	2.8 ± 0.6	24.5 ± 0.2	2.4 ± 0.3
3	Terekty canyon (Sarkand region)	67 ± 3.1	13 ± 0.2	2.4 ± 1.2	11.2 ± 0.1	0.5 ± 0.1
4	Terekty canyon deep, 2 km (Sarkand region)	80 ± 1.9	17 ± 1.8	3.0 ± 0.9	18.5 ± 0.1	1.0 ± 0.2
5	Dry river 61 km from the Terekty canyon (Sarkand region)	57 ± 3.1	10 ± 0.9	1.1 ± 0.4	18.5 ± 0.2	0.5 ± 0.1
6	Kzyl-Tal mountains near the river (Sarkand region)	51 ± 2.9	10 ± 1.9	1.0 ± 0.3	21.7 ± 0.7	0.4 ± 0.2
7	Kordaysky Pass (Zhambyl region)	57 ± 3.9	12 ± 0.4	2.0 ± 0.5	26.0 ± 0.5	0.3 ± 0.1
8	Turgen canyon	74 ± 3.9	14 ± 0.3	3.1 ± 1.0	21.0 ± 0.1	
9	Ketmen Mountains	45 ± 1.2	9.1 ± 0.5	2.0 ± 0.1	17.0 ± 0.2	

bush 0.3 kg. The plants of this genus seem to grow better under high mountain conditions as their morphometric parameters exceed those of the plants growing in the lowlands. For the plants growing in the Kordai Pass, the morphometric indices are lower than for the samples from the Dzungarian Alatau (Table 10.1). Analysis of the accumulation of ephedrine has shown that its content does not exceed 0.9%.

10.2.2 Licorice

Licorice is one of the most valuable economically important plants in the flora of Kazakhstan as well as the world. Its export in the form of raw material has increased lately, the reason being the high glycyrrhizic acid content in the roots. The latter is widely used in developed countries as a raw material for the production of antispasmodic, antiinflammatory, anti-allergic, anti-tumor drugs (Altay et al. 2016; Karahan et al. 2016; Ozturk et al. 2017a, b). It is intensively used as an emulsifier in the food and cosmetic industry, and as a blowing agent in the preparation of fire-fighting equipment and other industries (Ozturk et al. 2017a). The presence of large reserves of this plant in Kazakhstan makes it a very promising crop, both for domestic industrial use and for export. Much work has been done on the biochemical composition of various licorice species in Russia. Already in 1937 it was reported that the roots of this plant contain the sweet substance glycyrrhizin: a salt of glycyrrhizic acid (up to 22%), flavonoids (up to 4.3%), organic acids (up to 4.6%), mono- and disaccharides (up to 20%), together with other pharmacologically active ingredients. The flavonoids with antiinflammatory, antimicrobial, hypolipidemic, and antioxidant properties are extracted from its root, and organic extracts from the root have a pronounced antimicrobial and antioxidant effect (Wu et al. 2011; Karahan et al. 2016; Ozturk et al. 2017a). Flavonoids from licorice can reduce the side effects of certain drugs and enhance their pharmacological effect (Ozturk et al. 2017a).

The genus *Glycyrrhiza* belongs to the family Fabaceae. It is represented in the world by 30 species. The medicinal uses of licorice and its preparations, in which licorice came out on top among flowering plants, is diverse (Ozturk et al. 2017a). Its application in about 12 pharmaco-therapeutic groups for the prevention and treatment of various diseases has been recorded. Licorice species have attracted the attention of researchers as sources of plant raw materials used to produce valuable medicines, food, and other products. Biologically active substances (BAS) of the plants from this taxon belong to two large and sufficiently studied groups of natural substances—triterpenoids and phenolic compounds. Tremendous work has been conducted by the phytochemists of Russia, the Central Asian Republics, Japan, China, India, and other countries. A total of 346 individual substances of phenolic nature have been isolated and characterized (Ozturk et al. 2017a). As for Kazakhstan, it is currently sold abroad only as a raw material in the form of roots and syrup.

The export statics of licorice during the past 5 years shows that the quotas for 2015 for export are about 36,404 tons. In 2013 it was only 30,000 tons, bringing a revenue of 22,666,000 US dollars (USD) and 57 tons of licorice extract worth 262,000 USD in 2015. Nearly 17,513 tons of root worth 10,407,000 USD and 77 tons of extract worth 334,000 USD were exported to America.

The main types in the southeast of Kazakhstan are locally known as licorice naked and Urals. The Urals is characterized by a curved, wrinkled form of fruit (01A07), and the naked has straight and unwrinkled type (01A30). Some populations possess small slightly curved fruits (01A24). This form looks like a hybrid between the licorice of the Urals and the naked. The phenotype is close to Korzhinsky's malt. The original species and hybrids were investigated based on biochemical characteristics. As a biochemical marker, the composition of flavonoids and the nucleotide sequence of the rbcL gene were considered.

Licorice naked and Uralsk both have hair-bearing fruits and non-hairy fruits. In the Ural form the leaf is oval (01A10), and in the nude linearly oval (more elongated, 01A12). In the hybrid forms we see intermediate morphs between these two forms (01A11 and 01A13). In the uralskaya, the plants basically accumulate isokvetsitin, and others have rutin. The hybrid form also accumulates more rutin, but the spectrum of flavonoids is specific, although it is close to the licorice naked spectrum. In the same methanol extract, the accumulation of triterpenes in the roots was determined. It was found that glycyrrhizin, glabridine, and glycicoumarin accumulate in the roots; glabridin accumulates only at the head. Glycicumarin is from the Urals. The thinner the root, the more glycocumarin present in the plant (Table 10.2).

Table 10.2 Influence of growth conditions (low and high thickening) on the pigment content and peroxidase activity in plants from different populations (mg of pigment per 1 g of dry matter)

Type of association	Control	Intensity of associations	
		Low	High
1	2	3	4
Chlorophyll "a"			
Liquorice	6.6 ± 0.1	5.9 ± 0.7	5.8 ± 0.9
Cane-liquorice	8.8 ± 0.2	6.9 ± 0.9	7.9 ± 1.1
Azhrekovo-licorice	8.0 ± 0.1	5.3 ± 0.6	7.2 ± 0.9
Chingilovo-licorice	4.3 ± 0.2	4.1 ± 0.1	3.7 ± 0.6
Vein-and-Liquorice	5.0 ± 0.1	3.8 ± 0.4	2.9 ± 0.3
Sweet and Liquorice	5.4 ± 0.1	3.7 ± 0.9	3.9 ± 0.4
Chlorophyll "b"			
Liquorice	3.7 ± 0.2	2.5 ± 0.1	2.1 ± 0.1
Cane-liquorice	5.6 ± 0.5	3.5 ± 0.1	3.7 ± 0.2
Azhrekovo-licorice	7.1 ± 0.8	3.4 ± 0.2	2.8 ± 0.1
Chingilovo-licorice	1.2 ± 0.1	1.3 ± 0.1	1.5 ± 0.1
Vein-and-Liquorice	2.6 ± 0.1	2.6 ± 0.1	1.4 ± 0.2
Sweet and Liquorice	2.1 ± 0.1	1.2 ± 0.1	18 ± 0.1
Chlorophyll "ab"			
Liquorice	10.3 ± 0.2	8.5 ± 0.2	7.9 ± 0.2
Cane-liquorice	14.5 ± 0.4	10.3 ± 0.3	11.7 ± 0.4
Azhrekovo-licorice	15.1 ± 0.3	8.6 ± 0.2	10.0 ± 0.2
Chingilovo-licorice	6.2 ± 0.1	5.4 ± 0.1	5.1 ± 0.1
Vein-and-Liquorice	7.3 ± 0.1	4.4 ± 0.1	6.4 ± 0.2
Sweet and Liquorice	7.5 ± 0.2	4.9 ± 0.1	5.7 ± 0.2
Carotenoids			
Liquorice	5.3 ± 0.1	6.6 ± 0.2	1.5 ± 0.1
Cane-liquorice	3.1 ± 0.1	5.5 ± 0.1	2.5 ± 0.1
Azhrekovo-licorice	4.2 ± 0.1	2.4 ± 0.1	2.3 ± 0.1
Chingilovo-licorice	1.1 ± 0.05	0.9 ± 0.05	1.01 ± 0.08
Vein-and-Liquorice	1.4 ± 0.05	0.9 ± 0.05	0.87 ± 0.05
Sweet and Liquorice	1.2 ± 0.05	0.81 ± 0.04	1.18 ± 0.05
Peroxidase activity			
Liquorice	0.9 ± 0.04	0.2 ± 0.04	0.03 ± 0.01
Cane-liquorice	23.9 ± 0.6	29.6 ± 0.7	26.7 ± 0.7
Azhrekovo-licorice	118.4 ± 8.0	203.1 ± 11.3	169.5 ± 8.3
Chingilovo-licorice	50.6 ± 2.5	51.6 ± 2.6	50.1 ± 2.7
Vein-and-Liquorice	13.8 ± 0.2	8.2 ± 0.2	11.2 ± 0.3
Sweet and Liquorice	104.7 ± 5.7	49.3 ± 2.1	73.5 ± 4.3

The accumulation of glycyrrhizin varies from 1.2 to 6% on a dry weight basis of the root; more glycyrrhizin accumulates in the roots.

A comparison of the nucleotide sequence (primary structure) of the rbcL gene (ribulose-bisphosphate carboxylase/oxygenase) of licorice nude, Ural, and hybrid revealed that previously the difference between the two species in the sequence of this enzyme is two nucleotides. On the basis of our data, the nucleotide sequence in the gene of the large subunit of rbcL was designated as the Ural G-A type, with the naked type A-T type. The hybrid form contains both G-A and A-T types of sequences. Although the sequence of the rbcL gene is species specific, in hybrid forms it is not the same. The sequence of the rbcL gene cannot be used to determine the degree of kinship of the hybrid forms to one or another type. Therefore, both licorice naked and Ural show species-specific flavonoids, triterpenes, and the nucleotide sequence of the rbcL gene. In a hybrid form, the set of these markers is also specific, although it is not sufficient to assign it to a particular species. The hybrid is closer to the nude externally but is closer to the Urals on the basis of a number of biochemical features, which indicates the possibility of classifying this form as a hybrid between the mountain form and the Urals.

The effect of external factors such as salinity, and thickening of the association also affects these plants and leads to a change in the content of various forms of pigments and peroxidase activity (Table 10.2). Of the chlorophylls, the content of form "a" decreases most strongly. Moreover, in the case of the branch and other associations, this is more active, by 66–68%. The content is reduced by 9% in the association. The activity of peroxidase in leaves under the action of salinity and other unfavorable factors varies. In plants growing with sufficient soil moisture and in a homogeneous population, the activity of enzyme is low. The conditions of mixed populations and salinization cause an increase in the activity. The highest level of activity is characteristic for some plant associations. An analysis of the effect of thickening on the activity of the enzyme shows that low and high thickening reduces the activity level of the enzyme, and increases in some associations.

The peroxidase activity and its level, the chlorophyll content, and carotenoids in different populations are not the same. There is a change in these indicators, depending on the degree of plant density in the population, salinity level in the soil, hydration, and several other factors. In conditions of multifactorial field experiment it is impossible to see the strongest influence of any particular damaging factor of the environment. The absence of data on the rate of change in these indicators in plants of a particular population does not allow us to speak of the degree of disrepair of the population. However, the data obtained indicate the specificity of plant metabolism in various associations.

Currently, there is a very favorable situation for export of licorice abroad, but in recent years its quantity has sharply decreased. The emerging difficulty was from Afghanistan as well as China. The Chinese as a monopolist built a liquorice plant in Urumchi and decreased the cost of dry root. They are selling 1 kg of root for 3 USD whereas Kazakhstan sells 1000 kg for nearly 100 USD. The pharmaceutical companies in southeast Asia desire to buy licorice directly from Kazakhstan, but several reasons hinder an increase in the exports. Most important is the lack of a strategy for the production of natural thickets of licorice. Despite the long history of exploitation of natural populations, there is no multifaceted strategy for its large-scale production. Deep plowing of floodplain meadows and steppe zones leads toward a destruction of biocenoses formed over the centuries. Moreover, the places of harvesting are not properly leveled; pits are left, and soil erosion is the result.

Most of the good-quality licorice grows in the floodplains of the Rivers Ili and Syr Darya. A study of the licorice associations was conducted during 1960–1980. During the past two decades much has changed. The Syr-Darya River was shallow; the water level in the Lake Balkhash fell. Data on previously known licorice stocks, suitable for harvesting, are not dependable. Currently we do not know the general stocks of licorice and, most importantly, their distribution. In this situation, it is necessary first to determine the total reserves of licorice, and then the annual quota. It is also necessary to go for artificial plantings of licorice; the technology has been available for decades, but it is not used. There is one essential element in the cultivation technology for licorice seed production. It is mentioned that only the seeds of local plants should be used, but the quality of seeds and the ability to synthesize glycyrrhizic acid is questionable. It varies from generation to generation, from population to population; the plant is perfectly adapted to arid conditions.

10.2.3 Ferula

The genus *Ferula* has more than 180 species. It is one of the most polymorphic genera of the family Apiaceae. The main range of representatives of this genus is Central Asia and Kazakhstan (Safina 2012). During 1878–2012 much work was done on the species of the genus *Ferula* naturally distributed in Central Asia and Kazakhstan, notably Regel, Fedchenko, Kozo-Polyanskii, Korovin, Baitenov, Pimenov, Safina, Rechinger, Chamberlain, and Kamelin (Mukumov 1993; Safina 2012). The most comprehensive classification of *Ferula* was proposed by Korovin (1939, 1940, 1947) and later modified in the *Flora of the USSR* (Korovin et al. 1951). Korovin distinguished 6 subgenera and 8 groups, some of which, on the basis of differences in the characteristics of leaves and flowers, were further subdivided. Safina and Pimenov (1984) reject the subgenus of Korovin in the "*Ferula* of Kazakhstan," and instead 12 groups are singled out. The results of their phenotype cluster analysis involve 90 species of *Ferula* using 33 morphological, physiological, and phytochemical characteristics. These findings contradict the classification of Korovin (Safina and Pimenov 1984). Camberlin and Rechinger in their edition of the *Flora of Iran* retain only the subgenus of Korovin, but do not refer to any of the 4 new species (Rechinger 1987); this however does not in any way ignore the contributions of Korovin to the taxonomy and botany of the genus *Ferula*. The interpretation of Safina and Pimenov (1984) was subsequently adopted by Pimenov and Klyuykov (2002) in the flora of umbellate plants in Kyrgyzstan.

Fifty-two species of this genus grow in Kazakhstan (Markova and Medvedeva 1965), the highest number in Central Asia (Safina 2012). The main exported species is the stinking *Ferula*, which is widespread in the deserts of Iran and Central Asia, the Kopet-Dag, the small Balkhans, in Kazakhstan: the Embene Plateau, the Western

Melkosopochnik, the Aral Sea, the Moyinkum, the Balkhash-Alakul, the Kyzylkum, Turkestan, the Chu-Ili mountains, in the western part of Kazakhstan, and Karatau, on the Peninsula Mangyshlak, in the Northern and Southern Ustyurt. In these areas, the species is found on sandy habitats, as well as on clayey soils, in the foothill of deserts, on less steep forest slopes, river terraces, along creeks, and fixed and semi-fixed sands; it occurs in the belts of wormwood and saltwort deserts, often as a dominant or co-dominant in the communities (Korovin 1939, 1940, 1947; Markova and Medvedeva 1965; Safina and Pimenov 1984; Baytenov 2001; Safina 2012).

In the natural flora of Mangistau there are four species of *Ferula*: *Ferula foetida* (smelly), *F. dubjanskyi*, *F. lehmannii*, and *F. nuda* (naked). Ecological confinement of the first species to the sandy terrain allows us to class it among the xerophytes (sclerophytes), heliophytes, and psammophytes. It is drought resistant and able to germinate in places with a dry hot climate. Periodic droughts, characteristic of deserts and semi-deserts, do not influence the development of *F. foetida* because of its powerful root system (Safina and Pimenov 1984; Baytenov 2001; Safina 2012). The genus occupies an exceptional place in the flora of Central Asia. The representatives of this genus have wide practical application in the form of fodder, as aromatics, and as essential oil plants. Some species have been used in the form of fuel, honeycombs, and building material, as well as for landscaping for decorative purposes, and as source material for musical instruments. The economical importance of the plants from this genus shows that they contain resinous substances, essential oils, starchy products, and gum and possess medicinal and fodder properties (Markova and Medvedeva 1965; Mukhtubaev 2010; Zubaydova et al. 2014).

The products from *F. foetida* (Fig. 10.3) are used in various industries (chemical, textile, rubber, medicine), and alcohol is obtained from its starch (Mukhtubaev 2010; Kareparamban et al. 2012). It is a perennial, powerful plant up to 1.5–2 m tall. The central trunk is thick, branched at the top, and has large leaves. The underground part is very massive and complex, consisting of a root and a stalk of a cedar (kaduksa); the root is swollen and oval. The caudex is branched, consists of lignified basal parts of monocarpic shoots, metamorphosed hypocotyl, and the basal part of

Fig. 10.3 General appearance of *Ferula foetida* (Bunge) Regel in nature

the root. In the bark of the blossoming stem, schizogenic receptacles are found, in which gum resin accumulates (Kareparamban et al. 2012). At the end of development the cortex slides, which is a distinctive feature of the species. The stinky *Ferula* blooms and fruits only once, and then dies, which is typical for this species. Nutrients accumulate in the root and are directed to the formation of the stem and generative organs. In the first year of vegetation the main root reaches a depth of 30–50 cm; in 3-year plants the root system grows well. As a result, an oval thickening is formed on the main root, which penetrates into the soil to a depth of 2.5 m. Every year in spring the plant produces three or four separate leaves. The full stem, which grows after 5–9 years, carries a powerful inflorescence, a complex umbrella located at the end of the stem. The stem grows very intensively (Mukhtubaev 2010). The generative shoot reaches 20–25 cm; it completes its life cycle in 6 weeks and forms fruits in the form of viscous plants, then dries up. The resin is radically destroyed and the root becomes fibrous after its death. The stem is thick, stocky, in the upper third branched into a thick, spherical panicle; the lower branches are regular, the upper branches are somewhat collected together. Leaves are soft, usually naked on top, always slightly pubescent from below; the basal ones possess short and thick petioles (Mukhtubaev 2010).

Radical leaves are triple, complexly dissected, with a petiole, which in the lower part passes into an inflated vagina. The plate of the leaf is triangular, mesomorphic; their lobes are large, oblong up to 15 cm long and 5 cm wide, following its growth in spring. Stem leaves are smaller with a vagina, the upper leaves are without plate, oval, represented only by flat vaginas, covered with hairs from the outer side. Hairs on the surface of leaves are unicellular, with a folded surface. The vagina of the leaves is swollen, soft, pubescent, and the upper leaves are deflected downward, thinly membranous (Mukhtubaev 2010; Ashena et al. 2014; Imanbayeva et al. 2015).

Inflorescence is in the form of an umbel, located in the form of a panicle at the end of the stem. Each panicle, in turn, terminates with 1 central umbel and 3–6 side umbels on long peduncles. The central umbels have 25–30 rays, 15–20 cm in diameter, with bisexual flowers, lateral with staminate flowers. The rays of the umbel are pubescent, with 12–15 flowering stalks, 3–5 mm long, without wrapper, calyx absent. There are five stamens, the pistil with a lower two cavities, hairy ovary, two elongated posts, and stigmas. Petals are flat, with a blunt apex, not bent inside. Flowers are yellow, showing cross-pollination. Pollen grains have three furrow-pores, oblong-trihedral in form. The ovary is semi-inferior, of two carpels from the surface, pubescent. Fruit is oval, thin, flat, two-seeded or viscous, as two half-fruits; flat, elliptical with 5 filiform ribs, straw-yellow, hairy, notched at the apex, 16–30 mm long. The weight of 1000 seeds (half-fruit) is 49 g; flowering in March, fruiting in April–May. The plants multiply by seeds. The number of chromosomes in *Ferula* stink is $2n = 22$, which is typical for all species of the genus. The plant requires a combination of cold and snowy winters with a warm early spring. However, the abundance of sprouts also depends on the edaphic conditions of the region (Mukhtubaev 2010; Safina 2012; Ashena et al. 2014).

Industrially valuable thickets of *Ferula* occur in the Tuiesu sands. The area is estimated to be around 600 ha with a yield of 3 tons/ha. The operational reserve of underground parts is calculated at the level of 1800 tons; the volume of possible collections is 18 tons. A comparative study of the interspecies and population variability of the genomic DNA of *F. foetida* has been carried out on specimens from natural communities of the desert zone of Mangistau. The genomic DNA of plants was evaluated by collecting samples of two species of the genus *Ferula*, growing in the Mangistau desert: *Ferula foetida* and *Ferula nuda*. Polymerase chain reaction (PCR) evaluation with genomic DNA of plants from genus *Ferula* was undertaken. In all 30 arbitrary decimal primers were tested, differing in the nucleotide sequence and GC composition. Of these, five primers were selected as the most informative, flanking reversed repeats in the genome of plants of the genus *Ferula*. As a result, interspecies and population variability of *Ferula* was determined, depending upon the conditions of growth of species in varying soil and water regimes of the region.

The genomic DNA of *F. foetida* and *F. nuda* revealed that clustering of RAPD (random amplification of polymorphic DNA) spectra shows the population studies form two main clusters, 6 and 10. The distance between the root of cluster 11 and cluster 10 is 37, 42 units. High rates of genetic distances between clusters 6 and 10 indicate that the populations forming these clusters are genetically significantly distant from each other, which indicates that the plants belong to the same genus, but to different species; that is, cluster 10 had one species of *F. foetida*, and cluster 6 *F. nuda*. Cluster 11 combined both species into one cluster, because both species belong to the same genus, *Ferula*. A study was thus conducted of the composition of the essential oil of *Ferula foetida*, which grows in two different ecological populations in the Mangyshlak Peninsula: on loamy plains of the Tynybai Shoka Upland and on Tuyusu Sands. In the samples studied, gas chromatography-mass spectrometry was used. The components identified were 31–47, constituting 88.2–92.5% of the total essential oil. The essential oil in *F. foetida* is a yellow liquid with a specific garlic-scented smell from the presence of sulfide compounds. The essential oil of the two populations of the Tuiesu sands and the west of the Tynimbai Upland are different in terms of the content of some components, depending on the phase of development. For all, the presence of 2,5-diethylthiophene, 3,4-diethylthiophene, bulnesol, guaiol, caryophyllene oxide, β-*trans*-caryophyllene, and myristicine was noted, which possesses acaricidal, antitussive, and expectorant activity, as well as anxiolytic and antidepressant properties (Kareparamban et al. 2012; Zubaydova et al. 2014).

Information on the content and composition of organic acids in the smelly *Ferula* in the literature is lacking, so we used capillary electrophoresis to determine them. In the roots 11 organic acids were identified: oxalic, fumaric, acetic, lactic, benzoic, sorbic, succinic, formic, propionic, citric, and malic. The predominant organic acid is lactic acid. It was established that the most important of the identified organic acids is lactic acid (0.16–2.28%), the lowest is fumaric acid (0.005–0.08%). Oxalic, fumaric, acetic, lactic, sorbic, and succinic acids are found mainly in the roots of plants of Tuiesu populations. Formic, propionic, citric, and malic acids are found in large amounts in the roots of plants from populations west of the Tynimbai Shoka Upland. The benzoic acid content in all three populations is at the same level.

The composition of phenolic compounds in the leaves, seeds, and roots of the smelly *Ferula* from the natural populations of the Mangistau desert was first studied by HPLC. Up to 20 phenolic compounds in the roots and 27 in the aerial organs were detected. Among these, in the above- and underground organs, umbelliferone, chlorogenic, and ferulic acids were identified. In the leaves and seeds of the plant coffee acid, luteolin, diasmetin, and five substances of flavon nature were also detected. The concentrations of chlorogenic acid, umbelliferone, and ferulic acid in the roots are higher than in the aerial organs. The content of the identified phenolic compounds in the roots is higher during the growing season, in contrast to the fruiting period. The number of phenolic compounds in the roots is greatest in populations growing on sandy-loam soils, in average numbers in the populations from sandy massifs, and the lowest in the populations from loamy soil. In the seeds, the smallest number of components is determined in plants growing in the region of the Tynimbai Upland (to the west and south of the elevation).

The results of phytocoenotic descriptions and areas of the identified thickets testify to the possibility of practical exploitation of the natural thickets of smelly *Ferula*. The communities of stinking *Ferula* successfully flourish on different types of soils, from sandy to loamy-stony.

The following types of communities are described: saxaul-ferul-santolinno-sagebrush; Ferul-herbage-saxaul; Kamyanodolinovo-Ferulovoy-herbage-saxaul; saxaul-ferul-teresken; saxaul-santolinol-weed-mixed-herb-ferulic; sedge-wormwood-ferul-saxaul; wormwood-ferul-nanophyte; wormwood-ferul-saxaulchikovoye; wormwood-ferul-shrub; and wormwood-ferul-cereal.

In all the investigated communities the smelly *Ferula* is either dominant or codominant with enough abundance, and good vitality. The coeno-populations can enter both the lower grassy vertical part (in the virgin period), and into the medium shrubby-grassy section (in the breeding period). The most active in most morphological isomers, including the height of the plants and the depth of the roots, the smelly *Ferula* develops in the first 3–4 years. There is a gradual slowdown in their growth after 5–6 years and even termination occurs, regardless of the edaphic conditions of natural populations. The results obtained show and testify the 7-year life cycle of the mangystau smelly *Ferula*. However, according to Tajik botanist Rakhimov Safarbek, in *Ferula* a 17-year life cycle is also possible. However, the difficulty lies in the reliable determination of the age. Analysis of 17 aboveground and 16 underground growth indices covering the entire spectrum of the age interval, from 1 to 7 years, showed the determining effect of habitat conditions on the morphology of the *Ferula*. Stronger habitat conditions reduce the intensity of growth processes. With regard to the exports of *Ferula*, one can note its illegal sales around Turkestan. The underground part of *Ferula* is harvested, whereas juice is prepared in Tajikistan. The culture of harvesting and processing of *Ferula* raw materials in Kazakhstan is still very low. Mainly Afghans are engaged in its harvesting. The presence of large natural reserves of *Ferula* make it promising for future export.

10.2.4 Cistanche

Cistanche is one of the most valuable important plants in the flora of Kazakhstan (Kobayashi et al. 1984; Abdulina 1998; Sarsenbayev et al. 2013). The reason is the high content of various polysaccharides, iridoids, biologically active compounds. In developed countries it is widely used as a raw material for the production of many biologically active compounds of a wide spectrum action such as; increase of energy, in potency, and as an antioxidant (Kobayashi et al. 1984). The genus belongs to the family Orobanchaceae. All the species from this group are perennial root parasites and are completely devoid of chlorophyll. The main features of this family are the peculiar appearance of plants, the absence of real roots, scaly leaves, and very small and numerous seeds with a very weakly reduced embryo. Seeds easily penetrate into the soil layer after rains, and the seedlings come into contact with the roots of the host plants (Sarsenbayev et al. 2013). After reaching the root of the host plant, the tip of the embryonic root of the broomrape is transformed into a haustorium (sucker) and is actively introduced into it by means of proteolytic enzymes. Further development of the sprout appears only after the contact of its haustorial cells with the conducting root system of the host plant. At the site of implantation from the part of the sprout approximately corresponding to the hypocotyl, a tuberous formation, usually called a nodule, develops. The nutrients are stored in this nodule, and fertile shoots as well as secondary haustorium-forming organs are formed to ensure vegetative reproduction of the parasitic plant (Stefanova et al. 2011).

The broomrape family includes 13 genera and about 200 species, distributed widely, but highly unevenly. Nearly 90% of all species occur in Europe, Asia, and northern Africa. These species are especially numerous in Eurasia, from the Canary Islands and the Iberian Peninsula to the Himalayan Mountains. In Kazakhstan this genus is represented by three species: **1**. yellow (*C. flava*), **2.** ts. Solscha (*C. salsa*), **3**. *C. dubious* (*C. ambigua*), a synonym for *Cistanche deserticola* (Abdulina 1998). *Cistanche deserticola* (a synonym for *C. ambigua*) is a perennial herbaceous plant, not cultivated, chlorophilic, glandular-hairy. In Kazakhstan, it is a parasite living on the roots of saxaul (*Haloxylon*), whose forests occupy an area of about 15 million ha.

This parasitic species attaches to the roots of saxaul, zhuzguna, or tamarix and sucks the nutrients. It has a fruit body or stem with flowers. The term root or grass is used in some cases (in an unscientific way) meaning a stem or stalk of the plant. Plants grow in the desert, preferring low elevations between 225 and 1150 m, under the continental climate, on sandy habitats with enough sun.

Cistanche grows in Mongolia, in China in the provinces of Kansu, Tsinghai, Sinkiang, and Xinjiang. The harvesting of plants, mainly *Cistanche deserticola,* is carried out in spring. This species is listed in many Chinese "type" collections in the literature. *C. salsa* can also be collected but it is not as widespread, although it is also used in Chinese medicine. *C. salsa* can be used to produce a yellow pigment. It is a non-green, chlorophilic, glandular-hairy parasite living on the roots of saxaul, with a simple unbranched stem with naked appearance or almost bare scales, which end in a spicate inflorescence.

Saltwater *Cistanche* is a perennial, 10–40 cm in height, more or less hairy, stalk thick, in the middle part 5–20 mm in thickness, inflorescence shortly cylindrical, or cylindrical, sometimes strongly shortened, covering scales ovate or oblong-lanceolate, on the backside and along the margin woolly-hairy, up to 2–3.5 cm in length, bracts linearly oblong, obtuse, almost equal to calyx, woolly at the edge, calyx 9–14 mm long. Corolla obtuse-campanulate, 25–35 mm long, with light yellow tube and purple bend, sometimes all light yellow. Stamens at base hairy, anthers 3–4 mm long, strongly hairy. The stigma is thick, slightly notched; the capsule opens into 2, rarely 3, leaves. It grows on clay solonetz desert steppes; it parasitizes the plants from the genera *Anabasis* and *Salsola*, but prefers *Saxaul*.

The anatomical structure of the parasite and the roots of *Cistanche* and saxaul join into one knot. *Cistanche* is a dubious long-term root parasite (although indicated as an annual in some references). Seeds germinate in the spring, reach the root of the host, penetrate into it, and begin to absorb nutrients, gradually gaining biomass. It hibernates and in the spring, in the presence of sufficient biomass, a flower spike is produced. In bad weather the stolon can be retained in the soil until the next year. It has a short, powerful stem, studded with regular scales, lying in the soil, only the flowering spike is above the surface. As a parasite, the roots are mutated into so-called haustoria, which penetrate into the roots of the host. This function is performed by the primary root of the sprout. At the site of implantation from the part of the sprout approximately corresponding to the hypocotyl, a tuberous formation, usually called a nodule, develops. In the nodule, nutrients are stored, and fertile shoots and secondary haustorium-forming organs are formed to ensure vegetative reproduction of the parasite. In the next year, the formation of an underground stalk takes place. The transverse section of the peduncle is oval, cells of the epidermis are covered with cuticle, elongated, pressed from both sides, and the lateral walls are lignified. Cortical parenchyma is loose, with wide-lobed intercellular spaces. Conductive tissues of the central cylinder are located under the endodermis. As the pedicel is erect, all the vascular bundles are the same and are arranged in a ring. They are poorly developed; only one or two vessels can be seen on the xylem portion, the cambium is absent (procambium), and central core of the pedicel towards the middle is hollow.

In the stem, reduction of leaf blades leads to a significant decrease in the evaporating surface and a weak development of vascular fibrous bundles in *Cistanche*. The stem turns into an organ adapted for the accumulation of water. The outer part of the epidermis is covered with a thick cuticle. There are no mechanical tissues. Cortical parenchyma is permeated with vascular bundles of pedicels. The tissue of stems is uniform and mainly consists of a thin-walled parenchyma, permeated with underdeveloped vascular bundles. The vessels in each bundle are few; all have a narrow cavity and primitive structure, which is the result of the lack of leaves, which stimulates the formation of fibrous vascular bundles of the stem. The core consists of thin-walled parenchyma cells.

The rhizome is a perennial shoot and usually thicker than aboveground part. On the outside it is covered with periderm. As a main function, the storage of reserve substances occurs in the rootstock, and the core develops. It consists of round, thin-walled parenchymal cells, with small intercellular spaces between them. The core in volume and in thickness predominates over other tissues of the rhizome. In the latter, cells with inclusions not found in other *Cistanche* tissues.

The tuber shows a predominance of parenchyma in the nodule over other cells. It is even more distinct than in the rhizome. Just as in the stems of succulent plants, the entire mass of the tuber, with the exception of the integumentary (periderm), is represented by a storage cover, permeated with underdeveloped and primitively constructed bundles of the conducting system.

No significant difference is seen in the anatomical structure of plants from different populations and the two species studied, which is probably associated with the same parasitic way of life.

The vegetation period of the aboveground part is 10–20 days in April–May. After the flowering, the stalk becomes brittle and the plant quickly dies. In the body of a dead stolon, we find many worms. On the surface for a long time, dry stems of *Cistanche* with seeds are visible. It was not clear if this is an annual or a perennial plant. It is impossible to imagine that such a small fruit, seed, and flowering stem can be formed from such a small seed within a short time. The presence of large reserves of *Cistanche* in Kazakhstan makes it a very promising crop, both for domestic use in the food industry and for healthcare. At the same time, quantitative studies on the detection of biochemical, physiological, and genetic features of natural populations of *Cistanche* in Kazakhstan were not been conducted earlier. Of special scientific and practical interest is the study of the population polymorphism of accumulation of BAS in various habitats and regions of Kazakhstan, and the chemical composition of stolons; research on the genotypes most enriched with BAS is needed.

The aboveground and underground parts of the *Cistanche* dubious were studied by us. For this purpose, 30 *Cistanche* plants were taken from different populations in April, near the Moinkum village of the Zhambyl region. The plants were parasitized on the roots of saxaul. The vegetation here is rather monotonous, but the soil where the studied populations grew was not the same in chemical composition. At the same time, samples of *Cistanche* from 18 populations in Mangistau on solonchak were doubtful. Differences between Moinkum and Mangistau populations on biomass accumulation and morphometric parameters revealed that, in Mangistau, unlike Moinkum, the *Cistanche* has low productivity and growth processes. Plants of these species in Mangistau parasitize on the roots of *Tamarix*. It was found that the germination of seeds begins in June; the plants attach to the roots of the host plant and begin to powerfully collect biomass. By the end of the growing season, *Cistanche* collects a sufficient mass for the next spring to form the peduncle. *Cistanche* is at least biannual, rather than an annual plant. It is interesting that, by dying as a *Cistanche* stone, a new plant forms on the site of attachment to the roots of the host. From it, a full-fledged flowering plant is formed the next year. There are great differences in the structure of the stolon between the questionable *Cistanche* of Mangyshlak and Moinkumov. In Moinkum, it is loose, and tastes sweet; in Mangishlak, the structure of the stolon is fibrous.

For 2000 years this plant has been used as an additive in herbal collections for the treatment of various diseases, which was the result of the small reserves of raw materials of this plant in China, Mongolia. The people in the southeast of Kazakhstan used this plant in its pure form for the treatment of various diseases. It was applied in raw form and after removing the peel it resembled a cucumber, although it contained 65% water. Kobayashi et al. (1984) and Sarsenbayev et al. (2013) showed the presence of a variety of high molecular weight organic compounds in the *Cistanche* as a desert plant: alkaloids, iridoids, kankanoside, orrobanchin, 6-methyl indole, 3-methyl-3ethylhexane, 2,6-bis (1,1-dimethylethyl)-4-methyl phenol, heptadecane, 2-methyl-5-propyl nonane, nonadecane, eicosane, and henicosane. The water-soluble fraction contains *N,N*-dimethyl glycine methyl ester, betaine, sitosterol, daucosterol, triacontanol, acteoside, 8-epiloganic acid, stearic acid, 2-nonacosanone, and bis-2-ethyl-hexyl-phthalate. The proven pharmacological actions include stimulation in sexual activity, as well as anticancer activity. The following compounds have been determined; 2′-acetylac teoside; acteoside; bicyclo [2,2,2] oct-5-en-2-ol; 2,6-bis (1,1-dimethylethyl)-4-methyl phenol; bis-2-ethyl-hexyl-phthalate cistachlorin; cistanin; cistanoside B; cistanoside C; cistanoside D; cistanoside E; cistanoside G; cistanoside H; daucosterol; 4,6-dimethyl dodecane; *N,N*-dimethyl glycine methyl ester; 3,6-dimethyl-undecane; (2,5-dioxo-4-imidazolidinyl) carbamic acid; echinacoside; eicosane; 8-epiloganic acid; geniposidic acid; laxative; heneicosane, heneicosanic acid; heptadecane; leonuride; liriodendrin; 2-methyl-5-propyl nonane; 3-methyl-3-ethylhexane; 6-methyl indole; 2-nonacosanone; *n*-nonadecane; phenylalanine; β-sitosterol; stearic acid; succinic acid; *n*-triacontanol; and tubuloside B.

The evidence regarding the presence of physiological activity in the extracts of *Cistanche* is not great. Dong et al. (2007) found an increase in immunity in humans after taking two water-soluble polysaccharides from *Cistanche.* The effect of reducing the rate of aging in mice after treatment with phenylethanoid glycosides of *Cistanche* has been studied by Xian and Liu (2008). Studies by Wu et al. (2011) and by Byeon et al. (2012) on extracts and polysaccharides from licorice and ginseng have shown that they activate rodent macrophages through activation of such transcription factors as NF-κB and AP-1 and enzymes ERK and JNK. In addition, the stimulation of stress proteins is possible. Similar action has also been observed with the administration of *Cistanche*. The intake of *Cistanche* powder inhibits the aging of rats, the formation of cataracts, and the noninflammatory lesion of the retina of the eyeball (retinopathy). Similar to most desert plants, *Cistanche* has many biologically active compounds. The physiological effect in *Cistanche* is similar to that of mesophytic ginseng. It appears to be related to a group of compounds. However, it can be said with certainty that after identification and study of individual components, the *Cistanche* itself can act as a medicament, and not with the composition of herbal compositions. The positive side in this connection is medicament is natural not artificial. The investigations carried out by us with sufficient reliability showed the physiological activity of powders, tinctures, and extracts from *Cistanche*.

In a 2016 review of the chemical composition of *Cistanche* stolons, it is noted that the component composition of *Cistanche* has been actively studied from 1980 onward. Among the nonvolatile *Cistanche* compounds, more than 100 have been currently isolated and identified. These compounds are mainly PhGs (phenylethanoid glycosides), iridoids, lignans, alditols, oligosaccharides, and polysaccharides. PhG were well studied as an important class of Herba *Cistanche* compounds. To date, in cystange 34, PhG compounds have been successfully isolated, including 22 disaccharide glycoside, 10 trisaccharide glycosides, and 2 monosaccharide glycosides (Collective Authors 2016).

Fundamental research devoted to an objective assessment of the therapeutic potential of raw materials from this plant is negligible; nevertheless, *Cistanche* is used in the composition of herbal preparations in China, Korea, and America. In culture on the roots of saxaul, the *Cistanche* does not grow in China, and in Russia it has long been included in the Red Book. Kazakhstan has sufficient and renewable reserves, but *Cistanche* populations are practically not used, although industrial harvesting has been carried out for several decades. More than 200 tons of dried stolons are illegally exported annually (the ratio of dry raw material to raw material is 1:8). In 2009, in connection with the creation of the Customs Union, the export of the roots of the desert *Cistanche* was banned. Export of products from *Cistanche* is permitted, but this is not available in Kazakhstan; this can lead to a complete halt in exports, that is, because of the lack of processing technologies, the country does not receive profit from the sale of this valuable plant in the flora of Kazakhstan. As per Jiang et al. (2009) *Cistanche* becomes extremely fashionable and promising in pharmacology. Kazakhstan possesses enormous resources of this plant and must rationally use this property with profit and benefit to the people.

In traditional Chinese, Korean, and Japanese medicines, *Cistanche deserticola* has been widely used for the past 2000 years (Li et al. 2016). Modern studies have shown that this species has wide therapeutic functions, especially when using hormones, aperitifs, immunomodulating, neuroprotective, antioxidant, antiapoptotic, antinociceptive, antiinflammatory, fractures, and bone formation for regulation (Choi et al. 2011; Li et al. 2016).

According to the Pharmacopoeia of China, Herba *Cistanche* refers only to the species *Cistanche deserticola*. However, other species of *Cistanche* (*salsa*, *tubulosa*, etc.) are also sold as Herba *Cistanche*. The most valuable is only *C. deserticola*, which is included in the Red Book in the Russian Federation and China and is protected by the Convention on International Trade in Endangered Species of Wild Fauna and Flora (CITES). Annex II. The annual demand in China for *Cistanche* since 1995 has increased much; the estimated demand was 550 tons, and in 2002 3500 tons, all because of the improvement of living standards in China and the concurrent increase in demand for tonics (Tan et al. 2004). As for the external demand for *Cistanche*, it was estimated at 1000 tons per year (Tan et al. 2004).

Cistanche exports from Kazakhstan continue, and according to the Forestry Committee of the Ministry of Agriculture, the license for export of *Cistanche* from Kazakhstan in 2015 was issued for 390 tons of dried stems. In 2014 it was 240.7 tons, and in 2013 it was 375.9. The export data are not easy to find, because the international export code HS1992–1211908500, used for the export of *Cistanche* from Kazakhstan, is generalized to plants and parts thereof, pharmacology, perfume, and insecticides included.

There is an inventory of raw materials for *Cistanche* around the village of Bakanas. On the territory of the Forest area, with an area of 3333.4 hectares, there are 840.0 tons of raw materials. According to unofficial data, only 100–150 tons of fresh raw materials were harvested in the territory of Akzhar in 2012, which implies a high level of depletion of the thickets of the plant. Collection of *Cistanche* adversely affects the environment, leading to rapid depletion of plants, followed by soil erosion, desertification, and damage to *Haloxylon ammodendron* and *H. persicum*, commonly known as saxaul forests. In Kazakhstan, forests cover 4.6% (or 12.3 million ha) of the territory, and the saxaul forests account for almost half the forest area. The collectors leave "traces" in the form of deep depressions or pits that need to be filled to prevent the drying of the roots of such host plants as *Haloxylon*, *Calligonum*, and *Tamarix*.

10.3 Conclusions

The biodiversity of Kazakhstan in general and plant diversity in particular is very rich. This rich diversity is attracting more importance because of the presence of a large number of endemics in its flora. The species of *Haloxylon* are systematically cut for tasty meat-grilling. If this activity continues, in the near future this genus will be found only in nature reserves. One of the hardy elm species possesses 20-m-long deep-penetrating roots and is useful as a windbreak in areas sensitive to erosion. The species of *Tamarix* flourish on large areas with showy flowers. The *Ephedra* species together with sand-thorn bushes survive on poor soil cover. The charming berries of the latter adorn the yellow-brown landscape. A large number of bulbous plants are distributed in the steppes, flowering during the months of April and May after the soils become wet after the water from snow melting reaches these habitats. In addition to the small steppe tulips, crocuses, and anemones, one also comes across medicinally valuable *Cistanche* and *Ferula* species. During May, the common poppy is distributed all over the vast steppe like a purple carpet. The high mountainous slopes are studded with elegant Tien Shan spruce and the valleys are covered with juniper woods, whereas high-lying pine tree forests occur on the Ertis. Wild apple, pear, cherry, and apricot trees are generally distributed at the lower elevations of the mountains. Species of marigold, primrose, edelweiss, and gentian are distributed in the alpine meadows below the glaciers. The alpine zone of Tien Shan mountains has twice the number of plant species as compared to the Alps. The riverbanks are covered by Torgay woods including Turanga poplar, ash, tamarisk, and reeds. A mixed willow and ash wood from the last Ice Age has been preserved. Most impressive are the following taxa: *Tulipa*, *Stipa*, *Allium*, *Jurinea*, *Ferula*, *Oxytropis*, *Astragalus*, *Adonis*, and others (Republic of Kazakhstan 2009). All these need full protection. Currently, natural vegetation is facing a significant change in species composition in natural cenoses. There is a reduction in the area of medicinal plants, and wild stocks of raw materials are decreasing (Shtephan 2012). The system of conservation of biodiversity including both the flora as well as fauna in the Republic of Kazakhstan needs sufficient protected areas such as national natural reserves, national parks, and state natural reserves. The total area of specially protected areas is currently 5678.7 ha, which includes 10 state nature reserves, 1612 thousand ha; 11 State National Parks, 2249.3 thousand ha; and four state natural reserves, 1817.4 thousand ha (Nyssanbayev 2013). However, there is an urgent need to increase the area of these protected places to save the impressive nature of the country.

Acknowledgments My special thanks go to my Ph.D. student M. Sagindikova for providing valuable data.

References

Abdulina SA (1998) In: Kamelin RV (ed) List of vascular plants of Kazakhstan. Academy of Sciences of Kazakhstan, Almaty, p 187

Altay V, Karahan F, Ozturk M, Hakeem KR, Ilhan E, Erayman M (2016) Molecular and ecological investigations on the wild populations of *Glycyrrhiza* L. taxa distributed in the East Mediterranean area of Turkey. J Plant Res 129(6):1021–1032

Ashena F, Jafari A, Shahrokhabady KN (2014) Comparative anatomical study on *Ferula* L. species in NE Iran. Greener J Biol Sci 4(4):103–110

Baytenov MS (2001) Flora of Kazakhstan. T.2. Gylym, Almaty, p 280s

Byeon SE, Lee J, Kim JH, Yang WS, Kwak YS, Kim SY, Choung ES, Rhee MH, Cho JY (2012) Molecular mechanism of macrophage activation by red ginseng acidic polysaccharide from Korean red ginseng. Mediators Inflamm 2012:732860

Choi JG, Moon M, Jeong HU, Kim MC, Kim SY, Oh MS (2011) Cistanches Herba enhances learning and memory by inducing nerve growth factor. Behav Brain Res 216(2):652–658

Collective Authors (1969) Illustrated determinant of plants in Kazakhstan. Alma-Ata. T.1: 40–41

Collective Authors (1980) Atlas of areals and resources of medicinal plants in the USSR. M, C 334–335

Collective Authors (2016) Biomedical industry of the Republic of Kazakhstan. RFCA Ratings

Dong Q, Yao J, Fang JN (2007) Structural characterization and immunological activity of two cold-water extractable polysaccharides from *Cistanche deserticola* Y.C. Ma. Carbohydr Res 342:1343–1349

Egeubaeva R (2002) Essential oil plants in South-East Kazakhstan and path management. Almaty

Egeubaeva RA, Bizhanova GK, Aydarbaeva DK, Kuzmin EV (2006) Distribution and reserves of ephedra horsetail in Almaty region. In: Proceedings of the State National Natural Park "Altyn-Emel". Вып.1.РИЦ "Asia" Almaty, 2006, pp 62–66

Gubanov A, Sinitsin GS (1966) Biological and ecological features of ephedra horsetail. Tr Inst Botany. AN KazSSR. C., pp 1–20

Imanbayeva AA, Sarsenbayev KN, Sagyndykova MS (2015) Anatomical organization of above and underground organs of *Ferula foetida* (Bunge) Regel in Mangistau natural populations. Contemp Prob Ecol 8(6):743–753

Ivanov A, Buhteeva A (1981) Most important forage plants in Kazakhstan. Works Appl Bot Genet Breed 54(2):3–17

Jiang Y, Li SP, Wang YT, Chen XJ, Tu PF (2009) Differentiation of Herba *Cistanches* by fingerprint with high-performance liquid chromatography-diode array detection-mass spectrometry. J Chromatogr A 1216(11):2156–2162
Karahan F, Avsar C, Ozyigit I, Berber I (2016) Antimicrobial and antioxidant activities of medicinal plant *Glycyrrhiza glabra* var. *glandulifera* from different habitats. Biotechnol Biotechnol Equip 30(4):797–804
Kareparamban JA, Nikam PH, Jadhav AP, Kadam VJ (2012) *Ferula foetida* "Hing": a review. Res J Pharmaceut Biol Chem Sci 3(2):775–786
Kleptsova I, Volkotrub L, Karavayev N (2001) Features of technogenic pollution of medicinal plants. Pharmacy [Farmaciya] 1:28–29 (in Russian)
Kobayashi H, Karasawa H, Miyase T, Fukushima S (1984) Studies on the constituents of Cistanches Herba. II: Isolation and structures of new iridoids, cistanin and cistachlorin. Chem Pharm Bull 32(5):1729–1734
Korovin EP (1939) The main lines in the taxonomy of the genus *Ferula* (Tourn) L. Bul. MOIP., Dep. Biol. T.48., vol. 5–6, C 5–15
Korovin EP (1940) The differentiating role of conditions of existence in the evolution of plants. The genus *Ferula* L. Plant Environ T1:237–274
Korovin EP (1947) Illustrated monograph of the genus *Ferula* (Tourn.) L. Izd-vo AN UzSSR, Tashkent, p 91
Korovin EP, Koroleva KM, Krishtofovich AN, Mandenova IP, Poyarkova AI, Shishkin BK (1951) Flora of the USSR. M.-L.: Izd. AN SSSR. T.17. 390 sec
Kukenov M (1996) Medicinal plants in Kazakhstan and their uses. Almaty
Kurkin K (1998) The relationship of plants in the meadow plant communities: characteristics, types, mechanisms. Ekologiya 6:419–423
Li Z, Lin H, Gu L, Gao J, Tzeng CM (2016) Herba Cistanche (Rou Cong-Rong): one of the best pharmaceutical gifts of traditional Chinese medicine. Front Pharmacol 7:41
Markova LP, Medvedeva LI (1965) Materials on the distribution, intraspecies variability and economic value of some species of *Ferula* from the subgenus *Peucedanoides* (Boiss.) Korov. Regetative Raw Mater 13:42–99
Mukhtubaev SK (2010) On the current trends in the use of *Ferula foetida* L. in southern Kazakhstan. Vestn Pavlod Gos Univ. Chem Biol 1:87–91
Mukumov IU (1993) Resourceological characteristics of some species of the genus *Ferula* L. sources of biologically active compounds, Tashkent. PhD Thesis
Nyssanbayev EN (2013) Current state of forestry in Kazakhstan and the main priorities of further development. Materials of the International Scientific and Practical Conference, dedicated to the 55 anniversary of the Kazakh Research Institute of Forestry. Kazakh Agro Technical University, Shchuchinsk, October, vol 12–13, pp 4–7
Ozturk M, Secmen O, Gork G (eds) (1996a) Plant life in southwest and Central Asia, vol 1. Ege University Press, Izmir, 499 p
Ozturk M, Secmen O, Gork G (eds) (1996b) Plant life in southwest and Central Asia, vol 2. Ege University Press, Izmir, 1093 p
Ozturk M, Altay V, Hakeem KR, Akçiçek E (2017a) Liquorice: from botany to phytochemistry. Springer International, Cham
Ozturk M, Altay V, Karahan F (2017b) Studies on trace elements distributed in *Glycyrrhiza* taxa in Hatay-Turkey. Int J Plant Environ 3(02)
Pimenov MG, Klyuykov EV (2002) Umbelliferae of Kyrgyzstan, 288s. ISBN 5-87317-113-0. M. KMC Science Press
Rechinger KH (1987) Flora Iranica. Academische Druck und Verlagsanstalt, Austria T. 162. 426 c
Republic of Kazakhstan (2009) The fourth National Report on progress in implementation of the convention on biological diversity. Ministry of Environmental Protection, Astana, 91 p
Safina LK (2012) Ferul of Central Asia and Kazakhstan (carpenter's review). T.18 (3). Almaty, p 243e
Safina LK, Pimenov MG (1984) *Ferula* of Kazakhstan. Science, Alma-Ata, 200 p

Sarsenbaev KN, Polymbetova FA (1986) The role of enzymes in plant resistance. Nauka, Almaty, p 184

Sarsenbaev KN, Kozhamzharova LS, Bekbayeva LK, Misni S, Tuymebayeva BE (2013) Genetic signs of interspecific polymorphism of *Ephedra* L. species in Kazakhstan flora. World Appl Sci J 21(3):428–432

Sarsenbayev KN, Sarsenbayeva AB, Balmukhanov K (2013) Chemical composition of low-molecular weight organic compounds (LMWOC) of water extracts from *Cistanche deserticola* stolones depending on treatment. World Appl Sci J 25(1):28–35

Shaushekov Z (2004) Botanical study of the floristic regions of Central Kazakhstan and introductions of species. Karaganda

Shtefan G (2001) Conservation and utilization of genetic diversity of medicinal plants in northern Kazakhstan. Herald Agric Sci Kazakhstan 11:23–25

Shtefan G (2004) Introduction of medicinal plants in the north of Kazakhstan. International Conference: Adaptive strategy of breeding field crops due to global climate change. Saratov 3:81–84

Shtephan G (2012) Wild medical plants in the phytocenoses of the northern Kazakhstan. Med Health Sci J 13:128–132

Sokolov N, Zamotaev J (1991) Handbook of medicinal plants (phytotherapy). Kazakhstan

Stefanova NA, Fursova AZ, Sarsenbaev KN, Kolosova NG (2011) Effects of *Cistanche deserticola* on behavior and signs of cataract and retinopathy in senescence-accelerated OXYS rats. J Ethnopharmacol 138(2):624–632

Tan DY, Guo QS, Wang CL (2004) Effects of the parasite plant (*Cistanche deserticola*) on growth and biomass of the host plant (*Haloxylon ammodendron*). For Res 17:472–478

Tkachenko K (2006) Wild medicinal, fodder and edible plants in the gene pool of useful plants. Conference on the role of botanic gardens in the conservation of biodiversity of flora in Asiatic Russia: present and future. Novosibirsk, pp 284–285

Urazaliev R, Kenenbaev S, Esimbekova M (2004) Plant genetic resources in Kazakhstan: status and prospects, international conference key directions of development of agricultural science in Kazakhstan: breeding, biotechnology, plant genetics. Almaty, pp 45–50

Vavilov N (1926) Centres of origin of cultivated plants. Works Appl Bot Genet Breed 16(2):1–248

Vavilov N (1965) Selected works, vol 5. Selkhozizdat, Moscow

Wu TY, Khor TO, Saw CLL, Loh SC, Chen AI, Lim SS, Park JHY, Cai L, Kong ANT (2011) Anti-inflammatory/anti-oxidative stress activities and differential regulation of Nrf2-mediated genes by non-polar fractions of tea *Chrysanthemum zawadskii* and licorice *Glycyrrhiza uralensis*. AAPS J 13(1):1–13 www.fito.nnov.ru/special/alkaloids/alkaloids.phtml

Xian GD, Liu CQ (2008) Research on the effect of phenylethanoid glycosides (PEG) of the *Cystanche deserticola* on anti-aging in an old age induced by D-galactose. Zhong Yao Cai 31:1385–1388

Zubaydova TM, Jamshedov JN, Isupov SJ, Zagrebelny IA, Davlatkadamov SM, Sodikov J, Sukhrobov PSh (2014) On the pharmacological study of different species of the genus *Ferula* L. in medicine of the twentieth century. Bulletin of the Tajik National University (scientific journal). Series of natural sciences no. 1/3 (134). Dushanbe, Sino, p 225

Chapter 11
Causes and Impacts of Land Degradation and Desertification: Case Study from Kazakhstan

Aigul Tokbergenova, Gulnara Nyussupova, Mehmet Arslan, and Shnar Kairova Lyazat Kiyassova

Abstract The problem of desertification is a serious threat that is causing a variety of negative social and economic effects. It is always accompanied by soil degradation as the result of human overuse of the land. During the past few decades this process has accelerated because of population growth, and the forecast is that this growth will continue to increase, which necessitates increased food production. Estimations depict that nearly 50,000–70,000 km^2 of fertile land becomes unusable annually. The main reason for this disastrous phenomenon is desertification. Natural factors are contributing toward the desertification processes in Kazakhstan as well, as it is a landlocked country. The country experiences a continental and arid climate together with the scarcity and uneven distribution of water resources. All these factors are responsible for the spread of sandy habitats and saline soils. The land degradation processes are enhanced by the seasonal changes in the characteristics of soils as well as impacts from drought. Nearly 75% of the countries are exposed to a high risk of ecological destabilization, and Kazakhstan is no exception. Natural habitats in the country show a weak environmental resistance to anthropogenic influences. The degradation is most severe in the pastures adjacent to the rural settlements, milking machines, wells, and distant pasturing territories. Several workers have been studying this phenomenon to identify the agricultural lands most vulnerable to desertification. These workers have proposed measures for the prevention of desertification. Some of the key measures are forest plantation and a sustainable use of pasture ecosystems through conservation programs for restoring the functional integrity of steppe ecosystems.

Keywords Ecosystems · Pastures · Drought · Land degradation · Desertification

A. Tokbergenova (✉) · G. Nyussupova · M. Arslan · S. K. L. Kiyassova
Department of Geography, Land Management and Cadastre, Al-Farabi Kazakh National University, Almaty, Kazakhstan

D. Egamberdieva, M. Öztürk (eds.), *Vegetation of Central Asia and Environs*,
https://doi.org/10.1007/978-3-319-99728-5_11

11.1 Introduction

One of the most serious problems of our time is land degradation or desertification. According to the terminology used widely by international organizations, "desertification" means land degradation in arid, semiarid, and dry sub-humid areas that results from various factors, including climatic variations and human activities. It is significantly different from the observed phenomenon of cyclic fluctuations in the bioproductivity of plants on the border of the desert related to the enlargement or reduction of the desert: the latter is shown by satellite data and is linked to variations in climate.

Agricultural activities have a tremendous impact on the ecological balance of the Earth's surface. The limited natural potential of land used for agricultural production leads to the expansion and development of steppes, which in turn disrupt the natural balance, increase degradation processes, and reduce soil fertility. Efforts to improve the agro-ecological condition of the land are based on the use of man-made technologies, adding to the destruction of land resources, and pollution of our air, water resources, and forests. Water erosion and the subsidence of soils develop rapidly and add to desertification. Production decreases, and land quality is reduced.

Land degradation and desertification are global phenomena of our age, concerning most countries around the world. International organizations [Food and Agriculture Organization (FAO), United Nations Environment Programme (UNEP), International Council for Research in Agroforestry (ICRAF)] are paying great attention to measures to combat this dangerous phenomenon. Active and sometimes irrational human activities in arid regions, which occupy about 30% of the land area, pose a real threat to ecological balance (Abdieva et al. 2013).

11.2 Study Area

The land resources of Kazakhstan are vast, but from the 1960s until now the territory has been exposed to desertification, increasing 10–12%. Earlier, mostly arid and sub-arid areas, generally semi-desert and desert zones, and areas of intensive economic use were exposed to desertification. Today, desertification has moved to the north toward the main grain sowing areas of Kazakhstan, which are occupied by the forest-steppe and steppe zones.

In view of this, attempts were made to collect information based on work supported by statistics. Stock materials and data on land degradation and desertification problems were pooled from the Institute of Geography, Statistics Agency, Committee of Land Resources Management, Department of Environmental Monitoring, and the Ministry of Agriculture, as well as monographs, books, and scientific publications. The "Ratification of UN Convention to Combat Desertification affiliated to the Program to combat desertification in the Republic of Kazakhstan for 2005–2015," a National Action Program and Strategy to Combat Desertification, and other documents have been recently adopted. These issues have been investigated by Chupahin, Chigarkin, Vilesov, Medeu, Mamutov, Borovskyi, Akiyanova, Skorinceva, and others.

The problem of desertification and degradation of lands was evaluated by following such methods as comparative geographic, cartographical, and spatial analysis, together with economic and statistical analysis.

11.3 Observations

The Convention to Combat Desertification in Kazakhstan was signed in 1994 and ratified in 1997. It included a number of commitments; application of integrated approach to the planning and the implementation of activities to combat desertification and its relationship with the fight against poverty; promotion of awareness and participation of the local population; encouraging the exchange of information; the transmission, acquisition, and adaptation of technologies to combat desertification.

According to the data pooled up to November 1, 2015, including information on the structure of land in the country, the area of reserve land is around 100.1 million ha (38.3%) and agricultural land comprises 100.8 million ha (38.6%) (Fig. 11.1).

During the process of reformation of agricultural enterprises in 1991–2005, the area of agricultural land in the republic decreased by 136.2 million ha. In subsequent years the area increased every year and the total increase from 2005 to 2015 was around 18.6 million ha (Summary analytical report on the state and use of the lands of the Republic of Kazakhstan for 2015) (Astana 2016). Despite the fact that part of the agricultural land has been transferred to a variety of nonagricultural purposes, this category of land in 2015 compared to 2014 has increased to 2.2 million ha, mainly because of the development of reserve lands (Fig. 11.2).

The main increase in 20 years has taken place in Aktobe, East Kazakhstan, Karaganda, Kostanay, and North Kazakhstan oblasts. At the same time, because of

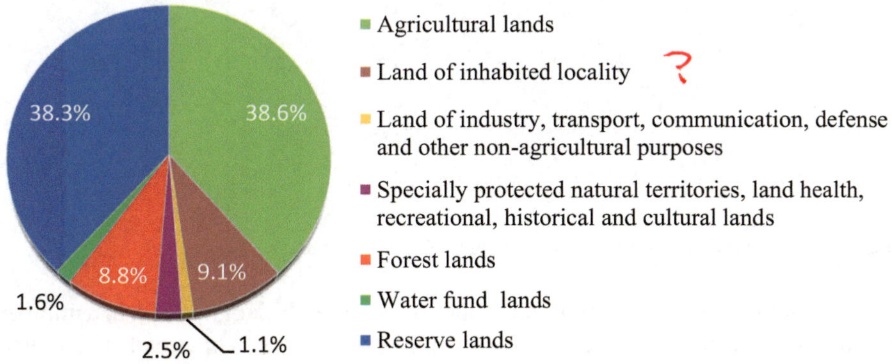

Fig. 11.1 Land resources of Kazakhstan (1 Nov 2015)

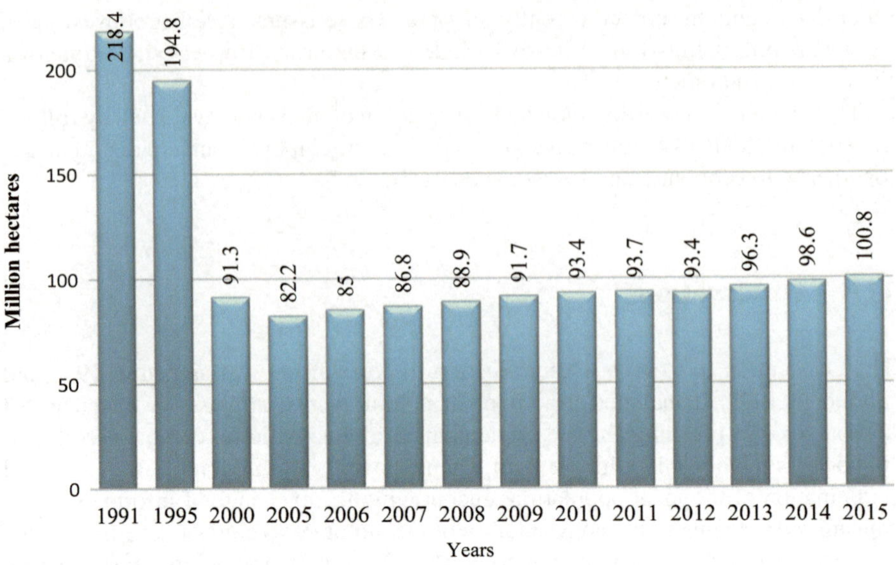

Fig. 11.2 Agricultural land area changes in Kazakhstan

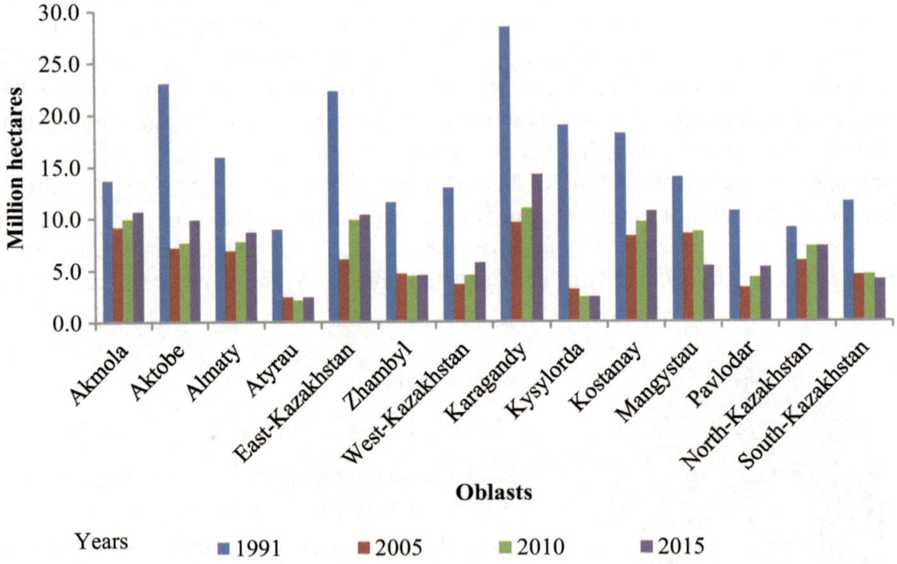

Fig. 11.3 Agricultural land area changes in the oblasts of Kazakhstan

non-use of the land for agricultural production and their lack of demand in a number of areas these lands have been converted into reserve lands. This trend is observed in Atyrau, Zhambyl, Kyzylorda, and South Kazakhstan oblasts (Fig. 11.3). The proportion of land in the reserve land resources of the republic as a whole is 38.3%. In

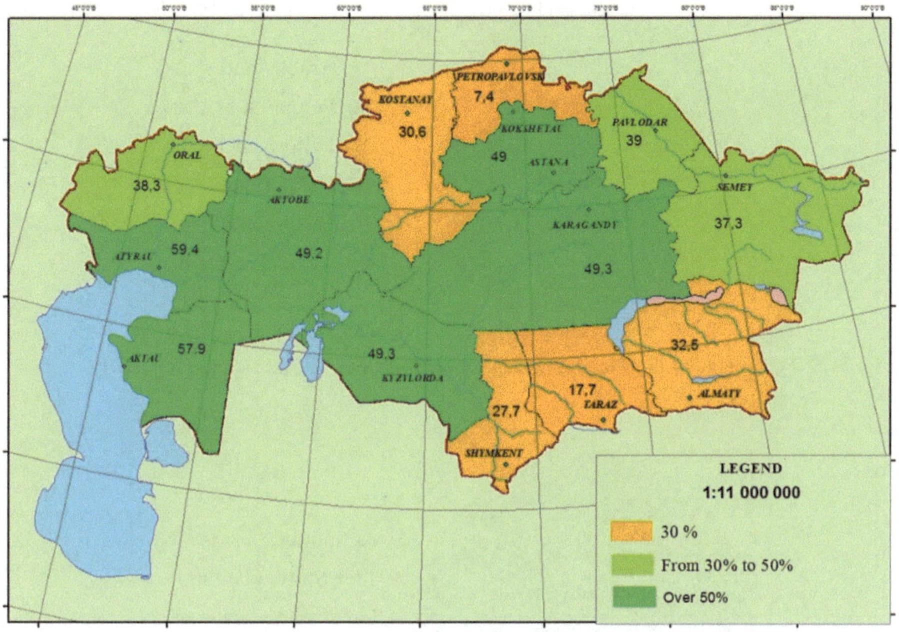

Fig. 11.4 Reserve land percentages in the oblasts of Kazakhstan (1 Nov 2015)

the structure of land reserves in oblasts, it ranges from 7.4%, in North Kazakhstan, to 58.4% in Atyrau oblast (Fig. 11.4).

A total of 26.5% of the pasture area is degraded, which has been happening for more than 10 years, but these pastures continue to be used and should be removed from this use (Fig. 11.5). A general trend toward deterioration of pasture condition is still present, but there are successful examples of reduction in degradation resulting from back-circulation of a number of pasturing areas and dispersal of livestock from settlements. Currently, the state program to restore the watering wells on pasturing areas with equity participation of land users has been launched, which will organize turnover in pastures and mitigate impacts on biodiversity. The degradation of pasture areas increases the effect of extending the processes of desertification, accelerated by the the climate change already in progress. Climate change makes winters warmer, the distribution of rainfall throughout the year more uneven, and summers drier. In mountainous areas, the risk of natural disaster floods is increasing, which can also cause local destruction of existing landscapes and ecosystems (http://www.zakon.kz/4746870-kazgidromet-na-planete-ozhidaetsja.html).

The total area of the land resources of Kazakhstan is around 272.5 million ha: the distribution is shown in Fig. 11.6. In the structure of farmlands, arable farmland is 24,934.7 thousand ha (11.2%), including irrigated lands, —1597.0 thousand ha (0.7%); fallow land, 4798.4 thousand ha (2.2%); and hayfields, 5131.1 thousand ha (2.3%). Natural pastures include 186,526.6 thousand ha (84.2%). These categories are dominated mostly by desert and semi-desert types. As a result of the loss of land,

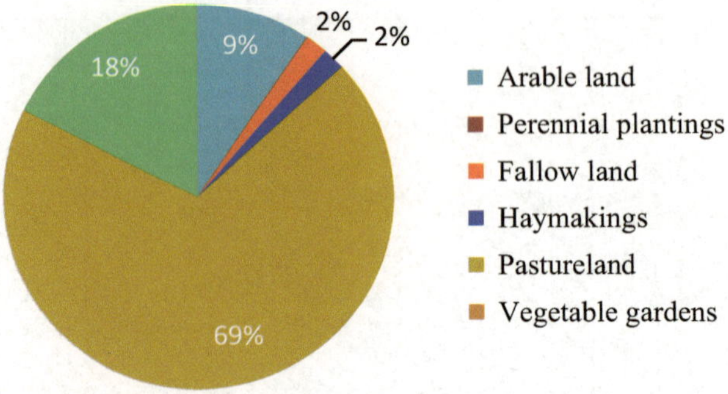

Fig. 11.5 Percentage of farm land in the agricultural land of Kazakhstan (1 Nov 2014)

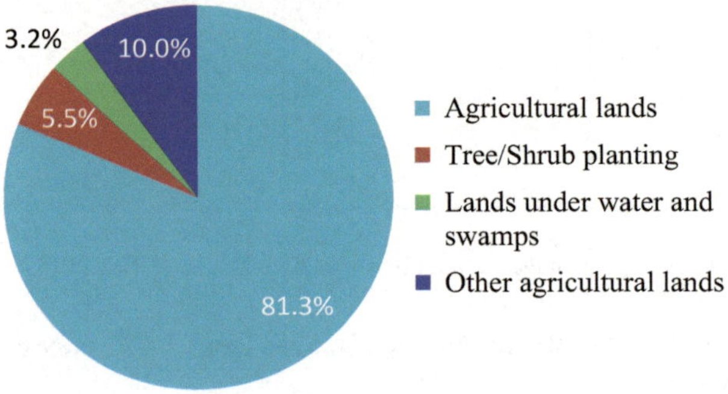

Fig. 11.6 Percentage of land resources in Kazakhstan (1 Nov 2015)

the area under major crops has been reduced from 35.21 million ha in 1990 to 21.5 million ha in 2015, including grain production, from 23.4 to 15.2 million ha, respectively. Approximately 83.0 million ha of land has been transferred into reserve lands; 10.2 million ha into part of forest land, and 16 million ha into the category of settlement lands for pasture use. Up to 10 million ha of arable land has been transferred into fallow land as a consequence of low productivity. The process of natural recovery of these lands without reclamation will require at least 20–30 years.

Erosion is one of the most dangerous types of land degradation, causing destruction of soil and blowing off the top layer of the humus-accumulative horizon, with loss of the land fertility. In many cases, erosion arises and develops as the result of anthropogenic impacts. Along with the erosion of soil, the dehumification of soil is the most common of all types of degradation in the country. Erosion causes huge economic and environmental damages, as it threatens the very existence of the soil as the main means of agricultural production and the independent component of the biosphere. The development of erosion processes is caused by both a suite of natural conditions (climate, landscape, soil texture) and the degree of human impact on the land, including the intensity of land use, primarily for agriculture.

Depending on the soil destruction and loss of its fertility, these lands suffer from water and wind erosion, which could be either normal or rapid. The former is always in the presence of any kind of drainage; it develops more slowly than soil formation and does not lead to noticeable changes in the level and shape of the land surface. The latter is faster than soil formation, leads to soil degradation, and is accompanied by a marked change in the landscape. There are also natural and human-induced types of erosion: human-induced erosion is not always rapid, and vice versa (Gusakov 2010).

The published data on the qualitative characteristics of land resources in Kazakhstan reports more than 90 million ha of eroded and erosion-exposed lands. Nearly 29.3 million ha are reported to have actually eroded, and 24.2 million ha (11.3%) of the agricultural land is also exposed to wind erosion. The degree of manifestation of the deflation process of land is subdivided into three subgroups:

- Slightly deflated, including slightly deflated homogeneous soil contours and their complexes moderately, and highly deflated area of 10–30 and 30–50% of sand. The total area is 2.2 million ha (9.1%).
- Moderately deflated, including soil with moderately deflated homogeneous contours, their complexes with medium level, highly deflated from 30 to 50% and 30–50% of sand and sandy soil plain area of light-brown, brown, and gray-brown zones and subzones. The total area is 4.9 million ha (20.2%).
- Highly deflated, including the soil with highly deflated homogeneous contours, their complexes with their predominance, moderately deflated and highly deflated soil complexes from 30 to 50% and all sand. The total area is 17.1 million ha (70.7%).

Eroded lands are one of the largest groups that need reclamation. These groups are adversely affecting the quality of land and its productivity. Wind erosion manifests itself in the form of deflation of sand and automorphic soils, saline soils, and dust storms. Besides natural factors such as light texture, active wind activity, and a few others, anthropogenic factors have a significant role in the development of soil deflation. Uncontrolled grazing, cutting down shrubs, and vehicular traffic on the roads all promote intensification of deflationary processes that alter the structural composition, bulk density, and humus content of the soil, causing land degradation and loss of fertility. The strongest negative impact of wind erosion can be shown during years of drought, when there is an acute shortage of soil moisture. Most active erosion processes are in the vast tracts of sand of Kyzylkum, Moyunkum, large and small Barsukov, Saryishikotrau, in the regions that are in the desert, semi-desert, and steppe zones with carbonate soils and the soils with light mechanical composition.

The main areas of agricultural land exposed to wind erosion are located in Almaty oblast, about 5 million ha, Atyrau and South-Kazakhstan, 3.1 million ha each, Kyzylorda, 2.8 million ha, and Zhambyl and Aktobe, more than 2.0 million ha. The largest proportion of eroded agricultural land (more than 30% of the total area) is located in Almaty, Atyrau, and South Kazakhstan oblasts. The smallest proportion of eroded land (5%) as a part of agricultural land is registered in Akmola,

Karaganda, Kostanay, and North Kazakhstan oblasts. Soils exposed to water erosion (eroded) of the total area of eroded lands cover an area of 4.9 million ha or 2.3% of agricultural land. Water erosion is observed in all regions of the country, and the intensity of its development is affected by the nature of relief (steepness and length of slope, size and shape of the watershed), the amount and intensity of precipitation, the type and texture of the soil, carbonate content, salinity, water permeability, and the nature of the use of land. The largest areas of eroded soils within the resources of agricultural land are located in South Kazakhstan (1.0 million ha), Almaty and Mangistau (0.8 million ha each), and Akmola (0.6 million ha) oblasts (Summary analytical report on the state and use of the lands of the Republic of Kazakhstan for 2015) (Astana 2016).

To fulfill the obligations of the UN Convention to Combat Desertification and to prevent degradation and desertification of land, Kazakhstan is taking certain measures. One of the most effective steps of Kazakhstan in this direction is the implementation of the regional integrated program "Central Asian Countries Initiative for Land Management (CACILM),", aimed at combating desertification and drought in the context of the UNCCD and supporting the productive functions of land resources.

The problem of desertification is of global importance (Ozturk 1995, 1999; Ozturk et al. 1996, 2006, 2011a, b, 2012; Chedlly et al. 2008; Feoli et al. 2003; Nurlu et al. 2008). Nearly 70% of the world's arid lands or nearly 3.6 billion ha has been degraded. In Central Asia, the total area affected by desertification amounts to more than 1073 thousand km^2, which is caused by the extensive use of land and leads to its degradation (Abdullayev 2010). Its main economic impact includes reduced crop yields and pasture productivity, and decrease in livestock and the productivity and export potential of agriculture. Identification of the factors of degradation and desertification of Kazakhstan is an urgent task (Chupahin 2010).

In the absence of cooperation with agricultural enterprises to improve the recovery of agriculture and soil fertility, Kazakhstan by 2025 is expected to lose up to 50% of its farmland because of soil erosion and degradation. The country is characterized by desert, semi-desert, and steppe, and the combination of the arid and continental climate makes the ecosystems vulnerable to desertification and land degradation. Another factor influencing the aggravation of the problem is human activities. Of 272.5 million ha of the country's territory, nearly 180 million ha of land at present is facing desertification, which is 66% of the total amount of land. More than 20 million ha of arable land and 25 million ha of pastures face wind erosion. Water erosion effects 19.2 million ha, and if we add to this man-made desertification caused by industrial activities, the loss of humus in soils, and salinization of irrigated lands, the problem becomes quite serious. During the past 40 years, the humus content in the soil has decreased by 20–30%, and the total damage in the country is estimated at $2.5 billion (http://news.caravan.kz/news).

Major alteration of ecosystems in Kazakhstan took place more than 50 years ago after a massive plowing of steppe and forest steppe zones. The plowing of grassland steppes on the plains has been around 90%. Until the late 1980s there has been rapid degradation on other types of landscape such as pastures suitable for grazing. The restoration of natural ecosystems on fallow lands as well as abandoned pastures began around the 1990s and is continuing strongly: previously abandoned areas have been brought back into the economy. The overgrazing near settlements has increased because of increased livestock. According to the data from 2015, up to 15% of agricultural lands are used inefficiently, about 125 million ha of pastures are not irrigated and not used, and 9 of more than 20 million ha of pasture areas adjacent to the settlements are classified as degraded. Following the lack of adequate control in Northern and Central Kazakhstan, 5.6 million ha of arable land is affected by water erosion and the yield of grain crops has been reduced by 20–30%. Degradation of agricultural lands in 9 of 14 oblasts of Kazakhstan, including grasslands, is 30–50%, and greater in some places.

Degradation impacts are visible on more than 90% of the arable land in the country. The problems of irrational use of land and livestock grazing economy are aggravated by numerous small agro-industrial and livestock units that do not have sufficient resources for a full management of territories (Fifth National Report of the Republic of Kazakhstan on Biological Diversity for 2011.) (Astana 2012). The extent of desertification and land degradation of Kazakhstan is presented in Fig. 11.7. The natural factors of land degradation are sharp continental climate, salt deposits in the alluvial plains, deflation, erosion and mudslides, and salt and dust aerosols from the dried bottom of the Aral Sea.

Industrial sources of land degradation include liquid and solid emissions from industrial plants and the oil and gas sector, emissions of transport and radiation and chemical pollution, waste of military space complex, greenhouse and ozone-depleting gases, waste in the areas of mining, oil, and gas, and construction of linear and point structures not accompanied by remediation activities.

The landlocked position of the country is effective in the arid climate, with scarcity and uneven distribution of water resources and widespread sands (up to 30 million ha) and saline soils (more than 93 million ha). These features are responsible for the poor resistance of the environment to human impact, such as illegal logging, fires, uncontrolled recreation, soil and groundwater contamination (Regional Action Plan on the protection of the environment, approved by the decision of the Interstate Commission on Sustainable Development.) (Astana 2001).

About 43% of the population lives in rural areas, the majority being dependent on income directly or indirectly related to the agricultural sector and land use.

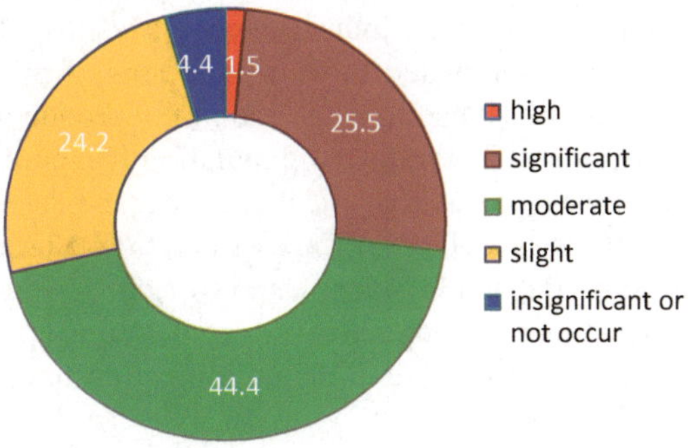

Fig. 11.7 Extent of desertification/land degradation percentages in Kazakhstan

Table 11.1 Key indicators of degradation (desertification) of pastures

Type of desertification	Degree of desertification				Causes of desertification
	Slight	Moderate	High	Very high	
Degradation of vegetation	Signs of degradation in drought years	Reducing the size and productivity of plants, rare plants replacement, appearance of weeds	Replacement of main types of plants by unwanted annuals, productivity reduction, reducing the seasonality of use	Strong thinning and blocking up with unwanted species	Grazing, cutting of shrubs, littering
Soil degradation	Slight surface disturbance, recovery possible	Surface consolidation, salinity increase, drying	Signs of deflation, silt removal	Strong salinity, waterlogging caused by flooding	Grazing, plowing, pollution, the impact of wild animals

Source: Kurochkina et al. (2005)

Kazakhstan ranks sixth in the world in terms of its pasture resources (188 million ha).The total area of degraded pasture lands was more than 48 million ha on January 1, 2010 (nearly 26%) (Resolution of the Government of the Republic of Kazakhstan 2010). There are 180.2 thousand ha of degraded land; the main reasons for pasture degradation are presented in Table 11.1.

The Aral and Ile-Balkash regions are the most susceptible to desertification. Water shortages caused by overregulation of river flow, cessation of flooding, lowering of groundwater levels, and increase in the saline land area has decreased livestock numbers per capita: all these factors have had a negative impact on economic development and the living conditions of the people. Habitat conditions for wild animals and fish have deteriorated. In the Northern and Central Kazakhstan, 5.6 million ha of arable land has been affected by water erosion, and the yield of grain crops has decreased as much as 20–30%. In the Caspian Sea region, 357 thousand ha of fertile coastal pastures and hayfields have become flooded. The land around industrial centers has been contaminated by the emissions from industrial enterprises. About 10 million ha of pasture and arable land has gone out of circulation around the military–industrial complex sites. The damage from desertification in the country is estimated to be around tens of million US dollars. Depletion of water resources has lead to a decline in the production, job cuts have been effective, the standards of living of the population has gone down, and the migration of people from the areas of ecological crisis has increased.

Of the 14 oblasts of Kazakhstan, only 5 (Aktobe, Mangistau, North Kazakhstan, Karaganda, and Kostanay), show negative signs (i.e., desertification) on 30% of forage land. In the other areas, degradation is up to 30–50% and higher. On the arable lands, soil erosion and dehumidification have intensified everywhere. The humus content has decreased by 25–30%. Therefore, soil fertility as well as the productivity of grain crops has decreased; 17 million ha of arable land has been taken out of use as fallow lands and pastures. The fallow lands are covered by tall weeds, adding to an increase in the populations of insects as crop pests. There has been soil salinization, water and wind erosion, reduction of humus, and salinization with discharges of water after irrigation on neighboring territories on more than 90% of the republic's territory. Problems with the irrational use of land and livestock grazing economy have been aggravated by numerous small agro-industrial and livestock units that are not able to provide a cost-effective use, or purchasing equipment, fertilizers, veterinary services for livestock, its relocation, providing insurance of stocks to feed them, and processing of agricultural products. The socioeconomic problems of reorientation of farms are also increasing. Under these conditions, the unreasonable use of high-grade natural land has been enhanced, without taking into account the norms of resource withdrawals, that is, their degradation. In view of all these impacts, the degree of desertification is amplified.

11.4 Concluding Remarks

- Most of the territory of the country is located in a zone very vulnerable to anthropogenic desertification: the environmental situation is thus rapidly deteriorating. The main areas of environmental stress and soil degradation are the Aral and Caspian regions, as well as abandoned land in the northern regions. Wind erosion (deflation) of sandy, light-textured, and carbonate-rich soils is widespread in Kazakhstan (40.4% of agricultural land) and more than 11% of farmland is affected by water erosion.
- Economic development indicators show a steady increase within GDP growth. Positive trends in the industrial and financial sectors of the economy are visible, but these are increasing the pollution load in their surroundings, areas occupied by industrial and domestic wastes together with the amount of wastewater and emissions of toxic pollutants.
- Desertification is accompanied by contamination of soil, groundwater, and surface waters, and finally reduction of the biological potential of the entire region. The main economic impacts of desertification/land degradation include decline in crop yields and crop production, decline in livestock productivity, reduction in the export potential in agriculture, a slower development of food and other small-scale industries, and a sharp decline in tax revenues from processing and agriculture sectors.
- The evaluations presented here have revealed that major economic and social losses caused by land degradation are not connected with a decrease in the volume of total product produced in the region, but with a decrease in its natural potential.

References

Abdieva ZB, Dyusenova GB, Botabekova GT (2013) Human impact on natural ecosystems and desertification/land degradation problems: materials of the Republican scientific-theoretical conference "Seifullin reading IX: a new vector of development of higher education and science" dedicated to the Day of the First President of the Republic of Kazakhstan, vol 1(2), pp 104–105

Abdullayev AK (2010) Problems of land degradation as a result of unsustainable agricultural use and ways to improve the situation. http://www.caresd.net/land/o1.html

Chedlly A, Ozturk M, Ashraf M, Grignon C (eds) (2008) Biosaline agriculture and high salinity tolerance. Birkhauser Verlag (Springer Science), Basel, 367 pp

Chupahin VM (2010) Regional environmental scheme to combat desertification. Nauka, Moscow

Feoli E, Giacomich P, Mignozzi K, Ozturk M, Scimone M (2003) Monitoring desertification risk with an index integrating climatic and remotely-sensed data: an example from the coastal area of Turkey. Manag Environ Qual 14(1):10–21

Fifth National Report of the Republic of Kazakhstan on Biological Diversity for 2011: Informational and Analytical Center of Environmental Protection. Astana, 2012

Gusakov VG (2010) The main directions of improvement of agrarian strategy and policy. Belarus State University Press, Minsk

http://www.zakon.kz/4746870-kazgidromet-na-planete-ozhidaetsja.html

http://news.caravan.kz/news

Kurochkina LY, Karibaeva KN, Baijumanov AB, Mishchenko AB, Toylybaev AJ, Masimov AK, Isabaeva AB, Usen K (2005) Manual for farmers (pasture management). OST-XXI Century, Almaty

Nurlu E, Erdem U, Ozturk M, Guvensen A, Turk T (2008) Landscape, demographic developments, biodiversity and sustainable land use strategy: a case study on Karaburun Peninsula, Izmir, Turkey. In: Petrosillo I et al (eds) Use of landscape sciences for the assessment of environmental security. Springer, The Netherlands, pp 357–368

Ozturk M (1995) Recovery and rehabilitation of Mediterranean type ecosystem—a case study from Turkish maquis. In: Rapport D, Gaudet CL, Calow P (eds) Evaluating and monitoring the health of large-scale ecosystems, NATO-ARW. Springer, Dordrecht, pp 319–332

Ozturk M (1999) Urban ecology and land degradation. In: Farina A (ed) Perspectives in ecology. Backhuys, Leiden, NL, pp 115–120

Ozturk M, Secmen O, Gork G (eds) (1996) Plant life in Southwest and Central Asia, vol II. Ege University Press, Izmir, 1093 pp

Ozturk M, Waisel Y, Khan MA, Gork G (eds) (2006) Biosaline agriculture and salinity tolerance in plants. Birkhauser Verlag (Springer Science), Basel, 205 pp

Ozturk M, Mermut A, Celik A (2011a) Land degradation, urbanisation, land use & environment. NAM S. & T., Delhi, 445 pp

Ozturk M, Okmen M, Guvensen A, Celik A, Gucel S (2011b) Land degradation, urbanisation and biodiversity in the Gediz Basin-Turkiye. In: Ozturk M et al (eds) Urbanisation, land use, land degradation and environment: NAM proceedings. Daya Publishing, Delhi, pp 74–93

Ozturk M, Kebapci U, Gucel S, Cetin E, Altundag E (2012) Biodiversity & land degradation in the lower Euphrates subregion of Turkey. J Environ Biol 33:311–323

Regional Action Plan on the protection of the environment, approved by the decision of the Interstate Commission on Sustainable Development on April 12, 2001. http://www.ecoportalca.kz/temy/degradaciya_zemel/

Resolution of the Government of the Republic of Kazakhstan dated by September 10, 2010 on approval of sectoral Program "Zhasyl Damu" for 2010–2014

Summary analytical report on the state and use of the lands of the Republic of Kazakhstan for 2015. Astana, 2016, pp 80–139

Chapter 12
Assessment of the Current Plant Diversity Status in Kazakhstan

Maulken Askarova, Alikhan Medeu, Akhmetkal Medeu, and Mehmet Arslan

Abstract Biological diversity is a main source of satisfying the human needs and serves as the basis for adaptation to changing environmental conditions. The practical value of biodiversity lies in the fact that it is an inexhaustible source of biological resources. At present, the number of species on the Earth is a rapidly decreasing. Among the main environmental issues, reduction of biodiversity takes a special place. This chapter presents an analysis of the plant resource biodiversity in the Republic of Kazakhstan. Attempt has been made to explore such natural zones in the ecosystems such as the forest-steppe, desert, semidesert, foothill, and mountainous. All include a detailed description of communities and dominant plants, endemics, and relict taxa. Furthermore, has been carried out the assessment of current biodiversity status, as well as degradation degree of the steppe ecosystems subjected to a large-scale plowing in the 1960s too. The aim here has been to present a quantitative assessment of biodiversity parameters as indicators of ecological safety based on multivariate statistical analysis. According to the indices, the biodiversity is determined as high, acceptable, medium, low, and critical.

Keywords Kazakhstan · Plant diversity · Plant communities · Ecological safety

12.1 Introduction

Kazakhstan is located in the center of the Eurasian continent and is characterized by large species, genetic, ecosystem, and terrain diversity. The area is 2724.9000 km^2. This is about 2% of our global surface. It stretches from the east of the Caspian Sea and the Volga plains to the Altai mountains, from the foothills of the Tien Shan in the south and southeast to the West Siberian lowland in the north. Ecosystems and

M. Askarova (✉) · M. Arslan
Department of Geography, Land Management and Cadastre, Al-Farabi Kazakh National University, Almaty, Kazakhstan

A. Medeu · A. Medeu
Institute of Geography, Almaty, Kazakhstan

D. Egamberdieva, M. Öztürk (eds.), *Vegetation of Central Asia and Environs*,
https://doi.org/10.1007/978-3-319-99728-5_12

terrain complexes are very unique: from deserts to highlands to inland aquatic ecosystems. The country has a full range of subzonal vegetation variants comprised of steppes, deserts, and mountain belts. These in fact are the characteristic features of Central Asia. More than 75% of these are occupied by arid and subhumid territories, containing more than 40% of species composition of the entire biodiversity of the country (Republic of Kazakhstan 2010).

Plains compose 60% of total land area of the republic, 30% small hills, and 10% mountainous areas. The length of the territory from the west to east exceeds 3000 km, from south to north 1700 km. In the southeast, there is a mountainous range named as Khan Tengri, which is 7000 m high. Kazakhstan is the ninth largest territory in the world (National Atlas 2010).

Flora of Kazakhstan includes more than 13,000 species, including more than 5754 species of higher vascular plants, about 5000 fungi, 485 lichens, more than 2000 algae, and about 500 bryophytes. Among the plants, 14% of the species are endemic, including many tertiary and quaternary relics. The centers of floral endemism are the Mount Karatau and Western Tien Shan (National Report 2015). Unique natural pine forests on sands (Arakaragai, Aman-Karagai, Naurzum) and forests and steppes on the lowlands of Central Kazakhstan; desert communities of Betpak-Dala, Southern Balkhash area, and Ili basin representing original floral composition; forest, shrub, and steppe communities in the Southern Altai, the Kalbin Mountains, and Tarbagatai, the midlands of the Dzungarian Alatau and Tien Shan with coniferous fir forests and fragments of apple forests, as well as wetland-swamp ecosystems of the lower reaches of the Urals, Torgay hollows, Tengiz, and Alakol lakes; and floodplain forests (tugai) of the Syr Darya, Ili, and Charyn.

The country is located inside the continent, within temperate climatic zone. Basic botanical and geographical zones such as forest-steppe, steppe zone, semidesert, and desert zones stretch along its flat part, which in turn is divided into number of subzones. The analysis of current ecological status of bioresources reveals that as a result of urbanization and intensive agricultural development, ecosystems are experiencing a heavy pressure leading to a decrease in the biodiversity. The ecosystems like steppes and vegetation around the foothills are under very high anthropogenic activity. Our major objective here is to analyze the biodiversity status of the country's ecosystems, evaluate their current ecological status, and determine their ecological safety level, based on methods of multidimensional statistical analysis.

12.2 Plant Diversity

12.2.1 Forest-Steppe Zone

This zone occupies a small territory in the Northern Kazakhstan, around Petropavlovsk and Kokshetau cities, represented by forests (0,7 million ha) and is rich in herbal-transformed steppes. Two forest-steppe subzones are clearly outlined in this zone (4th National Report 2010):

1. *Southern weak-wet moderately warm tree cover* (1% of the territory of Kazakhstan) is represented by aspen-birch and aspen forests in small plantations, surrounded by steppe, in the undergrowth of which we come across species like *Salix alba*, *Salix berberifolia*, *Padus racemosa*, *Rosa acicularis*, *Rhamnus cathartica*, and *Viburnum opulus*, together with the graminaceous species like *Brachypodium pinnatum*, *Calamagrostis epigejos*, *Elymus repens*, and few other important taxa such as *Sanguisorba*, *Filipendula*, *Vicia cracca*, and *Peucedanum*.
2. *Moderately dry herbaceous cover* (1.04%) is represented by rich motley grass-cereal steppes. The motley grass includes mesophylls: *Filipendula ulmaria*, *Pulsatilla patens, and Galium boreale*. The cereal steppes are dominated by *Stipa rubens*, *Stipa pennata*, and *Fescue sulcata*; also in small numbers we find species like *Koeleria gracilis*, *Poa angustifolia*, *Phleum phleoides*, *Bromus inermis*, *Elymus repens*, and *Helictotrichon hookeri* ssp. *schellianum*. In the forest areas, birch trees (*Betula pendula*, *Betula pubescens*, *Betula kirghisorum*) flourish together with willows (*Salix triandra*, *Salix caprea*, *Salix rosmarinifolia*, *Salix fragilis*, *S.alba*). The bushes include *Rosa acicularis*, *Rosa spinosissima*, *Spiraea crenata*, *Spiraea hypericifolia*, *Cerasus fruticosa*, and *Cotoneaster melanocarpa*.

The Artemisia-fescue-feather grass steppes with *Artemisia lercheana* are distributed in the saline depressions. The meadow cereal-mixed grassland is observed on the flat interfluves on wet habitats. Cereals include a dense cover of *Elymus repens*, *Calamagrostis epigejos*, *Bromus inermis*, and mesophyllous taxa like *Filipendula stepposa* and *Sanguisorba officinalis*. The saline habitats are covered by *Galatella punctata*, *Glycyrrhiza uralensis*, *Silaum silaus*, and *Artemisia tenuisecta*.

Steppes occur in complex formations on solonetzes, solonchaks, and saline meadows along the depressions on the flat interfluve and weakly drained valleys. In the northern part of the forest-steppe zone on flat interfluve, which is characterized by most humid habitats, aspen-birch forests are widespread, which are found in the depressions of the south.

The centers of *minor depressions* are characterized by wet and even marshy meadows. Solonchaks are inhabited by the mixed grass-cereal floodplain meadows, widely used with a predominance of *Hordeum brevisubulatum*. Swampy sedge meadows on meadow-bog soils are often haggard due to the abundance of the *Carex omskiana*, *Carex caespitosa*, *Carex acuta*, *Carex disticha*, and *Carex physodes* together with *Juncus ambiguus*, *Comarum palustre*, *Filipendula ulmaria*, *Lythrum virgatum*, and *Lythrum salicaria*.

The forest-steppe zone is characterized by solonetzes and solonchaks with predominance of halophytic vegetation. On dry solonetzes, *Camphorosma monspeliaca* and *Artemisia pauciflora* with *Limonium gmelinii*, *Kochia prostrata*, and a number of annual *Salsola* taxa prevail. *Atriplex verrucifera* and *Puccinellia dolicholepis* grow on more humid crustal solonetzes and solonchaks; *Plantago maritima* prevails in the transition zone on gray meadow solonchaks. On wet solonchaks thickets of the *Salicornia europaea* grow densely (Fig. 12.1).

Fig. 12.1 A view of *Salicornia europaea*. Source: http://www.petrovne.ru/flora_fauna/mediterranean_wildlife/

12.2.2 Steppe Zone

The steppe zone covers an area of 121 million ha (45,5% of the territory of the Republic) between 48° and 52° north latitude, where the entire spectrum of zonal steppe types is observed. The climate of the steppes is characterized by cold winters and hot summers. Annual amount of precipitation is 350–400 mm, and in the southern regions, their amount drops to 250 mm. Vegetation period in the steppes lasts up to 190 days (from late March to late September). In the eastern part of the zone, due to increasing continental feature and longer winters, the vegetation period is shorter and lasts about 175–160 days.

Since steppe plants during the vegetation period are influenced by high temperatures, combined with insufficient precipitation, main vegetation cover is composed of xeromorphic and turf grasses. From north to south, their share in the plant cover increases. In addition, as the ratio of living forms changes, the number of ephemerals and ephemeroids increases. Ephemeral plants with a very short vegetation period grow in early spring and autumn, when there is a lot of moisture in the soil. These plants grow early in the spring and don't exceed a few centimeters in height and include small spring groove, spring veronica, etc. They are distinguished by a very short period of growth—fast and are ephemeral. In a short time period, they sprout, bloom, give seeds, and die down.

The perennial plants with short vegetation period are hibernating in the form of bulbs, nodules, or rhizomes. These are called ephemeroids and include *Tulipa* (Fig. 12.2), *Iris*, *Gagea*, *Hyacinthus*, some species of *Carex*, etc. (Rachkovskaya et al. 1999). The steppes in Kazakhstan, being a unique part of the Eurasian steppe region,

Fig. 12.2 A general look of *Tulipa greigii* Regel. Source: http://www.phytology.ru/cveti-v-sadu/nemnogo-interesnoie-informacii-o-stepnix-tyulpanax.html

represent a transitional zone from the boreal type of vegetation to arid (4th National Report 2010). In the vegetation cover of the steppe zone of the country, xerophilous turf grasses are dominant: *Stipa rubens*, *Stipa lessingiana*, *Stipa capillata*, *Stipa sareptana*, as well as *Festuca valesiaca*. The western part of the steppe zone is characterized by the *Stipa stenophylla* and partly *Stipa ucrainica* and the east by the *Stipa kirghisorum*. The distinctive feature of the eastern part of the steppe zone in comparison with the western part is the presence of shrub steppes (especially with the taiga and caragana), as well as the presence of such oriental species as *Iris scariosa*, *Chamaerodos erecta*, *Potentilla acaulis*, *Orostachys spinosa*, *Astragalus fruticosus*, *Ferula songorica*, etc. in many types of steppes.

The steppe zone is divided into three subzones (Ivashchenko 2009):

1. *Moderate dry rich mixed grass-cereal steppes* distributed on normal and southern black soils (24 million ha or 9%). The basis of the herbage includes dense-turf cereals with a large number of motley herbs on fertile soils. The mixed grasses include *Phlomis tuberosa*, *Salvia stepposa*, and *Medicago lupulina*. These are the most important agricultural regions of the country, which have been subjected to the largest economic development.
2. *Moderate dry steppes* occupying dark chestnut and chestnut soils (53 million ha or 20%) and represented by the feather grass, tyrsa, and fescue as well as ephemerals and ephemeroids—*Poa bulbosa*, desert sedge (*Carex physodes*), *Spiraea hypericifolia*, and *Halimodendron halodendron* which grow in *minor depressions*, in the river floodplains—aspen, and poplar forests with hips, honeysuckle (*Lonicera tatarica*). Most part of it is plowed up.

3. *Desert steppes* on light chestnut soils (44 million ha or 16.5%). This subzone is a transition zone between steppe and desert. It is characterized by the predominance of wormwood-cereal-thin vegetation. Solonets and solonchaks predominate here. The vegetation is similar to the forest-steppe zone. Deserted steppes are distinguished by great complexity of vegetation cover.

The specific feature of these steppes is that the largest number of endemic species are mostly concentrated in the northeast of Kazakhstan. These include the *Amygdalus ledebouriana*, *Calophaca tianschanica*, herbs—*Stipa iljinii*, *Agropyron tarbagataensis*, *Silene karkaralensis*, *Isatis frutescens*, *Clausia kasakhorum*, *Astragalus trautfetteri*, *Cachris macrocarpa*, *Hyssopus ambiguus*, and *Hyssopus macranthus*—*Craniospermum subfloccosum*, etc. The number of endemics decreases in southern direction and especially in western parts. Some endemic species like *Craniospermum echioides*, *Astragalus subarcuatus*, *Astragalus mugodsharicus*, and *Jurinea mugodsharica* are recorded from Ulytau, Mugodzhary, and Western Kazakhstan.

The turf grasses include mesophilic grasses. Meadow-steppe plants such as the *Salvia*, *Phlomis*, *Gypsophila*, *Syzygium*, *Libanotis*, and *Laserpitium* are common in depressions and on the northern slopes of the gullies. Half-shrubs such as the *Artemisia*, *Pyrethrum*, and *Linosyris* play a significant role in the grass stands.

12.2.3 Semidesert Zone

Semideserts are a transition zone between steppes and deserts—stretching from the banks of Zhaiyk to the Altai Mountains, 2900 km in width from 30 to 300 km. They occupy 10.8% of the territory of the country. The southern boundary of the zone passes 48° northern latitude.

The climate of semideserts is much drier than in the northern zones. Precipitation is low; average annual amount varies between 180 and 300 mm. The Most precipitation falls at the end of the spring—the beginning of summer and least in winter and midsummer. The summer in the whole territory is characterized by hot season. The average temperature in July is +22, +24 °C, and sometimes it rises to +40 °C. Winter is severe; clear frosty days prevail. The average temperature of January is −15 to −17 °C; the lowest temperature −50 °C is observed in the east.

In the desert-steppe zone, both steppe and desert plants are common. The vegetation cover mainly consists of the *Festuca valesiaca*, *Artemisia*, *Matricaria*, and *Stipa capillata*. Often *Artemisia* occupies large areas, creating a dull monotonous picture. In some places, among *Artemisia* grow *Kochia*, *Ceratocarpus*, *Krascheninnikovia ceratoides*, and *Atriplex*. Thickets of *Achnatherum splendens* are represented in places where groundwaters approach close to the surface of the earth, on solonchak soils.

According to Gvozdetskiy and Nikolaev (1971), two subzones are distinguished in the semidesert zone:

1. *Desert* (*semishrub-turf-cereal*s *steppes*—northern semidesert with a dominance of turf-cereal steppe, which are combined with desert communities.
2. Steppe deserts—the southern semidesert. Desert communities dominate in the southern semidesert: cereal-*Artemisia* deserts are combined with turf-cereal steppe communities, *Artemisia*, and *Salsola* deserts. So, in both cases the complex nature of vegetation cover is typical with a combination of steppe and desert communities where the first ones dominate in the northern semidesert and the second in the southern. In comparison with vegetation of the steppe and dry steppe zones, the vegetation cover of the semidesert is characterized by depleted species composition, thinness, and a lower height of grass stand (10–15 cm). Moreover, vegetation in desert communities is especially thin. Drought-resistant annual plants predominate: *Stipa sareptana*, *Festuca sulcata*, and *Artemisia. Salsola* is common on saline soils. Ephemerals and ephemeroids such as the *Poa bulbosa* and *Tulipa* are widely represented. The seasonal vegetation development is sharply expressed. In spring and early summer, when there is enough moisture in the soil, the plants grow rapidly, but in the second half of the summer, most of them dry up and burn out. In autumn, when temperatures go down and rainfalls begin, the plants come to life again for a short time.

Complex feature is the most characteristic of the semidesert vegetation cover with a combination of steppe (turf-cereal) and desert (*Artemisia-Salsola*) communities. The *Festuca sulcata*, *Stipa*, *Agropyron desertorum*, and *Agropyron sibiricum* grasses dominate in the steppe communities and in the desert—the *Agropyron sibiricum*. The *Artemisia lercheana*, *Artemisia lessingiana*, and *Artemisia sublessingiana* are common on light chestnut soils and *Artemisia pauciflora* on solonetzes. The *Pyrethrum achilleifolium* and *Kochia prostrata* play a significant role in the vegetation cover of the southern subzone. Except for *Artemisia pauciflora*, *Anabasis salsa*, and *Atriplex cana*, *Camphorosma monspeliaca* grow widely on crockery solonetzes. *Nanophyton erinaceum* is found on salinized gravelly soils. Solonchaks are home to various succulent *Sálsola*. The meadows are dominated by *Lasiagrostis splendens*, *Aeluropus littoralis*, and *Puccinellia distans*. These prevail in river valleys and lake depressions and mostly on saline soils. On low-capacity crashed stone light chestnut soils in Kazakh small hills, the sparse vegetation cover consists mainly of *Stipa capillata* and *Artemisia sublessingiana*, and in the most stony areas grow *Artemisia frigida*, *Stipa orientalis*, shrub thickets of *Spiraea*, *Cytisus biflorus*, and rose hips (*Rosa*).

12.2.4 Desert Zone

Deserts (area of 124.6 million ha) cover the regions of the Caspian lowland, the peninsula of Mangyshlak, the Ustyurt Plateau, the southern part of the Turgai and Kazakh small hilly areas (eastern Betpak-Dala and Balkhash area), Turan Lowland (Aral area), Kyzylkum, Moyinkum sands, Alakol depression, and Ili basin and in the south reach the foothills of Northern Tien Shan, Dzhungar Alatau, and Tarbagatai mountains.

The climate is characterized by very strong radiation, exceptional aridity of the entire vegetation period, very high temperatures, and extremely low precipitation (100–200 mm/year). In the northern regions of Central Asian and Caspian deserts, winter is quite severe: the average January temperature is −15 °C, and absolute minimum temperature in winter reaches −40 °C with very low snow cover. Spring is frosty. The summer is very hot; average July temperatures range from +24 to +27 °C. Precipitation falls mainly in autumn, winter, and early spring. In order to survive under such conditions, desert plants develop some adjustments such as reduction of evaporating surface by reducing the leaf surfaces (up to their reduction) and developing a thick cuticle layer. As a rule, all desert plants have powerful underground organs, which are much bigger than the mass and volume of the aboveground parts. Along with roots that go deep into the soil, many plants have a large surface area and very thin roots, with the help of which the plants catch moisture from condensed dew (Ogar 2003).

The vegetative component of the desert ecosystems (main communities) is represented by semishrubs and shrubs. It is characterized by small species diversity, small projective cover, and absolute dominance of drought-resistant xerophytes and hyperxerophytes. Three subzones of the desert ecosystems are represented on the plains, northern, middle, and southern, as well as special climate type of foothill deserts (The State cadastre of plants 2006).

1. *Northern* (*steppe deserts*) on brown desert soils (area of 40.0 million ha). They form a huge arc, based on the northern coast of the Caspian Sea in the west and on the foothills of the Tien Shan and Dzhungar Alatau in the east, including the north of Ustyurt Plateau, the Northern Aral area (up to the Syr Darya River), Betpak-Dala desert, southern Balkhash area, and Moyinkum. They include not only the *Artemisia*-semishrubby, *Artemisia-Salsola* deserts, but also sand deserts in the Kyzylkum, dominated by the *Haloxylon* and shrub deserts. They are characterized by semishrub communities, mostly sagebrush and rarely perennial *Salsola* (Toderich et al. 2010). The specific feature of plant communities is the presence of steppe grasses *Stipa sareptana*, *Stipa kirgisorum*, *Stipa richteriana*, and *Agropyron fragile* on the sands.
2. *Middle deserts* (area 51.2 million ha) on gray-brown desert, freezing soils. In the middle deserts, perennial *Salsola* dominate—*Anabasis salsa* (Fig. 12.3), *Salsola arbusculiformus*, *Salsola orientalis*, *Nanophyton erinaceum*, and *Artemisia* species such as *Artemisia terrae-albae* and *Artemisia turanica. Saxaul* white (*Haloxylon persicicum*) and black (*Haloxylon aphyllum*) are widely spread on the sand and psammophilous shrubs and semishrubs (*Calligonum*, *Ephedra*, *Ammodendron*).
3. *Southern deserts* (area of 30.3 million ha) occupy the southern part of the arid-denudational Ustyurt Plateau composed of limestone and marls and Kyzylkum sandy area, as well as part of the territory of Southern Kazakhstan within the Atyrau, Kyzylorda, and Shymkent regions. Due to uneven terrain and different salinity levels of the substrate, vegetation cover of these desert communities is heterogeneous and differs in complexity, i.e., combination of different plant micro groups in small areas and in sandy deserts—sand relief and their firmness.

Fig. 12.3 A general view of *Anabasis salsa*. Source: http://silkadv.com/ru/node/800

Fig. 12.4 A general view of *Haloxylon aphyllum*. Source: http://www.plantarium.ru/page/image/id/221341.html

In addition, semishrubs and shrubs dominate in these deserts, but their species composition varies accordingly. The communities of *Salsola gemascons* and *Artemisia kemrudica* prevail. The phytocoenotic role of ephemerals and ephemeroids is significantly increased on the sands, especially of the *Carex physodes* (Figs. 12.4 and 12.5) and juzgun plants (Novikova 1990).

Fig. 12.5 A general view of the *Haloxylon* white (*Haloxylon persicum*). Source: http://nahman.livejournal.com/213044.html

This *Haloxylon* (Chenopodiaceae) is named as black because of its crown which is painted in a rather dark green color. This color is preserved in spring and summer. In autumn the crown becomes orange-brown. It grows on soils containing many salts. This is what distinguishes it from most of the accompanying plants, which die from excess salts in the soil. As a result, its notched young branches taste salty-acidic. For the normal growth, black *Haloxylon* requires heavily wet soil; it grows on lowlands, the places where rivers used to flow and groundwater approaches close to the soil surface or where the wind forms deep basins. Black *Haloxylon* forest is often located in close proximity to oases and protects them from moving sands.

4. *Piedmont desert* (area 14.8 million ha). Mountain systems play a barrier role and, thus, contribute to the distribution of desert communities which is due to latitudinal changes in hydrothermal conditions. After approaching the mountains, the amount of precipitation on the foothills increases due to the intensive thermal convections and active atmospheric fronts. The leading factor in the formation of ecosystems in foothill areas (plains, small hills, and sand areas) is a significant increase in the precipitation due to the effect of piedmont moistening, which forms a "humid-foothill" zone.

In the foothill areas, precipitation is two to three times higher than on the plains which are beyond the influence of the mountains. Mountain massifs serve as significant barrier traps for the northwestern air masses. Piedmont deserts occur at the foot of all mountain systems of Kazakhstan from the Tarbagatai to Karatau and the Western Tien Shan. The main soil types in the ecosystems of foothill deserts are light gray soils (northern and southern). The vegetation of the foothill deserts is characterized by the presence of semishrubs and shrub communities in their composition, as well as well-marked tier of ephemeroids formed by *Poa bulbosa* and *Carex pachystilis*.

In many respects it resembles black *Haloxylon*, but the color of the crown is lighter and slightly whitish. This impression is due to the fact that branches of the previous year are almost white and the shoots of the current year are light green. The trunk is rough, curved, and covered with grayish bark. White *Haloxylon* differs from black by the presence of quite discernible leaves, although poorly developed. They resemble small scales, which transfer upward in a rather long taper. These scales are located on the shoots in pairs, opposite and closely adjacent to the surface of the stem. Difference between black and white *Haloxylon* is that the black shoots taste salty or sour-salty and white ones have bitter taste. *Haloxylon* white grows mainly in the sandy desert and defines the indigenous terrain of the Central Asian sand deserts, emphasizing its originality.

12.3 Mountain Ecosystems

12.3.1 Classification of Vertical Zoning by Semyonov-Tyan-Shansky

The vertical zoning means change in vegetative cover in the mountains and is determined by the position of mountains on the earth's surface, general climatic conditions of certain terrain, and altitude above the sea level. The vertical zoning in the Tien Shan mountains is especially expressed and, accordingly, well explored, which flora is extremely rich and diverse, including almost half of the species diversity of Kazakhstan. It is two to three times higher than the flora of any of the European regions. The number of endemic species is up to 10% of the flora. The endemism is progressive type with neo-endemic characters (Volkova 2003). In addition, Popov has found relict representatives of tertiary mesophilic forests in Tien Shan. His aforementioned "ginkgo flora" includes thickets of wild apple (*Malus sieversii*), apricot (*Armeniaca vulgaris*), and maple (*Acer semenovii*) in the lower belts of Dzungar and Ile (Trans-Ili) Alatau, as well as herbaceous species such as the remarkable *Corydalis semenovii*, *Adonis chrysocyathus*, and others.

Undoubtedly, the plantations of the *Fraxinus potamophila* should also be referred to the same forest relics in the valley of the Charyn River, described by Bykov (1979). In the ash forests (the lower reaches of Charyn River), Bykov (1979) has found a kind of mesophilous orchid (*Neottia kamtschatica*). In general scheme of this system zoning, six vertical zones have been outlined (Chupakhin 1987).

Semyonov-Tyan-Shansky has presented a general scheme of this system zoning, where he distinguished six vertical zones:

1. *Steppe*, stretching from the Ili River valley to the foot of the Trans-Ili Alatau, between 153 and 612 m above the sea level. This zone, now rather called desert, is divided into two tiers. The lower tier stretches from 153 to 306 m above the sea level and is characterized by the presence of *Halóxylon*, solonchak plants, *Calligonum*, *Tamarix*, *Alhagi*, and *Halimodendron*. The upper tier of this zone is

located between 306 and 612 m and is characterized by a vegetation cover of *Artemisia* with some admixture of European species.

2. *Zone of shrub-motley grass steppe* occupies the plumes and foothills of the Trans-Ili Alatau, starting from a height of 612 m and ending near the lower boundary of the coniferous forest on the northern slope at an altitude of 1377 m and on the southern slope in the valley of the Issyk-Kul lake, at an altitude of 1530 m. This zone is rich and diverse, in both herbaceous and tree-shrubby vegetation. This zone is characterized by a wild apple tree (*Malus sieversii*) and apricot (*Armeniaca vulgaris*), which Popov considers a native of China. Also *Crataegus songorica*, *Crataegus altaica*, *Atraphaxis muschketovii*, *Rhamnus cathartica*, *Padus racemosa*, *Acer semenovii*, and a lot of thorn bushes, which usually remain after cutting down as the main components, *Berberis heteropoda*, *Rosa platyacantha*, *Rosa spinosissima*, and others characterize this belt as a zone of hardwood species.
3. *Zone of coniferous forests* or subalpine zone is located along the mountain slopes at altitudes from 1377 m to 2325 on northern exposures and 2448 m on southern slopes. Its upper boundary is the upper limit of forest vegetation. The dominant species is the Tien Shan spruce (*Picea schrenkiana*). There are hardwood species in the subordinate state: *Populus tremula*, *Betula*, and *Sorbus*, in addition to a number of shrubs, including seven species of *Lonicera*. Significant areas of subalpine meadows are found. Ecologically, the Tien Shan spruce differs significantly from other species of this genus. Firstly, it is much more photophilic than all other spruces, and, secondly, it needs less moisture. They usually grow dissociated, rarely forming light rare forests with lots of glades and clearings, often interspersed with rocks, gravelly areas, or stony placers. Seed reproduction of spruce occurs not under the canopy but in strongly lighted places: on fringes, glades, etc.
4. *Zone of the lower Alpine or Alpine bushes begins above the coniferous forest, i.e., above 2448 m and ends at an altitude of 2754 m.* The characteristic shrubs are *Juniperus turkestanica*, *Juniperus sibirica*, *Caragana jubata*, *Spiraea*, *Salix*, *Potentilla fruticosa*, *Potentilla salessovi*, currants, honeysuckle, etc. In addition to shrubs, this zone is also rich in excellent meadows—summer pastures "zhailau."
5. *Zone of the upper Alpine or Alpine grasses* extends over the Alpine shrubbery and up to the lower boundary of the eternal snows located at the altitudes of 3213–3366 m. Various herbaceous alpine forms are mainly spread in this zone, including not more than 25% of European alpine-polar plant types. About 75% of the rest is typical for the Alps, Altai, and Sayan Mountains as well as Polar Siberia, a small part of Himalayan, and a whole number of actually Tien Shan species. However, in the main and the most extensive area is typical grouping of low-grass cereal-mixed grass Alpine meadow. This is low, only 25–30 cm squat grass which fully covers the soil and densely and is unusually flowery, because all dicotyledons of its composition bear bright and large flowers which are disproportionate with the height. Above them stand the single stems of several peculiar alpine cereals and sedges: *Poa pratensis* and *Poa alpina*, *Festuca kirilovii*,

Fig. 12.6 A general view of *Viola altaica*. Source: http://m.cvetki.org/cvetki_viola.php

Trisetum spicatum, and *Anthoxanthum odoratum*. *Carex melanantha* prevails within the sedges. From the representatives of cereals *Polygonum nitens*, *Polygonum viviparum*, *Viola altaica* (Fig. 12.6), *Anemone protracta* and especially bright orange *Erigeron aurantiacus* grow widely. New species also join here: *Aster flaccidus*, *Pedicularis violascens*, *Pedicularis songorica*, *Gentiana algida*, and *Gentiana kaufmanniana*.

The upper limit of vegetation is a scattered and disconnected plant cover of large-scale moraines, terminating around the glaciers at altitudes above 3000 m, which serve as a transition to the final plant free glacial zone: eternal snows, ice, and stone. At the lower edge of the moraines, one can find some Himalayan species, for example, *Viola kunawarensis* or *Lonicera glauca*. On the moraine itself, usually many species of *Poa lipskyi* and *Poa dschungarica* grow together with *Festuca tianschanica*, *Carex griffithii*, *Taraxacum lilacinum*, and *Saussurea glaciales*.

6. *Zone of eternal snow is* represented by peaks and high ridges, lying above 3366 m. In this zone only some areas open up from snow for a short time, and rare representatives of plants from the upper-alpine zone can be seen here.

12.4 Current Biodiversity and Ecological Status

Conservation and careful use of biodiversity is a key element of sustainable development. The species composition of any ecosystem is formed as a result of long evolution. In the ecosystem, all species are in close relationship with each other and with their habitat, performing certain functions, and therefore the whole system is well balanced. Loss of one species can lead to an imbalance as a result of loss of certain functions performed by this species. This does not pass without consequences not only for the ecosystem but for the entire biosphere (Ozturk et al. 2010). In a report published in early 2010, the UN Commission on Biodiversity draws attention to catastrophic changes in the wildlife on global basis. Currently, more than 40% of all living species on the Earth are threatened with extinction. If these rates of extinction continue or accelerate, the number of endangered species in the next decades will be counted in millions (http://web.unep.org/annualreport/2016/index.php).

The variety of natural conditions of Kazakhstan determines the wealth and diversity of its biological resources. The country's biological resources are vital for its economic and social development. Biological diversity is an asset of great value to the present and future generations. In Kazakhstan as in the rest of the world, the decrease of biodiversity is taking place. Major reasons are a loss of habitat, excessive exploitation of resources, environmental pollution, and displacement of natural species by introduced exotic species (Ozturk et al. 2010). Apparently, in all cases, these reasons have anthropogenic character. Among the globally threatened (CR, EN, VU, NT) categories in the flora of Kazakhstan, there are 15 species, including five critically endangered (*Berberis karkaralensis*, *Calligonum triste*, *Lonicera karataviensis*, *Populus berkarensis*, *Sibiraea tianschanika*), eight endangered, and two vulnerable. The list of rare and endangered plant species at the national level contains 387 plant species (5th National Report 2015).

As a result of large-scale plowing of land which took place more than 50 years ago, the total changes in majority of ecosystems have occurred, especially in the steppe and forest-steppe zones. This is fully observed in the rich mixed feather grass, 8.5 million ha, and mixed feather grass steppe, 13.6 million ha. The transformation of territories on the plains reaches 90%, in small hills areas up to 30%. The dry steppes in the plains have changed by 50–60% and in small hills areas by 10–15%. The depletion of biodiversity in the desert subareas is different. In northern deserts, predominantly sagebrush, the local overgrazing around wintering grounds, settlements, a linear trend is observed along the livestock trail routes. In the middle and southern deserts (perennial *Salsola*), except for overgrazing, the violations are associated with technogenic impacts and not well-planned road network, regulation of river flow, and illegal cutting of *Halóxylon*. Vegetation of sandy deserts has been strongly disturbed. Severe disturbance of the vegetation cover has occurred due to urbanization and intensive agricultural development around the foothill zones. The areas in the original Kazakhstan ephemeroid-wormwood deserts have practically been destroyed. Especially sharp changes in meadow vegetation have taken place in the floodplains of the Ili, Syr Darya, Shu, and Talas rivers. Moreover, highly productive floodplain communities have almost completely got degraded. In connection with limited river flow, meadow vegetation has degraded everywhere, and meadows have disappeared as a type of vegetation. Their diversity has reduced; yield capacity has also reduced 10–15 times (from 15–40 to 1.5–2 ha).

In northwest Kazakhstan floodplain forests (oak forests, ash trees, maple trees, elms, willows, etc.), there has been degradation due to violation of hydrological regimes. Mountain forests are also under the pressure of livestock overgrazing, which has resulted in the degradation of live surface cover and destruction of natural renewal. Logging has resulted in a decrease in the density of plantations and a change of coniferous species to deciduous and shrubby ones (East Kazakhstan region). Lower boundary of the spruce belt has gone up to about 200 m in the Ili Alatau during the last 100–150 years and in the Dzungar (Zhetysu) Alatau fir belt up to 100 m (5th National Report 2015).

12.5 Evaluation of Biodiversity and Ecological Safety Levels

Ecological safety under current conditions is considered as an integral and important part of the security of the "individual, society, and the state." The main subject of ensuring environmental safety is the state, which performs its functions in this area through the bodies of national legislative, executive, and judicial authorities. The main objects of environmental safety are:

- Individuals have a right to live a healthy life in a favorable environment.
- Society and its material and spiritual values, depending on ecological status of the country.
- Natural resources and natural environment are the basis for sustainable development of society and welfare of future generations.

Environmental safety is a sustainable environmental status, which provides an opportunity for improving the life quality of the society and the state as well. Ecological safety means environmental impacts, which can result in the changes in the environment and, as a consequence, changes in the conditions for the existence of human and society.

The issues of ensuring environmental safety and issues of sustainable development are becoming especially relevant today and are treated as the most important for each state. Currently, in the context of globalization, environmental safety for any country is considered an obligatory, necessary, and the most important part of the general state policy. Ecological safety of each state separately determines the overall international political stability and global security.

The national report on biodiversity of the Republic of Kazakhstan (5th National Report 2015) notes the depletion of biodiversity and ecosystem degradation in the 66% of the country's area. Reduction of biodiversity components is caused, first of all, by anthropogenic influences. Main threats include desertification, economic activities, environment pollution, natural disasters, forest fires, and illegal logging on the territory of the state forest resources, the impact of introduced species (biological pollution) (5th National Report 2015). It is especially necessary to note that Kazakhstan, as an agricultural country, has practically mono crop agriculture which has adverse effects on biological diversity.

Threats to environmental safety include conditions and factors that represent a threat to environmental resources, natural balance, human-friendly natural environmental conditions, and economic value of natural resources. A threat to ecological well-being can be a result of the excessive extraction of natural resources and, as a result, a decrease in biodiversity. Biodiversity is the basis of life on the Earth, is one of the most important life resources, and is considered to be the main factor determining the stability of biogeochemical cycles of the substance and energy in the biosphere (Ozturk et al. 2010). Based on the importance of this issue, we calculated the level of biodiversity ecological safety.

12.6 Methods Used

Indicators of biodiversity have been estimated by many researchers, but none has primarily focused on quantitative values but qualitative parameters of the ecosystem. Such indicators cannot be estimated in view of environmental safety management; therefore we propose a system of quantitative assessment on biodiversity parameters, which determines the indicator of biodiversity ecological safety. The principle of quantitative evaluation is oriented on the fact that populations of species depend on the average life expectancy of each representative and more precisely on the life cycle of population. However, the dependence for life forms is nonlinear but rather quadratic; therefore there is a slight increase in the population life cycle, and population increases significantly. In nature such phenomena are considered anomalous. To express the Ginni index, the larger the anomaly, this index changes more rapidly (Russell 2013). Therefore, to determine the biodiversity safety coefficient (B_{SC}), we used the following formula:

$$B_{SC} = \frac{\left|1 - \sum_{i=2}^{n} (Y_i - Y_{i-1})\right|}{\left|1 - \sum_{i=2}^{n} (Z_i - Z_{i-1})\right|} \times \frac{\sum_{i=1}^{n} Z_i}{\sum_{i=1}^{n} Y_i}$$

where n is the number of animal species inhabiting Kazakhstan, Y_i, Z_i is population number of i (fauna species) and population life cycle. The sets (Y_1, Y_2, …, Y_k, Y_{k+1}, …, Y_n) and (Z_1, Z_2, …, Z_k, Z_{k+1}, …, Z_n) go in strictly descending order. In high biodiversity of the study area, the safety index tends to be 1. Disturbance of ecological balance occurs in dominance of the populations of the species at approximately the same stage in the population life cycle; in this case the safety index tends to be 0.

In general, such innovative approach on biodiversity requires additional large-scale researches and refinements, but following the logic of our research, we tried to assess the safety levels on a five-point scale given in Table 12.1.

Table 12.1 Biodiversity safety levels

Index	Safety level	
I	More than 0,7	High
II	0,7–0,6	Accessible
III	0,6–0,5	Average
IV	0,5–0,4	Low
V	Less than 0,4	Critical

12.7 Conclusions and Recommendations

As a result of the large data base analysis, this study includes an assessment of the bioresources of Kazakhstan on the basis of botanical-geographical zones like forest-steppe, steppes, deserts, semideserts, and mountain zones. They characterize the country's rich biodiversity. Reduction of biodiversity as a result of strong anthropogenic pressure on the environment is more than urgent for Kazakhstan, as well as around the globe. The estimation of current ecological condition here shows the degree of ecosystem degradation caused by climate change, habitat loss, excessive exploitation of resources, environment pollution, and displacement of natural species by introduced exotic species. Moreover, we propose the formula to calculate the indicator of biodiversity ecological safety level, which is defined according to five indices as high, acceptable, average, low, and critical.

The following recommendations are proposed for effective biodiversity management:

- Inventory assessment of plant communities and determination of new boundaries of natural zones in the context of the climate change and development of desertification processes
- Further expansion of specially protected natural areas linked by ecological corridors
- Inclusion of the republic's specially protected natural areas in the UNESCO World Natural and Cultural Heritage List and Biosphere Territories under the program "Human and Biosphere"
- Conservation of natural populations of rare, endangered, relict, and endemic species through their artificial ex situ reproduction and restoration in disturbed areas

The analysis of the rich fauna status in ecosystems of Kazakhstan, agrobiodiversity, diversity of aquatic ecosystems, as well as specially protected natural areas (SPNA), which occupy 8.6% of the total territory, has not been included in this study, as these are a subject of separate research study.

References

Bykov BA (1979) Essays on the history of flora in Kazakhstan and Central Asia. Alma-Ata, Nauka

Chupakhin VM (1987) Altitude and zonal geosystems of Central Asia and Kazakhstan. Alma-Ata, Nauka, p 256

Gvozdetskiy NA, Nikolaev VA (1971) Kazakhstan: essay on nature. Moscow, p 295
Ivashchenko AA (2009) The flora of Kazakhstan. Almaty, p 228
Russell J (2013) Gini coefficient, p 88
Kazakhstan's 4th National Report for Biodiversity Convention (2010)
Kazakhstan's 5th National Report for Biodiversity Convention (2015)
National Report on the state of environment and natural resources of the Republic of Kazakhstan for 2011 - 2014 (2015)
Novikova SS (1990) Botanical and geographical patterns of distribution of vegetation in Northeastern Balkhash area and Alakol depression: Abstract of Thesis, Alma-Ata
Ogar NP (2003) The vegetation of river valleys. Botanical geography of Kazakhstan and Central Asia (within the desert area), St. Petersburg
Ozturk M, Aksoy A, Gucel S, Altundag E (2010) Biodiversity & conservation-a case study from Europe & Turkey. In: Bhatti R, Soomro SA (eds) Proceedings of second international conference on "biodiversity is our life" (IC Biour-Life). Center for Biodiversity & Conservation, Shah Abdul Latif University, Khairpur, pp 1–12
Rachkovskaya EI, Ogar NP, Marynich OV (1999) Rare plant communities of the steppes of Kazakhstan and their protection. Steppe Bull Novosibirsk 79:63–70
The National Atlas of the Republic of Kazakhstan, Vol 1 Natural Conditions and Resources (2010)
The Republic of Kazakhstan (2010) Natural conditions and resources vol 1, p 505
The State cadastre of plants in Mangistau region in 3 books, Main Botanical Garden of the Republic of Kazakhstan, p 368, 2006
Toderich KN, Shuyskaya EV, Ozturk M, Juylova A, Gismatulina L (2010) Pollen morphology of some Asiatic species of genus *Salsola* (Chenopodiaceae) and its taxonomic relationships. Pak J Bot, Special Issue (SI Ali Festschrift) 42:155–174
Volkova EA (2003) The vegetation cover of the mountains. Botanical geography of Kazakhstan and Central Asia, St. Petersburg

Chapter 13
Ecology and Environmental Aspects of "Makmalzoloto" Gold Mining Area in Kyrgyzstan

Nazgül Imanberdieva, Nazima Chukunkyzy, Zeki Severoğlu, and Zheenbek Kulenbekov

Abstract This investigation enlightens the original data on the ecological situation of Makmal Valley and neighboring villages Makmal and Chetbulak Toguz-Toro in "Makmalzoloto" gold mining area in Kyrgyzstan. The study was undertaken to put forward the possibilities for preservation of rich biodiversity and ecological state in these villages. The area is under pressure from the activities of gold extraction. Even today environmental pollution and high levels of radiation are observed. The extraction activity has seriously damaged and produced a negative impact on pastures, livestock, as well as biodiversity in this region. The dosimetric studies have revealed that in the village Chetbulak, radiation levels lie around 38–40 mcR/h. Since the first extraction from the plant, the tailing pond is lying in the open with no protection or warning signs. The cattle grazes in the area openly all year-round; children lead the cattle herds and play freely around the tailing pond.

Keywords Gold mine · Plants · Heavy metals · Kyrgyzstan

Abbreviations

CUC	Common-use channel
GRF	Gold recovery factory

N. Imanberdieva (✉)
Biology Department, Kyrgyzstan-Turkiye Manas University, Bishkek, Kyrgyzstan

N. Chukunkyzy
Kyrgyz State University, Bishkek, Kyrgyzstan

Z. Severoğlu
Biology Department, Kyrgyzstan-Turkiye Manas University, Bishkek, Kyrgyzstan

Biology Department, Faculty of Science and Arts, Marmara University, Istanbul, Turkey

Z. Kulenbekov
American University of Central Asia, Bishkek, Kyrgyzstan

D. Egamberdieva, M. Öztürk (eds.), *Vegetation of Central Asia and Environs*,
https://doi.org/10.1007/978-3-319-99728-5_13

HPP	Heat and power plant
MAE	Maximum allowable emission
SES	Sanitation and epidemiological station
SNiP	Construction Norms and Rules

13.1 Introduction

According to Jenkins and Yakovleva (2006), concerns about the sustainability and social responsibility of gold mining are becoming an increasingly high-profile issue in many countries. These authors have presented recent trends in the impacts and issues in the global mining industry. Gold mining has created several environmental problems in different parts of our globe. Not much has been published on the effects of an increased load of sediments in aquatic ecosystems. The water courses affected by gold mining are generally reported to show high concentration of suspended sediments together with higher concentrations of mercury. The sediment loads of the polluted stream have been reported to be around 95.6% produced by the eroding goldfields (Mol and Ouboter 2004a, b). The fishes found in the streams affected by gold mining are reported to show low species diversity, low proportion of young fishes, and high proportion of midchannel surface-feeding fishes, and fishes adapted to low light, low proportion of visually orienting fishes (Mol and Ouboter 2004a, b). The erosion originating from the gold mining reduces fish diversity and shifts community structure. The sediments stored in the streambed will pose a pollution problem long after the gold mining stops (Mol and Ouboter 2004a, b).

Studies undertaken on the Madeira River Basin around the gold mining area (Malm et al. 1990) have shown that gravimetric separation of heavy and fine particles amalgamated with mercury and gold is recovered by heating the amalgam, and nearly 5–30% of the mercury is lost during this process which goes to the rivers, and 20% is released to the atmosphere by burning. These authors have reported 292 and 3.2 μg/m^3 atmospheric mercury concentrations in the vicinity of the mining area. They have reported higher concentrations in water, suspended particles, and bottom sediments in forest streams adjacent to the river. The values reported for forest streams are 3.05 ± 4.5 μg/L water, 1.21 μg/g dry weight in suspended particles, and 10.2 ± 30.3 μg/g dry weight in bottom sediment; in the river these values were 1.56 ± 2.9 μg/L, 0.50 ± 0.13 μg/g, and 0.13 ± 0.11 μg/g, respectively. Concentrations in fish have been reported to range from 0.21 to 2.7 μg/g wet weight.

Veiga et al. (2006) report that in over 50 industrializing countries, mercury is used by small-scale gold miners. Its release into the environment affects human health. Significant releases of mercury are associated with inefficient amalgamation techniques, estimated to lie between 800 and 1000 tons a year. Nearly 200–250 tons are released in China, 100–150 tons in Indonesia, and 10–30 tons each in Bolivia, Brazil, Colombia, Peru, the Philippines, Venezuela, and Zimbabwe. It enters these countries legally as imported material from countries in Europe who paradoxically continue to produce virgin Hg at government-owned mines. Toguz-Torouz Rayon is

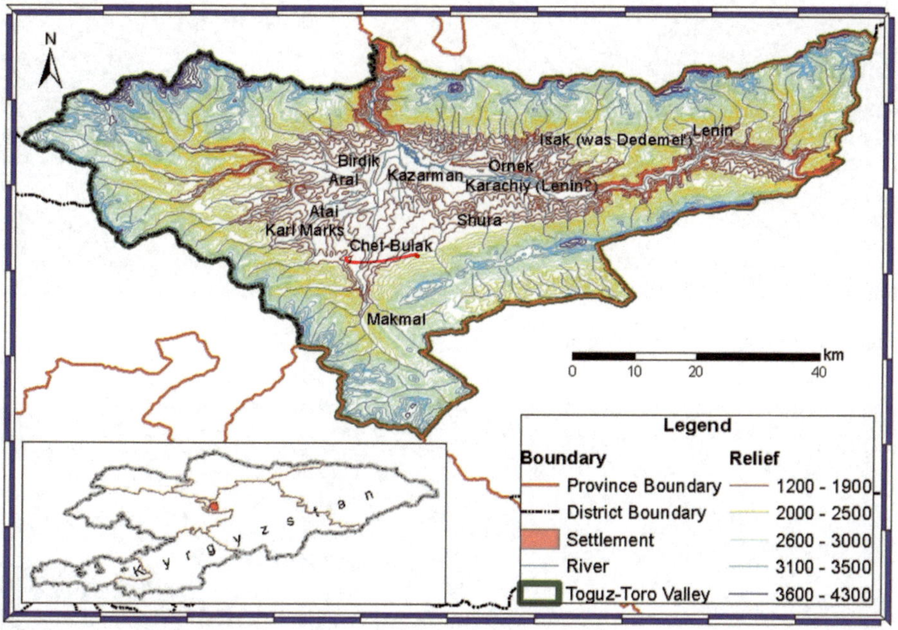

Fig. 13.1 Physiographic map of Toguz-Toro Valley

an administrative area of Jalal-Abad region in Kyrgyzstan. It is located within the southwest part of the country; the area makes 3962,8 km^2. The region is highly mountainous, at an altitude of 1200–2000 m above sea level. From the north it borders with Naryn region, from the west with Jumgal Rayon, from the east with Ak-Talaa Rayon, and from the south with Uzgen Rayon of Osh Region. Kazarman village is in the center of this region. The Naryn River separates the region into two parts. Toguz-Toro Valley is related to the southwest part of Internal Tien Shan, and from different directions it is surrounded with mountains such as Fergana-Too, Kekerim-Too, Kabak-Too, and Moldo-Too (Fig. 13.1). The Toguz-Toro Valley is rich in ore deposits (Yusupov 1961) like Tash-Tuz deposit (Besh-Kul, Ken-An, Tunuk-Tuz), a construction material (granite, clay, lime, sand, crushed stone, etc.), polymetallic ores (Batysh, Akshyirak, Kabak-Too, Kok-Yirim-Too), gold (Makmal), mercury (Baidamtal), wollastonite (Makmal), coal (Kek-Kyiya, Minteke) and many others (Popova 1946; Lebedev 1972; Asanov 2007).

From all ores mentioned above, the Makmal Altyn deposit holds a special place. It is located within the western part of Akshyirak ridge (Chaar-Tash), at the confluence of Fergana-Too Ridges, at the Chaar-Tash massive. The quarry is located at an altitude of 2500 m above sea level, 45 km to the south from Kazarman village, within the Makmal River basin. The ore processing is carried out since 1986. Distance between Makmal and Chetbulak villages is around 5 km (Fig. 13.2).

Activity of Makmal gold ore mining plant causes severe damage to the population and biodiversity of the region (Fig. 13.3). The mining activity and high radiation

Fig. 13.2 Mining and processing plant

Fig. 13.3 A general view of the "Makmalzoloto" Plant

have a negative impact on the ecology and biodiversity of the area, and this issue is of global urgency. The scientific data published to date reveals that the readings of radioactivity in the Chetbulak Village territory lie around 38–40 mcR/h (source). Similar findings have been reported by Viers et al. (2000) from Nyong River basin, Cameroon, as well as by Zhaoyong et al. (2015) in the surface water in the Tian Shan Mountains of China. The tailing pond of the mining plant is near the settlements and is uncovered, an open area, without any protection. The livestock always graze around the tailing pond (Fig. 13.4).

13.2 Study Area

The primary task here was to assist local residents to increase their awareness toward saving the biodiversity as well as make decisions on how to come up with a solution of the existing environmental situation. The impact of radioactive substances in plants leads to negative mutational changes. It is known that similar environmental problems would lead to breaking the balance of the whole ecological

Fig. 13.4 Tailing pond in "Makmalzoloto"

system. Studies were conducted at three most common sites for Chetbulak and Makmal villages of Toguz-Torouz Valley, with different degrees of effects originating from the Makmalzoloto gold ore mine on the vegetation and soil resources within the area.

Sites of Survey

1. Vegetation and Soils of Makmal Village
2. Vegetation and soils around the tailing pond
3. Vegetation and soils at gold mining mountain section

The spectral analysis were performed in the central laboratory of the State Geological Agency of the Kyrgyz Republic. The collection of plants was carried out at the stage of flowering. In the plant determinations, the herbarium materials and flora books like *Kirghiz SSR Flora*, volumes I–XI (1950–1965), and *Plants of Central Asia*, volumes I–X (1968–1993), were used. Differentiation of grass stands by tiers was made based on major biomorphs, by sub-tiers – taking into account the height of active assimilating parts of adult components rather than generative sprouts. For geographical parameters, altitudes and GPS-12 were considered. Determination of horizontal and vertical composition of grass stands, total projective cover of soil with vegetation layer, and partial covering with certain types of species was implemented based on generally accepted method of sketches using the framing square and by eye.

13.3 Data Evaluation

No data on the biodiversity of Toguz-Torouz Rayon has been published till now. The soils and vegetation of Toguz-Torouz valley differ from other valleys of Internal Tien Shan because it is bordered by Fergana-Too Ridges from one side, where basically ephemeral-sagebrush deserts and savannah-type steppes exist (Yusupov 1961). Climatic conditions of the area are more or less favorable. Summers are humid and wet, but winters are less cold and more snowy. The amount of atmospheric precipitation is much more. Soils are rather diverse. At lower altitudes various gray desert soils are distributed in combination with alkaline lands, particularly in the valleys and at the foothills. With increase in the altitude, these soils are replaced by mountain steppe and mountain meadow soils, among which the brown wood soils with small spots are located. At the slopes and sliding rocks, the soil cover is in the stage of formation.

The vegetation is varying, but steppes are dominating almost everywhere, with highest distribution on the landscape; meadows and bushes show a lesser distribution. Deserts and swamps too are met with. Deserts are poorly developed and mainly dominated by wormwood communities. They occupy small areas on dry foothills of southern and adjacent slopes, e.g., *Festuca*, *Stipa*, *Ferula*, *Andropogon*, and other steppes, mixed up at places with *Acantholimon korolkovii* and *Onobrychis echidna* plants. Small spots of saltwort deserts are met on halophytic habitats. In the wormwood deserts, the grass stand is composed of *Artemisia maritime*, *A. serotina*, *A. terrae-albae*, and several other species. We also come across such taxa as *Kochia prostrata*, *Ceratoides papposa*, *Stípa capillata*, *Ferula ovina*, *Ceratocarpus arenarius*, *Salvia sclarea*, *Achillea micrantha*, *Bromus kalmia*, *Trigonella orthoceras*, *Dodartia orientalis*, *Centaurea squarrosa*, *Acantholimon lepturoides*, and the genera like *Galatella* and *Onosma*. Herbaceous cover is thin. The cover is around 30–40%.

In the geobotanical territory described here, the steppes are landscaped. In Toguz-Toro Rayon, they are widespread. Steppes occupy flat foothills, slopes, and smooth hilly tops, reaching up to subalpine elevations. Toguz-Torouz steppes show communities, many of which are not typical for the Internal Tien Shan but are typical for Western Tien Shan. These plant covers are represented by *Agropyron cristatum*, *Ferula*, *Prangos*, *Inula*, *Andropogon*, and others (Popova 1946). For Toguz-Toro steppes the representative communities are generally composed of *Festuca valesiaca*, *Stipa pennata*, and *S. capillata*, together with *Stipa sareptana*, *S. consanguinea*, and the taxa belonging to genera *Caragana* and *Achnatherum*. *Festuca valesiaca* dominates the steppes. Other taxa typical for the Western Tian Shan like *Koeleria cristata*, *Phleum pratense*, *Galium verum*, *Salvia sclarea*, *Stípa capillata*, *Artemisia serotina*, *Dipsacus fullonum*, *Ferula*, and *Astragalus* also grow here. Many different taxa belonging to the genera *Prangos*, *Ferula*, *Euphorbia*, and *Medicago* are also distributed here. Feather grass steppes show a fragmentary distribution. They are diverse by their floristic composition and cenological structure. There are massifs dominated by *Stípa capillata*, mixed up with such species as *Salvia sclarea*, *Centaurea squarrosa*, *Dodartia orientate*, *Scutellaria pseudocoerulea*, and other species from the genera *Thymus*, *Ziziphora*, and *Lagochilus* (Lebedev 1972). In some areas, especially riverbanks, there is a domination of violet *Stipa*. These steppes are more monochromatic and get violet-reddish shade during fruitification of *Stipa*. Areas of steppes with domination of *Stipa sareptana* are not large and usually occupy washed-out rubbish soils. Massifs where *Achnatherum* is distributed at the foot of the mountains are not large and monotonous as far as their floristic composition is concerned.

Ferula steppes also grow widely in this area, mainly dominating the rest of the territory. They cover flat mountain slopes, especially in the southwestern part. These steppes are quite original, against a background of different types of grasses. The vigorous bushes of *Ferula* plants grow higher with a height of 250–305 cm. *Prangos* steppes are less widespread here than *Ferula* steppes. *Ferula* and *Prangos* steppes sharply differ by their floristic composition, cenological structure, and developmental rhythms as compared to the bunch steppes. Plants of ephemeral type dominate their grass stands. *Astragalus sewerzowii*, *Eremurus*, *Agropyrum*, *Galium verum*, *Lagochilus*, *Euphorbia ferganensis*, *Artemisia serotina*, *Scaligeria allioides*, *Carex turkestanica*, *Trigonella*, *Festuca valesiaca*, *Scabiosa*, *Hedysarum*, and many other taxa grow here (Lebedev 1972). The coverage of soil by plants lies around 60–70%.

There are also *Andropogon ischaemum* steppes formed by *Bothriochloa ischaemum*. Their floristic composition is poor and monotonous: *Bromus tectorum*, *B. oxyodon*, *Alyssum desertorum*, *Arenaria serpyllifolia*, *Onobrychis chorassanica*, *Veronica agrestis*, and *Trigonella* species.

Elytrígia steppes are less spread. The grass stands here are formed by *Agropyron glaucum*, *Onobrychis pulchella*, *Hedysarum montanum*, *Scaligeria alaica*, *Artemisia ferganensis*, *Convolvulus pseudocantabrica*, and the species of *Peucedanum*. In spring time the ephemeral species of *Bromus*, *Borago*, *Malcolmia*, and *Arenaria* are prominent (Lebedev 1972).

There are tall grasses and subalpine and alpine meadows found in this area. *Ligularia* meadows have a significant place among these meadows. They include *Ligularia thompsoni*, together with *Rumex confertus*, *Potentilla*, *Allium aflatunense*, *Alopecurus pratensis*, *Fagopyrum songaricum*, *Paeonia intermedia*, *Pedicularis*, *Dactylis glomerata*, *Solenanthus tortousum*, *Heracleum sibíricum*, *Cerastium latifolium*, *Alchemilla retropilosa*, *Aquilegia karelinii*, *Geranium pratense*, *Myosotis suaveolens*, *Poa pratensis*, and the species of *Potentilla*, *Pedicularis*, as well as Alliaceae and Poaceae meadows. Floristic composition is rich and diverse, although there are few plants valuable as forage. *Polygonum coriarium* shows a floristic composition very similar to some meadows (Lebedev 1972). Plants of Toguz-Torouz geobotanic region are mainly used as spring, summer, and autumn pastures. Some meadows are used as haymaking areas.

During the gold extraction process at Makmalzoloto Plant, the activity affects the environment due to the emission of harmful chemicals into atmosphere, as well as wastewater discharge and solid waste disposal. The production and other activities in the area are carried out at two sites: the mine and the gold recovery factory (GRF).

The first source of pollution is related to pollutants emitted to atmosphere by the factory in the form of solids and gases. Most aggressive compounds are listed in Table 13.1 together with their aggressiveness coefficients.

The biggest emission is of hydrogen cyanide, mainly from the tailing pond area (Table 13.2). The gold recovery factory (GRF) contributes around 10.3% of total hydrogen cyanide emissions, and it shows one of the highest rates of aggression. It shows instability in the atmosphere. After entering the atmosphere, it gets

Table 13.1 Atmospheric pollutants emitted in the gold mine area

Compounds	A = aggressiveness coefficient
Lead	A = 3333
Manganese oxides	A = 1000
Welding spray	A = 1000
Anhydrous hydrogen fluoride	A = 200
Fluorides	A = 200
Cyanogen hydrogen	A = 100
Silicone oxid	A = 50
Chlorine	A = 33
Nitrogen dioxide	A = 25
Inorganic dust Sio 220–7%	A = 10
Suspended substances	A = 6,6
Hydrocarbons	A = 0,66
Carbone monoxid	A = 0,33

Table 13.2 Annual emission of pollutants into the atmosphere

Lead	0,0009 t/year or 0,002% of total emission mass
Iron-welding spray	0,3007 t/year or 0,068% of total emission mass
Manganese oxide	0,0089 t/year or 0,02% of total mass
Anhydrous hydrogen fluoride	0,0093 t/year or 0,021% of total mass
Fluorides	0,0054 t/year or 0,0121% of total mass
Hydrogen cyanide	1.3364 t/year or 2,96% of total mass
Chlorine	0.2410 t/year or 0,53% of total mass

decomposed into different components, which is due to catalytic character. The length of hydrocyanic acid stay in the atmosphere does not exceed 5–10 min. The annual amount of harmful substance emissions into the atmosphere in general for Makmalzoloto Mining Plant is around 45.10 mln t/year.

In Table 13.2 the maximum allowable quantities of emissions are presented. Different substances do not have the high values of aggressiveness, being 10 and 0.33, respectively. Contribution of substances with high aggressiveness coefficient is insignificant—3.613% of the total emissions. These substances on the basis of estimations of dispersions completely settle down on the industrial territory of gold recovery factory (GRF) and mine. Out of these the most aggressive are lead 0,0009 tons/year or 0.002% of the total emission mass; iron welding spray 0.3007 t/year or 0,068% of the total mass; manganese oxide 0,0088908 t/year or 0,02% of the total mass; hydrogen fluoride 0.0093 t/year or 0.021% of the total mass; fluorides 0.0054 t/year or 0.0121% of the total mass; hydrogen cyanide 1.3364 t/year or 2.96% of the total mass; and chlorine 0.2410 t/year or 0.53% of the total mass.

The calculations of dispersion as maximum allowable emissions show that the active pollution zone from pollutants emitted into the atmosphere is related to industrial areas and it is within the sanitary protection zones of each site. The Quarry C33 is 1000 m, for the gold recovery factory (GRF) C33 500 m, and for tailings 500 m.

The second source of pollution is wastewater generated during the process of "Makmalzoloto" Plant. There are three types of wastewaters: household-domestic, industrial, and surface. Domestic wastewater is produced during the operation of taps, dining rooms, showers, and other facilities related to the household human activities within the territory of industrial sites. Water is discharged following mechanical and biological treatment facilities located in the mine area, around the gold recovery factory (GRF) and Kazarman village. After treatment the discharge of domestic wastewaters is carried out from the mine site to Makmal, from the gold recovery factory (GRF) site, to the tailing pond and from the Kazarman village to the Naryn River.

The total volume of wastewater entering the treatment facility is approximately around 198 K m^3/year. Constant work to maintain the treatment effectivity is also carried out at wastewater treatment facilities, and also control of treatment quality is made both in the existing mine laboratory and at the Rayon sanitation and epidemiological station (SES) laboratory. For domestic fecal sewages, the concrete septic tanks/cesspools have been constructed at the industrial site of the mine. Fecal sewerage volume formed is estimated in accordance with the SNiP 2.04.03-85 (Construction Norms and Rules) and makes 133 m^3/year.

The third source of pollution is solid wastes. These are generated within the plant territory; they are domestic solid wastes including waste paper, food waste, and others, which are stored at the mine in special pits that are closed on top with floor slabs, and at the factory the wastes are disposed at special partitioned areas. Volume of solid waste generated is accepted according to the norms of SNiP 01-03-95 KR that makes 91,77 tons/year. Household wastes are transported to dump site of Kazarman village, located on the southern outskirts of the village.

Wastes generated after processing of ore and entering as pulp to the tailing pond are a commercial product, as the gold content is sufficiently high and further processing of such wastes is possible. Solid part of pulp is represented by ore crushed up to class of 0,074 mm mining at the Makmal deposit. The liquid pulp part is presented by water containing chemical reagents used at the factory. The liquid pulp part after settling is fully taken into the water circulation system of the gold recovery factory (GRF). The total industrial product stored in lies around approximately 430,000.

The rocks that were used in 1986 as the dump sites in the quarry are not wastes as their disposals are temporary. After working off the deposit, the rocks will be used for recultivation activity. These rocks are not toxic and do not contain acid-forming elements.

Barren rocks in the quarry are also not wastes of the plant, their location is temporary as today, and in the future the rocks will be used as the raw material for the gold recovery factory (GRF).

For solid waste disposal from the gold recovery factory (GRF) (scrap metal, sawdust, ash and slag wastes, barrels, and other containers after using reagents), the landfill site has been organized near the GRF territory and warehouse. Volume of waste disposed lies around 566 tons, but the volume is reduced slowly, because noncompleted lime in the plant is used in the technological cycle.

Following the calculations made to determine the hazard category of the enterprise of the mining complex (common-use channel (CUC) = 125.1), we come across the level IV hazard category. The processing complex (common-use channel (CUC) = 1447.6) has level III hazard category. Our studies have enlightened the fact that at the implementation of analysis for ash collected from plants and vegetation in the vicinity of the tailings and from the Makmal Village, the elemental contents are different. In the ash of plants gathered at the tailing pond, such elements like cobalt (Co), tungsten (W), zirconium (Zr), niobium (Nb), tantalum (Ta), antimony (Sb), bismuth (Bi), arsenic (As), zinc (Zn), and cadmium (Cd) were not detected, and in the plants collected from Makmal, such elements like chromium (Cr) and silver (Ag) too were missing in addition to those listed above.

In the ash of plants around the tailing pond, the strontium (Sr) content was around 100 that exceeds the dose by ten times. The same element in the ash of Chetbulak Village plants exceeded the dose by three times and lithium (Li) by seven times. In Makmal Village manganese content (Mn) was higher by three times. The remaining elements were normal or below normal. The results of spectral analysis of plants around the tailing pond showed that strontium (Sr) exceeds the dose almost double than the prescribed levels. The remaining elements are normal. In the analysis of soil, it was found that manganese (Mn) in the Makmal Village exceeds the normal level by 1.5 times and lithium (Li) by seven times; availability of beryllium (Be) is little more than normal.

Green and Renault (2008) have carried out a study on the suitability of adding papermill sludge to neutral/alkaline gold mine tailings to improve the establishment of *Festuca rubra*, *Agropyron trachycaulum*, and *Medicago sativa* and *Festuca rubra* root and shoot biomass. They report that *A. trachycaulum* shoot biomass increased with papermill sludge amendment. The addition of papermill sludge and fertilizer has increased the shoot and root biomass of *M. sativa* (20–30 times), while *A. trachycaulum* and *F. rubra* have shown a moderate increase in growth. Their study suggests that addition of papermill sludge and adequate fertilization can alleviate some of the adverse conditions of neutral/alkaline gold mine tailings.

13.4 Conclusions

Gold mining has a negative image for being highly polluting industry, but it has many benefits, especially in the industrializing countries through employment and foreign exchange earnings (Kumah 2006; Telmer and Veiga 2009; Mosquera et al. 2009). Anti-mining activists are often tepid on this issue to avoid being blamed for lack of sensitivity (Ali 2006). The rudimentary nature of small-scale gold mining activities often generates a legacy of extensive degradation and social conditions (Vieira 2006). The most popular form of small-scale gold mining is called "land dredging," which requires application of large volumes of water for both mining and mineral processing. The preferred method employed by these miners for gold recovery is using mercury—a dangerous pollutant. There has been a growing focus on the impact of small-scale gold mining on water resources over the last few decades. It is an important source of income for rural areas and drives economic development of the countries involved (Mol and Ouboter 2004a, b; Seccatore et al. 2014; Bashwira et al. 2014). However, it is a potential threat to the aquatic and terrestrial environment. Some countries have seen great progress in mining lately (Bakia 2014; Omang et al. 2014; Asaah et al. 2015; Vishiti et al. 2015). The gold mine sites have been exploited in a way which predisposes the water resources in these areas to serious environmental pollution (Bakia 2014; Djibril et al. 2017). To refine gold, companies often use toxic chemical products, such as cyanide and mercury (CAPAM 2016; Fashola et al. 2016). Rakotondrabe et al. (2018), Pahimi et al. (2015), Tehna et al. (2015), and Leopold et al. (2016) have studied the influence of gold mining on the water quality in Cameroon, Ibrahima et al. (2015) in Senegal, and Huang et al. (2010, 2016) in Tibet. The high concentrations of trace elements have been found here as a result of gold mining activities as well as digging of riverbeds, excavation, and gold amalgamation. Health effects too have studied in Nigeria (Taiwo and Awomeso 2017).

Gold prices have increased 360% lately (Swenson et al. 2011; Comercio 2009a, b). Poverty, ineffective institutions, and environmental regulations hinder the organization of the sustainable extraction of gold, and evidence is even available from space images (Butler 2009). The study undertaken by Swenson et al. (2011) in Peru reveals that mining deforestation has increased nonlinearly alongside a constant annual rate of increase in international gold price. Similar environmental destruction can be seen in Kyrgyzstan endowed with rich gold deposits. Some countries have reached environmental accords with gold mining companies, not using mercury. However, in remote areas no environmental impact analysis is done due to lack of education and awareness (Stenson 2006; Sousa and Veiga 2009). Major environmental threats caused by gold mining include deforestation, acid mine drainage, and air and water pollution from arsenic, cyanide, and mercury contamination (Fraser 2009). The miners are directly exposed to liquid mercury and vapors during gold processing. The gold price continues to set new records; as a response, informal gold mining has risen along with grave environmental and health consequences (Keane 2009).

Toguz-Toro Valley is surrounded by mountains and relatively small area. Here, many species of animals are distributed because of high altitude and suitable landscape. The region is rich in minerals. Topography of the area has the features of ancient glacial sculpture and is located at altitude of 2800–3000 m above sea level. Around 40 large and small rivers, which originate from the glaciers, are registered within the territory of the Toguz-Toro Valley. The flora of the valley is rich and diverse. The ecological situation of the rayon is getting worse due to the negative impact of radioactive substances from the natural ore processing. It is an urgent problem for the whole biodiversity of the valley and is expected to lead to disruption of the natural balance in the region. The protection of living organisms and prevention of global catastrophic actions are the responsibility of all of humanity. Makmal Altyn today is one of the largest sites in the gold mining sector of Kyrgyzstan. It is the place, where local population lives and cattle are grazing year-round. The tailing pond lacks any sort of fencing around it; there are no barriers and clearly marked-out warning signs. Analyses of plants and soils from the area have shown that limits of certain heavy metals exceed the normal safety levels. The impact from the activity of Makmal Altyn gold mining plant to ecological situation and biodiversity of the whole region is negative. There is need for a quick restoration process in the area.

References

Asaah AV, Zoheir B, Lehmann B, Frei D, Burgess R, Suh CE (2015) Geochemistry and geochronology of the ~620 Ma gold associated Batouri granitoids, Cameroon. Int Geol Rev 57:1485–1509

Asanov (2007) Toguz-Toro Region: Encyclopedia/Main edition-284

Bakia M (2014) East Cameroon's artisanal and small-scale mining bonanza: how long will it last? The futures of small-scale mining in sub-Saharan Africa. Futures 62:40–50

Bashwira MR, Cuvelier J, Hilhorst D, Van der Haar G (2014) Not only a man's world: women's involvement in artisanal mining in eastern DRC. Res Policy 40:109–116

Butler R (2009) High gold prices, army collaboration, play role in mining invasion in southern Venezuela. Mongabay.com. http://news.mongabay.com/2009/1124-venezuela.html

CAPAM (2016) Mission de suivi environnemental des sites d'exploitations minières artisanales dans les régions de l'Est et de l'Adamaoua. Technical repport (44p)

Comercio EL (2009a) Buried in Mercury. (30 Nov 2009). Lima Peru. http://elcomercio.pe/impresa/notas/sepultados-mercurio/20091110/366858. Accessed 5 Dec 2009

Comercio EL (2009b) Rising Gold Price Boosts Informal Mining in La Selva] (29 December). Lima Peru. http://elcomercio.pe/impresa/notas/madre-dios-problema-ecologico-sin-visos-solucion/20091229/387199. [In Spanish] Accessed 10 Jan 2010

Fashola MO, Ngole-Jeme VM, Babalola OO (2016) Heavy metal pollution from gold mines: environmental effects and bacterial strategies for resistance. Int J Environ Res Public Health 13:1047

Fraser B (2009) Peruvian gold rush threatens health and the environment. Environ Sci Tech 43:7162–7164

Green S, Renault S (2008) Influence of papermill sludge on growth of Medicago sativa, Festuca rubra and Agropyron trachycaulum in gold mine tailings: a greenhouse study. Environ Pollut 151(3):524–531

Huang X, Sillanpää M, Gjessing Egil T, Peräniemi S, Vogt Rolf D (2010) Environmental impact of mining activities on the surface water quality in Tibet: Gyama valley. Sci Total Environ 408:4177–4184

Huang X, Deng H, Zheng C, Cao G (2016) Hydrogeochemical signatures and evolution of groundwater impacted by the Bayan Obo tailing pond in northwest China. Sci Total Environ 543:357–372

Ibrahima M, Moctar D, Maguette DN, Diakher MH, Malick NP, Serigne F (2015) Evaluation of water resources quality in Sabodala gold mining region and its surrounding area (Senegal). J Water Resour Prot 7:247–263

Jenkins H, Yakovleva N (2006) Corporate social responsibility in the mining industry: exploring trends in social and environmental disclosure. J Clean Prod 14(3):271–284

Keane L (2009) Rising prices spark a new gold rush in Peruvian Amazon. 19 December. 10. Washington, DC, USA: Washington Post. http://www.washingtonpost.com/wp-dyn/content/article/2009/12/18/AR2009121804139.html

Djibril KNG, Cliford TB, Pierre W, Alice M, Kuma JC, Flore TDJ (2017) Artisanal goldmining in Batouri area, East Cameroon: impacts on the mining population and their environment. J Geol Min Res 9(1):1–8
Kumah A (2006) Sustainability and gold mining in the developing world. J Clean Prod 14(3–4):315–323
Lebedev LP (1972) Popular classifier of plants in Kirgizia, Frunze
Leopold EN, Sabine DD, Philemon ZZ, Jung MC (2016) Physical and metals impact of traditional gold mining on soils in Kombo-Laka area (Meiganga, Cameroon). Int J Geosci 7:1102–1121
Mol JH, Ouboter PE (2004a) Downstream effects of erosion from small-scale gold. Conserv Biol 18:201–214
Mosquera C, Chavez ML, Pachas VH, Moschella P (2009) Diagnostic study of artisanal mining activities in Madre de Dios Lima, Peru: Cooperación, Caritas, Conservation International. ISBN: 978-9972-865-07-7. http://www.minam.gob.pe/mnilegal/images/files/estudio_diagnostico_mineria_artesanal_madredios.pdf. Accessed 15 Dec 2009
Malm O, Pfeiffer WC, Souza CMM, Reuther R (1990) Mercury pollution due to gold mining in the Madeira River basin. Brazil Ambio 19(1):11–15
Mol JH, Ouboter PE (2004b) Downstream effects of erosion from small-scale gold mining on the instream habitat and fish community of a small neotropical rainforest stream. Conserv Biol 18(1):201–214
Omang BO, Bih CV, Fon AN, Embui V, Suh CE (2014) Regional geochemical stream sediment survey for gold exploration in the upper Lom Basin, Eastern Cameroon. Int J Geosci 5:1012–1026
Pahimi H, Panda CR, Ngassoum MB, Tchameni R (2015) Environmental impacts of mining in the volcano-sedimentary basins of Cameroon: case study of artisanal gold mine tailings (Betare Oya, East-Cameroon). Int J Energy Sustain Environ Eng 2:5–15
Popova LI (1946) Report on fodder and geobotanical situation in Toguz-Toro Rayon of Tien-Shan Oblast of Kyrgyzstan. Academy of Science of Kirgizistan, Frunze
Rakotondrabe F, Ngoupayou JRN, Mfonka Z, Abolo AJN, Ako AA (2018) Water quality assessment in the Bétaré-Oya gold mining area (East-Cameroon): multivariate statistical analysis approach. Sci Total Environ 610–611:831–844
Ali SH (2006) Gold mining and the golden rule: a challenge for producers and consumers in developing countries. J Clean Prod 14(3–4):455–462
Seccatore J, Veiga M, Origliasso C, Marin T, De Tomi G (2014) An estimation of the artisanal small-scale production of gold in the world. Sci Total Environ 496:662–667
Sousa RN, Veiga MM (2009) Using performance indicators to evaluate an environmental education program in artisanal gold mining communities in the Brazilian Amazon. Ambio 38:40–46
Stenson J (2006) Disaster management as a tool for sustainable development: a case study of cyanide leaching in the gold mining industry. J Clean Prod 14(3–4):230–233
Swenson JJ, Carter CE, Domec JC, Delgado CI (2011) Gold mining in the Peruvian Amazon: global prices, deforestation, and mercury imports. PLoS One 6(4):e18875. https://doi.org/10.1371/journal.pone.0018875
Taiwo AM, Awomeso JA (2017) Assessment of tracemetal concentration and health risk of artisanal gold mining activities in Ijeshaland, Osun State Nigeria-part 1. J Geochem Explor 177:1–10
Telmer KH, Veiga MM (2009) World emissions of mercury from small scale and artisanal gold mining. In: Pirrone N, Mason R (eds) Mercury fate and transport in the global atmosphere: emissions, measurements and models. Springer, New York, pp 131–172
Tehna N, Nguene FD, Etame J, Medza Ekodo JM, Noa Tang S, Suh EC, Bilong P (2015) Impending pollution of Betare Oya opencast mining environment (Eastern Cameroon). J Environ Sci Eng B 4:37–46
Veiga MM, Maxson PA, Hylander LD (2006) Origin and consumption of mercury in small-scale gold mining. J Clean Prod 14(3–4):436–447
Vieira R (2006) Mercury-free gold mining technologies: possibilities for adoption in the Guianas. J Clean Prod 14(3–4):448–454
Viers J, Dupre B, Braun JJ, Deberdt S, Angeletti B, Ngoupayou JR, Michard A (2000) Major and trace element abundances, and strontium isotopes in the Nyong basin rivers Cameroon

constraints on chemical weathering processes and elements transport mechanisms in humid tropical environments. Chem Geol 169:211–241
Vishiti A, Suh CE, Lehmann B, Egbe JA, Shemang EM (2015) Gold grade variation and particle microchemistry in exploration pits of the Batouri gold district, SE Cameroon. J Afr Earth Sci 111:1–13
Yusupov T (1961) Toguz-Toro Basin Soils-Soils of Naryn Basin Middle Part, Frunze
Zhaoyong Z, Abuduwaili J, Fengqing J (2015) Heavy metal contamination, sources, and pollution assessment of surface water in the Tianshan Mountains of China. Environ Monit Assess 187:33

Chapter 14
Floodplain Forest Mapping with Sentinel-2 Imagery: Case Study of Naryn River, Kyrgyzstan

Akylbek Chymyrov, Florian Betz, Ermek Baibagyshov, Alishir Kurban, Bernd Cyffka, and Umut Halik

Abstract The article presents the results of studies on the use of Sentinel-2 satellite data and the application of SNAP and ArcGIS software for the classification and mapping of forest cover of the Naryn river floodplain. Available inventory maps of the Kyrgyz Forestry Administration are outdated and do not meet the current requirements and need to be updated with the use of satellite images from different systems. High-resolution Sentinel-2A multispectral imagery has been used to study the supervised forest cover classification of the floodplain areas of the Naryn River in Kyrgyzstan for contributing to forest inventory and general analysis of the floodplain forest ecosystems. Using such high-resolution images in this study was due to the peculiar properties of classification and mapping of small vegetation areas of the unstable floodplains of mountain rivers. Supervised classification was performed using S2A MSI and WorldView-2 satellite images through SNAP software and field investigation data. Level-1C S2A multispectral images are processed to the Level-2A using Sen2Cor for the atmospheric corrections and further classification.

A. Chymyrov (✉)
Department of Geodesy and Geoinformatics, Kyrgyz State University of Construction, Transport and Architecture, Bishkek, Kyrgyzstan
e-mail: akylbek.chymyrov@acagiscience.org

F. Betz · B. Cyffka
Faculty of Mathematics and Geography, Catholic University Eichstaett-Ingolstadt, Ingolstadt, Germany
e-mail: florian.betz@ku.de; bernd.cyffka@ku-eichstaett.de

E. Baibagyshov
Department of Ecology, Naryn State University, Naryn, Kyrgyzstan

A. Kurban
Xinjiang Institute of Ecology and Geography, Chinese Academy of Sciences, Beijing, China
e-mail: alishir@ms.xjb.ac.cn

U. Halik
Key Laboratory of Oasis Ecology, College of Resources & Environmental Sciences, Xinjiang University, Urumqi, Xinjiang, China
e-mail: halik@xju.edu.cn

D. Egamberdieva, M. Öztürk (eds.), *Vegetation of Central Asia and Environs*,
https://doi.org/10.1007/978-3-319-99728-5_14

The research results show the usefulness of high-resolution Sentinel-2 imagery for land use and land cover classification as well as the best freely available tool for thematic mapping of riparian forests.

14.1 Introduction

In the semiarid climate of Kyrgyzstan, the rivers and their floodplains have a high ecological importance for biodiversity as well as for the supply of people with relevant ecosystem services (Karthe et al. 2015). For instance, riparian forests along the Naryn River provide timber, harvestable fruits, or pastoral land. Furthermore, they are important for mitigating erosion or—from a global perspective—for the climate system via storage of carbon (Betz et al. 2015). Despite the importance of these ecosystems, there is no modern monitoring. Available information is outdated and does not meet the requirements of a modern, effective forest inventory. For instance, there are no recent maps of the forest coverage (Betz et al. 2016). The need for monitoring the floodplain forests along a more than 250 km long stretch of the Naryn River together with a lack of infrastructure in the Naryn catchment makes remote sensing (RS) the method of choice for assessing the state and dynamics of riparian areas along the Naryn River.

Over the past decades, a wide range of algorithms has been developed for the processing of remotely sensed imagery and land cover classification (Gomez et al. 2016; Vorobyev et al. 2015). There are a number of satellite systems, such as Landsat, ALOS, and SPOT, which provide low-, medium-, and high-resolution images suitable for mapping and monitoring of the forested large areas. Improvements have been made in the assessment of riparian zones from satellite remote sensing. For the European Union, Clerici et al. (2013) presented a continent-scale delineation of floodplain vegetation combining digital terrain analysis with remote sensing and existing datasets. In a similar approach, Betz et al. (2016) estimated the extent and spatial distribution of riparian ecosystems along the Naryn River and its major tributaries. Despite the interesting insights of this Landsat-based assessment, the medium resolution of 30 m did only allow distinguishing between vegetation, non-vegetation, and water as the Naryn River shows partly narrow floodplain sections.

The use of the very high-resolution WorldView-2 image in land use/land cover (LULC) classification is investigated for mapping of the part of study area (Chymyrov et al. 2016). This research has developed large-scale and high-quality thematic maps (1:1000–1:10,000) for the forest management unit and ecosystem service management project EcoCAR (Ecosystem Assessment and Capacity Building for sustainable Management of Floodplains along the Central Asian Rivers Tarim and Naryn). But the extremely high cost of such satellite imagery limits their wider use in forest mapping in conditions of Kyrgyzstan.

For management issues, however, information about the exact land cover classes are required. Since 2015, the Sentinel satellites of the European Space Agency (ESA) offer a new remote sensing tool for environmental monitoring (Drusch et al.

2012). The sensors of the optical Sentinel-2 offer a wide range of spectral bands with spatial resolutions between 10 and 60 m suitable for manifold tasks of land observations (Drusch et al. 2012). Especially the 10 m resolution in the visible and near-infrared bands makes Sentinel-2 imagery an interesting dataset for monitoring small landscape features such as forest patches or narrow river channels (Radoux et al. 2016; Ozdogan and Kurban 2012). The available resolution has a high potential for thematic mapping of floodplain areas with a suitable degree of detail to support research and management of ecosystem services along the Naryn River.

The goal of this study is to create a land cover mapping system for the floodplain areas along the Naryn River using Sentinel-2 imagery. This up-to-date land cover information is the objective basis for further research on ecosystem services of the floodplain forests and can be used as baseline data for future land cover change analysis.

14.2 Material and Methods

14.2.1 Study Area

The Naryn River originates in the inner Tian Shan Mountains and flows westward across Kyrgyzstan toward the arid Fergana Valley. The major share of this 807 km long stream is in a near natural state without dams or extensive embankments. There is a number of hydropower stations located in the river basin including the largest Toktogul Reservoir with 19.5 cubic km capacity. The river catchment upstream from the Toktogul Reservoir has a size of 52,130 km^2. The discharge regime is glacial to nival-glacial with one single peak in July. Climate is highly continental with a high-temperature amplitude between summer and winter as well as in generally low annual precipitation of 300 mm (Betz et al. 2016).

Riparian vegetation is mainly composed by Salicaceae like *Salix* spp. and *Populus* spp. Further frequently occurring species are *Tamarix* spp. and *Hippophae rhamnoides*. The relevant ecosystem services belong to the category of provisioning services according to the classification of the Millennium Ecosystem Assessment (MA 2005). People get fuel and construction wood or collect the berries of the sallow thorn. Furthermore, the floodplain forests are important pastoral land and are used for recreation (Betz et al. 2015, see also Fig. 14.1).

For our remote sensing analysis, we focus on the central part of the Naryn Basin as here the most extensive floodplain areas can be found due to the geomorphological character of the region (Betz et al. 2016, 2018). The area of interest (AOI) for the remote sensing data collection is defined with geographic coordinates (N41.450°, E75.031°; N41.352°, E75.031°; N41.450°, E75.691°; N41.352°, E75.691°) and has a total area of about 700 km^2 (Fig. 14.2). The Naryn River floodplain with a total area of 69.5 km^2 is selected for land cover classification and further investigations (Fig. 14.3a).

Fig. 14.1 (**a**) Grazing in floodplain forests; (**b**) cutting of trees for firewood (Photos: F. Betz (l) and B. Cyffka (r))

14.2.2 Satellite Imagery Acquisition and Preprocessing

Sentinel-2A images for the vegetation season 2016 were downloaded from the scientific data hub of the ESA Sentinel mission (ESA 2017a). Satellite imagery from this platform is available as top-of-atmosphere (TOA) data. A detailed list of the used S2A scenes is given in Table 14.1. Atmospheric correction has been carried out by using the Sen2Cor processor, a Python tool for conducting atmospheric correction as well as cloud, cloud shadow, and snow detection using the algorithms of Zhu et al. (2015). For the processing, the Sentinel Application Platform (SNAP) offers a comfortable open-source solution, and this desktop software is available for download from the ESA website (ESA 2017b).

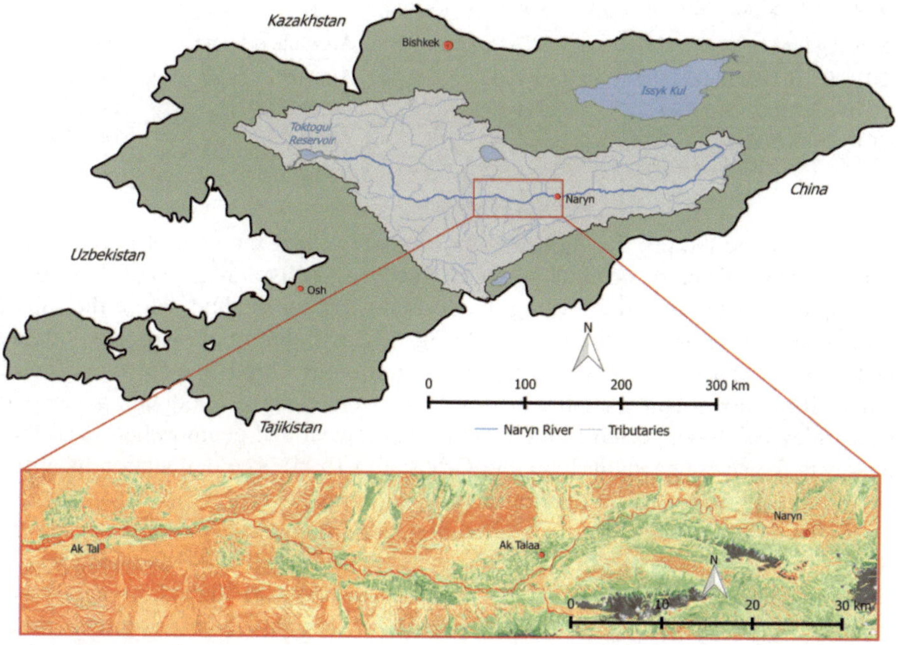

Fig. 14.2 Overview over the study area: the detailed map shows a NDVI composite of the area of interest

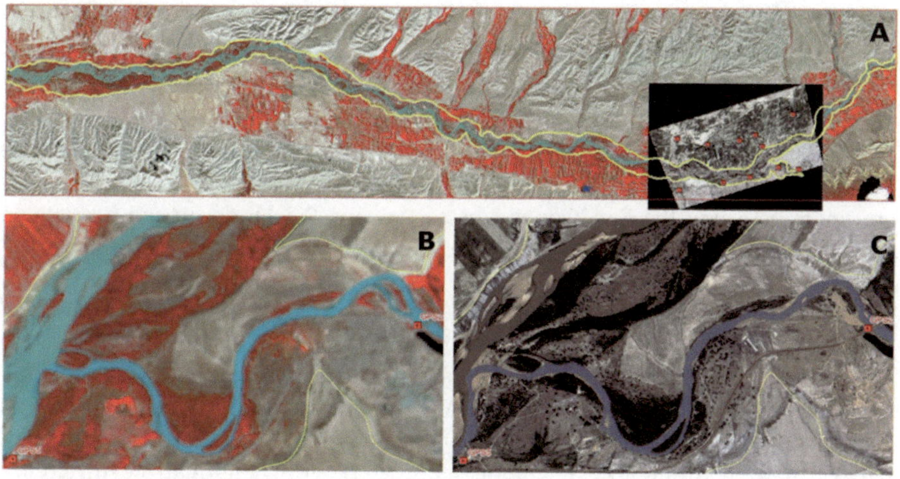

Fig. 14.3 (**a**) Multispectral satellite imagery for the area of interest (the floodplain is delineated with a yellow line); (**b**) false color image (S2A, 2016, 10 m resolution); (**c**) natural color image (WV-2, 2014, 0.5 m resolution)

Table 14.1 Sentinel-2A MSI granules used in this study

Sensing time	Processing level	Absolute orbit no.	Tile no.
2016-06-10T05:52:06.158Z	Level-1C	A005050	T43TEF
2016-07-20T05:51:55.331Z	Level-1C	A005622	T43TEF
2016-08-09T05:54:34.076Z	Level-1C	A005908	T43TEF

To minimize the processing time, the band subsets of the area of interest have been extracted from the original granules. The 20 and 60 m image bands have been resampled to 10 m resolution using the nearest neighbor method before the subset processing for the area of interest (AOI) (Fig. 14.3a, b). Preliminary product quality check shows that all three single-tile S2A images are cloud-free for the selected Naryn River floodplain section. For this study, we define the floodplain as the part of a valley being connected to the recent hydrological and geomorphological processes in the river channel (Hupp and Osterkamp 1996). The delineation has been performed via the visual interpretation of satellite imagery.

For calibrating a supervised classification, a training dataset has been generated during field surveys in the 2015–2016 vegetation seasons and via visual interpretation of a WorldView-2 image (14JUN29060129-S2AS) with 50 cm horizontal resolution. Georeferencing control and correction of this very high-resolution image were realized by using 11 ground control points (GCPs) positioned by a geodetic class Trimble R-8 GNSS receiver (Fig. 14.3b). These GCPs have been used to confirm the declared 12.5–20 m geo-location accuracy of Sentinel-2 imagery (ESA 2012).

14.3 Level-2A Image Processing

The Level-2A (L2A) processing includes an atmospheric correction applied to TOA S2A Level-1C products. L2A main output is an orthoimage bottom-of-atmosphere (BOA)-corrected reflectance product. Atmospherically corrected Level-2A images are produced using the Sen2Cor processor (version 2.3.1), developed by ESA to perform atmospheric, terrain, and cirrus correction of top-of-atmosphere Level-1C input data (ESA 2017c).

The scene classification algorithm allows the detection of clouds, snow, and cloud shadows and generation of a classification map, which consists of different classes for clouds, shadows, cloud shadows, vegetation, bare soil or desert, water, and snow (Fig. 14.4). The classification algorithm is based on a series of threshold tests that use as input TOA reflectance of spectral bands, band ratios, and indexes like normalized difference vegetation index (NDVI) and normalized difference snow and ice index (NDSI). Preliminary analysis of the automated L2A classification shows that further image processing is needed to develop more detailed and specific land use/land cover classification.

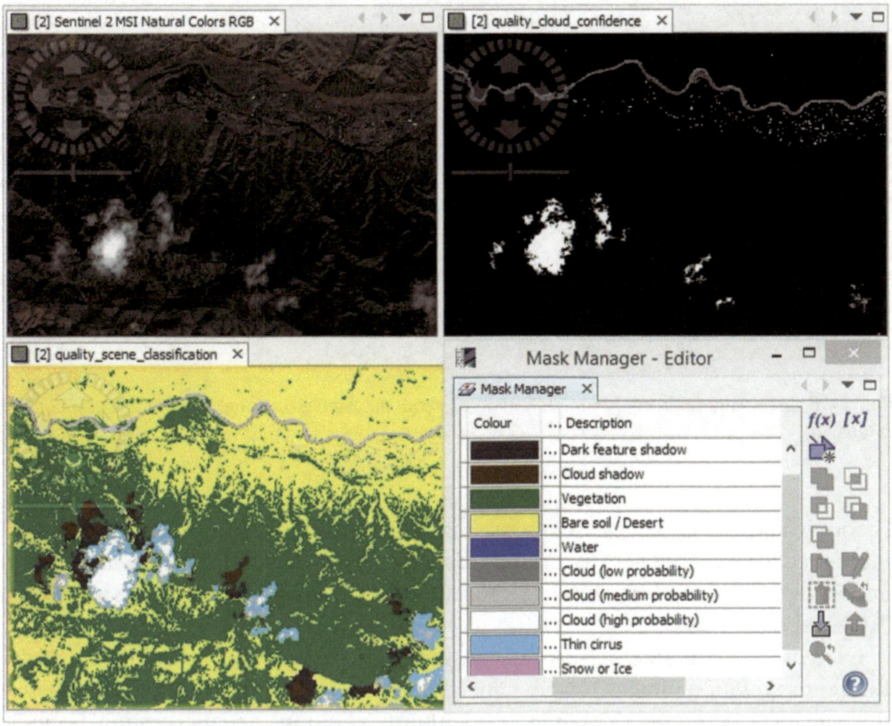

Fig. 14.4 Level-2A product with RGB image, quality indicators, and classification map

The normalized difference vegetation index (NDVI) is a numerical indicator that uses the visible and near-infrared bands of the electromagnetic spectrum adopted to analyze remote sensing materials and assess live green vegetation content. NDVI has been used to estimate vegetation health, crop yields, pasture conditions, and rangeland-carrying capacities, among others. NDVI can be derived using the near-infrared and red spectral bands that are most sensitive to vegetation information.

The NDVI algorithm subtracts the red reflectance values from the near-infrared and divides it by the sum of near-infrared and red bands:

$$\mathrm{NDVI} = (\mathrm{NIR} - \mathrm{RED}) / (\mathrm{NIR} + \mathrm{RED}).$$

This formulation allows coping with the fact that two identical patches of vegetation could have different DN values if one were, for example, in bright sunshine and another under a cloudy sky. The bright pixels would all have larger values and therefore a larger absolute difference between the bands. This is avoided by dividing the sum of the reflectance values and the NDVI ratio for Sentinel-2 is outlined below:

$$\mathrm{NDVI} = (\mathrm{Band8} - \mathrm{Band4}) / (\mathrm{Band8} + \mathrm{Band4}).$$

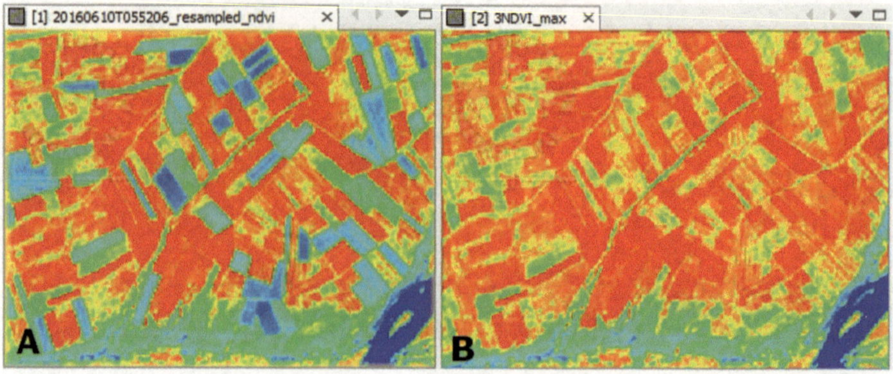

Fig. 14.5 S2A image from June 10, 2016 (**a**), and the stacked image with maximum NDVI values (**b**)

Theoretically, NDVI values are represented as a ratio ranging in value from −1 to 1, but in practice extreme negative values represent artificial surfaces and water, values around zero represent bare soil, and values over 0.6 represent dense green vegetation (Akkartal et al. 2004).

Each of the three Sentinel-2A images from a different sensing time demonstrates different vegetation indices for the same land plot because of the strong dependence of vegetation mass from seasons, agricultural planting, and harvesting periods. Three different NDVI images received on June 10, July 20, and August 9, 2017, give different NDVI values for the same land plot and individually may lead to the wrong land cover classification (Fig. 14.5a). Stacking of these three NDVI images with calculated maximum pixel NDVI values gives new raster with leveled NDVI for the improved agricultural land use classification (Fig. 14.5b).

The preliminary analysis of three different NDVI images shows that calculated minimum pixel NDVI values are more suitable for the floodplain forestry classification. It can be explained by more stable vegetation biomass of forested areas in comparison to the grass vegetation more vulnerable in the dry and wet vegetation periods of the year. Classification of NDVI image from June 10 has many land plots misclassified as mixed forest areas (Fig. 14.6a), and the off-season NDVI image with calculated maximum pixel NDVI value has removed cropped fields or grass vegetation from the mixed forest areas (Fig. 14.6b).

14.4 Supervised Classification of NDVI Imagery and Raster Generalization

In this study, supervised land cover classification of SNAP-processed NDVI imagery, with combined minimum values from three S2A scenes, is realized using field survey datasets, available WorldView-2 image, and ArcGIS 10.4 software.

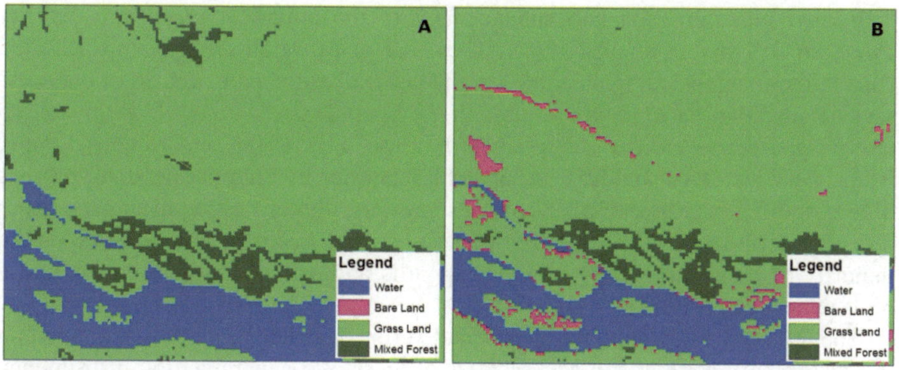

Fig. 14.6 Result based on the NDVI image from June 10, 2016 (**a**), and based on the stacked image with minimum NDVI values (**b**)

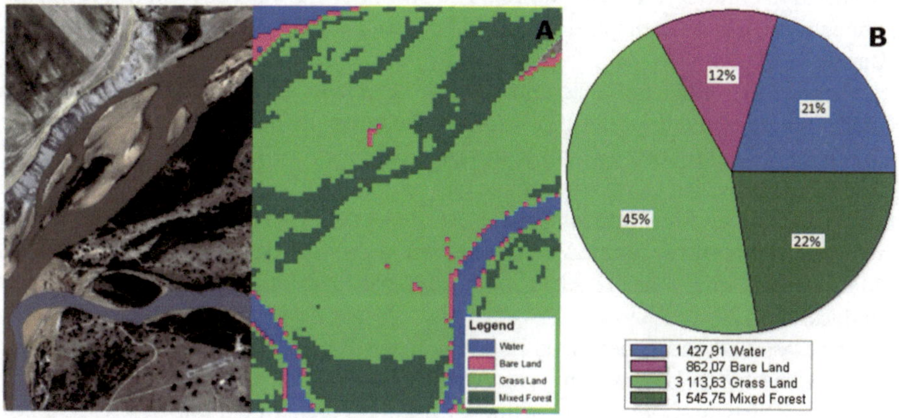

Fig. 14.7 (**a**) Land cover classification control; (**b**) floodplain land cover types (the numbers in the legend gives the area in ha)

For classification of each of the three Level-2A Sentinel-2A images, the study proceeded in the next steps:

- Calculation of NDVI values for each of images and calculation of the combined maximum and minimum NDVI values
- Supervised classification of the NDVI image into LULC classes
- Enhancement of the classification and creation of a final LULC map

Four LULC classes were selected for the supervised classification of the study area. Urban land and artificial surfaces are not significant and not included in the floodplain classification, rock surfaces and gravel/sand areas are classified as bare land, and mixed forest areas include trees and shrubs. A verification of the classification was performed by using field data and the visual interpretation of a WV-2 image. Along with the map output, a pie chart was used to display the pattern of the different land cover types (Fig. 14.7a, b).

Further generalization and enhancement of the classified image are realized using ArcGIS spatial analysis tools. There are many small areas of misclassified cells in the image needing cleaning up and generalizing data to get rid of unnecessary details. Two basic tools available in the Spatial Analyst module applicable to removing artifacts from raster images are Boundary Clean and Majority Filter. Boundary Clean tool smooths the jagged boundaries by eliminating small patches of less significant groups of pixels of the same class, therefore simplifying the structure of raster images. The Majority Filter tool replaces cells in a raster based on the majority of their contiguous neighboring cells and satisfies two criteria before a replacement can occur.

The Majority Filter tool has been selected for the raster generalization based on the preliminary comparison and analysis of these two generalization instruments. The number of neighboring cells of a similar value must be large enough (either by being the majority or half of all the cells), and those cells must be contiguous around the center of the filter kernel. The second criterion concerning the spatial connectivity of the cells minimizes the corruption of cellular spatial patterns (Bartuś 2014).

The majority filtering is determined by the results of observations of the cell values found around the central cell. The procedure selection is realized by defining the number of neighboring pixels involved in the analysis. The variable is set to "4," which means that the calculations have only involved pixels adjacent to the edges of the central cell and the values of pixels located in the corners of the neighborhood are not be taken into account. After four iterations with majority replacement threshold, the raster image has reached the acceptable level of stability without further changes.

The classified and generalized image in raster format was converted into vector format shapefile polygons using the ArcGIS Conversion. Cartographic smoothing with the PAEK smoothing algorithm is used for generalization of LULC class polygons to improve its aesthetic quality. The final versions of the thematic maps are prepared and printed out for the Naryn forestry department in different scales (Fig. 14.8) and for the ecosystem service research purpose within the project "EcoCAR" (Betz et al. 2015).

14.5 Results and Discussion

Sentinel-2A Level-1C images for the 3 months of 2016 vegetation season were downloaded and preprocessed applying the Sentinel Application Platform (SNAP). Output Level-2A (L2A) products have atmospheric, terrain, and cirrus correction of top-of-atmosphere L1C input data. SNAP has been used for the resampling of 20 and 60 m image bands to 10 m resolution and further subset processing of all bands for the area of interest (AOI).

The use of multi-temporal Sentinel-2A images results in different vegetation indices for sample plots and LULC classifications, with maximum and minimum NDVI values facilitating an increase in classification accuracy depending on the

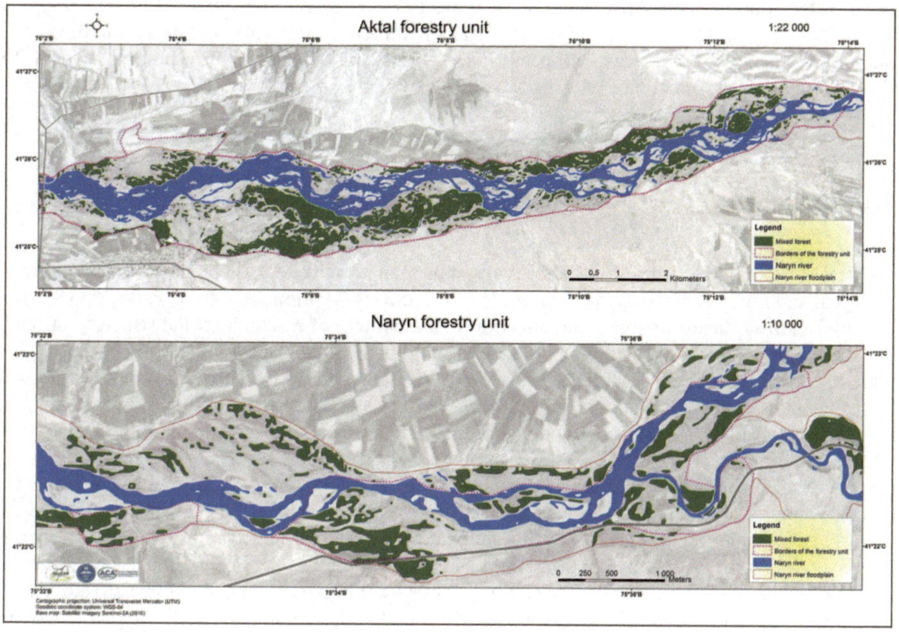

Fig. 14.8 Thematic maps for the forestry department

analysis purposes. Classification of the stacked image with maximum NDVI cell values will reduce the misclassification between crop and bare land due to the artificially impacted seasonal vegetation. Application of the stacked image with minimum NDVI values will improve the identification of mixed floodplain forest areas by excluding the grassland more affected by the seasonal drought.

The supervised classification is used to identify four LULC classes for the study area as water surface, bare land, grassland, and mixed forest areas with trees and shrubs. The final classification results are converted into polygon shape files for further map design and land use/land cover analysis.

The classification accuracy assessment results for the study area indicate that the Sentinel-2A multispectral imagery can be used efficiently in forest studies, providing the fundamental data source for examining LULC changes as well as for the mapping and monitoring purposes. Such satellite image has high classification accuracy (90–93%) for the investigation area with 25 sample plots. The major problem causing relatively low accuracy compared to the very high-resolution imagery classification was the misclassification between grass vegetation and mixed forest due to their complex vegetation stand structure and species composition and between initial succession (other vegetation) and mixed agro-pasture due to the lack of a clear boundary between them.

Application of the high-resolution Sentinel-2A images, enhanced by using different image correction and improvement tools and algorithms, makes it as one of the most advantageous and accessible multispectral optic satellite system for forest mapping and monitoring.

Acknowledgments This study is part of the project *Ecosystem Assessment and Capacity Building along the Central Asian Rivers Tarim and Naryn* funded by the Volkswagen Foundation. The authors would like to thank the European Space Agency (ESA) for the provision of Sentinel-2A data and Sentinel Application Platform.

References

Akkartal A, Türüdü O, Erbek FS (2004) Analysis of changes in vegetation biomass using multitemporal and multisensor satellite data. In: ISPRS archives—volume XXXV part B8, pp 181–185

Bartuś T (2014) Raster images generalization in the context of research on the structure of landscape and geodiversity. Geol Geophys Environ 40(3):271–284

Betz F, Cyffka B, Halik U, Kurban A, Chymyrov A, Baibagyshov E, Ding J (2015) GIS for environmental management along the Central Asian rivers Tarim and Naryn. In: Proceedings of the GIS in Central Asia conference—GISCA 2015 "geospatial management of land, water and resources", May 14–16, 2015, TIIM, Tashkent, Uzbekistan, pp 13–18.

Betz F, Rauschenberger J, Lauermann M, Cyffka B (2016) Using GIS and remote sensing for assessing riparian ecosystems along the Naryn River, Kyrgyzstan. Int J Geoinform 12(4):25–30

Betz F, Lauermann M, Cyffka B (2018) Delineation of the riparian zone in data-scarce regions using fuzzy membership functions: an evaluation based on the case of the Naryn River in Kyrgyzstan. Geomorphology 306:170–181. https://doi.org/10.1016/j.geomorph.2018.01.024

Chymyrov A, Baibagyshov E, Degembaeva N, Ismailov T, Urmambetova T, Aiypov B (2016) Study of the Naryn river floodplain forest ecosystem by using very high-resolution satellite imagery. Forest ecosystems under climate change: biological productivity and remote monitoring. Compendium of research papers/Executive editor Prof. E.A. Kurbanov. Volga State University of Technology, Yoshkar-Ola, pp 100–106

Clerici N, Weissteiner CJ, Paracchini ML, Boschetti L, Baraldi A, Strob P (2013) Pan-European distribution modelling of stream riparian zones based on multi-source Earth Observation data. Ecological Indicators 24:211–223

Drusch M, Del Bello U, Carlier S, Colin O, Fernandez A, Gascon F, Hoersch B, Isola C, Laberinti P, Martimort P (2012) Sentinel-2: ESA's optical high resolution mission for scientific observations of ocean, cryosphere and land. Remote Sens Environ 120:91–101

ESA (2012) Sentinel-2: ESA's optical high-resolution mission for GMES operational services (ESA SP-1322/2 March 2012)

ESA (2017a) Copernicus open access hub. https://scihub.copernicus.eu/dhus. Accessed 10 May 2017b

ESA (2017b) Sentinel application platform (SNAP). http://step.esa.int/main/download. Accessed 10 May 2017

ESA (2017c) Sen2Cor. http://step.esa.int/main/third-party-plugins-2/sen2cor. Accessed 10 May 2017

Gomez C, White JC, Wulder MA (2016) Optical remotely sensed time series data for land cover classification: a review. ISPRS J Photogramm Remote Sens 116:55–72

Hupp CR, Osterkamp WR (1996) Riparian vegetation and fluvial geomorphic processes. Geomorphology 14:277–295

Karthe D, Chalov S, Borchardt D (2015) Water resources and their management in Central Asia in the early twenty first century: status, challenges and future prospects. Environ Earth Sci 73(2):487–499

MA (Millenium Ecosystem Assessment) (2005) Ecosystems and human well-being: synthesis. Island Press, Washington, DC

Ozdogan M, Kurban A (2012) The effects of spatial resolution on vegetation area estimates in the lower Tarim basin, NW China. In: Chen J, Wan S, Henebry G, Qi J, Gutman G, Sun G,

Kappas M (eds) Dryland East Asia (DEA): land dynamics amid social and climate change. HEP-De Gruyter, Berlin
Radoux J, Chome G, Jaques DC, Waldner F, Bellemans N, Matton N, Lamarche C, d'Andrimont R, Defourny P (2016) Sentinel-2's potential for sub-pixel landscape feature detection. Remote Sens 8:488. https://doi.org/10.3390/rs8060488
Vorobyev O, Kurbanov E, Gubaev A, Demisheva E, Воробьев ОН (2015) Method of stepwise classification of satellite images for the thematic mapping of forest cover. Bull Volga State Univ Technol Russia 4(28):57–72
Zhu Z, Wang S, Woodcock CE (2015) Improvement and expansion of the Fmask algorithm: cloud, cloud shadow, and snow detection for Landsats 4-7, 8, and Sentinel 2 images. Remote Sens Environ 159:269–277

Chapter 15
Potential Impacts of Climate Change on Plant Diversity of Sary-Chelek Biosphere Reserve in Kyrgyzstan

Nazgül Imanberdieva, Bazaraliev Imankul, Zeki Severoğlu, Volkan Altay, and Münir Öztürk

Abstract The Sary-Chelek Nature Reserve established in 1959 with an area of 239 km^2 and declared as a UNESCO-MAB Biosphere Reserve in 1978 is located on the southeast-facing slope of the northern Chatkal Range in central Jalalabad province. It is centered around the scenic Sary-Chelek Lake. The mountains immediately surrounding the lake abound in exceptionally diverse flora and fauna, what may be the widest variety found in the Tian Shan. The area is spread over varied ecosystems and habitats, which include the forests of spruce and juniper together with the forests of wild fruits and nuts and riparian forests as well as a variety of steppe and meadow ecosystems. The total number of vascular plants growing in the study area is 668 taxa. These belong to 69 families and 313 genera. 531 taxa are from the dicotyledonous group, whereas 127 taxa are from the monocotyledonous group. This chapter presents an analysis of the potential impacts of climate change on plant diversity of this biosphere reserve.

Keywords Sary-Chelek · Biosphere reserve · Biodiversity · Climate change

N. Imanberdieva
Biology Department, Kyrgyzstan-Turkiye Manas University, Bishkek, Kyrgyzstan

B. Imankul
Sary-Chelek Government Biosphere Reserve, Jalal-Abad Oblast, Kyrgyzstan

Z. Severoğlu
Biology Department, Kyrgyzstan-Turkiye Manas University, Bishkek, Kyrgyzstan

Biology Department, Faculty of Science and Arts, Marmara University, Istanbul, Turkey

V. Altay
Biology Department, Faculty of Science and Arts, Hatay Mustafa Kemal University, Antakya, Hatay, Turkey

M. Öztürk (✉)
Vice President "The Islamic World Acadmey of Sciences" Amann-Jordan and Centre for Environmental Studies-Botany Department, Ege University, Izmir, Turkey

D. Egamberdieva, M. Öztürk (eds.), *Vegetation of Central Asia and Environs*,
https://doi.org/10.1007/978-3-319-99728-5_15

15.1 Introduction

Biosphere reserves are established by the UNESCO directly to bring together biodiversity, cultural diversity, and ecosystem services, in order to promote the ecological security and models for sustainable development (Bridgewater and Babin 2017). After an adoption of several strategies, the number of sites in the world biosphere network (WNBR) has shown a rapid growth in the extent and across biodiversity hotspot countries. This depicts the promising progress in the recognition of the value of the biosphere program for providing achievable models for conservation and sustainable use of natural resources (Ishwaran 2012; Brenner and Job 2012; Coetzer et al. 2013; Van Cuong et al. 2017). After the establishment of first biosphere reserves in 1976, WNBR has included 669 sites from 120 countries with a total area of more than 600 million ha in the list (UNESCO 2015; Van Cuong et al. 2017; Lowell 2017). The aim behind the UNESCO-MAB Biosphere Reserve program is to preserve natural and cultural landscapes of global importance including both unique ecosystems and traditional cultures (UNESCO 2015). Our study area "The Sary-Chelek" set aside as a "State Nature Reserve" was designated as a biosphere reserve in 1978 by UNESCO. Its exceptionally high biodiversity resulting from the valleys is unique with lush green forest ecosystem full of spruce and fir trees together with wild fruits and nut forests occurring together (UNESCO 2015). Our objective in this chapter has been to analyze potential impacts of climate change on plant diversity of Sary-Chelek Biosphere Reserve Area.

15.2 Brief Information on Biosphere Reserve

This reserve is found in the central Jalalabad province and reached via the town of Tashkomur, located on the Bishkek-Jalalabad highway, on the northern fringe of the Ferghana Valley. It was first established in 1959 with an area of 239 km^2. The reserve is centered around the scenic Lake Sary-Chelek. The reserve is full of exceptionally diverse flora and fauna. This biodiversity is found all over the mountains immediately surrounding the lake. In 1978 the entire area was declared as a UNESCO-MAB Biosphere Reserve, with administrative headquarters in the town of Arkyt, located inside the reserve about 8 km south of the Lake Sary-Chelek (Farrington 2005). The elevations range from approximately 1200 m on the Khodja-Ata River lying on the reserve's southern boundary to 4247 m in the Chatkal Range lying on the northern boundary of the reserve. The lake is located at an elevation of 1876 m in the center of reserve, with an area of approximately 13.9 km^2. Out of the lake waters, the cliffs and steep mountain slopes rise abruptly (Farrington 2005).

The climate in the study area is humid and warm, the reason being high mountains surrounding the reserve which protect it from the cold air masses flowing from the north. In winter the temperature in the reserve is higher than in the surrounding valleys, and in summer it is not lower than in other parts of southern Kyrgyzstan.

The average annual rainfall is 993 mm, mostly falling during the months of March to May, and average monthly precipitation is around 120–150 mm; least rainfall occurs in August and September with an average of 37 mm per month. Average annual temperature is 7.9 °C, and hottest month is August with an average temperature of 19 °C, while the coldest month is January with an average temperature of −5.3 °C (Magnuszewski et al. 2015).

The area before delineation as a reserve was used intensively by the villagers of Arkyt for collecting walnuts, timber cutting for fuel and lumber, as well as for grazing livestock. Following establishment of the reserve, restrictions were placed on these subsistence activities; however, Sary-Chelek became a popular destination for Soviet tourists. The growing tourism industry is expected to help support the families who reside in the village of Arkyt today (Farrington 2005).

We come across two zones in the reserve; 90% of reserve territory forms the core zone with an area of 215 km^2 which is nominally closed to the entrances and touristic visits, and a 10% of reserve territory with an area of 24 km^2 forms the multiuse zone, where villagers of Arkyt living in the area are permitted to utilize natural resources at a subsistence level (Mambetaliev and Shukurov 2002; Farrington 2005).

There are some sacred sites in the reserve. These include the tomb of Toskol-Ata situated on the south shore of the lake, a much respected local spiritual person. The facilities for recreation include homestays in the village of Arkyt and horse and foot trekking. A 3–5 days trekking visit takes a start around the Talas River valley, near the town of Kyzyl Oktyabr. The tour group first crosses the crest of the Talas Range, then a spur of the Chatkal Range, and finally descends to the Lake Sary-Chelek (Farrington 2005). The basic conservation issues in the study area resulting from the activities of the local residents of Arkyt include mainly over cutting of trees for firewood and overgrazing by livestock. As far as the excessive tourist pressure in the reserve, it hardly totals about 1000 persons a year. The program includes trampling of lake-shore vegetation, picking of endangered medicinal plants and wildflowers, and widespread littering (Mambetaliev and Shukurov 2002; Farrington 2005).

15.3 Plant Diversity of Sary-Chelek Biosphere Reserve Area

The Reserve is situated on the southeast-facing slope of the northern Chatkal Range, widely recognized for its unique ecological values as a UNESCO Biosphere Reserve (Grisa et al. 2008). Some of the taxa growing here include the walnuts which have been placed in the Red Book of Kyrgyzstan (Eastwood et al. 2009). The reserve has the widest variety of flora and fauna found in the Tian Shan area, a wide variety of ecosystems and habitats spread all over the reserve (Grisa et al. 2008). These ecosystems include the forests of spruce and juniper, the wild fruit and nut forests, riparian forests, and a variety of steppe and meadow ecosystems.

The fauna in the Sary-Chelek reserve includes the famous snow leopard listed as an endangered species, 35 species of mammals, 157 species of birds, brown bear,

Eurasian lynx, badger, golden eagle, lammergeyer, saker falcon, and peregrine falcon (Mambetaliev and Shukurov 2002). Floral list of the reserve consists of 668 taxa of higher plants, including Schrenk's spruce and wild apple and walnut trees, as well as several types of wild geraniums (Mambetaliev and Shukurov 2002). Floristic data published by Alimbaeva and Goncharova (1971), Borlakov and Golovkora (1971), Sultanova and Lazkov (1987), Borlakov (1996), Ozturk et al. (1996a, b), Mambetaliev and Shukurov (2002), Farrington (2005) and Eastwood et al. (2009) has been evaluated in the light of latest findings, together with other plant biodiversity investigations. The vascular plants whose status or name have changed, or have become synonyms, or have been included under new combinations have been corrected following the "www.plantlist.org."

The total number of vascular plants growing in the study area includes 668 taxa belonging to 69 families and 313 genera. 531 taxa are placed in the dicotyledonous group, whereas 127 taxa belong to the monocotyledonous group (Table 15.1). The following families are represented by the largest number of taxa: Poaceae (71 taxa, 10.63%), Asteraceae (62 taxa, 9.28%), Lamiaceae (58 taxa, 8.68%), Rosaceae (55 taxa, 8.23%), and Fabaceae (54 taxa, 8.08%) (Table 15.2). The genera represented by the largest number of taxa are *Astragalus* (16 taxa), *Carex* (13 taxa), *Allium* (11 taxa), and *Poa* (10 taxa) (Table 15.3).

In addition to this, as far as the floristic structure of the Sary-Chelek Biosphere Reserve, coniferous trees and shrubs, deciduous trees and shrubs, wild fruit and nut trees, potential economic wildflowers, and medicinal plants are listed below.

Table 15.1 A general account of the floristic data from Sary-Chelek

			Angiospermae		
	Pteridophyta	Gymnospermae	Dicotyledonous	Monocotyledonous	Total number
Family	1	3	53	12	69
Genera	1	4	252	56	313
Taxa	2	8	531	127	668

Table 15.2 The richest families with numbers of taxa and their percentages

Family	Number of taxa	%
Poaceae	71	10.63
Asteraceae	62	9.28
Lamiaceae	58	8.68
Rosaceae	55	8.23
Fabaceae	54	8.08
Brassicaceae	41	6.14
Apiaceae	34	5.09
Boraginaceae	25	3.74
Caryophyllaceae	22	3.29
Ranunculaceae	20	2.99

Table 15.3 The numbers and percentages of the genera on the basis of taxa

Genera	Number of taxa	%
Astragalus	16	5.11
Carex	13	4.15
Allium	11	3.51
Poa	10	3.19
Silene	9	2.88
Potentilla	9	2.88
Rosa	9	2.88
Nepeta	9	2.88
Draba	8	2.56
Euphorbia	8	2.56
Scutellaria	8	2.56

The endemic plants found in the study area:

Abies sibirica ssp. *semenovii, Tulipa greigii, Allium oreoscordum, Silene fetissovii, Pseudosedum ferganense, Pyrus korshinskyi, P. regelli, Malus niedzwetzkyana, Astragalus aflatunensis, Oxytropis susamyrensis, Hedysarum chaitocarpum, H. santalaschi, H. turkestanicum, Bupleurum ferganense, Elaeosticta ferganensis, Seseli fasciculatum, S. unicaule, Phlomoides urodonta, Phlomis hypoleuca, Salvia schmalhausenii, Onosma ferganensis, Rindera tschotkalensis, Lactuca alaica, Kovalevskiella kovalevskiana, Hypacanthium echinopifolium, and Rhaponticum namanganicum (*Fig. 15.1*).*

The coniferous trees and shrubs found in the study area:

Abies sibirica ssp. *semenovii, Picea schrenkiana, Juniperus semiglobosa, J. polycarpos* var. *seravschanica*, and *J. pseudosabina.*

The deciduous trees and shrubs found in the study area:

Salix alatavica, S. iliensis, S. tianschanica, S. turanica, S. niedzwieckii, S. wilhelmsiana, Populus alba, Betula pendula, B. tianschanica, Celtis australis ssp. *caucasica, Spiraea pilosa, S. hypericifolia, Exochorda racemosa, Cotoneaster oliganthus, C. multiflorus, Sorbus tianschanica, S. turkestanica, S. persica, Rosa albertii, R. beggeriana, R. laxa, R. fedtschenkoana, R. maracandica, R. nanothamnus, R. platyacantha, R. kokanica, R. canina, Acer tataricum* ssp. *semenovii, A. platanoides* ssp. *turkestanicum, Rhamnus cathartica, Elaeagnus rhamnoides, Fraxinus sogdiana, Abelia corymbosa, Lonicera microphylla, L. stenantha, L. olgae, L. webbiana, L. korolkowii*, and *L. nummulariifolia.*

The major wild fruit and nut trees found in the study area are *Juglans regia, Berberis heteropoda, B. nummularia, Pyrus korshinskyi, P. regelli, Malus sieversii, M. kirghisorum, M. niedzwetzkyana, Crataegus songarica, C. pseudoheterophylla* ssp. *turkestanica, C. azarolus* var. *pontica, Prunus cerasifera, P. domestica, P. armeniaca, P. petunnikowii, P. mahaleb, P. erythrocarpa, P. verrucosa, Cerasus tianshanica*, and *Pistacia vera.*

Fig. 15.1 Some of the endemic plants in the study area. (**a**) *Malus niedzwetzkyana* (www.alchetron.com); (**b**) *Allium oreoscordum* (www.plantarium.ru); (**c**) *Oxytropis susamyrensis* (www.plantarium.ru); (**d**) *Hedysarum chaitocarpum* (www.plantarium.ru); (**e**) *Pyrus korshinskyi* (www.stories.rbge.org.uk); (**f**) *Tulipa greigii* (www.morena.kz)

The potential economically important wildflowers in the study area are the species of *Allium*, *Aquilegia*, *Anemone*, *Arum*, *Campanula*, *Colchicum*, *Eremurus*, *Fritillaria*, *Iris*, *Ixiolirion*, *Paeonia*, *Primula*, and *Tulipa*.

The medicinal plants distributed in the area are:

Equisetum arvense, *Juniperus semiglobosa*, *J. polycarpos* var. *seravschanica*, *Ephedra equisetina*, *Betula pendula*, *Oxyria digyna*, *Polygonum aviculare*, *Persicaria hydropiper*, *Capsella bursa-pastoris*, *Hypericum perforatum*, *H. scabrum*, *Elaeagnus rhamnoides*, *Carum carvi*, *Plantago major*, *Cichorium intybus*, *Tussilago farfara*, and *Inula helenium*.

15.4 Potential Impact of Climatic Changes on Plant Biodiversity

Multiple scales of widespread impacts on the biodiversity are expected, including genus, species, communities, and ecosystems, due to climate change (Bellard et al. 2012). The responses to climate change from the biological side widely vary among species and populations. Some of these responses are positive, leading to increased growth rates or range expansions; others are negative, resulting in localized or widespread declines (Montoya and Raffaelli 2010). Many species are shifting their geographical ranges, distributions, and phenologies rapidly than thought previously as a result of climate change. However, these rates are not uniform across the species (Ahmad et al. 2015). The backbone of ecosystem structure and function is biodiversity, which is fundamental to and controls the broad spectrum of goods and services from natural systems derived by us (Naeem 2009; Mace et al. 2012). The ecosystem function, persistence, and services are dependent to a large extent on the decrease or loss of any aspect of biodiversity, and it has direct as well as indirect impacts (Hooper et al. 2005). The keystone of the carbon cycle are plants. They are at the same time major regulators of global climate events; therefore if any plant species disappears, it will disproportionally affect the rurals, a large majority of whom rely on wild natural resources for daily uses as well as medicinal uses (Ahmad et al. 2015). An alteration of the abiotic conditions influencing biological systems is also affiliated to the climate change. The biological responses to its effects depend on several factors such as the rate magnitude, together with the character of the change, ecological sensitivity, and adaptive capacity to environmental change. All levels of biodiversity, like distribution, organization, and interactions, among the biota are shifting over spatial and temporal scales by a combined effect of these factors (Walther 2010; Ahmad et al. 2015).

The scientific community has come to a growing consensus that climate change is occurring. It is occurring and having impacts on biodiversity, and we have now an unequivocal evidence the Intergovernmental Panel on Climate Change (IPCC 2007). The absolute magnitudes of predicted changes are uncertain, but a high degree of confidence is existing in the direction of changes. There is a recognition that climate change effects will continue for many centuries (Ahmad et al. 2015). A conclusion reached by the "The United Nations IPCC" is that the global atmosphere is warming. The average global surface temperature has increased by nearly 1 °C over the past century and is likely to rise by another 1.4–5.8 °C over the next century (IPCC 2001). The atmospheric warming is affecting other aspects of the climate system such as pressure and composition of the atmosphere; temperature of surface air, land, water, and ice; water content of air, clouds, snow, and ice; wind and ocean currents; ocean temperature, density, and salinity; and finally the physical processes like precipitation and evaporation (Ahmad et al. 2015). This change is recognized as a major global threat (Stern 2007). It is expected to have significant impacts on many aspects of biodiversity in the near future (Campbell et al. 2009). The change takes place on a day-to-day basis. It will continue affecting biodiversity and will lead to biodiversity loss. Latter will produce negative effects on human well-being and natural systems (EEA 2010).

The climate changes will definitely threaten the sustainability of existing protected biodiversity components as well as their habitats in different ways and with different magnitude (Normand et al. 2007; Andrade et al. 2010). The expectations are that the species composition will be affected. It will also reduce the richness of taxa and will modify the functionality of many ecosystems substantially (Andrade et al. 2010). According to Araújo et al. (2011), several studies and projections claim that about 58% of the European terrestrial plants and vertebrate species will no longer find climatic conditions suitable for survival within the current protected areas by 2080. Several potential impact classes of climate change on habitats in protected areas have been reported by Sârbu et al. (2014) such as:

(a) Seasonality (changes of mean and maximum temperature, precipitation, frost, and snow days)
(b) Hydrology (decrease of precipitation during vegetation period, change in precipitation intensity and variability)
(c) Soil (change of soil structure, nutrients, and chemistry)
(d) Sea level rise (local coastal flooding)
(e) Extreme events (heavy rains, floods, drought, wildfire, storm)
(f) CO_2 concentration (increasing concentration)
(g) Cumulative effects (the shift in species composition and abundance, the invasion of aliens, pest development, land use changes)

A major impact of climate change on the coastal biodiversity in particular halophytic habitats will be in related to the sea level rise, which induces erosion of coastlines followed by impacts on soil structure (Sârbu et al. 2014). In the near future we can expect effects due to flooded seacoast on the wetland habitats with vulnerable halophyte vegetation (Gafta and Mountford 2008; Gaştescu and Ştiuca 2008; Strojan and Robic 2009; Sârbu et al. 2014). The complex habitats including pioneer terophyte and lichen communities growing on barren lands like old and decalcified inland dunes will get their full share from these changes. In particular changes of soil structure together with the hydrological changes can be considered as significant impacts (Gafta and Mountford 2008; Sârbu et al. 2014).

According to Sârbu et al. (2014), for standing and running water habitats, most significant impacts are hydrological changes affecting water regime and water level. In the case of latter waters, these factors can induce an irreversible alteration of floodplain areas, riverbed modification, or drying out of the river. The changes in temperature regime, which influence the development pattern of aquatic organisms and changes in bank ecotone conditions, will be the other relevant impacts. The water quality will deteriorate with a negative effect on microorganisms, benthic invertebrate, plankton species, and for different categories of aquatic macrophytes because of high water temperatures (Campbell et al. 2009). The changes regarding the structure of aquatic biocoenosis and the expansion of alien species will have cumulative effects (Sârbu et al. 2014).

The periodic variations will have significant impacts on temperate heaths and scrubs in the alpine zone as a result of the changes in temperature and precipitation regime, frost and snow days. The torrents and storms associated with erosion, ruptures, and habitat fragmentation will occur more often as extreme events. A cumulative effect from all these changes will lead to a potential loss of plant species strongly depending on humidity and low temperature, together with degradation and potential loss of the habitats dominated by mesophytes or mesohygrophytes and psichro-thermophytes. The temperature rise and humidity decrease due to climate change are expected to affect the dominant species, significantly reduce their diversity, and endanger the quality and integrity of such habitats. The human activities like tourism, construction works, transportation, or eutrophication together with erosion are all factors intensifying this development (Sârbu et al. 2014).

The increasing temperatures (mild winters, warm spring, and summer), decreasing precipitation (droughts in summers, less snow in winter), and more frequent extreme events like torrents and heavy rains associated with soil erosion will be responsible for the highly significant climate change impacts in the alpine natural grasslands. The changes in the composition of grasslands are likely in response to climate change (Campbell et al. 2009). In the alpine zone, a change in environmental conditions will determine the disappearance of plant species sensitive to drought and high temperature. These will favor the expansion of drought-tolerant taxa and lead to an increase of scrub abundance, which in turn will modify the sequences of succession (Sârbu et al. 2014).

The frequency, duration, and period of flooding are significant hydrological preconditions for a sustenance of seminatural tall-herb humid meadows. This necessary environmental setting due to severe drought and increasing mean temperatures will interfere in this action. All these are often associated with a significant frequency of wildfires, together with the effects like mineralization of peat and accumulation of organic material, with impacts on soil structure and nutrients (Sârbu et al. 2014).

Significant impacts in the mesophyll grasslands will originate from the hydrology, e.g., changes in flooding regime and groundwater level, rising temperature, change in precipitation pattern, or extreme weather events like heavy rains associated with soil erosion. Their cumulative effects will lead to the changes in the sequences of succession, expansion of alien species, habitat degradation, and possible loss of species with high conservation value, in particular the endangered and rare plants (Ozturk and Pirdal 1988, 1991; Ozturk et al. 2012; Sârbu et al. 2014).

Many different impacts can be considered for the raised bogs, mires, and fens such as changes of groundwater level, precipitation pattern, temperature increase and drought, wildfires, or soil quality changes like erosion, soil leaching, mineralization of peat, and accumulation of organic material. Cumulative effects will be in the form of changes in succession sequences, increasing tree cover, and spread of reed and alien species, all affecting bryophytes and other specialized plants (Sârbu et al. 2014).

The significant impacts in the case of rocky habitats and caves are related to temperature increases and precipitation decreases associated with permafrost melting and glacier retreat. Some consideration is to be given to the soil erosion as well as the frequency of avalanches as extreme events. The loss of the colder climatic zone at high altitudes and a significant loss of plant species strongly depending on vernalization can be expected as cumulative impacts (Sârbu et al. 2014).

The changes in seasonality will especially produce a negative effect on the forest tree taxa, the temperature will increase, and changes in precipitation pattern will add to these impacts (Milad et al. 2011; Profft and Frischbier 2009; Sârbu et al. 2014). The forest trees are also sensitive to increased drought stress associated with fluctuation in frequency, amplitude, and moment of drought appearance and to changes in groundwater level vis-a-vis the flooding regime. The heavy impacts can be listed as torrents, storms, and heavy rains. All these will lead to the loss of relevant and sensitive species, changes in the structure of the forest community, drying out of forest areas, an increased advantage of the propagation of pests, and improved conditions for the development of alien plants as a result of cumulative impacts (Sârbu et al. 2014). The mesophytes and mesohygrophytes in the temperate coniferous forests are dominated by trees depending on water supply and low temperatures (microthermophytes). They will get affected by changes in temperature and precipitation pattern as well as in groundwater level. Other significant impacts for forest life will be heavy rains, storms, and torrents, because these can have a great impact on stand stability, causing sudden and dramatic ecological and economic losses. The forest ecosystems are put at a high risk by the extent and rate of climate change. There is a strong dependency on existing site conditions that cannot be modified, existing stand conditions exposed to nitrogen deposition and acidification, and multiple demands and expectations by society; all these will have a direct influence on the management strategies of forestry in general. These will create big difficulties on the management system of forests. However, on the basis of short and medium terms, a positive effect on forest growths can be expected due to the changes in mean climatic variables. But, on long-term basis, increasing drought and extreme weather events will be great risk factors for forest sustainability (Lindner et al. 2010; Sârbu et al. 2014).

A well-known fact is that vegetation as the backbone of biodiversity will definitely suffer from climate change effects, and these will become noticeable everywhere. The vegetation groups will react according to the vital requirements of species composing the community. According to Sârbu et al. (2014), the impacts will be at a slower or faster rate, but vegetation will react in a non-favorable direction. Generally protected areas and biosphere reserves already face a number of climate change impacts leading to a wide variety of effects from the biodiversity loss, habitat quality degradation, and potential loss of certain habitat types. All these strongly depend on actual climatic conditions. There are several opinions regarding the level of potential vulnerability of different habitat categories, but it is difficult to point out which will be ultimately more profoundly affected by climate change impacts.

15.5 Conclusions

Many protected areas are already facing biodiversity loss as the first consequence of climate change. The spring now starts earlier in the year; as a result of this, timing of seasonal events like the first flowering date for plants and the breeding dates of birds are advancing. The geographical distribution of several plant taxa is moving to northward or to higher altitudes. In view of this situation, there is a disruption in time or in space of typical ecological interactions like hatching of offspring and availability of food sources. Moreover the pattern and intensity of extreme events like floods, heavy rains, the heat waves, and dry seasons are changing. As a result of this, the individual species and habitats are affected severely. The character of habitats and ecosystems is expected to change due to altered water regimes or other abiotic conditions. A further acceleration will be observed because of the projected future climate trends, and these will end up with changes in the distribution and abundance of endangered species and ecosystems, intensifying overall biodiversity loss. Undoubtedly the mitigation of climate change is very important, but conservation management is to be adapted to this change; otherwise its impacts will lead to the degradation of habitats, extinction of species, and the loss of ecosystem services, which are essential for human well-being (Rannow et al. 2014).

An adjustment in ecological, social, or economic systems to prevent or reduce harm or benefit from potential opportunities is defined as an adaptation to climate change (Smit and Pilifosova 2001). An adjustment in management practices, decision-making processes, and organizational structures will mean adaptation of conservation management practices (Welch 2005). There is great need to start the adaptation process now, but it must be planned as a long-term process. Its success will depend only if as many institutions and stakeholders as possible are actively involved and are willing to support it. Although to facilitate effective implementation of adaptation actions local communities and decision-makers are essential, however, the development of adaptation strategies needs a helping hand from the scientists which is very important (Rannow et al. 2014).

For a transparent and understandable decision-making process, the basis will be the expertise and data provided by research, but scientific results need to be translated and presented in a form that is accessible to professionals and decision-makers and local stakeholders (Welch 2005). It must be transparent in the process of production, authorized and trusted by the people involved, and meet as well as address the needs, knowledge, and language of local communities, because they have to implement the management practices, which is a major challenge for many scientists in climate research. The little awareness on the climate change as a main driver of biodiversity loss currently is not acceptable. Its significance should increase much more in the future. Neither management authorities nor land users and stakeholders generally have enough information, knowledge, or incentives to plan and negotiate necessary adaptations to climate change; they need to be fully involved (Rannow et al. 2014).

The administrations of National Parks or Biosphere Reserves are a crucial component within the adaptive capacity of local institutions. Any lack of expertise, methods, and tools for climate adaptation as well as limited resources will prevent proper management and adaptation. The protected areas and their institutional setting have capacity and willingness to respond to new challenges and opportunities. This capacity is very, very important for biodiversity conservation at the local level as the biological capacity of species to adapt at the level of individuals, populations, or species (Rannow et al. 2014).

The snowcapped peaks, forests, and meadows of Kyrgyzstan Tian Shan area form a remarkable ecological bridge across the arid heart of Asia that connects the mountain ecosystems of Mongolia and Siberia with those of Tibetan Plateau, Himalayas, and other areas of West and East Asia (Ozturk et al. 1996a, b; Farrington 2005). Kyrgyzstan has remarkably high levels of species diversity in spite of the fact that it is a small, temperate nation bounded by vast deserts and arid grasslands. The preservation of its diverse ecosystems is of utmost importance for the ecological well-being of Inner Asian living beings. The old system of protected areas and protected area management inherited by the Kyrgyz Republic is slowly expanding. It has been modernized after the republic gained independence in 1991. However, due to economic crisis which began in the late 1980s and got intensified during the 1990s, the conditions are not expected to improve dramatically in short time. The protected areas in Kyrgyzstan are mainly protected weakly; many units are used as public commons for grazing, woodcutting, hunting, and the collection of wild plants (Farrington 2005).

These areas are staffed by dedicated rangers, but wages are so low that they can no longer support their living and must spend much time engaged in subsistence activities, such as herding livestock in protected area multiuse zones. In view of this, not much time is spent time for patrolling parks and reserves. As a result of this, the poaching of wildlife and illegal harvesting of plants and trees has increased in protected areas throughout the republic. Currently many rare animal and plant species are threatened with extinction, and the nation's already extremely limited forest cover is decreasing. Other major threat to the future integrity of the nation's ecosystems comes from the accelerated rate of retreat of glaciers in the Tian Shan, which are the source of much of the water for the Central Asian republics and the extremely arid Xinjiang in China (Farrington 2005).

Undoubtedly the economic hardships of independence have had a bad effect on the state of conservation in the Kyrgyz Republic. The old economic collapse has provided new opportunities for establishment of protected areas in the republic, primarily through the large-scale depopulation of the nation's remotest grazing lands and by opening the nation's borders to adventurous foreign visitors, who form the basis of a significant tourist economy. This may help support and promote Kyrgyzstan's protected areas. However, with the opening of borders other economic pressures have come on both existing and potential conservation areas in the form of expanded mineral exploration and extraction activities; proposals for creation of international highways, railroads, and pipelines; as well as increased international trafficking in endangered species. Hopefully, the conservationists in Kyrgyzstan will succeed in their efforts to preserve the nation's fragile ecosystems in the face of new economic pressures. Currently the deterioration of Kyrgyzstan's national conservation program is unlikely to be reversed without substantial support from the international conservation community (Farrington 2005).

Some species or populations, especially endemics, rare, and medicinal plants, may already be in decline due to loss of suitable habitat. They may be sink populations occupying marginal habitat at the edge of their range, or remnant populations. In these cases, protection of habitat will not be sufficient on its own, as there is very low possibility of the species or population dispersing there through space and time. More drastic intervention is required, such as assisted translocation. Taking account of the timing and spatial variation across the landscape of the climate change impact will be critical in such cases (Ozturk et al. 2011, 2015, 2016).

Conservation choices made to address climate change will be different according to which factors are taken into account, and some of the challenges we described may be more important than others. In the case of climate extremes, completely different management decisions should be made when only shifts in climate averages are considered compared with when climate extremes and extreme events are taken into account (Ozturk et al. 2011, 2015, 2016).

Climate change is operating on a long time scale, and some facets will not have large impacts in the near- or midterm, but it is clearly crucial to take into account the fact that climate change is already having an impact in some places. It is of course difficult to draw meaningful conclusions over the short term given the large uncertainty associated with short-term climate projections, which decreases with longer-term forecasts. In terms of human responses and synergistic and feedback interactions, the distinction between local (human-driven change) and global (climate change) drivers, and the choices taken to manage them, is critical to ensure that actions operate at the appropriate scale (Ozturk et al. 2011, 2015, 2016).

Acknowledgment Münir Öztürk and Volkan Altay are highly thankful to the Kyrgyz-Turkish Manas University, Bishkek, Kyrgyzstan, for providing facilities and permitting us to collaborate with them during our trip to the biosphere reserve in September 2015.

References

Ahmad M, Bano A, Zafar M, Sultana S, Rashid S (2015) Interdependence of Biodiversity, Applied Ethnobotany, and Conservation in Higher Ecosystems of Northern Pakistan Under Fast Climatic Changes. In: Oztürk M et al (eds) Climate change impacts on high-altitude ecosystems. Springer, Cham, Switzerland, pp 455–489

Alimbaeva PK, Goncharova AV (1971) Wild medicinal plants in Kirgizia (pharmacological studies). Bishkek, Kyrgyzstan, Frunze

Andrade PA, Herrera FB, Cazzola GR (eds) (2010) Building resilience to climate change: ecosystem-based adaptation and lessons from the field. IUCN, Gland

Araújo MB, Alagador D, Cabeza M, Nogués-Bravo D, Thuiller W (2011) Climate change threatens European conservation areas. Ecol Lett 14(5):484–492

Bellard C, Bertelsmeier C, Leadley P, Thuiller W, Courchamp F (2012) Impacts of climate change on the future of biodiversity. Ecol Lett 15:365–377
Borlakov HU (1996) Flora of Sary-Chelek Reserve. Works of Sary-Chelek State Reserve, 2nd edn. Bishkek, Kyrgyzstan, Frunze
Borlakov HU, Golovkora FG (1971) Plants of Sary-Chelek Reserve. Works of Sary-Chelek State Reserve, 2nd edn. Bishkek, Kyrgyzstan, Frunze
Brenner L, Job H (2012) Challenges to actor-oriented environmental governance: examples from three Mexican biosphere reserves. Tijdschr Econ Soc Geogr 103:1e19
Bridgewater P, Babin D (2017) UNESCO-MAB Biosphere Reserves already deal with ecosystem services and sustainable development. Proc Natl Acad Sci 114(22):E4318
Campbell A, Kapos V, Scharlemann JPW, Bubb P, Chenery A, Coad L, Dickson B, Doswald N, Khan MSI, Kershaw F, Rashid M (2009) Review of the literature on the links between biodiversity and climate change: Impacts, adaptation and mitigation. Secretariat of the Convention on Biological Diversity, Montreal, Technical series No. 42, 124p
Coetzer KL, Witknowski ETF, Erasmus BFN (2013) Reviewing Biosphere Reserves globally: effective conservation action or bureaucratic label? Biol Rev Camb Philos Soc 89(1):82–104
Eastwood A, Lazkov G, Newton A (2009) The Red List of Trees of Central Asia. Fauna & Flora International, Cambridge, UK
EEA (2010) The European environment—State and outlook 2010: Synthesis. European Environment Agency, Copenhagen
Farrington JD (2005) A Report on Protected Areas, Biodiversity, and Conservation in the Kyrgyzstan Tian Shan with Brief Notes on the Kyrgyzstan Pamir-Alai and the Tian Shan Mountains of Kazakhstan, Uzbekistan, and China. U.S. Fulbright Program, Environmental Studies Section, Bishkek
Gafta D, Mountford O (coord) (2008) Manual de interpretare a habitatelor Natura 2000 din România. Cluj-Napoca: Editura Rosprint
Gaştescu P, Ştiuca R (2008) Delta Dunării Rezervaţie a Biosferei. Editura CD Press, Bucureşti
Grisa E, Venglovsky B, Sarymsakov Z, Carraro G (2008) Forest typology in the Kyrgyz Republic. Practice oriented document for field assessment and sustainable management of forest stands. Bishkek
Hooper DU, Chapin FS, Ewel JJ, Hector A, Inchausti P, Lavorel S, Lawton JH, Lodge DM, Loreau M, Naeem S, Schmid B, Setala H, Symstad AJ, Vandermeer J, Wardle DA (2005) Effects of biodiversity on ecosystem functioning: a consensus of current knowledge. Ecol Monogr 75:3–35
IPCC (Intergovernmental Panel on Climate Change) (2001) Climate change 2001: The scientific basis. Contribution of Working Group I to the Third Assessment Report of the Intergovernmental Panel on Climate Change. In: Houghton JT et al (eds) . Cambridge University Press, Cambridge, UK and New York, 881 pp
IPCC (Intergovernmental Panel on Climate Change) (2007) Climate Change 2007: The Physical Science Basis. Contribution of Working Group I to the Fourth Assessment Report of the Intergovernmental Panel on Climate Change, Cambridge, UK and New York
Ishwaran N (2012) Science in intergovernmental environmental relations: 40 years of UNESCO's Man and the Biosphere (MAB) Programme and its future. Environ Dev 1:91–101
Lindner M, Maroschek M, Netherer S, Kremer A, Barbati A, Garcia-Gonzalo J, Seidl R, Delzon S, Corona P, Kolstro M, Lexer MJ, Marchetti M (2010) Climate change impacts, adaptive capacity, and vulnerability of European forest ecosystems. For Ecol Manag 259:698–709
Lowell K (2017) A socio-environmental monitoring system for a UNESCO biosphere reserve. Environ Monit Assess 189(12):601
Mace GM, Norris K, Fitter AH (2012) Biodiversity and ecosystem services: a multilayered relationship. Trends Ecol E 27:19–26
Magnuszewski M, Bijak S, Orozumbekow A, Howe B, Musuraliev K, Zasada M, Bronisz K, Bronisz A (2015) Different growth patterns of *Picea schrenkiana* subsp. *tianshanica* (Rupr.) Bykov and *Juglans regia* L. coexisting under the same ecological conditions in the Sary-Chelek Biosphere Reserve in Kyrgyzstan. Dendrobiology 73:11–20

Mambetaliev UA, Shukurov EJ (2002) Sary-Chelek State Biosphere Reserve—Management Plan 2001–2004. Central Asia transboundary West Tien-Shan Biodiversity Project

Milad M, Schaich H, Bürgi M, Konold W (2011) Climate change and nature conservation in Central European forests: A review of consequences, concepts and challenges. For Ecol Manag 261(4):829–843

Montoya JM, Raffaelli D (2010) Climate change, biotic interactions and ecosystem services. Philos Trans R Soc B Biol Sci 365:2013–2018

Naeem S (2009) Ecology: gini in the bottle. Nature 458:579–580

Normand S, Svenning J, Skov F (2007) National and European perspectives on climate change sensitivity of the habitats directive characteristics plant species. J Nat Conserv 15(1):41–53

Ozturk M, Pirdal M (1988) Ecology of grasslands in the State of Mugla. Turk J Bot 12(2):164–174

Ozturk M, Pirdal M (1991) Productivity and Nutrient Turnover in Isikeli Canakkale Grasslands. I. Turkish Grasslands-Fodder Congress, Izmir, (ed) Avcioglu R, pp 202–212

Ozturk M, Secmen O, Gork G (eds) (1996a) Plant Life in Southwest and Central Asia, vol I. Ege University Press, Izmir, Turkey, 499 pp

Ozturk M, Secmen O, Gork G (eds) (1996b) Plant Life in Southwest and Central Asia, vol II. Ege University Press, Izmir, Turkey, 1093p

Ozturk M, Gücel S, Altundağ E, Çelik S (2011) Turkish Mediterranean medicinal plants in the face of climate change. In: Ahmad A et al (eds) Medicinal plants in changing environment. Capital Publication, New Delhi, ISBN: 81-85589-14-3

Ozturk M, Altay V, Gucel S, Aksoy A (2012) Aegean Grasslands as Endangered Ecosystems in Turkey. Pak J Bot 44:7–17

Ozturk M, Hakeem KR, Faridah-Hanum I, Efe R (eds) (2015) Climate change impacts on high-altitude ecosystems. Springer, Cham, 696 pp

Ozturk M, Altay V, Aksoy A (2016) Ecology of some endangered endemic plant taxa of Turkiye in relation to climate change. Conservation and Sustainable use of genefund of plant world in Eurasia at the present stage—International Scientific conference within "Day of Kazakhstan", at EXPO, pp. 12–15, Antalya-Turkey

Profft I, Frischbier N (2009) Forestry in a changing climate—the necessity of thinking decades ahead. In: Feldmann F et al (eds) Crop plant resistance to biotic and abiotic factors. Deutsche Phytometrische Gesellschaft, Braunschweig, pp 66–73

Rannow S, Wilke C, Gies M, Neubert M (2014) Chapter 20: Conclusions and recommendations for adapting conservation management in the face of climate change. In: Rannow S, Neubert M (eds) Managing Protected Areas in Central and Eastern Europe Under Climate Change. Springer, Dordrecht, pp 291–303

Sârbu A, Janauer G, Profft I, Kaligarič M, Doroftei M (2014) Potential impacts of climate change on protected habitats. In: Rannow S, Neubert M (eds) Managing protected areas in central and Eastern Europe Under Climate Change, Advances in global change research, vol 58. Springer, Dordrecht, Netherlands, pp 45–60

Smit B, Pilifosova O (2001) Adaptation to climate change in the context of sustainable development and equity. Contribution of the working group to the third assessment report of the intergovernmental panel on climate change. Cambridge University Press, Cambridge, pp 879–912

Stern N (2007) The economics of climate change. The stern review. Cambridge University Press, Cambridge

Strojan I, Robic M (2009) Environmental indicators in Slovenia, sea level. [Slovenian]. http://kazalci.arso.gov.si/?anddata¼indicatorandihd_id¼146andmenu_group_id¼10

Sultanova BA, Lazkov GA (1987) List of higher plants of Sary-Chelek Reserve with identification of endemic, rare and valuable species

UNESCO (2015) World Network of Biosphere Reserves. http://www.unesco.org/new/en/natural-sciences/environment/ecological-sciences/biospherereserves/world-network-wnbr/wnbr/

Van Cuong C, Dart P, Hockings M (2017) Biosphere reserves: attributes for success. J Environ Manag 188:9–17

Walther GR (2010) Community and ecosystem responses to recent climate change. Philos Trans R Soc B Biol Sci 365:2019–2024
Welch D (2005) What should protected areas managers do in the face of climate change? The George Wright Forum 22(1):75–93

Web Sites

www.alchetron.com
www.morena.kz
www.plantarium.ru
www.plantlist.org
www.stories.rbge.org.uk

Chapter 16
Plant Diversity of Ala-Archa National Park in Kyrgyzstan with Emphasis on Its Economical Potential

Nazgül Imanberdieva, Zeki Severoğlu, Gulbubu Kurmanbekova, Volkan Altay, and Münir Öztürk

Abstract The national parks are the most important tools in biodiversity conservation. The generally include remote and rugged areas possessing a low possibility for human use. They are of lower economic value. Currently the Republic of Kyrgyzstan possesses a total of eight national parks. These cover approximately 1.4% of national territory. One of these national parks is the Ala-Archa National Park. During the period covering the months from May to October, every weekend hundreds of visitors pay a visit to this oldest national park of the country established in 1976. The 25 km long park has a total area of 194 km^2. This area encompasses the upper watershed of the Ala-Archa River. The park shows a microclimate of almost every ecosystem existing in the Republic of Kyrgyzstan, including the forests of mixed conifers, deciduous and dwarf juniper forests, scrubland, grassland steppe, sagebrush semidesert, river valleys, glaciers, and high alpine environments. The total number of vascular plants growing in the national park is 756 taxa. These belong to 73 families and 324 genera. Six hundred twenty-three taxa are from the dicotyledonous group, whereas 121 taxa are from the monocotyledonous group. This chapter presents an analysis of the economical potential of plant diversity of this national park.

Keywords Ala-Archa National Park · Plant diversity · Economic potential

---

N. Imanberdieva · G. Kurmanbekova
Biology Department, Kyrgyzstan-Turkiye Manas University, Bishkek, Kyrgyzstan

Z. Severoğlu
Biology Department, Kyrgyzstan-Turkiye Manas University, Bishkek, Kyrgyzstan

Biology Department, Faculty of Science and Arts, Marmara University, Istanbul, Turkey

V. Altay
Biology Department, Faculty of Science and Arts, Hatay Mustafa Kemal University, Antakya, Hatay, Turkey

M. Öztürk (✉)
Vice President "The Islamic World Acadmey of Sciences" Amann-Jordan and Centre for Environmental Studies-Botany Department, Ege University, Izmir, Turkey

D. Egamberdieva, M. Öztürk (eds.), *Vegetation of Central Asia and Environs*,
https://doi.org/10.1007/978-3-319-99728-5_16

16.1 Introduction

The national parks today together with other forms of protected areas (PAs) are of great importance for biodiversity conservation, although these were originally established to preserve monuments and wonders of nature (Margules and Pressey 2000; Hoffmann et al. 2010; Rauset et al. 2016). The successful results obtained from PAs toward conservation have spurred scientific effort into designing representative and persistent reserves (Rauset et al. 2016). However, scenic areas and wilderness often get priority in the protection efforts. Such places generally include remote and rugged areas, with little human impact as well as lower economic value (Pressey 1994; Soule and Sanjayan 1998; Margules and Pressey 2000; Rauset et al. 2016).

The main aim behind the national parks is to protect important ecosystems and rare wildlife, like nature reserves, but at the same time provide an opportunity to the human beings living around the area or at a short distance to come and enjoy the outdoor recreational activity in a scenic natural setting. As such, when compared to the nature reserves, national parks welcome all living beings (Farrington 2005). These parks usually possess three parts: (1) prohibited, (2) recreational, and (3) production zones. The first zone is usually located at a remote distance of the park. It is an area of high ecological importance—close to the human entrance as well as cattle, etc. But, the scientists undertaking research in such areas can enter together with the rangers on patrol. The zone number 2 is open for outdoor recreation, not much disruptive to the natural surroundings. The activities here are compatible with the objectives lying behind the establishment of the park, like hiking, camping, and picnic visits. The zone number 3 is a multipurpose zone in the park. The traditional subsistence activities, like grazing and limited logging, are permitted; however, such areas at the same time include tree nurseries and afforestation projects together with timber cutting (Farrington 2005). There are eight national parks in Kyrgyzstan covering approximately 1.4% of national territory (Farrington 2005). One of these national parks is the Ala-Archa National Park.

16.2 Brief Information on National Park

Ala-Archa is located in the highest section of the Kyrgyz Range, at a distance of 30 km in the south of Bishkek. It is the oldest and most visited national park. An approximately 25 km long park was established in 1976 on an area of 194 km^2 that encompasses the upper watershed of the Ala-Archa River and receives hundreds of visitors each weekend from May up to October. There first 10 km are a roughly paved road starting from the entrance gate to the park's small lodge and main picnic as well as the camping area (Farrington 2005). The park is centered around the steep mountain canyon of the upper Ala-Archa River, and the elevations range from approximately 1600 m at the entrance of the park and goes up to 4895 m at "Peak Semenova." A classic V-shaped mountain valley forms its lower part, but a series of several broad, glacially formed steps constitute the upper part, set at the bottom of a stunning amphitheater of sheer cliffs and rocky crags topped by dozens of glaciers (Farrington 2005).

Recreational features of the park include hiking and climbing. Most attractive trail is the one which leads to the Ak-Sai glacier, climbing roughly 1200 m over 6 km to a climbers' hut at the toe of the glacier. Approximately 13 km long trail leads up the main Ala-Archa Valley from the park lodge to a former ski lodge. This trail goes up to the edge of the ice fields near the southern boundary of the park, about 8 km up the valley (Farrington 2005).

The Republic of Kyrgyzstan recognized the Ala-Archa River gorge as its first "State National Nature Park" in 1976. It was only one of the series of parks classified on a scale similar to IUCN categories (International Union for Conservation of Nature (IUCN) 2014), from reserves with relatively low management to preserves that allowed for only scientific access. A general look reveals that the park embodies a rich biodiversity, distinguishing it as a little-known spot among national and international parks. A critical glacial and snowpack watershed, forests with highly diverse plant and animal diversity, towering alpine mountains, and ease of accessibility from the major center in Bishkek have been much advantageous for this park. The park has succeeded in maintaining its existence, in spite of the little management and policy enforcement. It is now rapidly emerging as a country's potential focus point for the flourishing tourism industry (Mukanbetov 2013; Fry 2017).

In view of the unique concentration of Central Asian flora and fauna within the park area as well as high ecosystem services, the park has successfully passed most disturbing periods during the post-Soviet period. The water resources together with an already-existing potential management issues, and proximity to the capital city center, it is an ideal area for the development and application of low-cost, high-effect information and education programs, which concentrate on ecosystem services and socio-ecological systems (Fry 2017). The park is not a large one, but it defines itself with an array of plant as well as animal diversity and ecosystem services. It encompasses the highest peaks of the Kyrgyz Range of the larger Tian Shan Mountains in part; a cirque of towering peaks surround the upper reaches of the Ak-Sai glacier (Fry 2017). Although only 194 square kilometers large, it contains a microclimate of nearly every ecosystem that exists at large in Kyrgyzstan. These include the mixed conifers, deciduous forests, dwarf juniper forests, scrubland, alpine grassland, steppe, sagebrush semidesert, river valley, glaciers, and high alpine environments (Farrington 2005; Fet 2007; Matyas 2010; Fry 2017).

16.3 Data Analysis

This floristic data published by Komarov (1934-1969), Cherepanov (1995), Ozturk et al. (1996a, b), Sultanova et al. (1998), Farrington (2005), Lazkov (2006, 2007), Umralina and Lazkov (2008), and Sennikov et al. (2011) has been evaluated in this chapter in the light of latest findings, together with other plant diversity investigations. The vascular plants whose status or name has changed, or have become synonyms, or have been included under new combinations have been corrected following the "www.plantlist.org."

The existing potential of medicinal, food, and fodder plants has been evaluated taking into account the floristic structure of the study area. The potential pollen- and nectar-rich plants (honey bee plants) too have been identified, in order to contribute economically to the good-quality honey production in the region. The economic potential of the floristic data published by Golovkova and Chubarova (1988), Eisenman et al. (2013), and Zaurov et al. (2013) for medicinal plants; by Golovkova and Chubarova (1988) for food plants; by Golovkova and Chubarova (1987) and Ozturk et al. (2017a) for fodder plants; by Pelmenev (1985), Ozturk et al. (1989, 2012d), and Sorkun (2008) for honey plants; and by Farrington (2005) for other plants with alternative potential use has been investigated.

16.4 Plant Diversity

A small forest composed of introduced trees, like Siberian larch and non-native birch are found in the lower part of the canyon. The sparse juniper forest is distributed in the valley following the entrance area. The word "archa" means juniper in Kyrgyz language. The name of this park comes from this word. This is followed by pure stands of native Schrenk's spruce beyond the park lodge, which is distributed between 2100 and 2800 m. This narrow belt of plant cover is mainly found on shaded, north-facing slopes. All along the entire length of valley, we come across deciduous shrubs, which eventually are followed by alpine grasslands above 3000 m elevations. The wild flowers blooming in spring and summer are spread over the whole valley (Fig. 16.1) (Farrington 2005). This national park is located in the central part of northern macroslope of Kyrgyz Range of Tien Shan Mountains, stretching from east to west for 450 km. The valley of Ala-Archa is found on the highest part of the mountain range, Semenov-Tien-Shansky Peak. The elevation of this peak is 4875 m above sea level. The territory of the park includes meadow-steppe, forest-meadow, and subalpine and alpine vegetation zones, represented by 8 types and 22 formations (Ionov et al. 2000).

The total number of vascular plants distributed in the study area includes 756 taxa from 73 families. These belong to 324 genera. The number of dicotyledons taxa is 623 and that of monocotyledons 121 (Table 16.1). The following families are represented by the largest number of taxa: Asteraceae (131 taxa, 17.83%), Poaceae (61 taxa, 8.07%), Rosaceae (47 taxa, 6.22%), Brassicaceae (45 taxa, 5.95%), and Fabaceae (43 taxa, 5.69%) (Table 16.2). The following genera are represented by the largest number of taxa: *Allium* (18 taxa), *Potentilla* (15 taxa), *Astragalus* (14 taxa), *Veronica* (14 taxa), *Taraxacum* (13 taxa), and *Artemisia* (13 taxa) (Table 16.3).

Stellaria brachypetala

Geranium saxatile

Tulipa zenaidae

Tulipa heterophylla

Picea schrenkiana

Rosa beggeriana

Astragalus fedtschenkoanus

Myricaria germanica
(synonym: ***Myricaria bracteata***)

Fig. 16.1 Some of the plants from the study area

Table 16.1 The numbers regarding the floral diversity of Ala-Archa National Park

			Angiospermae		
	Pteridophyta	Gymnospermae	Dicotyledoneae	Monocotyledoneae	Total number
Family	4	3	55	11	73
Genera	4	3	276	41	324
Taxa	5	7	623	121	756

Table 16.2 The richest families with numbers of taxa and their percentages

Family	Number of taxa	%
Asteraceae	131	17.83
Poaceae	61	8.07
Rosaceae	47	6.22
Brassicaceae	45	5.95
Fabaceae	43	5.69
Caryophyllaceae	35	4.63
Ranunculaceae	32	4.23
Lamiaceae	30	3.97
Apiaceae	30	3.97
Boraginaceae	22	2.91

Table 16.3 The numbers and percentages of the genera on the basis of taxa

Genera	Number of taxa	%
Allium	18	2.38
Potentilla	15	1.98
Astragalus	14	1.85
Veronica	14	1.85
Taraxacum	13	1.72
Artemisia	13	1.72
Erigeron	12	1.59
Poa	10	1.32
Pedicularis	10	1.32
Carex	9	1.19
Draba	9	1.19

16.5 Economically Important Plants

16.5.1 Potential Medicinal Plants

A medicinal plant is the one which produces a definite curing physiological response in the treatment of various ailments in humans or seed (Hamayun et al. 2006). Various chemicals act collectively to reach equilibrium in the body as they do in the plant. In this way they produce gentle progressive healing in the tissues. From the earliest times, humans have used medicinal plants from the earliest times so as to cure diseases and relieve physical suffering (Hamayun et al. 2006). A total of 104 naturally distributed medicinal plant taxa have been determined in the study area (Table 16.4). Most common taxa used in traditional medicine are *Artemisia viridis*, *Elaeagnus rhamnoides* (syn. *Hippophae rhamnoides*), *Ephedra equisetina*, *Glycyrrhiza glabra* and *G. uralensis*, *Hypericum perforatum*, *Leonurus turkestanicus*, *Origanum vulgare*, *Plantago major*, *Rhodiola kirilowii* (syn. *Rhodiola linearifolia*), *Tussilago farfara*, *Urtica dioica*, and *Ziziphora clinopodioides*.

Many medicinal plants used in the traditional folk medicine in Kyrgyzstan have been reported to be used for many diseases such as gastrointestinal, tuberculosis, headache, different of cancer, rheumatism, gynecological disorders, skin diseases, diuretic, hypertension, heart diseases, anti-inflammatory, fever, and eye problems (Eisenman et al. 2013; Zaurov et al. 2013).

16.5.2 Potential Fodder Plants

A backbone of the cultivated plants is meadows and pastures. These are an important genetic source contributing much to the biodiversity. They serve as the areas of shelter for animals and shield land from erosion, in addition to serving as the most important natural sources of forage plants for animals (Aydın and Uzun 2002; Ozturk et al. 2012c; 2017a). Such habitats are also very important as a cheaper feed source for animals in terms of the quality of animal products. Therefore, pasture and meadow areas and their efficiency are of paramount importance (Kaya et al. 2001; Babalık and Sönmez 2009; Ozturk et al. 2017a). In Ala-Archa a total of 48 fodder plant taxa are distributed (Table 16.5). These belong to 28 genera and are placed in 9 families. These taxa include members of the families Fabaceae (17), Poaceae (14), Cyperaceae (6), and Asteraceae (5), which emphasizes the richness of the species diversity.

Table 16.4 A list of the medicinal plants distributed in Ala-Archa National Park

No	Botanical name	No	Botanical name
1	*Achillea asiatica*	53	*Hypericum scabrum*
2	*Achillea filipendulina*	54	*Impatiens parviflora*
3	*Achillea setacea*	55	*Inula britannica*
4	*Aconitum leucostomum*	56	*Inula helenium*
5	*Aconitum soongaricum*	57	*Juniperus sabina*
6	*Agrimonia eupatoria* subsp. *asiatica*	58	*Juniperus semiglobosa*
7	*Alcea nudiflora*	59	*Lagochilus platycalyx*
8	*Amaranthus retroflexus*	60	*Leonurus turkestanicus*
9	*Anagallis arvensis*	61	*Lithospermum officinale*
10	*Artemisia absinthium*	62	*Marrubium anisodon*
11	*Artemisia annua*	63	*Melilotus officinalis*
12	*Artemisia dracunculus*	64	*Mentha longifolia* var. *asiatica*
13	*Artemisia scoparia*	65	*Onopordum acanthium*
14	*Artemisia viridis*	66	*Origanum vulgare*
15	*Artemisia vulgaris*	67	*Patrinia intermedia*
16	*Astragalus sieversianus*	68	*Picea schrenkiana*
17	*Bidens tripartita*	69	*Plantago lanceolata*
18	*Capsella bursa-pastoris*	70	*Plantago major*
19	*Carum carvi*	71	*Polemonium caucasicum*
20	*Ceratocephala testiculata*	72	*Polygala comosa*
21	*Cichorium intybus*	73	*Polygonum aviculare*
22	*Clematis orientalis*	74	*Polygonum coriarium*
23	*Codonopsis clematidea*	75	*Prunus padus*
24	*Conium maculatum*	76	*Rhamnus cathartica*
25	*Convolvulus arvensis*	77	*Rhodiola kirilowii*
26	*Crataegus songarica*	78	*Roemeria refracta*
27	*Daucus carota*	79	*Rosa canina*
28	*Delphinium confusum*	80	*Rosa fedtschenkoana*
29	*Delphinium iliense*	81	*Rubus caesius*
30	*Descurainia sophia*	82	*Rubus idaeus*
31	*Dianthus superbus*	83	*Rumex tianschanicus*
32	*Dipsacus dipsacoides*	84	*Salvia deserta*
33	*Dodartia orientalis*	85	*Salvia sclarea*
34	*Elaeagnus rhamnoides*	86	*Scabiosa songarica*
35	*Ephedra equisetina*	87	*Sorbus tianschanica*
36	*Epilobium hirsutum*	88	*Tanacetum vulgare*
37	*Equisetum arvense*	89	*Taraxacum campylodes*
38	*Erodium cicutarium*	90	*Thalictrum foetidum*
39	*Fragaria vesca*	91	*Thalictrum isopyroides*
40	*Fumaria vaillantii*	92	*Thalictrum minus*
41	*Galium verum*	93	*Tribulus terrestris*
42	*Gentiana olivieri*	94	*Trifolium pratense*

(continued)

Table 16.4 (continued)

No	Botanical name	No	Botanical name
43	*Geranium collinum*	95	*Tussilago farfara*
44	*Geum rivale*	96	*Urtica dioica*
45	*Glaucium fimbrilligerum*	97	*Vaccaria hispanica*
46	*Glycyrrhiza glabra*	98	*Verbascum songaricum*
47	*Glycyrrhiza uralensis*	99	*Verbascum thapsus*
48	*Haplophyllum acutifolium*	100	*Vicia cracca*
49	*Helichrysum maracandicum*	101	*Viola suavis*
50	*Herniaria glabra*	102	*Xanthium strumarium*
51	*Hyoscyamus niger*	103	*Ziziphora clinopodioides*
52	*Hypericum perforatum*	104	*Ziziphora tenuior*

Table 16.5 Potential fodder plants from Ala-Archa National Park

No	Botanical Name	No	Botanical Name
1	*Aegilops cylindrica*	25	*Hordeum murinum* subsp. *leporinum*
2	*Alopecurus pratensis*	26	*Kobresia capillifolia*
3	*Artemisia serotina*	27	*Lathyrus gmelini*
4	*Artemisia tianschanica*	28	*Lathyrus pratensis*
5	*Astragalus* sp.	29	*Medicago falcata*
6	*Atriplex patula*	30	*Medicago lupulina*
7	*Bromus danthoniae*	31	*Medicago minima*
8	*Bromus japonicus*	32	*Medicago medicaginoides*
9	*Carex melanantha*	33	*Medicago sativa* subsp. *varia*
10	*Carex orbicularis*	34	*Melilotus albus*
11	*Carex songorica*	35	*Melilotus officinalis*
12	*Carex stenocarpa*	36	*Onopordum acanthium*
13	*Carex turkestanica*	37	*Phleum pratense*
14	*Cichorium intybus*	38	*Poa bulbosa*
15	*Convolvulus arvensis*	39	*Poa pratensis*
16	*Dactylis glomerata*	40	*Stipa lessingiana*
17	*Ferula* sp.	41	*Taraxacum* sp.
18	*Festuca rubra*	42	*Trifolium pratense*
19	*Festuca valesiaca*	43	*Trifolium repens*
20	*Festuca alatavica*	44	*Turgenia latifolia*
21	*Festuca olgae*	45	*Vicia cracca*
22	*Galium aparine*	46	*Vicia semenovii*
23	*Geranium collinum*	47	*Vicia tenuifolia*
24	*Hedysarum neglectum*	48	*Vicia tetrasperma*

16.5.3 Potential Food Plants

Important and cheap sources of protein, carbohydrates, fats, vitamins, and minerals are wild edible plants. Their dietary contribution increases as these are available during most of the seasons. They are found during the periods in a year when

Table 16.6 Potential food plants from Ala-Archa National Park

No	Botanical name	No	Botanical name
1	*Aegopodium* sp.	15	*Prunus padus*
2	*Allium atrosanguineum*	16	*Rheum wittrockii*
3	*Berberis heteropoda*	17	*Ribes janczevskii*
4	*Carum carvi*	18	*Ribes meyeri*
5	*Cerasus tianschanica*	19	*Rosa alberti*
6	*Cichorium intybus*	20	*Rosa canina*
7	*Crataegus songarica*	21	*Rosa platyacantha*
8	*Daucus carota*	22	*Rubus caesius*
9	*Elaeagnus rhamnoides*	23	*Rubus idaeus*
10	*Eremurus* sp.	24	*Rubus saxatilis*
11	*Fragaria vesca*	25	*Rumex* sp.
12	*Gentiana* sp.	26	*Sorbus tianschanica*
13	*Humulus lupulus*	27	*Urtica dioica*
14	*Oxyria digyna*		

conventional staples and vegetables are scarce. Most of the species used as vegetables are collected in the dry season when cultivated vegetables are scarce (Samant and Dhar 1997). Edible plants of the wild habitats contribute economically due to their indigenous uses. Most of the wild edible species provide a good root stock for the commercial cultivars of the fruit crops. They have wider adaptability, vigorous growth, and resistance to major diseases and pests (Samant and Dhar 1997). A total of 27 food plant taxa have been determined in the Ala-Archa National Park (Table 16.6). These include some tree and shrub taxa as well. Those at the forefront are *Rubus idaeus*, *R. caesius*, *R. saxatilis*, *Ribes meyeri*, *R. janczevskii*, *Crataegus songarica*, *Sorbus tianschanica*, *Cerasus tianschanica*, *Rosa alberti*, *R. canina*, *R. platyacantha*, *Berberis sphaerocarpa*, and *Elaeagnus rhamnoides*.

16.5.4 Potential Pollen- and Nectar-Rich Plants for Honey Production

A list available as a guide for the potential honey bee plants is of great value for the beekeepers as well as researchers. Their importance for honey bee colonies as a source of pollen, nectar, or both is very essential (Ozturk et al. 2012d; Abou-Shaara 2015). Such lists help in better understanding of the park flora suitable for honey bees. Identifying honey bee plants can help beekeepers in selecting suitable sites for their apiaries as well as in cultivating suitable plants for their colonies. Such a list can also guide researchers mainly during their melissopalynological studies and during their studies on bee flora in Kyrgyzstan (Pelmenev 1985; Ozturk et al. 1989, 2012d; Sorkun 2008; Abou-Shaara 2015). Over 100 naturally distributed honey plant taxa have been determined from the study area (Table 16.7).

Table 16.7 Potential plants for honey production in the Ala-Archa National Park

No	Botanical name	No	Botanical name
1	*Achillea*	53	*Mentha longifolia* var. *asiatica*
2	*Aegopodium kashmiricum*	54	*Myosotis*
3	*Alcea nudiflora*	55	*Nepeta cataria*
4	*Alchemilla krylovii*	56	*Nepeta nuda*
5	*Alfredia acantholepis*	57	*Onobrychis arenaria*
6	*Anagallis arvensis*	58	*Onobrychis* spp.
7	*Aster*	59	*Onopordum acanthium*
8	*Astragalus* spp.	60	*Origanum vulgare*
9	*Barbarea vulgaris*	61	*Phlomoides oreophila*
10	*Bidens tripartita*	62	*Plantago lanceolata*
11	*Capsella bursa-pastoris*	63	*Plantago major*
12	*Carduus nutans*	64	*Polygonum*
13	*Carum carvi*	65	*Potentilla reptans*
14	*Centaurea* spp.	66	*Prunella vulgaris*
15	*Cerasus tianschanica*	67	*Prunus padus*
16	*Chenopodium*	68	*Ranunculus*
17	*Cichorium intybus*	69	*Rosa canina*
18	*Cirsium*	70	*Rosa* spp.
19	*Convolvulus arvensis*	71	*Rubus* spp.
20	*Cotoneaster melanocarpus*	72	*Rubus idaeus*
21	*Cotoneaster multiflorus*	73	*Rumex*
22	*Crataegus songarica*	74	*Salix*
23	*Crepis*	75	*Salvia deserta*
24	*Daucus carota*	76	*Salvia sclarea*
25	*Draba* spp.	77	*Sanguisorba minor*
26	*Dracocephalum stamineum*	78	*Scabiosa*
27	*Echinops maracandicus*	79	*Solenanthus circinatus*
28	*Echium vulgare*	80	*Spirea hypericifolia*
29	*Epilobium*	81	*Stachys betoniciflora*
30	*Ferula* spp.	82	*Stellaria media*
31	*Fragaria vesca*	83	*Tanacetum vulgare*
32	*Galatella coriacea*	84	*Taraxacum* spp.
33	*Galium tricornutum*	85	*Thymus seravschanicus*
34	*Galium verum*	86	*Thymus pulegioides* subsp. *pannonicus*
35	*Geranium*	87	*Tragopogon*
36	*Glycyrrhiza glabra*	88	*Tribulus terrestris*
37	*Hedysarum*	89	*Trifolium pratense*
38	*Hypericum perforatum*	90	*Trifolium repens*
39	*Isatis*	91	*Turgenia latifolia*
40	*Lactuca*	92	*Tussilago farfara*
41	*Lamium album*	93	*Urtica dioica*
42	*Lamium amplexicaule*	94	*Valeriana fedtschenkoi*

(continued)

Table 16.7 (continued)

No	Botanical name	No	Botanical name
43	*Lathyrus* spp.	95	*Valeriana ficariifolia*
44	*Leonurus turkestanicus*	96	*Valeriana turkestanica*
45	*Linum*	97	*Verbascum*
46	*Lonicera* spp.	98	*Veronica*
47	*Malva neglecta*	99	*Vicia* spp.
48	*Marrubium*	100	*Vicia cracca*
49	*Medicago* spp.	101	*Xanthium strumarium*
50	*Melilotus albus*	102	*Ziziphora clinopodioides*
51	*Melilotus officinalis*	103	*Ziziphora tenuior*
52	*Mentha arvensis*		

16.5.5 Other Plants with Potential Alternative Uses

The bulbous plants and wild flowers have a great commercial value. However, their overharvesting together with other wild plants has led to a rapid disappearance of many plant taxa. A decrease in the occurrence of such plants is maximum in the neighborhood of villages (Farrington 2005). Most of the visitors roaming in the parks during spring and summer seasons have a very bad habit of picking wild flowers which is another problem. Some of these are rare or endangered (Farrington 2005). The wild flowering plants growing in the park with an economical potential are the species of *Allium*, *Anemone*, *Aster*, *Campanula*, *Crocus*, *Dianthus*, *Eremurus*, *Gagea*, *Gentiana*, *Iris*, *Ixiolirion*, *Paeonia*, *Phlomoides*, *Primula*, *Trollius*, *Tulipa*, and *Viola*. Moreover, among the economically important decorative plants, it is necessary to note that the species of trees and shrub which can be used in the gardens in the houses constructed in the urban and rural areas are *Acer tataricum* ssp. *semenovii*, *Betula tianschanica*, *Cotoneaster* spp., *Elaeagnus rhamnoides*, *Juniperus* spp., *Lonicera* spp., *Picea schrenkiana*, *Populus* spp., *Rhamnus cathartica*, *Rosa* spp., *Salix* spp., *Sorbus tianschanica*, and *Spirea* spp.

16.6 Conclusions

The threats to some of the local floral components in their natural habitat include a large number of adverse human activities like expansion of agricultural fields, overharvesting, overgrazing, uncontrolled fires, and woodcutting for fuel production. An examination of the facts based on interviews with key respondents during field survey, on the attitude of local residents living in the vicinity of such areas, regarding such threats to local ecological resources in particular collection of wild food and medicinal plant species has revealed that an achievement of economic payback from plant species can endorse the interest of locals in protection and preservation of significant and endangered species (Balemie and Kebebew 2006; Abbasi et al. 2015).

Global climate change is reported to have increased the glacial melting in the area. The scientists have documented a 27% loss in Kyrgyz glacial mass from 1935 up to 1985 (Hagg et al. 2005; Farrington 2005; Fet 2007). The mitigation potential of floods in the country is increased by the juniper forests of the park, which play an increasingly major role in this connection in an already susceptible area (Stucker et al. 2012). The role of juniper forests is like a "geomorphic glue" for hillslope soils, thereby mitigating the effects of flooding, erosion, and eventual landslides (Byers 2009). However, in spite of all this ecological potential, juniper forests too have relatively lost ground as woodlands have started recovering from overgrazing pressure experienced by the park before its establishment (Shukhurov and Domashov 2009). The country cannot afford more degradation of such ecosystems and their services. The suburbs of capital city Bishkek are located on the border of Ala-Archa gorge. Any deterioration of juniper groves will mean potentially higher risk for floods due to melting of glaciers or the moraine lake bursting. All these will have a devastating effect on the people living at the foothill (Fry 2017).

The park is also suffering a pressure from heavy recreation, including overcollection of wood for campfires in the lower parts of the park. Although the park has not been heavily grazed in recent times, preliberation era grazing damage is still evident. Generally mudslides in the country often result from overgrazing. On July 21, 2003, a catastrophic mudslide took place in the lower park area, destroying 25 summer huts near the park entrance, and there were many human casualties (Farrington 2005). Some people make sure that their traditional knowledge is used equitably—as per the limitations set by their traditions or requiring benefit sharing for its use (Ahmad et al. 2015). The role of indigenous knowledge in several sectors is being discussed during the recent years. It includes the use of traditional medicine in human healthcare, traditions and education in natural resource management, knowledge about seed varieties, soil conservation, intercropping techniques, pest control, crop diversity, breeding techniques of plant varieties, and irrigation together with water conservation (Ahmad et al. 2015). The contribution of indigenous knowledge in this connection has increased the interest of academicians and policymakers related to such matters. Several governmental/nongovernmental organizations, as well as international organizations like the World Bank, International Labor Office, UNESCO, and FAO, have started appreciating the role indigenous knowledge can play in achieving sustainable development (Ahmad et al. 2015).

Another important factor is "ecological sustainability." This is especially important for medicinal, aromatic, and other economically important plants. One cannot place a price on their great economic value (Costanza and Farber 2002; Farber et al. 2006; Ozturk et al. 2017a, b, c). The use of such plants is very important for humans; therefore, the correct identification of these taxa is imperative. A high priority also is needed in the conservation of economically important medicinal and other plants in this context. It is necessary during conservation to look at how these plants are utilized by local people. Like many countries in the world, protection of genetic resources is perhaps the highest priority in Kyrgyzstan (Pleskanovskaya et al. 2011; Ozturk et al. 2017a).

An effective application of the laws for "nature and species protection" is not possible unless alternatives are developed for their collection from nature. The best option is to start cultivation practices for such plants. Endangered species and those threatened with extinction should be considered in terms of their contribution to the natural environment and the economy of the country. A sustainable conservation of genetic resources from our natural wealth and for future research is also very important (Bayram et al. 2010; Ozturk et al. 2017a, c). During the next five decades, global population will be around 9 billion. This means a decrease in renewable natural resources and an increase in housing and farmland use. Both will mean a decrease in the number of species; area of fertile soils will go down; and deforestation will add to the species loss. A depletion of water resources due to climate change will add to this loss. All these will pose a great threat for future generations (Pleskanovskaja et al. 2011; Ozturk et al. 2017a). The availability of medicinal and aromatic plants together with other consumable herbals will suffer greater loss (Ozturk and Ozcelik 1991; Ozturk et al. 2011, 2012a, b, c, 2017a, b, c). In the case of medicinal and aromatic plants, also other economically important plants, the relevant stakeholders and industry will suffer much, and they need to prepare a long-term plan as a prerequisite. In the case of economically important plants, most important action should be that we study their behavior under future climate change scenarios, together with drought, flooding, erosion, other natural disasters, ecosystem viability, and sustainable land management. There is a great need to work on the market preferences and demand trends of genetic resources and biodiversity, their varietal development, organic products, as well as the planning related to their production based on industry issues. We should create research inventories, and collaboration platforms are a must. Finally each organization needs to cooperate closely with others. There should be good communication among themselves and local inhabitants (Bayram et al. 2010; Ozturk et al. 2017a).

Acknowledgment Münir Öztürk and Volkan Altay are highly thankful to the Kyrgyz-Turkish Manas University, Bishkek, Kyrgyzstan, for providing facilities and permitting us to collaborate with the staff during our trip to the Ala-Archa National Park in September 2015.

References

Abbasi AM, Shah MH, Khan MA (2015) Ethnobotanical aspects of wild edible vegetables. In: Abbasi AM, Shah MH, Khan MA (eds) Wild edible vegetables of lesser Himalayas. Chapter 4. Springer, Cham, pp 67–140

Abou-Shaara HF (2015) Potential honey bee plants of Egypt. Cercetari Agron Moldova 48(2):99–108

Ahmad M, Bano A, Zafar M, Sultana S, Rashid S (2015) Interdependence of biodiversity, applied ethnobotany, and conservation in higher ecosystems of Northern Pakistan under fast climatic changes. In: Oztürk M et al (eds) Climate change impacts on high-altitude ecosystems. Springer, Basel, pp 455–489

Aydın I, Uzun F (2002) Cayır-Mera Amenajmanı ve ıslahı. Textbook No. 9, Faculty of Agriculture, Ondokuz Mayıs University, Samsun

Babalık AA, Sönmez K (2009) Otlatılan ve korunan mera kesimlerinde bakı faktörünün toprakiistü biomass miktarı üzerine etkileri. Suleyman Demirel Universitesi Orman Fakultesi Dergisi Seri A 1:52–58
Balemie K, Kebebew F (2006) Ethnobotanical study of wild edible plants in Derashe and Kucha Districts, South Ethiopia. J Ethnobiol Ethnomed 2:53
Bayram E, Kırıcı S, Tansı S, Yılmaz G, Arabacı O, Kızıl S, Telci I (2010) Tıbbi ve Aromatik Bitkiler Uretiminin Arttırılması Olanakları. In: Turkey agricultural engineering technical conference proceedings, Ankara, pp 437–456
Byers A (2009) A comparative study of tourism impacts on alpine ecosystems in the Sagarmatha (Mt. Everest) National Park, Nepal and the Huascaran National Park, Peru. In: Hill J, Gale T (eds) Ecotourism and environmental sustainability. Ashgate Publishing Group, Abington, Oxon, pp 51–71
Cherepanov SK (1995) Vascular plants of Russia and neighbor countries. Saint-Petersburg, 990 p. (in Russian)
Costanza R, Farber S (2002) Introduction to the special issue on the dynamics and value of ecosystem services: integrating economic and ecological perspectives. Ecol Econ 41:367–373
Eisenman SW, Zaurov DE, Struwe L (eds) (2013) Medicinal plants of Central Asia: Uzbekistan and Kyrgyzstan. Springer, Berlin
Farber S, Costanza R, Childers DL (2006) Linking ecology and economics for ecosystem management. Bioscience 56:121–133
Farrington JD (2005) A report on protected areas, biodiversity, and conservation in the Kyrgyzstan Tien Shan. Fulbright Fellow-Environmental Studies Kyrgyzstan, Former Soviet Central Asia 2003–2004, 13–18
Fet GN (2007) Satellite image processing for biodiversity conservation and environmental modeling in Kyrgyz Republic National Park. MAI 46(4):81-2038
Fry NE (2017) Master's Project: The Ala-Archa ecological leadership project: evaluating the effectiveness of environmental adventure education in growing responsible environmental leaders in the Kyrgyz Republic. Rubenstein School Masters Project Publications 13
Golovkova AG, Chubarova AV (1987) Fodder Plants of Kyrgyzstan. Frunze (In Russian)
Golovkova AG, Chubarova AV (1988) Useful flora Kyrgyzstan. Mektep, Frunze 112 p
Hagg W, Braun LN, Weber M, Becht M (2005) Runoff modeling in glacierized Central Asian catchments for present-day and future climate. Nordic Hydrol 37(2):93–105
Hamayun M, Khan SA, Sohn EY, Lee IJ (2006) Folk medicinal knowledge and conservation status of some economically valued medicinal plants of District Swat, Pakistan. Lyonia 11(2):101–113
Hoffmann M, Hilton-Taylor C, Angulo A et al (2010) The impact of conservation on the status of the world's vertebrates. Science 330:1503–1509
International Union for Conservation of Nature (IUCN) (2014) IUCN Protected Areas Category System. http://www.iucn.org/about/work/programmes/gpap_home/gpap_quality/gpap_pacategories/
Ionov RN, Isakova TV, Lebedeva LP (2000) Vegetation of the National Park "Ala-Archa". Intern. University of Kyrgyzstan, Research Information Magazine 4(12):86–90 (in Russian)
Kaya I, Öncüler A, Ünal Y, Yıldız S (2001) Botanical composition and nutrient levels of the different forms of Kars herbaceous grassland taxa. In: National animal nutrition congress, Elazığ, pp 100–108
Komarov VL (1934-1969) Flora SSSR, T. pp. 1–30. Izd. AN SSSR. Moskva-Leningrad, RU
Lazkov GA (2006) The family Caryophyllaceae in the flora of Kyrgyzstan. KMK Scientific Press, Moscow (In Russian)
Lazkov GA (2007) New and rare species of vascular plants from Western and Central Tian-Shan in Kyrgyzstan. Novit Syst Pl Vasc 39:307–312 (In Russian)
Margules CR, Pressey RL (2000) Systematic conservation planning. Nature 405:243–253
Matyas C (2010) Forests and climate change in Eastern Europe and Central Asia. Food and Agriculture Organization of the United Nations, Rome

Mukanbetov S (2013) Presentation from unspecified US conference 2013. Mainstreaming sustainable development in national development strategies: Kyrgyz Republic. New York. https://sustainabledevelopment.un.org/content/documents/3911Kyrgyzstan.pdf

Ozturk M, Ozcelik H (1991) Useful plants of East Anatolia. Siskav, Ankara 196 pp

Ozturk M, Dalgıç R, Gemici Y (1989) Pollen and chemical analysis of honey from the Aegean region of Turkey. Journal of Faculty of Science Ege University, Series B 11(2):11–16

Ozturk M, Secmen O, Gork G (eds) (1996a) Plant Life in Southwest and Central Asia, vol I. Ege Univ. Press Izmir, Turkey 499 pp

Ozturk M, Secmen O, Gork G (eds) (1996b) Plant life in Southwest and Central Asia, vol II. Ege Univ. Press Izmir, Turkey 1093 pp

Ozturk M, Gucel S, Altundag E, Celik S (2011) Turkish Mediterranean medicinal plants in the face of climate change. In: Ahmad A et al (eds) Medicinal plants in changing environment. Capital Publishing Company, New Delhi, pp 50–71

Ozturk M, Gucel S, Altundag E, Mert T, Gork C, Gork G, Akcicek E (2012a) An overview of the medicinal plants of Turkey. In: Singh R (ed) Genetic resources, chromosome engineering and crop improvement: medicinal plants. CRC Press, LLC, Taylor & Francis, Boca Raton, pp 181–206

Ozturk M, Gucel S, Celik A, Altundag E, Mert T, Akcicek E, Celik S (2012b) *Myrtus communis* in phytotherapy in the Mediterranean. In: Singh R (ed) Genetic resources, chromosome engineering and crop improvement: medicinal plants. CRC Press, LLC, Taylor & Francis, Boca Raton, pp 923–934

Ozturk M, Kebapçı U, Gücel S, Çetin E, Altundag E (2012c) Biodiversity and land degradation in the lower Euphrates subregion of Turkey. J Environ Biol 33:311–323

Ozturk M, Dalgıç R, Güvensen A, Altay V, Gücel S (2012d) Honey-pollen-health: palinochemical analysis of honey from Turkey. Acta Hortic 1:83–92

Ozturk M, Altay V, Gücel S, Altundağ E (2017a) Plant diversity of the drylands in southeast Anatolia-Turkey: role in human health and food security. Chapter 5. In: Ansari AA, Gill SS (eds) Plant biodiversity: monitoring, assessment and conservation. CABI, Oxfordshire, pp 83–124

Ozturk M, Altay V, Gonenç TM (2017b) Herbal from high mountains in the East Mediterranean. In: Bhojraj S et al (eds) Drug discovery from herbs—approaches and applications, Chapter 24. Centre for Science & Technology of the Non-aligned and other Developing Countries (NAM S & T Centre), DAYA Publishing House, New Delhi, pp 327–367

Ozturk M, Altay V, Hakeem KR, Akçiçek E (2017c) Liquorice: from botany to phytochemistry. Springer Briefs in Plant Science. Springer, Cham

Pelmenev VK (1985) Honey bee plants. Russian Agricultural Publishing, 144 p, Moscow-Russia.

Pleskanovskaya S, Mamedova GA, Ozturk M, Gucel S, Ashyraliyeva M (2011) An overview of ethnobotany of Turkmenistan and use of *Juniperus turcomanica* in phytotherapy. In: Singh RJ (ed) Genetic resources, chromosome engineering and crop improvement: medicinal plants. CRC Press, LLC, Taylor & Francis, Boca Raton, pp 207–220

Pressey RL (1994) Ad hoc reservations: forward or backward steps in developing representative reserve systems. Conserv Biol 8:662–668

Rauset GR, Andrén H, Swenson JE, Samelius G, Segerström P, Zedrosser A, Persson J (2016) National parks in northern Sweden as refuges for illegal killing of large carnivores. Conserv Lett 9(5):334–341

Samant SS, Dhar U (1997) Diversity, endemism and economic potential of wild edible plants of Indian Himalaya. Int J Sustain Dev World Ecol 4(3):179–191

Sennikov AN, Lazkov GA, Uotila P, Weber HE (2011) Taxonomic corrections and new records in vascular plants of Kyrgyzstan. Memoranda Soc Fauna Flora Fenn 87:41–64

Shukhurov EJ, Domashov IA (2009) Potential forest vegetation and its geographic distribution in Kyrgyzstan. In: Laletin A et al (eds) Traditional forest related knowledge, biodiversity conservation, and sustainable forest management in Eastern Europe, Northern and Central Asia. Paper presented at International Union of Forest Researchers Organizations, Bishkek, Kyrgyzstan, 23–30 June, pp 56–61. IUFRO Headquarters, Vienna

Sorkun K (2008) Türkiye'nin nektarlı bitkileri, polenleri ve balları, Palme Yayıncılık, Ankara (In Turkish)

Soule ME, Sanjayan MA (1998) Ecology—conservation targets: do they help? Science 279:2060–2061

Stucker D, Kazbekov J, Yakubov M, Wegerich K (2012) Climate change in a small transboundary tributary of the Syr Darya calls for effective cooperation and adaptation. Mt Res Dev 32(3):275–285

Sultanova BA, Lazkov GA, Lebedeva LP, Ionov RN (1998) A preliminary list of endangered species proposed for inclusion in the Red Data Book of Kyrgyzstan. Sci New Technol 1998(2):119–227 (In Russian)

Umralina AR, Lazkov GA (2008) Endemic and rare plant species of Kyrgyzstan (Atlas). Institute of Biotechnology of the National Academy of Sciences of Kyrgyzstan, Bishkek (In Russian and English). www.plantlist.org

Zaurov DE, Belolipov IV, Kurmukov AG, Sodombekov IS, Akimaliev AA, Eisenman SW (2013) The medicinal plants of Uzbekistan and Kyrgyzstan. In: Medicinal plants of Central Asia: Uzbekistan and Kyrgyzstan. Springer, New York, pp 15–273